Operator Theory
Advances and Applications
Vol. 69

Editor
I. Gohberg

Well-Posedness of Parabolic Difference Equations

A. Ashyralyev
P.E. Sobolevskii

Translated from the Russian by A. Iacob

Springer Basel AG

Authors

A. Ashyralyev
Department of Mathematical Analysis
The Turkmen State University
31, Suparmyrat Turkmenbushy Shayoly
744000 Ashgabat
Turkmenistan

P.E. Sobolevskii
Institute of Mathematics
The Hebrew University
Givat Ram
91904 Jerusalem
Israel

A CIP catalogue record for this book is available from the Library of Congress, Washington D.C., USA

Deutsche Bibliothek Cataloging-in-Publication Data
Ašyral'ev, Allaberen:
Well-posedness of parabolic difference equations / A.
Ashyralyev ; P.E. Sobolevskii. Transl. from Russ. by A. Jacob.
– Basel ; Boston ; Berlin : Birkhäuser, 1994
  (Operator theory ; Vol. 69)
  ISBN 978-3-0348-9661-0   ISBN 978-3-0348-8518-8 (eBook)
  DOI 10.1007/978-3-0348-8518-8
NE: Sobolevskii, Pavel E.:; GT

© 1994 Springer Basel AG
Originally published by Birkhäuser Verlag in 1994
Softcover reprint of the hardcover 1st edition 1994
Printed on acid-free paper produced from chlorine-free pulp
Cover design: Heinz Hiltbrunner, Basel

9 8 7 6 5 4 3 2 1

TABLE OF CONTENTS

Chapter 2
THE ROTHE DIFFERENCE SCHEME

Chapter 3
PADÉ DIFFERENCE SCHEMES

Chapter 4
DIFFERENCE SCHEMES FOR PARABOLIC EQUATIONS

PREFACE

A well-known and widely applied method of approximating the solutions of problems of mathematical physics is the method of difference schemes. The main characteristics of difference schemes are their accuracy and stability. Modern computers allow the implementation of highly accurate difference schemes. Hence, a task of current interest is the construction and investigation of highly accurate difference schemes for various types of boundary value problems of mathematical physics. The present monograph is devoted to the construction of highly accurate difference schemes for parabolic boundary value problems, based on Padé approximation.

The convergence properties of these schemes and the estimates of their rate of convergence are determined by the type of stability they enjoy. The first stability estimates for such difference schemes were established in Hilbert space norms, which in applications led to L_2 estimates of stability and rate of convergence. However, in applications uniform estimates of the rate of convergence are far more interesting. Earlier such estimates were established on the basis of a maximum principle for the simplest difference schemes of low order of accuracy, and only for second-order equations. In the present monograph the study of the stability of difference schemes for parabolic equations is based on a new notion of positivity of a linear operator in a Banach space, which in applications allows one to deal with difference schemes of arbitrary order of accuracy and establish their stability and convergence in Hölder norms. Our investigations are based on the results of a new theory of interpolation of linear operators.

The type of stability that is the most interesting for application purposes is the so-called coercive stability. Establishing coercivity inequalities for solutions of difference schemes is important in applications because such inequalities allow one to obtain sharp, i.e., two-sided estimates of the rate of convergence. Such inequalities for highly accurate difference schemes in Banach space norms can be also proved based on the theory of interpolation of linear operators.

We note that the existence of coercivity inequalities is equivalent to the natural notion of well-posedness of difference problems.

Let us give a brief account of the contents of our monograph. It consists of four chapters.

In Chapter 1 we study the well-posedness of the abstract Cauchy problem

$$v'(t) + Av(t) = f(t), \quad 0 \le t \le 1, \quad v(0) = v_0,$$

in various spaces $F(E)$ of functions defined on $[0,1]$ with values in some Banach spaces. The chapter consist of five sections.

In Section 1 the Cauchy problem is studied in the space $C(E)$. It is shown that a necessary condition for well-posedness is that the semigroup $\exp\{-tA\}$ ($t \ge 0$) be analytic in E. In the general case this condition is not sufficient for well-posedness in $C(E)$.

The analyticity of the semigroup $\exp\{-tA\}$ is sufficient for well-posedness in $C(E)$ of the homogeneous Cauchy problem ($f \equiv 0$). Moreover, the solution of this problem enjoys additional smoothness for $t > 0$, and consequently the homogeneous Cauchy problem is well-posed in the weighted Hölder space $C_0^\alpha(E)$, which is smaller than $C(E)$. As it turns out, the general Cauchy problem, too, is well posed in $C_0^\alpha(E)$. This is discussed in Section 2.

In Section 3 we study the well-posedness of the Cauchy problem in the spaces $L_p(E)$, $p \in (1, \infty)$. A necessary condition for well-posedness is again the analyticity of the semigroup $\exp\{-tA\}$. This condition is also necessary for any $p \in (1, \infty)$ provided the Cauchy problem is well posed in $L_{p_0}(E)$ for some $p_0 \in (1, \infty)$. For example, if E is a Hilbert space one can take $p_0 = 2$.

A criterion for well-posedness of the Cauchy problem in $C(E)$ and $C_0^\alpha(E)$ is that the initial datum v_0 belong to the domain $D(A)$ of the operator A. In the case of the space $L_p(E)$ the element v_0 must belong to the so-called space of traces $E_{1-\frac{1}{p}}$. It turns out that the solution in $L_p(E)$ of the general Cauchy problem is a continuous function with values in $E_{1-\frac{1}{p}}$.

In Section 4 we isolate a rich family of spaces $E_{\alpha,q}$, $0 < \alpha < 1$, $1 \le q \le \infty$, for which one has continuous and dense embeddings $D(A) \subset E_{\alpha,q} \subset E$, such that the Cauchy problem is well posed in $L_q(E_{\alpha,q})$. From this and the results of

Section 3 we derive the well-posedness of this problem in the spaces $L_p(E_{\alpha,q})$ with $1 < p, q < \infty$.

Section 5 is devoted to the study of the well-posedness of the Cauchy problem in spaces of smooth functions $C_0^{\beta,\gamma}(E)$, $0 \le \gamma \le \beta$. A particular case of such spaces, obtained for $\gamma = \beta = \alpha$, appears earlier in Section 2. Although the spaces $C_0^{\beta,\gamma}(E)$ depend on two parameters, it turns out that the estimate in the corresponding coercivity inequality depends only on β. Consequently, one can choose γ arbitrarily in $[0, \beta]$. In particular, the Cauchy problem is well posed in the ordinary Hölder space $C^\alpha(E) = C_0^{\alpha,0}(E)$.

The study of difference Cauchy problems is initiated in Chapter 2. Therein we study the well-posedness of the difference Cauchy problem generated by an implicit difference scheme of first order of accuracy. Well-posedness of difference problems is understood as the existence of a coercive estimate that is uniform in the step τ of the difference grid. We succeed in establishing a sharp coercive estimate, i.e., a two-sided estimate of the solution error. Chapter 2 consists of six sections.

In Section 0 we give formulas and estimates for resolvents, and we establish the stability of the difference scheme in various spaces.

In Section 1 we study the well-posedness of the difference problem in the space $\mathcal{C}(E)$. From the well-posedness of the difference problem in $\mathcal{C}(E)$ follows the well-posedness of the differential Cauchy problem in $C(E)$, and hence the analyticity of the semigroup $\exp\{-tA\}$, $t \ge 0$. Since in the general case the differential problem is not well posed in $C(E)$, the difference problem cannot be stable uniformly in τ. This means that the coercive norm of the solution tends to infinity when $\tau \to 0+$. The study of the difference problem itself allows us to determine the order of growth of this norm to infinity.

In Section 2 we study the well-posedness of the difference problem in $C_0^\alpha(E)$. Here we generalize the results of Section 2 of Chapter 1. The proof is based on smoothness estimates of powers of the resolvent of the operator $-A$, which are equivalent to estimates of the semigroup $\exp\{-tA\}$, $t \ge 0$.

Section 3 is devoted to the well-posedness of the difference problem in the space $\mathcal{L}_p(E)$. As in the case of the differential problem (Section 3 of Chapter 1), the proof relies on the Benedeck-Calderón-Panzone extrapolation theorem. Here we use again the equivalence of the norms in the spaces of traces generated by the semigroup and the powers of the resolvent.

In Section 4 we study the well-posedness of the difference problem in the space $\mathcal{L}_p(E_{\alpha,q})$ and generalize the results of Section 4, Chapter 1. These results are based on the Cauchy-Riesz formula for functions of a strongly positive operator and on the introduction of an equivalent norm in the spaces $E_{\alpha,q}$ by means of the resolvent of this operator.

Finally, in Section 5 we are concerned with the well-posedness of the difference problem in difference analogues of spaces of smooth functions, and we generalize the results of Section 5, Chapter 1. The study relies on smoothness estimates of powers of the resolvent of a strongly positive operator, which are equivalent to the smoothness estimates for the semigroup $\exp\{-tA\}$, $t \geq 0$, given in Section 5 of Chapter 1.

Chapter 3 is devoted to the study of the well-posedness of the difference Cauchy problem generated by the Padé fractions $R_{j,l}(z)$ $(l-2 \leq j \leq l)$ approximating the exponential function. It consists of six sections.

In Section 0 we study the stability of the difference problem. The investigation of the stability and well-posedness of the Padé difference schemes relies on the properties of the rational functions $R_{j,l}(z)$ that generate them. Here we give formulas and establish estimates for $R_{j,l}(z)$. Then we construct Padé difference schemes and give criteria for their stability in the $\mathcal{C}(E)$-, $\mathcal{C}_0^\alpha(E)$-, and $\mathcal{L}_p(E)$-norms.

In Section 1 we study the well-posedness of the difference problem in the space $\mathcal{C}(E)$. Based on the estimates for $R_{j,l}(z)$ given in Section 0, we establish estimates for powers of the operator step $R_{j,l}(\tau A)$ $(l-2 \leq j \leq l)$ that are analogous to the estimates of powers of the resolvent. Let us note that the estimates of powers of $R_{l,l}(\tau A)$ have "worse" properties compared with those of the operator steps $R_{l-2,l}(\tau A)$ and $R_{l-1,l}(\tau A)$. The estimates obtained enable us to establish stability and almost coercivity inequalities for Padé difference schemes in $\mathcal{C}(E)$.

Section 2 is devoted to the well-posedness of the difference problem in the space $\mathcal{C}_0^\alpha(E)$. Based on the estimates of Section 1 we derive smoothness estimates for powers of the operator step $R_{j,l}(\tau A)$. The estimates obtained allow us to establish stability and coercivity estimates in $\mathcal{C}_0^\alpha(E)$ for difference schemes with $j = l-2$, $l-1$ or even $j = l$. For odd $j = l$ such inequalities are established in a space $\tilde{\mathcal{C}}_0^\alpha(E)$, smaller than $\mathcal{C}_0^\alpha(E)$.

In Section 3 we study the well-posedness of the difference problem in the space $\mathcal{L}_p(E)$. This study relies on the fact that the norm of the powers of the operator step $R_{j,l}(\tau A)$ decays exponentially. The latter holds for $j = l-2$, $l-1$ in the case in which the operator A is strongly positive with spectral angle $\phi(A, E) < \pi/(2l)$.

In Section 4 we study the well-posedness of the difference problem in the spaces $\mathcal{L}_p(E_{\alpha,q})$. We establish stability and coercivity estimates for the Padé difference schemes for any j, l such that $l - 2 \leq j \leq l$ in $\mathcal{L}_q(E_{\alpha,q})$, $1 \leq q \leq \infty$, in the case of an arbitrary strongly positive operator A, and for $j = l - 2$, $l - 1$ in $\mathcal{L}_p(E_{\alpha,q})$, $1 < p, q < \infty$, in the case of a strongly positive operator A with spectral angle $\phi(A, E) < \pi/(2l)$.

In Section 5 we study the well-posedness of the Cauchy problem in spaces of smooth functions. Here we generalize the results of Sections 2 and 4. The stability and coercivity estimates are now of two types. The first covers the case of Padé difference schemes with $j = l - 2$, $l - 2$ or $j = l$ an even number, while the second covers the case of Padé difference schemes with $j = l$ an odd number.

In Chapter 4 we obtain stability and coercivity estimates for highly accurate difference schemes for the approximate solution of the Cauchy problem for parabolic equations. The chapter consists of three sections.

In Section 1 we deal with elliptic difference operators with constant coefficients. We construct a difference operator A_h that approximates a given elliptic operator A of arbitrary order with constant coefficients in $\mathbf{R}^n$. For the operator A_h we study the Green function (fundamental solution) of the resolvent equation. We establish point estimates of the difference derivatives of the Green function, which go over into the corresponding estimates for the differential case when the grid step tends to zero. We should point out that here a new form of estimates for derivatives of the difference Green function for large absolute values of the spectral parameter is exhibited.

In Section 2 we study fractional spaces in the case of an elliptic difference operator. We prove the uniform-in-h positivity of the operator A_h in $L_p = L_p(\mathbf{R}_h^n)$, $1 \leq p \leq \infty$. We show that the $E'_{\alpha,p}(L_{ph}, A_h)$-norms are equivalent, uniformly in h, to the difference norms $W_p^{m\alpha}(\mathbf{R}_h^n)$ for $0 < \alpha < 1/m$. Finally, we prove the coercivity of the elliptic difference problem in the spaces $W_p^{m\alpha}(\mathbf{R}_h^n)$.

In Section 3 we show how the results about difference schemes obtained in Chapter 3 for abstract equations and the results of Sections 1 and 2 of Chapter 4 yield stability and coercivity estimates for parabolic difference equations.

The results presented in this monograph are due to the authors, and also to our colleagues A. E. Polichka, Kh. A. Alibekov, Hoàng Văn Lai, and Yu. A. Smirnitskiĭ. We gathered here only part of our investigations on the theory of difference equations, begun in 1967 in Voronezh, at the Department of Functional Analysis and Operator Equations of Voronezh University. A brief survey of all these investigations can be found in the "Comments on the Literature" that conclude the monograph.

CHAPTER 1

THE ABSTRACT CAUCHY PROBLEM

1. WELL-POSEDNESS OF THE DIFFERENTIAL CAUCHY PROBLEM IN $C(E)$

1. The Cauchy problem in a Banach space E. Definition of well-posedness in $C(E)$.

Consider the following Cauchy problem in an arbitrary Banach space E :

$$v'(t) + Av(t) = f(t), \quad 0 \le t \le 1, \quad v(0) = v_0. \tag{1.1}$$

Here $v(t)$ and $f(t)$ are the unknown and the given function, respectively, defined on $[0, 1]$ with values in E. The derivative $v'(t)$ is understood as the limit in the norm of E of the corresponding ratio of differences. A is a linear operator acting in E, with domain $D(A)$. Finally, v_0 is a given element of E.

A function $v(t)$ is called a *solution* of problem (1.1) if the following conditions are satisfied:

1) $v(t)$ is continuously differentiable on the segment $[0, 1]$. The derivative at the endpoints of the segment are understood as the appropriate unilateral derivatives.

2) The element $v(t)$ belongs to $D(A)$ for all $t \in [0, 1]$, and the function $Av(t)$ is continuous on $[0, 1]$.

3) $v(t)$ satisfies the equation and boundary conditions (1.1).

A solution of problem (1.1) defined in this manner will from now on be referred to as *a solution of problem* (1.1) *in the space* $C(E) = C([0, 1], E)$. Here

$C(E)$ stands for the Banach space of the continuous functions $\varphi(t)$ defined on $[0, 1]$ with values in E, equipped with the norm

$$\|\varphi\|_{C(E)} = \max_{0 \leq t \leq 1} \|\varphi(t)\|_E. \tag{1.2}$$

If $v(t)$ is a solution in $C(E)$ of problem (1.1), then the data of the problem must satisfy the following conditions:

a) $f(t)$ belongs to $C(E)$;

b) v_0 belongs to $D(A)$.

There arises the question of whether these necessary conditions are sufficient for the solvability of the problem (1.1) in $C(E)$. If $v(t)$ is a solution of problem (1.1) in $C(E)$, then $v(t)$ is a solution of the integral equation

$$v(t) = v_0 - \int_0^t Av(s)ds + \int_0^t f(s)ds, \tag{1.3}$$

with the property that the functions $v'(t)$ and $Av(t)$ are continuous on $[0, 1]$. Here in the left- and right-hand sides one has abstract Riemann integrals for continuous functions with values in the Banach space E. Conversely, any solution of equation (1.3) that possesses the indicated property is a solution in $C(E)$ of equation (1.1). The integral equation (1.3) is easy to study when A is a bounded linear operator $(D(A) = E)$.

In fact, equation (1.3) can be written in the operator form $v = Fv$, where

$$Fv = v_0 - \int_0^t Av(s)ds + \int_0^t f(s)ds.$$

Let us show by induction that for any nonnegative integer m,

$$\|F^m v_1(t) - F^m v_2(t)\|_E \leq \frac{(\|A\|t)^m}{m!} \|v_1 - v_2\|_{C(E)}. \tag{1.4}$$

It is readily proved that

$$\|Fv_1(t) - Fv_2(t)\|_E \leq \frac{\|A\|t}{1!} \|v_1 - v_2\|_{C(E)}$$

for all $t \in [0, 1]$. Hence, estimate (1.4) holds for $m = 1$.

Assume that (1.4) holds for some $m = n \geq 1$. Then

$$\|F^{n+1} v_1(t) - F^{n+1} v_2(t)\|_E \leq \|A\| \int_0^t \|F^n v_1(s) - F^n v_2(s)\|_E ds$$

$$\leq \|A\| \int_0^t \frac{\|A\|^n s^n}{n!} ds \, \|v_1 - v_2\|_{C(E)} = \frac{(\|A\|t)^{n+1}}{(n+1)!} \|v_1 - v_2\|_{C(E)},$$

i.e., (1.4) holds for $m = n + 1$. From (1.4) it follows that

$$\|F^m v_1 - F^m v_2\|_{C(E)} \leq \frac{\|A\|^m}{m!} \|v_1 - v_2\|_{C(E)}.$$

Thus, the power F^m of the operator F is contractive for sufficiently large m. Hence, applying the generalized contraction mapping principle we conclude that the Cauchy problem (1.1) has a unique solution in $C(E)$.

Since the operator A is bounded, the function $Av(t)$ is also continuous, and therefore $v(t)$ is a solution of problem (1.1) in $C(E)$. We see that in the case of a bounded operator A conditions a) and b) are not only necessary but also sufficient for the solvability in $C(E)$ of the problem (1.1).

It follows from the unique solvability of (1.1) that the solution $v(t)$ defines an operator $v(t; f(t), v_0)$, which acts from $C(E) \times E$ to $C(E)$. Here $C(E) \times E$ is understood as the Banach space of the pairs $(f(t), v_0)$, $f(t) \in C(E), v_0 \in E$, equipped with the norm

$$\|(f(t), v_0)\|_{C(E) \times E} = \|f\|_{C(E)} + \|v_0\|_E.$$

The unique solvability of problem (1.1) implies the additivity and homogeneity of the operator $v(t; f(t), v_0)$. Finally, by means of the theorem on integral inequalities one establishes the inequality

$$\|v(t; f(t), v_0)\|_{C(E)} \leq M(\|f\|_{C(E)} + \|v_0\|_E), \tag{1.5}$$

which shows that the operator $v(t; f(t), v_0)$ is continuous.

Definition 1.1. We say that the problem (1.1) is *well posed in* $C(E)$ if the following conditions are satisfied:

1) Problem (1.1) is uniquely solvable for any $f(t) \in C(E)$ and any $v_0 \in D(A)$. This means that an additive and homogeneous operator $v(t; f(t), v_0)$ is defined which acts from $C(E) \times D(A)$ to $C(E)$ and gives the solution of problem (1.1) in $C(E)$.

2) $v(t; f(t), v_0)$, regarded as an operator from $C(E) \times D(A)$ to $C(E)$, is continuous. Here $C(E) \times D(A)$ is understood as the normed space of the pairs $(f(t), v_0)$, $f(t) \in C(E), v_0 \in D(A)$, with the norm

$$\|(f(t), v_0)\|_{C(E) \times D(A)} = \|f\|_{C(E)} + \|v_0\|_E.$$

From linearity it follows that this property is equivalent to inequality (1.5).

2.   Examples of well-posed and ill-posed problems in $C(E)$.

Above it was established that problem (1.1) with a bounded operator is well posed in $C(E)$. The question arises of whether problem (1.1) can be well posed for unbounded operators A as well. First let us give an example of such a problem. Let E denote the Banach space of all real-valued functions $\varphi(\lambda)$ that are defined and continuous for all $\lambda \geq 1$ and tend to zero as $\lambda \to \infty$. The norm in this space is defined as

$$\|\varphi\|_E = \sup_{1 \leq \lambda \leq \infty} |\varphi(\lambda)|.$$

Next, define the operator A by the rule $(A\varphi)(\lambda) = \lambda\varphi(\lambda)$ on the functions $\varphi(\lambda) \in E$ for which $\lambda\varphi(\lambda)$ also belongs to E. Clearly, A is not bounded. Note that the space $C(E)$ obviously consists of all uniformly jointly continuous functions $f(t, \lambda)$ with the property that for each fixed t, $f(t, \lambda) \to 0$ as $\lambda \to \infty$. In the present case problem (1.1) becomes the Cauchy problem for an ordinary differential equation depending on a parameter λ:

$$\frac{\partial v(t, \lambda)}{\partial t} + \lambda v(t, \lambda) = f(t, \lambda), \quad 0 \leq t \leq 1, \quad v(0, \lambda) = v_0(\lambda). \tag{1.6}$$

Its solution will be a solution in $C(E)$ of the problem (1.1) whenever the functions $\partial v(t, \lambda)/\partial t$ and $\lambda v(t, \lambda)$ belong to $C(E)$.

    Suppose now that $f(t, \lambda)$ belongs to $C(E)$ and $v_0(\lambda) \in D(A)$. The solution of problem (1.6) (regarded as a Cauchy problem for an ordinary differential equation) obviously has the form

$$v(t, \lambda) = e^{-\lambda t}v_0(\lambda) + \int_0^t e^{-\lambda(t-s)} f(s, \lambda)ds.$$

    To show that $v(t, \lambda)$ is a solution in $C(E)$ of problem (1.1) it suffices to verify that the function $\lambda v(t, \lambda)$ is in $C(E)$. Now, the function

$$\lambda v_1(t, \lambda) = \lambda e^{-\lambda t}v_0(\lambda) = e^{-\lambda t}\lambda v_0(\lambda)$$

belongs to $C(E)$ since $v_0(\lambda) \in D(A)$. The function

$$\lambda v_2(t, \lambda) = \lambda \int_0^t e^{-\lambda(t-s)} f(s, \lambda)ds$$

also belongs to $C(E)$, since it obeys the inequality

$$|\lambda v_2(t, \lambda)| \leq (1 - e^{-\lambda}) \max_{0 \leq t \leq 1} |f(t, \lambda)|,$$

and $\max_{0 \leq t \leq 1} |f(t, \lambda)| \to 0$ as $\lambda \to \infty$. Finally, from the above estimates it follows that the solution depends continuously on the data of the problem. Thus, problem (1.6) is well posed in $C(E)$. It is readily seen that the analogous problem

$$\frac{\partial v(t, \lambda)}{\partial t} - \lambda v(t, \lambda) = f(t, \lambda), \quad 0 \leq t \leq 1, \quad v(0, \lambda) = v_0(\lambda)$$

is already not well posed in $C(E)$. This means that problem (1.1) is not necessarily well posed for an unbounded operator A. In connection with this it is of interest to isolate those unbounded operators A for which problem (1.1) is well posed.

3.   The homogeneous equation. Strongly continuous semigroups.

Problem (1.1) will be investigated here under the assumption that the set $D(A)$ is dense in E and the operator A has a bounded inverse A^{-1}. Note that the operator A in problem (1.6) satisfies these conditions. The assumption on the boundedness of the operator A^{-1} could be replaced by the assumption on the boundedness of the operator $(A + \mu I)^{-1}$ for some μ, since the substitution $v(t) = e^{\mu t} w(t)$ transforms problem (1.1) with an operator A into the same problem with the operator $A + \mu I$.

First let us consider the homogeneous problem

$$v'(t) + Av(t) = 0, \quad 0 \leq t \leq 1, \quad v(0) = v_0. \tag{1.7}$$

From the well-posedness of the general problem (1.1) it follows, of course, that this problem is uniquely solvable in $C(E)$ and that the operator $v(t; 0, v_0)$ is continuous as an operator from $D(A)$ to $C(E)$. Here $D(A)$ is regarded as a normed space with the norm of the space E. Problem (1.7) is studied in the theory of semigroups. It is established that this problem is well posed in $C(E)$ if and only if the operator $-A$ is the generator of a strongly continuous semigroup. Let us recall

Definition 1.2. A family $U(t), t \geq 0$, of bounded linear operators is a *strongly continuous semigroup* if the following conditions are satisfied:
  1) $U(t + \tau) = U(t)U(\tau) = U(\tau)U(t), \quad t \geq 0, \tau \geq 0; \quad U(0) = I$.
  2) For each fixed $v_0 \in E$ the function $U(t)v_0$ is continuous in t for $t \geq 0$.

From the strong continuity of the operator-function $U(t)$ it follows that its norm is uniformly bounded on any bounded segment $[0, T]$. Next, from the semigroup property it follows that when $t \to \infty$ this norm grows no faster than an exponential. Specifically, one has the estimate

$$\|U(t)\|_{E \to E} \leq Me^{\omega t}, \quad t \geq 0. \tag{1.8}$$

Definition 1.3. The operator $U'(0)$, defined by the formula

$$U'(0)v_0 = \lim_{\Delta t \to +0} \Delta t^{-1}[U(\Delta t) - I]v_0$$

on the elements $v_0 \in E$ for which the limit in the right-hand side exists, is called the *generator of the semigroup* $U(t)$.

The operator $U'(0)$ has a dense domain in E and for any $\mu > \omega$ in the case of a real space E, or any complex μ with $\operatorname{Re} \mu > \omega$ in the case of a complex space E, the operator $\mu I - U'(0)$ has a bounded inverse, i.e., $U'(0)$ is closed.

The membership of an operator B in the class of generators of strongly continuous semigroups can be characterized in terms of its resolvent.

Criterion. *Let B be an operator with dense domain acting in a complex Banach space E. In order for B to be the generator of a strongly continuous semigroup $U(t)$ satisfying the estimate (1.8), it is necessary and sufficient that any complex number λ with $\operatorname{Re} \lambda > \omega$ belong to the resolvent set of B and that the following estimates hold:*

$$\|(\lambda I - B)^{-n}\|_{E \to E} \leq M(\operatorname{Re} \lambda - \omega)^{-n}, \quad n = 1, 2, \dots. \tag{1.9}$$

Note that one may assume that the estimates (1.9) are satisfied only for some sequence λ_m such that $\operatorname{Re} \lambda_m \to \infty$ as $m \to \infty$.

In the case of a real Banach space E the estimates (1.9) must hold for all real $\lambda > \omega$.

In what follows a semigroup with generator $-A$ will be denoted by $\exp\{-tA\}$. By passing from the problem with the operator A to the problem with the operator $A + \mu I$ one can ensure that the norm of this semigroup decreases exponentially, i.e., the following estimate holds:

$$\|\exp\{-At\}\|_{E \to E} \leq Me^{-\delta t}, \quad M > 0, \delta > 0. \tag{1.10}$$

For each $v_0 \in D(A)$ and any $t \geq 0$ one has the equality

$$AU(t)v_0 = U(t)Av_0,$$

which follows from the identity

$$\frac{U(\Delta t) - I}{\Delta t}U(t)v_0 = U(t)\frac{U(\Delta t) - I}{\Delta t}v_0.$$

This means that the operator-function $U(t)$ maps the domain $D(A)$ of the operator A into itself. From the last identity it also follows that the function $v(t) = U(t)v_0$ admits a right derivative $v'_{\mathrm{r}}(t)$, and that

$$v'_{\mathrm{r}}(t) = -Av(t).$$

Thus, equation (1.7) for $t = 0$ is established. To show that the function $v(t)$ satisfies equation (1.7) for $t > 0$, too, we have to verify that its left derivative $v'_{\mathrm{l}}(t)$ exists and that $v'_{\mathrm{l}}(t) = v'_{\mathrm{r}}(t)$. This follows from the identity

$$\frac{v(t - \Delta t) - v(t)}{\Delta t} = U(t - \Delta t)\frac{U(\Delta t) - I}{\Delta t}v_0$$

for $t - \Delta t \geq 0$. Hence, the formula

$$v(t) = U(t)v_0 = \exp\{-tA\}v_0 \tag{1.11}$$

gives a solution of problem (1.7) in $C(E)$. Below, in the course of the investigation of a more general nonhomogeneous equation it will be shown, in particular, that any solution of problem (1.7) in $C(E)$ is given by formula (1.11).

4.   The nonhomogeneous equation. Analytic semigroups.

Let us consider now the nonhomogeneous Cauchy problem with initial condition zero,

$$v'(t) + Av(t) = f(t), \quad 0 \leq t \leq 1, \quad v(0) = 0. \tag{1.12}$$

It turns out that the well-posedness of this problem implies the membership of the semigroup $\exp\{-tA\}$ in the class of analytic semigroups, which is less wide than that of the strongly continuous semigroups. Let us recall the following

Definition 1.4. A strongly continuous semigroup $U(t)$ acting in a complex Banach space E is said to be *analytic* if it can be continued from the half-line $0 \leq t \leq \infty$ to an operator-function $U(z)$ that is analytic in some sector

$$S_\alpha = \{z : |\arg z| < \alpha, \ 0 < |z| < \infty\}, \quad 0 < \alpha \leq \pi/2,$$

and is strongly continuous in its closure $\overline{S_\alpha}$.

Let us show that the semigroup $\exp\{-tA\}$ is analytic. From the well-posedness of problem (1.1) in $C(E)$ it follows that the operator $v(t; f(t), 0)$ is continuous in $C(E)$, and the operator $Av(t; f(t), 0)$ is defined on the entire space $C(E)$. The operator A, which acts in the Banach space E with domain $D(A)$, generates via the formula $\mathcal{A}v = Av(t)$ an operator $\mathcal{A}$, which acts in the Banach space $C(E)$ and is defined on the functions $v(t) \in C(E)$ with the property that $Av(t) \in C(E)$. From the fact that the operator A^{-1} exists and is bounded it obviously follows that the operator $\mathcal{A}^{-1}$ exists and is bounded, and hence that $\mathcal{A}$ is closed in $C(E)$. Consequently, the operator $Av(t; f(t), 0) = \mathcal{A}v(\cdot\,; f, \cdot)$ is closed in $C(E)$. Next, by Banach's theorem, this operator is continuous, i.e., for any $f(t) \in C(E)$ one has the inequality

$$\|Av(t; f(t), 0)\|_{C(E)} \leq M\|f\|_{C(E)}, \tag{1.13}$$

where M does not depend on $f(t)$.

Now let us consider the function $v(t) = t \exp\{-tA\}v_0$. If $v_0 \in D(A)$, then obviously $v(t)$ is a solution in $C(E)$ of problem (1.12) with $f(t) = \exp\{-tA\}v_0$. Inequality (1.13) and estimate (1.10) yield the inequality

$$\max_{0 \leq t \leq 1} \|tA \exp\{-tA\}v_0\|_E \leq M\|v_0\|_E.$$

Since $D(A)$ is dense in E, this implies that for $t > 0$ the operator $A \exp\{-tA\}$ is bounded and obeys the estimate

$$\|A \exp\{-tA\}\|_{E \to E} \leq Mt^{-1}. \tag{1.14}$$

This means that the operator-function $\exp\{-tA\}$ is strongly differentiable for $t > 0$. Finally, from the semigroup property it follows that this operator-function is infinitely differentiable in the operator norm for $t > 0$ and that its derivatives obey the estimates

$$\left\|\frac{d^n \exp\{-tA\}}{dt^n}\right\|_{E \to E} = \|A^n \exp\{-tA\}\|_{E \to E} \leq M^n n^n t^{-n} e^{-\delta t} \tag{1.15}$$

for some $M > 0$ and $\delta > 0$.

Let us expand the operator-function $\varphi(t) = \exp\{-tA\}$ in a Taylor series at the point t_0:

$$\varphi(t) = \sum_{k=0}^{\infty} (t - t_0)^k \frac{\varphi^{(k)}(t_0)}{k!}. \tag{1.16}$$

From the estimates (1.15) it follows that this series converges for $|t - t_0| < t_0(Me)^{-1}$. This means that the operator-function $\varphi(t)$ can be continued to an operator-function $\varphi(z)$, analytic in the sector S_α and strongly continuous in its closure $\overline{S_\alpha}$, for any α satisfying $0 \le \alpha < \arcsin(Me)^{-1}$. Note that the estimate (1.15) is not only sufficient but also necessary for the analyticity of the semigroup $\exp\{-tA\}$; this follows from the Cauchy formula for analytic functions.

A semigroup $\exp\{-tA\}$ acting in a real Banach space E will be said to be *analytic* if it can be expanded in a Taylor series (1.16) that converges for any $t > 0$ and $t_0 > 0$ such that $|t - t_0| < t_0\lambda$ for some $0 < \lambda < 1$. This means that the natural extension of the semigroup $\exp\{-tA\}$ to the complexification of the space E is an analytic semigroup in that complex space.

The foregoing argument shows that the analyticity of the semigroup $\exp\{-tA\}$ is a necessary condition for the well-posedness of problem (1.1) in $C(E)$ even for real spaces E.

Generators of analytic semigroups acting in a complex space E admit the following characterization in terms of their resolvents.

Criterion. *Let B be an operator with dense domain acting in a complex Banach space E. In order for B to be the generator of an analytic semigroup it is necessary and sufficient that there exist real numbers ω and $\tau > 0$ such that all complex λ satisfying $\operatorname{Re} \lambda \ge \omega$ and $|\lambda| \ge \tau$ belong to the resolvent set of B and the following estimate holds:*

$$\|(\lambda I - B)^{-1}\|_{E \to E} \le M|\lambda|^{-1}. \tag{1.17}$$

We shall not use this criterion to show that the well-posedness in $C(E)$ implies the analyticity of the semigroup $\exp\{-tA\}$, since below this reasoning will be carried out in the study of the well-posedness of problem (1.1) in other function spaces.

5.   Well-posedness in $C(E)$ of the general Cauchy problem.

Let us consider now the general problem (1.1). As a consequence of unique solvability, its solution $v(t)$ can be represented in the form

$$v(t) = v(t; 0, v_0) + v(t; f(t), 0).$$

Hence, by (1.10) and (1.13), we have the coercivity inequality

$$\|v'\|_{C(E)} + \|Av\|_{C(E)} \leq M\big[\|f\|_{C(E)} + \|Av_0\|_E\big], \qquad (1.18)$$

where M does not depend on $f(t)$ and v_0.

Let $v(t)$ be a solution in $C(E)$ of problem (1.1), and let $-A$ be the generator of a strongly continuous semigroup $\exp\{-tA\}$. Then for any $0 \leq s \leq t \leq 1$ we have the identity

$$\frac{d}{ds}\big[\exp\{-(t-s)A\}A^{-1}v(s)\big] = \exp\{-(t-s)A\}A^{-1}f(s).$$

Integrating here with respect to s from 0 to t, we obtain the formula

$$A^{-1}v(t) = \exp\{-tA\}A^{-1}v_0 + \int_0^t \exp\{-(t-s)A\}A^{-1}f(s)ds.$$

Since the operator A is closed, we have

$$v(t) = \exp\{-tA\}v_0 + \int_0^t \exp\{-(t-s)A\}f(s)ds. \qquad (1.19)$$

Hence, if problem (1.1) is well posed in $C(E)$, then formula (1.19) defines the operator $v(t; f(t), v_0)$. From the semigroup estimates (1.10) it follows that $v(t; f(t), v_0)$ satisfies condition 2) for the well-posedness of problem (1.1) in $C(E)$. The question arises whether formula (1.19) gives a solution of problem (1.1) in $C(E)$ for all $v_0 \in D(A)$ and all $f(t) \in C(E)$.

If the function $f(t)$ is not only continuous, but also continuously differentiable on $[0, 1]$, it is known that formula (1.19) gives a solution of problem (1.1) in $C(E)$. However, this is a result with "loss of regularity," since the stronger requirement of continuous differentiability rather than continuity is imposed on $f(t)$.

Suppose now that the coercivity inequality (1.18) holds for arbitrary solutions $v(t)$ of problem (1.1) with continuously differentiable right-hand sides $f(t)$. This means, in particular, that the operator $Av(t; f(t), 0)$, acting in $C(E)$ and defined on all continuously differentiable functions is bounded, i.e., inequality (1.13) holds. Since the set of continuously differentiable functions is dense in $C(E)$, the operator $v(t; f(t), 0)$ is continuous in $C(E)$, and since the operator A is closed in E, it follows that $Av(t; f(t), 0)$ is continuous in $C(E)$. From this one readily concludes that problem (1.1) is well posed in $C(E)$.

Thus, to prove the well-posedness of problem (1.1) in $C(E)$ it suffices to establish the coercivity inequality for the solutions that correspond to smooth initial data. Of course, here the operator $-A$ is required to be the generator of an analytic rather than merely strongly continuous semigroup.

Is the analyticity of the semigroup $\exp\{-tA\}$ not only a necessary but also a sufficient condition for the well-posedness of problem (1.1) in $C(E)$? As it turns out, problem (1.1) is not well posed for all such operators A. Let us give an example.

Let E be the Banach space of continuous scalar functions $f(x)$ on $(-\infty, \infty)$, satisfying $f(x) \to 0$ as $|x| \to \infty$, with the norm

$$\|f\|_E = \sup_{-\infty < x < \infty} |f(x)|.$$

Let A be the operator acting in E according to the rule $Av(x) = -v''(x)$, so that we also have $v''(x) \in E$. From Taylor's formula it follows that $v'(x) \in E$, too. For all complex λ such that $\operatorname{Re} \lambda > 0$ the equation $\lambda v + Av = f$ has a unique solution $v \in D(A)$ for each $f \in E$, and

$$v = v(x, \lambda) = \int_{-\infty}^{\infty} \frac{1}{2\sqrt{\lambda}} e^{-\sqrt{\lambda}|x-y|} f(y) dy.$$

From this it follows easily that estimate (1.17) holds, i.e., $-A$ is the generator of an analytic semigroup.

Now let us show that in the present case problem (1.1) is not well posed in $C(E) = C([0, 1], E)$. Set $v_0 = 0$. Then (1.1) turns into the Cauchy problem

$$\frac{\partial v(t, x)}{\partial t} - \frac{\partial^2 v(t, x)}{\partial x^2} = f(t, x), \quad 0 \le t \le 1, \quad v(0, x) = 0, \quad -\infty < x < \infty. \quad (1.20)$$

Suppose problem (1.1) is well posed in $C([0, 1], E)$. Then for any function $f(t, x) \in C([0, 1], E)$ there exists a unique solution $v(t, x)$ of problem (1.20) such that the derivatives $\partial v(t,x)/\partial t$ and $\partial^2 v(t,x)/\partial x^2$ belong to $C([0,1], E)$ and the coercivity inequality is satisfied:

$$\max_{0 \le t \le 1} \sup_{-\infty < x < \infty} \left| \frac{\partial v(t,x)}{\partial t} \right| + \max_{0 \le t \le 1} \sup_{-\infty < x < \infty} \left| \frac{\partial^2 v(t,x)}{\partial x^2} \right|$$

$$\le M \max_{0 \le t \le 1} \sup_{-\infty < x < \infty} |f(t,x)|. \tag{1.21}$$

Now let us consider the Cauchy problem

$$\frac{\partial v(t,x)}{\partial t} - \frac{\partial^2 v(t,x)}{\partial x^2} = f(t,x), \quad a \le t \le b, \quad v(a,x) = 0, \quad -\infty < x < \infty. \tag{1.22}$$

In equation (1.22) let us make the substitutions $\tau = (t-a)/(b-a)$ and $x = \sqrt{b-a}\, y$. We obtain

$$\frac{\partial v}{\partial \tau} - \frac{\partial^2 v}{\partial y^2} = (b-a)f(\tau(b-a) + a, \sqrt{b-a}\, y),$$

$$0 \le \tau \le 1, \quad -\infty < y < \infty, \quad v(a, \sqrt{b-a}\, y) = 0.$$

Inequality (1.21) yields

$$\max_{0 \le \tau \le 1} \sup_{-\infty < y < \infty} \left| \frac{\partial v(\tau(b-a) + a, \sqrt{b-a}\, y)}{\partial \tau} \right|$$

$$+ \max_{0 \le \tau \le 1} \sup_{-\infty < y < \infty} \left| \frac{\partial^2 v(\tau(b-a) + a, \sqrt{b-a}\, y)}{\partial y^2} \right|$$

$$\le M \max_{0 \le \tau \le 1} \sup_{-\infty < y < \infty} |(b-a)f(\tau(b-a) + a, \sqrt{b-a}\, y)|. \tag{1.23}$$

Since

$$\frac{\partial v(\tau(b-a) + a, \sqrt{b-a}\, y)}{\partial \tau} = \frac{\partial v(t,x)}{\partial t}(b-a)$$

and

$$\frac{\partial^2 v(\tau(b-a) + a, \sqrt{b-a}\, y)}{\partial y^2} = \frac{\partial^2 v(t,x)}{\partial x^2}(b-a),$$

inequality (1.23) yields

$$\max_{a \le t \le b} \sup_{-\infty < x < \infty} \left| \frac{\partial v(t,x)}{\partial t} \right| + \max_{a \le t \le b} \sup_{-\infty < x < \infty} \left| \frac{\partial^2 v(t,x)}{\partial x^2} \right|$$

$$\le M \max_{a \le t \le b} \sup_{-\infty < x < \infty} |f(t,x)|. \tag{1.24}$$

Inequality (1.24) means that the Cauchy problem (1.22) is well posed in the space $C([a, b], E)$. Here $C([a, b], E)$ consists of the E-valued functions $\varphi(t)$ that are defined and continuous on $[a, b]$, and is equipped with the norm

$$\|\varphi\|_{C([a,b],E)} = \max_{a \le t \le b} \|\varphi(t)\|_E.$$

Since the constant M in inequality (1.24) does not depend on a and b, letting $a \to -\infty$ and $b \to \infty$ in problem (1.22) we establish the well-posedness of the problem

$$\frac{\partial v(t, x)}{\partial t} - \frac{\partial^2 v(t, x)}{\partial x^2} = f(t, x), \quad -\infty < t < \infty, \quad -\infty < x < \infty \qquad (1.25)$$

in $\overset{\circ}{C}((-\infty, \infty), E)$. Here $\overset{\circ}{C}((-\infty, \infty), E)$ is the space of all E-valued continuous functions $f(t)$ on $(-\infty, \infty)$ with the property that $\|f(t)\|_E \to 0$ as $|t| \to \infty$. Therefore, problem (1.25) has a unique solution $v(t, x)$ such that both $\partial v(t, x)/\partial t$ and $\partial^2 v(t, x)/\partial x^2$ belong to $\overset{\circ}{C}((-\infty, \infty), E)$ and the coercivity inequality holds:

$$\sup_{-\infty < t < \infty} \sup_{-\infty < x < \infty} \left| \frac{\partial v(t, x)}{\partial t} \right| + \sup_{-\infty < t < \infty} \sup_{-\infty < x < \infty} \left| \frac{\partial^2 v(t, x)}{\partial x^2} \right|$$

$$\le M \sup_{-\infty < t < \infty} \sup_{-\infty < x < \infty} |f(t, x)|. \qquad (1.26)$$

This implies the well-posedness in $\overset{\circ}{C}((-\infty, \infty), E)$ of the problem

$$\frac{\partial z(t, x)}{\partial t} + \frac{\partial^2 z(t, x)}{\partial x^2} = f(t, x), \quad -\infty < t < \infty, \quad -\infty < x < \infty. \qquad (1.27)$$

Indeed, the substitution $t = -\tau$ reduces (1.27) to the problem

$$\frac{\partial z(-\tau, x)}{\partial \tau} - \frac{\partial^2 z(-\tau, x)}{\partial x^2} = -f(-\tau, x), \quad -\infty < \tau < \infty, \quad -\infty < x < \infty.$$

From inequality (1.26) it follows that

$$\sup_{-\infty < \tau < \infty} \sup_{-\infty < x < \infty} \left| \frac{\partial z(-\tau, x)}{\partial \tau} \right| + \sup_{-\infty < \tau < \infty} \sup_{-\infty < x < \infty} \left| \frac{\partial^2 z(-\tau, x)}{\partial x^2} \right|$$

$$\le M \sup_{-\infty < \tau < \infty} \sup_{-\infty < x < \infty} |f(-\tau, x)|.$$

Since $\partial z(-\tau, x)/\partial \tau = -\partial z(t, x)/\partial t$, the last inequality yields

$$\sup_{-\infty < t < \infty} \sup_{-\infty < x < \infty} \left| \frac{\partial z(t, x)}{\partial t} \right| + \sup_{-\infty < t < \infty} \sup_{-\infty < x < \infty} \left| \frac{\partial^2 z(t, x)}{\partial x^2} \right|$$

$$\leq M \sup_{-\infty<t<\infty} \sup_{-\infty<x<\infty} |f(t,x)|. \tag{1.28}$$

Finally, let us consider the problem

$$\frac{\partial^2 v(t,x)}{\partial t^2} - \frac{\partial^4 v(t,x)}{\partial x^4} = f(t,x), \quad -\infty < t < \infty, \quad -\infty < x < \infty \tag{1.29}$$

in $\mathring{C}((-\infty,\infty),E)$. This problem splits into a system of two problems of the forms (1.25) and (1.27),

$$\frac{\partial v(t,x)}{\partial t} - \frac{\partial^2 v(t,x)}{\partial x^2} = u(t,x), \quad -\infty < t < \infty, \quad -\infty < x < \infty,$$

$$\frac{\partial u(t,x)}{\partial t} + \frac{\partial^2 u(t,x)}{\partial x^2} = f(t,x), \quad -\infty < t < \infty, \quad -\infty < x < \infty.$$

From the inequalities (1.26) and (1.28) it folllows that

$$\sup_{-\infty<t<\infty} \sup_{-\infty<x<\infty} \left|\frac{\partial^2 v(t,x)}{\partial t^2}\right| \leq M_1 \sup_{-\infty<t<\infty} \sup_{-\infty<x<\infty} |f(t,x)|.$$

In conjunction with (1.29) this inequality yields

$$\sup_{-\infty<t<\infty} \sup_{-\infty<x<\infty} \left|\frac{\partial^2 v(t,x)}{\partial t^2}\right|$$

$$+ \sup_{-\infty<t<\infty} \sup_{-\infty<x<\infty} \left|\frac{\partial^4 v(t,x)}{\partial x^4}\right| \leq M_2 \sup_{-\infty<t<\infty} \sup_{-\infty<x<\infty} |f(t,x)|.$$

This last inequality means that problem (1.29) is well-posed in $\mathring{C}((-\infty,\infty),E)$. But problem (1.29) cannot be well posed, since it is a (weighted) elliptic problem! A counterexample, due to S. L. Sobolev, is provided by the function

$$v(t,x) = \varphi[r(t,x)]\ln|\ln r(t,x)|, \quad r(t,x) = \frac{t^2}{2!} + \frac{x^4}{4!},$$

where $\varphi(u)$ is a smooth function equal to u near the point $u = 0$ and equal to zero for $|u| \geq 1$. Thus, the analyticity of the semigroup $\exp\{-tA\}$ is a necessary but not a sufficient condition for the well-posedness of problem (1.1) in $C([0,1],E)$.

2.   WELL-POSEDNESS OF THE CAUCHY PROBLEM IN $C_0^\alpha(E)$

1.   The homogeneous problem. The space $C_0^\alpha(E)$.

Necessary and sufficient conditions for the well-posedness of problem (1.1) can be established if one considers this problem in certain spaces $F(E)$ of smooth

E-valued functions on $[0, 1]$. A function $v(t)$ is said to be a *solution of problem* (1.1) *in* $F(E)$ if it is a solution of this problem in $C(E)$ and the functions $v'(t)$ and $Av(t)$ belong to $F(E)$. If $v(t)$ is a solution of problem (1.1) in $F(E)$, then obviously $v_0 \in D(A)$ and $f(t) \in F(E)$. As in the case of the space $C(E)$, one gives the following

Definition 2.1. Problem (1.1) is *well posed in* $F(E)$ if the following two conditions are satisfied:

     1) For any $f(t) \in F(E)$ and $v_0 \in D(A)$ there exists a unique solution $v(t) = v(t; f(t), v_0)$ of (1.1).

     2) The operator $v(t; f(t), v_0)$ is continuous as an operator from $F(E) \times D(A)$ to $F(E)$. Here $F(E) \times D(A)$ is equipped with the norm

$$\|(f, v_0)\|_{F(E) \times D(A)} = \|f\|_{F(E)} + \|v_0\|_E.$$

First let us consider the homogeneous problem (1.7). From the well-posedness of problem (1.7) in $F(E)$ follows, in particular, its well-posedness in $C(E)$. Hence, in order for problem (1.7) to be well posed in $F(E)$ it is necessary that the operator $-A$ be the generator of a strongly continuous semigroup. From formula (1.11) it follows that the solution of the homogeneous problem (1.7) has the form $w(t) = \exp\{-tA\}v_0$. Therefore, the well-posedness of problem (1.7) in $F(E)$ leads to the estimate

$$\|\exp\{-tA\}v_0\|_{F(E)} \leq M\|v_0\|_E. \tag{2.1}$$

This estimate and the well-posedness of problem (1.12) in $C(E)$ imply the analyticity of the semigroup $\exp\{-tA\}$ for $t > 0$. Note that the estimate (2.1) fails if we set $F(E)$ equal to $C^\alpha(E)$, $0 < \alpha < 1$, the Banach space obtained by completion of the set of smooth E-valued functions $v(t)$ on $[0, 1]$ in the norm

$$\|v\|_{C^\alpha(E)} = \|v\|_{C(E)} + \sup_{0 \leq t < t+\tau \leq 1} \frac{\|v(t+\tau) - v(t)\|_E}{\tau^\alpha}.$$

It turns out that the estimate (2.1) holds if we set $F(E) = C_0^\alpha(E)$ $(0 < \alpha < 1)$, the Banach space obtained by completion of the set of E-valued smooth functions $v(t)$ on $[0, 1]$ in the norm

$$\|v\|_{C_0^\alpha(E)} = \|v\|_{C(E)} + \sup_{0 \leq t < t+\tau \leq 1} t^\alpha \frac{\|v(t+\tau) - v(t)\|_E}{\tau^\alpha}.$$

Moreover, the analyticity of the semigroup $\exp\{-tA\}$ is a necessary and sufficient condition for the well-posedness in $C_0^\alpha(E)$ of the general Cauchy problem (1.1.).

First of all let us prove a lemma that will be needed in the sequel.

Lemma 2.1. *For any $0 < t < t + \tau \leq 1$ and $0 \leq \alpha \leq 1$ one has the inequalities*

$$\|\exp\{-tA\} - \exp\{-(t+\tau)A\}\|_{E \to E} \leq M\frac{\tau^\alpha}{t^\alpha}, \tag{2.2}$$

$$\|A(\exp\{-tA\} - \exp\{-(t+\tau)A\})\|_{E \to E} \leq M\frac{\tau^\alpha}{t^{\alpha+1}}, \tag{2.3}$$

where M does not depend on α, t, and τ.

Proof. Let us use the formula

$$\exp\{-tA\} - \exp\{-(t+\tau)A\} = \int_t^{t+\tau} A\exp\{-sA\}ds. \tag{2.4}$$

By inequality (1.14), we have

$$\|\exp\{-tA\} - \exp\{-(t+\tau)A\}\|_{E \to E} \leq M\int_t^{t+\tau}\frac{ds}{s} \leq M\frac{\tau}{t}. \tag{2.5}$$

The inequality

$$\|\exp\{-tA\} - \exp\{-(t+\tau)A\}\|_{E \to E} \leq M \tag{2.6}$$

follows from (1.10) and the triangle inequality. Interpolating (2.5) and (2.6) we obtain inequality (2.2). Further, using formula (2.4) and inequality (1.14) we obtain

$$\|A(\exp\{-tA\} - \exp\{-(t+\tau)A\})\|_{E \to E} \leq 4M^2\int_t^{t+\tau}\frac{ds}{s^2} \leq M_1\frac{\tau}{t^2}. \tag{2.7}$$

The inequality

$$\|A(\exp\{-tA\} - \exp\{-(t+\tau)A\})\|_{E \to E} \leq M_1\frac{\tau}{t}. \tag{2.8}$$

is an obvious consequence of (1.14) and the triangle inequality. Finally, (2.7) and (2.8) yield (2.3). The lemma is proved.

2.   Well-posedness in $C_0^\alpha(E)$ of the general Cauchy problem.

As in the case of the space $C(E)$, from the well-posedness of the Cauchy problem (1.1) one derives the coercivity inequality

$$\|v'\|_{C_0^\alpha(E)} + \|Av\|_{C_0^\alpha(E)} \leq M(\alpha)\Big[\|f\|_{C_0^\alpha(E)} + \|Av_0\|_E\Big]. \tag{2.9}$$

Conversely, if one can establish inequality (2.9) for continuously differentiable functions $f(t)$ and for $v_0 \in D(A^2)$, then, as in $C(E)$, one can show that problem (1.1) is well posed in $C_0^\alpha(E)$.

Now let us show that from the well-posedness of problem (1.1) in $C_0^\alpha(E)$ it follows that the semigroup $\exp\{-tA\}$ is analytic. Here problem (1.1) will be considered in a complex Banach space E. For the problem in a real space we need first to complexify.

Since the semigroup $\exp\{-tA\}$ is strongly continuous, by the strong continuitycriterion for semigroups, the operator $\lambda I + A$ has a bounded inverse for all complex λ with $\mathrm{Re}\,\lambda > \omega$. This means that for any $\varphi \in E$ the equation $\lambda u + Au = \varphi$ has a unique solution $u = (\lambda I + A)^{-1}\varphi$. Clearly, the function $v(t) = \exp(t\lambda)u$ is a solution in $C_0^\alpha(E)$ of problem (1.1) with $f(t) = \exp(t\lambda)\varphi$ and $v_0 = u$. From the coercivity inequality (2.9) it follows that

$$|\lambda|\,\|u\|_E + \|Au\|_E \le M(\alpha)\Big[\|\varphi\|_E + \|\exp(\lambda t)\|_{C_0^\alpha(\mathbf{R}^1)}^{-1}\|Au\|_E\Big],$$

where

$$\|\exp(\lambda t)\|_{C_0^\alpha(\mathbf{R}^1)} = \max_{0 \le t \le 1} |\exp(\lambda t)| + \sup_{0 \le t < t + \Delta t \le 1} t^\alpha \left|\frac{\exp(\lambda(t + \Delta t)) - \exp(\lambda t)}{\Delta t^\alpha}\right|.$$

Clearly, $\|\exp(\lambda t)\|_{C_0^\alpha(\mathbf{R}^1)} \to \infty$ as $\mathrm{Re}\,\lambda \to \infty$. Consequently, for sufficiently large $w_1 \ge w$ and any λ with $\mathrm{Re}\,\lambda \ge w_1$ we have

$$\|(\lambda I + A)^{-1}\|_{E \to E} \le M|\lambda|^{-1}.$$

By the analyticity criterion, this implies the analyticity of the semigroup $\exp\{-tA\}$.

Thus, the analyticity of the semigroup $\exp\{-tA\}$ is a necessary condition for the well-posedness of problem (1.1) in $C_0^\alpha(E)$. It turns out that this condition is also sufficient for well-posedness.

Theorem 2.1. *Let $-A$ be the generator of an analytic semigroup. Then problem (1.1) is well posed in $C_0^\alpha(E)$ and the coercivity inequality*

$$\|v'\|_{C_0^\alpha(E)} + \|Av\|_{C_0^\alpha(E)} \le M\Big[\|Av_0\|_E + \frac{1}{\alpha(1-\alpha)}\|f\|_{C_0^\alpha(E)}\Big]$$

holds, where M does not depend on α, v_0, and f.

Proof. If $v(t)$ is a solution in $C_0^\alpha(E)$ of problem (1.1), then it is a solution in $C(E)$ of this problem. Hence, by (1.19), we have the representation

$$v(t) = \exp\{-tA\}v_0 + \int_0^t \exp\{-(t-s)A\}f(s)ds \equiv w(t) + g(t). \tag{2.10}$$

To prove the theorem we must show that $w(t)$ is a solution in $C(E)$ of problem (1.7), that $g(t)$ is a solution in $C(E)$ of problem (1.12), and then that the functions $Aw(t)$ and $Ag(t)$ belong to $C_0^\alpha(E)$, and finally obtain estimates for the norms of these functions. Here we address only the last task. Using (1.10), we deduce that $w(t) \in D(A)$ and

$$\|Aw(t)\|_E = \|\exp\{-tA\}Av_0\|_E \le M\|Av_0\|_E, \quad 0 \le t \le 1.$$

Further, applying (2.2) we show that, for $0 < t < t + \tau \le 1$,

$$\|Aw(t+\tau) - Aw(t)\|_E = \|[\exp\{-(t+\tau)A\} - \exp\{-tA\}]Av_0\|_E \le M\|Av_0\|_E\frac{\tau^\alpha}{t^\alpha}.$$

Thus, we have proved that

$$\|Aw\|_{C_0^\alpha(E)} \le M\|Av_0\|_E. \tag{2.11}$$

Now let us estimate $Ag(t)$ in the norm of $C_0^\alpha(E)$. Using formula (2.10), we obtain the identity

$$g(t) = A^{-1}(I - \exp\{-tA\})f(t) + \int_0^t \exp\{-(t-s)A\}(f(s) - f(t))ds. \tag{2.12}$$

Using (1.10), (1.14), and the fact that the operator A is closed, we conclude that $g(t) \in D(A)$ and

$$\|Ag(t)\|_E \le M\left[\|f(t)\|_E + \int_0^t \|f(s) - f(t)\|_E\frac{ds}{t-s}\right]$$

$$\le M\|f\|_{C_0^\alpha(E)}\left(1 + \int_0^t \frac{ds}{(t-s)^{1-\alpha}s^\alpha}\right), \quad 0 \le t \le 1.$$

Since

$$\int_0^t \frac{ds}{(t-s)^{1-\alpha}s^\alpha} = \int_0^1 \frac{d\tau}{(1-\tau)^{1-\alpha}\tau^\alpha} = \frac{\pi}{\sin\pi\alpha},$$

it follows that for $0 \leq t \leq 1$,

$$\|Ag(t)\|_E \leq \frac{M_1}{\alpha(1-\alpha)}\|f\|_{C_0^\alpha(E)}. \tag{2.13}$$

Now let us estimate the difference $Ag(t+\tau) - Ag(t)$ for $0 < t < t+\tau \leq 1$. We shall consider separately the cases $t \leq 2\tau$ and $t > 2\tau$. If $t \leq 2\tau$, then (2.13) yields

$$\|Ag(t+\tau) - Ag(t)\|_E \leq \|Ag(t+\tau)\|_E + \|Ag(t)\|_E$$

$$\leq \frac{2M_1}{\alpha(1-\alpha)}\|f\|_{C_0^\alpha(E)}\tau^\alpha\tau^{-\alpha} \leq \frac{2^{1+\alpha}M_1}{\alpha(1-\alpha)}\|f\|_{C_0^\alpha(E)}\tau^\alpha t^{-\alpha}.$$

Now let $t > 2\tau$. From identity (2.12) it follows that

$$Ag(t+\tau) - Ag(t) = [f(t+\tau) - f(t) + \exp\{-tA\}f(t) - \exp\{-(t+\tau)A\}f(t+\tau)]$$

$$+ \int_{t-\tau}^{t+\tau} A\exp\{-(t+\tau-s)A\}(f(s) - f(t+\tau))ds$$

$$+ \int_{t-\tau}^{t} A\exp\{-(t-s)A\}(f(t) - f(s))ds$$

$$+ \int_{0}^{t-\tau} A\exp\{-(t+\tau-s)A\}ds\,(f(t) - f(t+\tau))$$

$$+ \int_{0}^{t-\tau} A[\exp\{-(t+\tau-s)A\} - \exp\{-(t-s)A\}](f(s) - f(t))ds$$

$$= I_1 + I_2 + I_3 + I_4 + I_5.$$

Using the estimates (1.10) and (2.2), we obtain

$$\|I_1\|_E \leq M\|f\|_{C_0^\alpha(E)}\tau^\alpha t^{-\alpha}.$$

Further, by (2.2), we have

$$\|I_2\|_E \leq \int_{t-\tau}^{t+\tau} \|A\exp\{-(t+\tau-s)A\}\|_{E\to E}\|f(s) - f(t+\tau)\|_E ds$$

$$\leq M \int_{t-\tau}^{t+\tau} \frac{ds}{(t+\tau-s)^{1-\alpha}s^\alpha}\|f\|_{C_0^\alpha(E)}$$

$$\leq M\|f\|_{C_0^\alpha(E)}\frac{1}{(t-\tau)^\alpha}\int_{t-\tau}^{t+\tau}\frac{ds}{(t+\tau-s)^{1-\alpha}} = \frac{2^\alpha M}{\alpha}\|f\|_{C_0^\alpha(E)}\frac{\tau^\alpha}{(t-\tau)^\alpha}.$$

Since $t - \tau = \frac{t}{2} + \frac{t}{2} - \tau > \frac{t}{2}$, we conclude that

$$\|I_2\|_E \leq \frac{M_1}{\alpha} \|f\|_{C_0^\alpha(E)} \frac{\tau^\alpha}{t^\alpha}.$$

In a similar manner we can show that

$$\|I_3\|_E \leq \frac{M_1}{\alpha} \|f\|_{C_0^\alpha(E)} \frac{\tau^\alpha}{t^\alpha}.$$

Next, using the identity

$$I_4 = [\exp\{-(t+\tau)A\} - \exp\{-2\tau A\}](f(t+\tau) - f(t))$$

and estimate (1.10), we obtain the estimate

$$\|I_4\|_E \leq M \|f\|_{C_0^\alpha(E)} \frac{\tau^\alpha}{t^\alpha}.$$

Finally, using (2.3) with $\alpha = 1$ we obtain the estimate

$$\|I_5\|_E \leq \int_0^{t-\tau} \|A[\exp\{-(t+\tau-s)A\} - \exp\{-(t-s)A\}]\|_{E\to E} \|f(s) - f(t)\|_E ds$$

$$\leq M \|f\|_{C_0^\alpha(E)} \int_0^{t-\tau} \frac{\tau ds}{(t-s)^{2-\alpha} s^\alpha}.$$

Since for $t > 2\tau$ we have the bound

$$\int_0^{t-\tau} \frac{\tau ds}{(t-s)^{2-\alpha} s^\alpha} \leq \frac{2^{2+\alpha}}{1-\alpha} \frac{\tau^\alpha}{t^\alpha},$$

we conclude that

$$\|I_5\|_E \leq \frac{M_1}{1-\alpha} \|f\|_{C_0^\alpha(E)} \frac{\tau^\alpha}{t^\alpha}.$$

Thus, we have shown that for any $0 < t < t+\tau \leq 1$ the following inequality holds:

$$\|Ag(t+\tau) - Ag(t)\|_E \leq \frac{M}{\alpha(1-\alpha)} \|f\|_{C_0^\alpha(E)} \frac{\tau^\alpha}{t^\alpha}. \tag{2.14}$$

The estimates (2.13) and (2.14) give

$$\|Ag\|_{C_0^\alpha(E)} \leq \frac{M}{\alpha(1-\alpha)} \|f\|_{C_0^\alpha(E)},$$

which in conjunction with estimate (2.11) yields the inequality

$$\|Av\|_{C_0^\alpha(E)} \leq M \left[\|Av_0\|_E + \frac{1}{\alpha(1-\alpha)} \|f\|_{C_0^\alpha(E)} \right].$$

By the inequality triangle, this last inequality and equation (1.1) yield

$$\|v'\|_{C_0^\alpha(E)} \leq M\Big[\|Av_0\|_E + \frac{1}{\alpha(1-\alpha)}\|f\|_{C_0^\alpha(E)}\Big].$$

Theorem 2.1 is proved.

3.   WELL-POSEDNESS OF THE CAUCHY PROBLEM IN $L_p(E)$

1.   Definition of the well-posedness of the Cauchy problem in $L_p(E)$.

Let us now enlarge the space in which we seek the solution of problem (1.1). We denote by $L_p(E) = L_p([0,1], E)$, where $1 \leq p \leq \infty$, the Banach space of all strongly measurable E-valued functions $v(t)$ on $[0,1]$ for which the norm

$$\|v\|_{L_p(E)} = \Big(\int_0^1 \|v(t)\|_E^p dt \Big)^{1/p}$$

is finite. Using Hölder's inequality one can show that for $p_1 \leq p_2$ the space $L_{p_2}(E)$ is imbedded in the space $L_{p_1}(E)$, i.e.

$$L_{p_1}(E) \supset L_{p_2}(E) \quad \text{and} \quad \|v\|_{L_{p_2}(E)} \geq \|v\|_{L_{p_1}(E)}, \quad v \in L_{p_2}(E).$$

A function $v(t)$ is said to be *absolutely continuous* if it has a derivative $v'(t)$ for almost every t such that $v'(t) \in L_1(E)$, and if the Newton-Leibniz formula

$$v(t) - v(\tau) = \int_\tau^t v'(s)ds$$

holds for all $t, \tau \in [0,1]$. Here the integral is understood in the sense of Bochner.

A function $v(t)$ is said to be a *solution of problem* (1.1) *in* $L_p(E)$ if it is absolutely continuous, the functions $v'(t)$ and $Av(t)$ belong to $L_p(E)$, equation (1.1) is satisfied for almost every t, and $v(0) = v_0$. From this definition it follows that a necessary condition for the solvability of problem (1.1) in $L_p(E)$ is that $f(t) \in L_p(E)$. It will be shown that in certain cases this condition is also sufficient for the solvability of problem (1.1). As concerns the initial element, in contrast to the situation considered earlier, from the solvability of problem (1.1) in $L_p(E)$ it follows only that $v_0 \in E$. In the case of an unbounded operator A this does not allow us to prove the solvability of problem (1.1). For this reason let us consider first the nonhomogeneous problem with null initial condition,

$$v'(t) + Av(t) = f(t), \quad 0 \le t \le 1, \quad v(0) = 0. \tag{3.1}$$

Definition 3.1. Problem (3.1) is *well posed in* $L_p(E)$ if the following conditions are satisfied:

    1) For any $f(t) \in L_p(E)$ it has a unique solution $v(t) = v(t; f(t), 0)$ in $L_p(E)$.

    2) The operator $v(t; f(t), 0)$ is continuous in $L_p(E)$.

From the unique solvability of problem (3.1) it follows that the operator $v(t; f(t), 0)$ is linear. From the continuity of this operator it follows that it is bounded.

As we did earlier, we establish that the operator $Av(t; f(t), 0)$ is bounded in $L_p(E)$. Further, from the solvability of problem (3.1) in $L_p(E)$ we infer, via the substitution $v(t) = w(t) + v_0$, the unique solvability of the general problem (1.1) for any $v_0 \in D(A)$ as well as the coercivity inequality for the solutions of this problem,

$$\|v'\|_{L_p(E)} + \|Av\|_{L_p(E)} \le M(p) \Big[\|f\|_{L_p(E)} + \|Av_0\|_E \Big]. \tag{3.2}$$

This inequality, like inequality (2.1) in the case of the space $C_0^\alpha(E)$, allows us to study the resolvent of the operator $-A$. In contrast to the case of $C_0^\alpha(E)$, here for the moment we do not know whether there exists a bounded inverse operator $(\lambda I + A)^{-1}$. Only for arbitrary complex λ satisfying $\operatorname{Re}\lambda \ge w_1$ and sufficiently large w_1 does inequality (3.2) allow us to derive the estimate

$$\|u\|_E \le M \|\varphi\|_E |\lambda|^{-1}$$

for the solutions of the equation $\lambda u + Au = \varphi$. This estimate shows that the equation $\lambda u + Au = \varphi$ cannot have more than one solution. The solvability of this equation for any $\varphi \in E$ can be established under the additional assumption that the operator A^{-1} is not only bounded but also compact in E. If this is the case, then we can pass to the equivalent equation $\lambda A^{-1}u + u = A^{-1}\varphi$. From the uniqueness theorem for this equation and the Leray-Schauder theory we obtain an existence theorem for it, and hence for the original equation.

The last inequality implies the estimate (1.17) for the resolvent of the operator $-A$. Thus, in the case of the space $L_p(E)$, too, the analyticity of the semigroup $\exp\{-tA\}$ is a necessary condition for the well-posedness of problem (3.1). This fact was established under the assumption that problem (3.1), rather than the general problem (1.1), is well posed. However, the proof was carried out under the stronger assumption that the operator A^{-1} is compact in E.

2.   A formula for the solution of the Cauchy problem in $L_p(E)$.

Let $v(t)$ be a solution in $L_p(E)$ of the Cauchy problem (1.1). Let us show that formula (1.19) for $v(t)$ remains valid in the present case. To this end we need the following assertion.

Lemma 3.1. *Let $\psi(t)$ and $\varphi(t)$ be a continuously differentiable operator-function and an absolutely continuous function, respectively. Then the function $\psi(t)\varphi(t)$ is absolutely continuous.*

Proof. Let $l \in E^*$. Consider the scalar function

$$\gamma(t) = l(\psi(t)\varphi(t)).$$

Since the functional l is linear, we have

$$\gamma(t) - \gamma(\tau) = l(\psi(t)\varphi(t) - \psi(\tau)\varphi(\tau)) = l([\psi(t) - \psi(\tau)]\varphi(t) + \psi(\tau)[\varphi(t) - \varphi(\tau)]).$$

From the boundedness of l, the continuous differentiability of $\psi(t)$, and the absolute continuity of $\varphi(t)$ it follows that

$$|\gamma(t) - \gamma(\tau)| \le \|l\|\left[M_1|t - \tau| + M_2\int_\tau^t \|\varphi'(s)\|_E ds\right].$$

Now, if $\{(\tau_k, t_k)\}_{k=1}^n$ is an arbitrary system of disjoint subintervals of $[0,1]$, then

$$\sum_{k=1}^n |\gamma(t_k) - \gamma(\tau_k)| \le \|l\|\left[M_1\sum_{k=1}^n |t_k - \tau_k| + M_2\sum_{k=1}^n \int_{\tau_k}^{t_k} \|\varphi'(s)\|_E ds\right].$$

Hence, by the definition of the absolute continuity, the scalar function $\gamma(t)$ is absolutely continuous. Consequently, we have

$$\gamma(t) - \gamma(\tau) = \int_\tau^t \gamma'(s)ds.$$

From this and the linearity of the functional l it follows that

$$l(\psi(t)\varphi(t) - \psi(\tau)\varphi(\tau)) = l \int_\tau^t (\psi(s)\varphi(s))' ds$$

for all $l \in E^*$ and all $t, \tau \in [0, 1]$. Therefore,

$$\psi(t)\varphi(t) - \psi(\tau)\varphi(\tau) = \int_\tau^t (\varphi(s)\psi(s))' ds.$$

Lemma 3.1 is proved.

Let $-A$ be the generator of an analytic semigroup $\exp\{-tA\}$. Then, using Lemma 3.1, we obtain for arbitrary $0 \le s \le t \le 1$ the identity

$$\frac{d}{ds}[\exp\{-(t-s)A\}A^{-1}v(s)] = \exp\{-(t-s)A\}A^{-1}f(s).$$

Integrating this identity with respect to s from 0 to t we obtain the relation

$$A^{-1}v(t) = \exp\{-tA\}A^{-1}v_0 + \int_0^t \exp\{-(t-s)A\}A^{-1}f(s)ds.$$

Since the operator A is closed and commutes with the semigroup $\exp\{-(t-s)A\}$, we obtain relation (1.19). This means, in particular, that problem (1.1) cannot have more than one solution in $L_p(E)$.

3. Spaces of initial data.

It follows from formula (1.11) that the solution of problem (1.7) has the form $w(t) = \exp\{-tA\}v_0$. Since the semigroup $\exp\{-tA\}$ is analytic, for any $v_0 \in E$ the function $w(t)$ is continuously differentiable for $t > 0$ and satisfies equation (1.7) for these values of t. Moreover, $w(t)$ is continuous at $t = 0$ and satisfies the initial condition (1.7). Hence, in order that $w(t)$ be a solution in $L_p(E)$ it is necessary and sufficient that the function $Aw(t) = Av(t; 0, v_0)$ belong to $L_p(E)$. The collection of all elements $v_0 \in E$ with this property is obviously a linear set containing $D(A)$. It becomes a Banach space, denoted by $E_{1-\frac{1}{p}}$, if one endows it with the norm

$$|v_0|_{1-\frac{1}{p}} = \|v_0\|_{E_{1-\frac{1}{p}}} = \left(\int_0^1 \|A\exp\{-tA\}v_0\|_E^p dt \right)^{\frac{1}{p}} + \|v_0\|_E. \tag{3.3}$$

The space $E_{1-\frac{1}{p}}$ is a particular case of the space $E_{\alpha,p}$ $(0 < \alpha < 1,\ 1 \le p \le \infty)$ with the norm

$$\|v_0\|_{E_{\alpha,p}} = \left(\int_0^1 \|t^{1-\alpha} A \exp\{-tA\} v_0\|_E^p \frac{dt}{t} \right)^{\frac{1}{p}} + \|v_0\|_E \quad (1 \le p < \infty),$$

$$\|v_0\|_{E_{\alpha,\infty}} = \sup_{0 \le t \le 1} \|t^{1-\alpha} A \exp\{-tA\} v_0\|_E + \|v_0\|_E. \tag{3.4}$$

Specifically, $E_{1-\frac{1}{p}} = E_{1-\frac{1}{p},p}$. Let us show that the second term $\|v_0\|_E$ in these norms can be discarded, since this leads to equivalent norms. To this end we shall use the identity

$$(I - \exp\{-A\}) v_0 = \int_0^1 A \exp\{-tA\} v_0 dt.$$

Since the semigroup $\exp\{-tA\}$ obeys the exponential decay estimate (1.10), the operator $I - \exp\{-A\}$ has a bounded inverse (because the spectral radius of the operator $\exp\{-A\}$ is smaller than 1). Hence, we have the identity

$$v_0 = (I - \exp\{-A\})^{-1} \int_0^1 A \exp\{-tA\} v_0 dt.$$

This implies the basic inequality

$$\|v_0\|_E \le \|(I - \exp\{-A\})^{-1}\|_{E \to E} \int_0^1 \|A \exp\{-tA\} v_0\|_E dt.$$

If $p = \infty$, the basic inequality yields

$$\|v_0\|_E \le \frac{1}{\alpha} \|(I - \exp\{-A\})^{-1}\|_{E \to E} \sup_{0 \le t \le 1} \|t^{1-\alpha} A \exp\{-tA\} v_0\|_E.$$

If $1 \le p < \infty$, the basic inequality and Hölder's inequality imply that

$$\|v_0\|_E \le \left(\frac{p-1}{\alpha p} \right)^{\frac{p-1}{p}} \|(I - \exp\{-A\})^{-1}\|_{E \to E} \left(\int_0^1 \|t^{1-\alpha} A \exp\{-tA\} v_0\|_E^p \frac{dt}{t} \right)^{1/p}.$$

Since $D(A)$ is dense in E, $D(A)$ is dense in $E_{\alpha,p}$, i.e., these spaces can be obtained by completion of $D(A)$ in the norms $\|\cdot\|_{E_{\alpha,p}}$.

Let $v(t)$ be a solution of the general problem (1.1). Since problem (3.1) has a solution $g(t)$ for any $f(t) \in L_p(E)$, the function $w(t) = v(t) - g(t)$ will be a solution of problem (1.7) in $L_p(E)$. Consequently, $v_0 \in E_{1-\frac{1}{p}}$. Therefore, for the solvability in $L_p(E)$ of the general problem (1.1) (under the assumption that problem (3.1) is well posed) it is necessary and sufficient that $v_0 \in E_{1-\frac{1}{p}}$.

4.   The values of the solution of the Cauchy problem in $L_p(E)$ for fixed t.

In the preceding subsection we have shown that the initial value v_0 of the solution of a problem (1.1) that is well posed in $L_p(E)$ must belong to the space $E_{1-\frac{1}{p}}$. It turns out that not only the initial value of the solution but also all its values belong to this space. If problem (1.1) is well posed, then its solution is given by formula (1.19) whenever $f(t) \in L_p(E)$ and $v_0 \in E_{1-\frac{1}{p}}$. As it turns out, it is always true that for arbitrary functions $f(t) \in L_p(E)$ and elements $v_0 \in E_{1-\frac{1}{p}}$ formula (1.19) defines a continuous function with values in $E_{1-\frac{1}{p}}$, and the operator defined in this way is bounded from $L_p(E) \times E_{1-\frac{1}{p}}$ to $C(E_{1-\frac{1}{p}})$. Here $L_p(E) \times E_{1-\frac{1}{p}}$ denotes the Banach space of the pairs $(f(t), v_0)$, $f(t) \in L_p(E), v_0 \in E_1 - \frac{1}{p}$ with the norm

$$\|(f(t), v_0)\|_{L_p(E) \times E_{1-\frac{1}{p}}} = \|f\|_{L_p(E)} + \|v_0\|_{E_{1-\frac{1}{p}}}.$$

Theorem 3.1. *For any $f(t) \in L_p(E)$ and $v_0 \in E_{1-\frac{1}{p}}$ formula (1.19) defines an $E_{1-\frac{1}{p}}$-valued continuous function on $[0,1]$ and*

$$\max_{0 \leq t \leq 1} |v(t)|_{1-\frac{1}{p}} \leq M \left[|v_0|_{1-\frac{1}{p}} + \frac{p^2}{p-1} \|f\|_{L_p(E)} \right]. \tag{3.5}$$

Proof. By (2.9),

$$v(t) = \exp\{-tA\}v_0 + \int_0^t \exp\{-(t-s)A\}f(s)ds = w(t) + g(t).$$

From the estimates (1.10) and the definition of the space $E_{1-\frac{1}{p}}$ it follows that

$$|w(t)|_{1-\frac{1}{p}} = \left(\int_0^1 \|\exp\{-(t+\tau)A\}Av_0\|_E^p d\tau \right)^{1/p}$$

$$\leq \|\exp\{-tA\}\|_{E \to E} |v_0|_{1-\frac{1}{p}} \leq M|v_0|_{1-\frac{1}{p}}. \tag{3.6}$$

Now let us consider the solution $g(t)$ of the nonhomogeneous problem (3.1). From the estimate

$$\|A \exp\{-tA\}\|_{E \to E} \leq M e^{-\delta t} t^{-1} \tag{3.7}$$

it follows that

$$\|A \exp\{-\tau A\}g(t)\|_E \leq M \int_0^t \frac{\|f(s)\|_E}{t-s+\tau} ds.$$

Hence,

$$|g(t)|_{1-\frac{1}{p}} \leq M\Big(\int_0^\infty \Big(\int_0^t \frac{\|f(s)\|_E}{t-s+\tau}ds\Big)^p d\tau\Big)^{1/p}.$$

Let us make the substitution $t - s = z$ and extend the function $f(z)$ by zero for $z \geq 1$ and $z \leq 0$. Then for the extended function $f^*(z)$ we obtain the inequality

$$|g(t)|_{1-\frac{1}{p}} \leq M\Big(\int_0^\infty \Big(\int_0^\infty \frac{\|f^*(t-z)\|_E dz}{\tau+z}\Big)^p d\tau\Big)^{1/p}.$$

Making the substitution $z = \tau r$ and using Minkowski's inequality, we have

$$|g(t)|_{1-\frac{1}{p}} \leq M\int_0^\infty \frac{1}{1+r}\Big(\int_0^\infty \|f^*(t-\tau r)\|_E^p d\tau\Big)^{1/p} dr.$$

Setting $r\tau = s$ in the inner integral, we obtain

$$|g(t)|_{1-\frac{1}{p}} \leq M\int_0^\infty \frac{dr}{(1+r)r^{1/p}}\Big(\int_0^\infty \|f^*(t-s)\|_E^p ds\Big)^{1/p}$$

$$\leq M\frac{\pi}{\sin(\pi/p)}\|f\|_{L_p(E)}. \tag{3.8}$$

The estimate (3.6) shows that $w(t) \in E_{1-\frac{1}{p}}$ for any $t \in [0,1]$ if $v_0 \in E_{1-\frac{1}{p}}$. If $v_0 \in D(A)$, then obviously $w(t) \in D(A)$ for any $t \in [0,1]$, and as a function with values in this space $w(t)$ is continuous. Consequently, $w(t)$ is continuous as a function with values in $E_{1-\frac{1}{p}}$ whenever $v_0 \in D(A)$. Now suppose $v_0 \in E_{1-\frac{1}{p}}$. Then (see Subsection 3.5) there exists a sequence $v_{0n} \in D(A)$ such that $v_{0n} \to v_0$ in $E_{1-\frac{1}{p}}$. Since the constant M in inequality (3.6) does not depend on v_0, the sequence of corresponding functions $w_n(t)$ converges uniformly in the norm of the space $E_{1-\frac{1}{p}}$. Hence, the limit function $w(t)$, which corresponds to $v_0 \in E_{1-\frac{1}{p}}$, is continuous in the norm of this space. The function $g(t)$ is dealt with in the same manner. The estimate (3.8) shows that its values belong to $E_{1-\frac{1}{p}}$ for all $t \in [0,1]$ and their norms are uniformly bounded in t. If $f(t) \in C_0^\alpha(E)$, then, as we have seen in Section 2, $g(t)$ is a continuous function with values in $D(A)$. Since $C_0^\alpha(E)$ is dense in $L_p(E)$, given a function $f(t) \in L_p(E)$ there exists a sequence $f_n(t) \in C_0^\alpha(E)$ such that $f_n(t) \to f(t)$ in $L_p(E)$. The corresponding functions $g_n(t)$ are continuous in the norm of $D(A)$, and hence in the norm of $E_{1-\frac{1}{p}}$. From the uniform (in t) estimate (3.8) it follows that the sequence of functions $g_n(t)$ converges uniformly to $g(t)$ in the norm of $E_{1-\frac{1}{p}}$. Therefore, $g(t)$ is a continuous function with values in $E_{1-\frac{1}{p}}$. Theorem 3.1 is proved.

5.   The coercivity inequality for the solutions in $L_p(E)$ of the general problem (1.1).

The boundedness of the operator $v(t; f(t), v_0)$ indicated in Theorem 3.1 allows us to sharpen inequality (3.2) and establish the following coercivity inequality for the solutions of problem (1.1) in $L_p(E)$:

$$\|v'\|_{L_p(E)} + \|Av\|_{L_p(E)} + \|v\|_{C(E_{1-\frac{1}{p}})} \le M(p)\left[\|f\|_{L_p(E)} + |v_0|_{E_{1-\frac{1}{p}}}\right]. \qquad (3.9)$$

The facts enumerated above concerning the well-posedness of the general problem (1.1) in $L_p(E)$ were derived from the well-posedness of problem (3.1) in $L_p(E)$. As we have seen, a necessary condition for the latter is the analyticity of the semigroup $\exp\{-tA\}$. There arises the question of whether this analyticity is also a sufficient condition for the well-posedness of problem (3.1). It turns out that here the following extrapolation result holds.

Theorem 3.2. *Suppose that problem* (3.1) *is well posed in* $L_{p_0}(E)$ *for some* p_0, $1 < p_0 < \infty$. *Then it is also well posed in* $L_p(E)$ *for any* p, $1 < p < \infty$, *and in* (3.9) $M(p) = M(p_0)p^2(p-1)^{-1}$.

The **proof** of this theorem is carried out according to the following scheme. The solution of problem (3.1) defines a convolution operator, given by the formula

$$v(t) = \int_0^t \exp\{-(t-s)A\}f(s)ds.$$

This operator extends in a natural manner to functions defined on the whole real line, and its investigation is carried out by means of an extrapolation theorem given in [11].

First, let us introduce the concepts and notations needed below. We denote by $L_p^\infty(E)$ the Bochner space of functions defined on the real line. Let $L_\infty^0(E)$ denote the class of bounded strongly measurable E-valued functions with compact support on the real line. We will address the problem of the boundedness of a certain operator in $L_p^\infty(E)$. To handle this problem we shall need the following extrapolation theorem (see [11]).

Theorem 3.3. *Let the operator* T *satisfy the following conditions:*

1) T *is a bounded linear operator in* L_{p_0} *for some* $p_0 \in (1, \infty)$.

2) *If the support of the function $f(t)$ is contained in an interval $|t - t_0| < \rho$ and $\int_{-\infty}^{\infty} f(t)dt = 0$, then there exist constants M_1 and M_2 such that*

$$\int_{|t-t_0|>M_1\rho} \|Tf(t)\|_E dt \le M_2 \int_{-\infty}^{\infty} \|f(t)\|_E dt = M_2\|f\|_{L_1^\infty}. \qquad (3.10)$$

Then T is a bounded linear operator in $L_p^\infty(E)$ for any $p \in (1, p_0)$. Moreover,

$$\|Tf\|_{L_p^\infty(E)} \le M(p)\|f\|_{L_p^\infty(E)},$$

where $M(p)$ depends only on M_1, M_2, and p:

$$M(p) = M(p_0)(p-1)^{-1}(p_0 - p)^{-1}.$$

Now let us define an operator K on functions $f(t) \in L_\infty^0(E)$ by the rule

$$Kf(t) = \int_{-\infty}^{\infty} \tilde{K}(t - s)f(s)ds,$$

where the kernel $\tilde{K}$ is given by the formula

$$\tilde{K}(\tau) = \begin{cases} \exp\{-\tau A\}, & \text{if } \tau > 0, \\ 0, & \text{if } \tau \le 0. \end{cases}$$

From the estimates (1.10) and (2.2) it follows that K is a bounded linear operator from $L_p^\infty(E)$ to itself for any $p \in [1, \infty)$, and from $L_p^\infty(E)$ to $C(E)$ for any $p \in [1, \infty)$. Next, let us define the superposition operator B in $L_p^\infty(E)$ by

$$B = AK. \qquad (3.11)$$

The domain $D(B)$ of B consists of the functions $f(t) \in L_p^\infty(E)$ with the properties that the function $\varphi(t) = Kf(t)$ takes values in $D(A)$ for almost every t and the function $\psi(t) = A\varphi(t) = AKf(t)$ belongs to $L_p^\infty(E)$. Since the operator A is closed and commutes with the semigroup $\exp\{-tA\}$, $D(B)$ includes the functions $f(t) \in L_p^\infty(E)$ such that $Af(t) \in L_p^\infty(E)$. The set of such functions $f(t)$ is dense in $L_p^\infty(E)$ because $D(A)$ is dense in E. Consequently, $D(B)$ is dense in $L_p^\infty(E)$. Obviously, the operator B defined as above is additive and homogeneous. Since K is bounded in $L_p^\infty(E)$ and A (regarded as an operator in $L_p^\infty(E)$) is closed, it follows that B is a closed operator in $L_p^\infty(E)$. The question for us is, when is B defined on all of $L_p^\infty(E)$, i.e., when is B bounded in $L_p^\infty(E)$?

Theorem 3.4. *Let the operator B be bounded in $L_{p_0}^\infty(E)$ for some p_0, $1 < p_0 < \infty$. Then B is bounded in $L_p^\infty(E)$ for any p, $1 < p < \infty$.*

Proof. First let us show that B satisfies the conditions of Theorem 3.3. This will imply that B is bounded in the spaces $L_p^\infty(E)$ with $p \in (1, p_0)$.

Suppose the support of the function $f(t)$ lies in the interval $|t - t_0| < \rho$ and $\int_{-\infty}^\infty f(t)dt = 0$. Then $f(t) \in L_p^\infty(E)$ for any p, $1 \le p \le \infty$. In particular, $f(t) \in L_{p_0}^\infty(E)$. Hence, $f(t) \in D(B)$ and $Bf(t) \in L_{p_0}^\infty(E)$. By the definition of the superposition operator,

$$Bf(t) = AKf(t) = A\int_{-\infty}^\infty \tilde{K}(t-s)f(s)ds = A\int_{t_0-\rho}^{t_0+\rho} [\tilde{K}(t-s) - \tilde{K}(t-t_0)]f(s)ds.$$

Since $\int_{t_0-\rho}^{t_0+\rho} f(s)ds = 0$ and $\tilde{K}(\tau) = 0$ for $\tau \le 0$, it follows that $Bf(t) = 0$ for $t < t_0 - \rho$, and

$$Bf(t) = A\int_{t_0-\rho}^{t_0+\rho} [\exp\{-(t-s)A\} - \exp\{-(t-t_0)A\}]f(s)ds$$

for $t > t_0 + \rho$. Therefore,

$$\int_{|t-t_0|\ge M_1\rho} \|Bf(t)\|_E dt = \int_{|t-t_0|\ge M_1\rho} \|A\int_{t_0-\rho}^{t_0+\rho} [\tilde{K}(t-s) - \tilde{K}(t-t_0)]f(s)ds\|_E dt$$

$$= \int_{|z|\ge M_1\rho} \left\|A\int_{t_0-\rho}^{t_0+\rho} [\tilde{K}(z+t-s) - \tilde{K}(s)]f(s)ds\right\|_E dz.$$

Let $M_1 \ge 2$. Then

$$\int_{|t-t_0|\ge M_1\rho} \|Bf(t)\|_E dt \le \int_{|t-t_0|\ge 2\rho} \|Bf(t)\|_E dt$$

$$= \int_{2\rho}^\infty \|A\int_{t_0-\rho}^{t_0+\rho} [\exp\{-(z+t_0-s)A\} - \exp\{-zA\}]f(s)ds\|_E dz$$

$$\le \int_{t_0-\rho}^{t_0+\rho} \int_{2\rho}^\infty \|A[\exp\{-(z+t_0-s)A\} - \exp\{-zA\}]\|_{E\to E} dz \, \|f(s)\|_E ds.$$

Let

$$Q = \int_{2\rho}^\infty \|A[\exp\{-(z+t_0-s)A\} - \exp\{-zA\}]\|_{E\to E} dz.$$

To show that the operator B satisfies the conditions of Theorem 3.3 it suffices to prove that Q is bounded for all $t_0 \in (-\infty, \infty)$ and $|s - t_0| \le \rho$. We shall examine separately two cases: $t_0 \leq s \leq t_0 + \rho$ and $t_0 - \rho \leq s \leq t_0$. In the first case inequality (2.3) with $\alpha = 1$ yields

$$Q \leq M \int_{2\rho}^{\infty} \frac{(s - t_0)dz}{(z + t_0 - s)^2} = \frac{M(s - t_0)}{2\rho + t_0 - s} = \frac{M\rho}{2\rho - \rho} = M.$$

In the second case inequality (2.3) with $\alpha = 1$ yields

$$Q \leq M \int_{2\rho}^{\infty} \frac{(t_0 - s)dz}{z^2} = \frac{M(t_0 - s)}{2\rho} \leq \frac{M\rho}{2\rho} = \frac{M}{2}.$$

Thus, the conditions of Theorem 3.3 are satisfied. We conclude that the operator B is bounded in L_p^{∞} for $1 < p < p_0$.

Let us extend this result for arbitrary $p \in (1, \infty)$. Since the operator B is bounded in $L_p^{\infty}(E)$, it has a bounded dual B^*, acting in $(L_p^{\infty}(E))^* = L_q^{\infty}(E^*)$, where $p^{-1} + q^{-1} = 1$. Let us find the explicit form of B^*. Denote by $\{x, y\}$ the value of the functional $y \in E^*$ on the element $x \in E$, and by $\langle f, g \rangle$ the value of the functional $g \in L_q^{\infty}(E^*)$ on the element $f \in L_p^{\infty}(E)$. We have

$$\langle Bf, g \rangle = \int_{-\infty}^{\infty} \left\{ A \int_{-\infty}^{\infty} \tilde{K}(t - s)f(s)ds, g(t) \right\} dt = \langle f, B^*g \rangle. \qquad (3.12)$$

Let us find the expression of B^*g. Assume first that $f(t) \in L_p^{\infty}(E)$ is such that $Af(t) \in L_p^{\infty}(E)$. Passing from the kernel $\tilde{K}(t - s)$ to the adjoint kernel $\tilde{K}^*(t - s)$ and switching the order of integration, we get

$$\int_{-\infty}^{\infty} \left\{ A \int_{-\infty}^{\infty} \tilde{K}(t - s)f(s)ds, g(t) \right\} dt = \int_{-\infty}^{\infty} \left\{ Af(s), \int_{-\infty}^{\infty} \tilde{K}^*(t - s)g(t)dt \right\} ds.$$

From this and (3.12) we obtain the equality $\langle Af, \psi \rangle = \langle f, B^*g \rangle$, where $\psi(s) = \int_{-\infty}^{\infty} \tilde{K}^*(t - s)g(t)dt$.

Since the right-hand side of the last equality is a continuous linear functional in f, the element ψ belongs to the domain of the dual A^* of A as an operator acting in $L_p^{\infty}(E)$. This implies that

$$B^*g(s) = A^*\psi(s) = A^* \int_{-\infty}^{\infty} \tilde{K}^*(t - s)g(t)dt.$$

Since the operator $\tilde{K}^*(\tau)$ obviously obeys the same estimates as $\tilde{K}(\tau)$, we see that B^* satisfies all the conditions of Theorem 3.3. Since B is bounded in $L_{p_0}^{\infty}(E)$, it follows that B^* is bounded in $L_{q_0}^{\infty}(E^*)$, where $q_0 = p_0(p_0 - 1)^{-1}$. By Theorem

3.3, B^* is bounded in $L_q^\infty(E^*)$ for any $q \in (1, p_0(p_0 - 1)^{-1})$. This implies that the operator B is bounded in $L_p^\infty(E)$, where $p = q(q-1)^{-1}$, i.e., B is bounded in $L_p^\infty(E)$ for an exponent $p \in (p_0, \infty)$. We conclude that the operator B is bounded in $L_p^\infty(E)$ for any $p \in (1, \infty)$. One can also verify that its norm is bounded by the quantity $M(p_0)p^2(p-1)^{-1}$. Theorem 3.4 is proved.

Now let us consider the operator G acting in the space $L_p(E)$ as

$$Gf(t) = A \int_0^t \exp\{-(t-s)A\}f(s)ds, \qquad (3.13)$$

which is generated by the solution of the problem (3.1). Let us show that the boundedness of G in $L_p(E)$ is equivalent to the boundedness of B in $L_p^\infty(E)$.

Theorem 3.5. *The operator G is bounded in $L_p(E)$ if and only if the operator B is bounded in $L_p^\infty(E)$.*

Proof. Using the definition (3.11) of the operator B and the expression for the kernel $\tilde{K}(\tau)$ we represent B as the sum

$$Bf(t) = A \int_{-\infty}^{t-1} \exp\{-(t-s)A\}f(s)ds + A \int_{t-1}^t \exp\{-(t-s)A\}f(s)ds$$

$$\equiv B_1 f(t) + B_2 f(t).$$

The operator B_1 is bounded in $L_p^\infty(E)$ for any p, $1 \le p \le \infty$. Indeed, using inequality (2.9) we have

$$\|B_1 f(t)\|_E \le M \int_{-\infty}^{t-1} \frac{e^{-\delta(t-s)}}{t-s} \|f(s)\|_E ds \le M \int_1^\infty e^{-\delta z}\|f(t-z)\|_E dz.$$

Next, using Minkowski's inequality we obtain for $\|B_1 f\|_{L_p^\infty(E)}$ the estimate

$$\|B_1 f\|_{L_p^\infty(E)} \le \left(\int_{-\infty}^\infty \left(M \int_1^\infty e^{-\delta z}\|f(t-z)\|_E dz \right)^p dt \right)^{1/p}$$

$$\le M \int_1^\infty e^{-\delta z} \left(\int_{-\infty}^\infty \|f(t-z)\|_E^p dt \right)^{1/p} dz \le M\delta^{-1}\|f\|_{L_p^\infty(E)}.$$

Hence, the problem of the boundedness of the operator B in $L_p^\infty(E)$ reduces to that of the boundedness of the operator B_2 in this space.

Suppose the operator G is bounded in $L_p(E)$. We shall need the identity

$$\|B_2 f\|_{L_p^\infty(E)} = \left[\sum_{k=-\infty}^{\infty} \int_k^{k+1} \|B_2 f(t)\|_E^p dt \right]^{1/p}$$

$$= \left[\sum_{k=-\infty}^{\infty} \int_0^1 \|B_2 f(t+k)\|_E^p dt \right]^{1/p}.$$

Using the expression of the kernel $\tilde{K}(\tau)$ and the triangle inequality we obtain

$$\|B_2 f\|_{L_p^\infty(E)} = \left[\sum_{k=-\infty}^{\infty} \int_0^1 \|A \int_{t-1}^0 \exp\{-(t-s)A\} f(s+k)ds\|_E^p dt \right]^{1/p}$$

$$\leq \left[\sum_{k=-\infty}^{\infty} \int_0^1 \|A \int_{t-1}^t \exp\{-(t-s)A\} f(s+k)ds\|_E^p dt \right]^{1/p}$$

$$+ \left[\sum_{k=-\infty}^{\infty} \int_0^1 \|A \int_0^t \exp\{-(t-s)A\} f(s+k)ds\|_E^p dt \right]^{1/p} = I_1 + I_2$$

First, let us estimate I_1. Changing the variable to $\tau = -s$ and using estimate (2.9) we get

$$\phi_k(t) \equiv \|A \int_{t-1}^0 \exp\{-(t-s)A\} f(s+k)ds\|_E \leq M \int_0^{1-t} \frac{\|f(-\tau+k)\|_E}{t+\tau} d\tau$$

$$\leq M \int_0^\infty \frac{\psi_k(\tau)}{t+\tau} d\tau; \qquad (*)$$

here $\psi_k(\tau) = \|f(-\tau+k)\|_E$ for $\tau \in [0,1]$ and $\psi_k(\tau) = 0$ for $\tau \notin [0,1]$. The last expression in $(*)$ is an operator of Hilbert type. Therefore, by Minkowski's inequality we have (after the substitution $\tau = rt$)

$$\left[\int_0^1 \phi_k^p(t) dt \right]^{1/p} \leq M \int_0^\infty \left[\int_0^1 \psi_k^p(rt) dt \right]^{1/p} \frac{dr}{1+r}$$

$$\leq M \frac{\pi}{\sin(\pi/p)} \left[\int_{-\infty}^\infty \psi_k^p(\tau) d\tau \right]^{1/p} = M \frac{\pi}{\sin(\pi/p)} \left[\int_0^1 \|f(-\tau+k)\|_E^p d\tau \right]^{1/p}$$

$$= M \frac{\pi}{\sin(\pi/p)} \left[\int_{k-1}^k \|f(s)\|_E^p ds \right]^{1/p}.$$

This finally yields

$$I_1 \le M \frac{\pi}{\sin(\pi/p)} \left[\sum_{k=-\infty}^{\infty} \int_{k-1}^{k} \|f(s)\|_E^p ds \right]^{1/p}$$

$$= M \frac{\pi}{\sin(\pi/p)} \|f\|_{L_p^\infty(E)} \le M_1 \frac{p^2}{p-1} \|f\|_{L_p^\infty(E)}.$$

Now let us estimate I_2. To this end we shall use the equality

$$I_2 = \Big(\sum_{k=-\infty}^{\infty} \int_0^1 \|A \int_0^t \exp\{-(t-s)A\}f(s+k)ds\|_E^p dt \Big)^{1/p}.$$

Let us estimate the integral

$$I_2^k = \int_0^1 \|A \int_0^t \exp\{-(t-s)A\}f(s+k)ds\|_E^p dt.$$

Since the operator G is bounded in $L_p(E)$, we have

$$I_2^k = \int_0^1 \|Gf(k+\tau)\|_E^p d\tau \le M_p^p \int_0^1 \|f(k+\tau)\|_E^p d\tau = M_p^p \int_k^{k+1} \|f(s)\|_E^p ds.$$

Consequently,

$$I_2 \le M_p \Big(\sum_{k=-\infty}^{\infty} \int_k^{k+1} \|f(s)\|_E^p ds \Big)^{1/p} = M_p \|f\|_{L_p^\infty(E)}.$$

The estimates for I_1 and I_2 yield

$$\|B_2 f\|_{L_p^\infty(E)} \le M(p) \|f\|_{|L_p^\infty(E)}, \quad M(p) = M_p + \frac{M_1 p^2}{p-1},$$

i.e., the operator B_2, and together with it the operator B, are bounded in $L_p^\infty(E)$.

Now suppose B is bounded. Let us extend the function $f(t)$ by zero to a function $f^*(t)$ defined on the whole real line. Clearly, $\|f\|_{L_p(E)} = \|f^*\|_{L_p^\infty(E)}$. Furthermore, we have the inequality

$$\|Gf\|_{L_p(E)} \le \|Bf^*\|_{L_p^\infty(E)}.$$

Consequently,

$$\|Gf\|_{L_p(E)} \le M \|f^*\|_{L_p^\infty(E)} = M \|f\|_{L_p(E)}.$$

This means that G is bounded in $L_p(E)$, and in fact its norm does not exceed the norm of B in $L_p^\infty(E)$. Theorem 3.5 is proved.

Proof of Theorem 3.2. From (3.9) with $p = p_0$ and formula (1.19) it follows that the operator G, defined as in (3.13), is bounded in $L_{p_0}(E)$. But then, by Theorem 3.5, the operator B is bounded in $L_{p_0}^\infty(E)$. Hence, by Theorem 3.4, it is bounded in $L_p^\infty(E)$ for any p, $1 < p < \infty$. Again using Theorem 3.5, we conclude that G is bounded in $L_p(E)$, and hence problem (3.1) is well posed in $L_p(E)$ for any p, $1 < p < \infty$.

From Theorems 3.1 and 3.5 we derive the following result.

Theorem 3.6. *Suppose problem* (1.1) *is well posed in* $L_{p_0}(E)$ *for some* p_0, $1 < p_0 < \infty$. *Then it is well posed in* $L_p(E)$ *for any* p, $1 < p < \infty$, *and* $M(p) = M(p_0)p^2(p-1)^{-1}$.

Note that the estimate (3.5) is established for arbitrary operators that generate analytic semigroups. In contrast, the full inequality (3.9) in $L_p(E)$ is established under the assumption that it holds in $L_{p_0}(E)$ for some p_0, $1 < p_0 < \infty$. When is the last condition satisfied? It turns out that it can be established in the case where $E = H$ is a Hilbert space.

Theorem 3.7. *Problem* (1.1) *is well posed in* $L_2(H)$.

Proof. As shown in Theorem 3.2, for the proof of this assertion it suffices to show that the operator B defined by formula (3.11) is bounded in $L_2^\infty(H)$. Let us denote by $\hat{f}(\lambda)$ the Fourier transform of the function $f(t) \in L_2^\infty(H)$; then

$$\hat{f}(\lambda) = \int_{-\infty}^{\infty} f(t)e^{-i\lambda t}dt.$$

Since

$$Bf(t) = AK * f(t) = A\int_{-\infty}^{\infty} \tilde{K}(t-s)f(s)ds,$$

we have

$$\widehat{Bf}(\lambda) = \widehat{AK}(\lambda)\hat{f}(\lambda). \tag{3.14}$$

Using the definition of the kernel $\tilde{K}(\tau)$, we deduce that

$$\widehat{AK}(\lambda) = A\int_{-\infty}^{\infty} \tilde{K}(t)e^{-i\lambda t}dt = A\int_0^{\infty} \exp\{-tA\}e^{-i\lambda t}dt = A(i\lambda I + A)^{-1}. \tag{3.15}$$

Using relations (3.14), (3.15) and the Parseval equality, we obtain

$$\|Bf\|^2_{L^\infty_2(H)} = 2\pi\|\widehat{Bf}\|^2_{L^\infty_2(H)} \leq 2\pi \sup_{-\infty<\lambda<\infty} \|A(i\lambda I + A)^{-1}\|^2_{H\to H}\|\hat{f}\|^2_{L^\infty_2(H)}$$

$$= \sup_{-\infty<\lambda<\infty} \|A(i\lambda I + A)^{-1}\|^2_{H\to H}\|f\|^2_{L^\infty_2(H)}.$$

Since the operator $-A$ is the generator of an analytic semigroup,

$$\sup_{-\infty<\lambda<\infty} \|A(i\lambda I + A)^{-1}\|^2_{H\to H} \leq M_0^2.$$

We see that

$$\|Bf\|_{L^\infty_2(H)} \leq M_0\|f\|_{L^\infty_2(H)}.$$

Theorem 3.7 is proved.

4.   WELL-POSEDNESS OF THE CAUCHY PROBLEM IN $L_p(E_{\alpha,q})$

In Section 3 we have shown that well-posedness in $L_p(E)$ implies (under the assumption that the operator A has a dense domain in E and a nonempty resolvent set) the analyticity of the semigroup $\exp\{-tA\}$. Presently it is not known whether the analyticity of this semigroup is sufficient for well-posedness in $L_p(E)$ in the case of arbitrary E and A.

It turns out that a Banach space E can be restricted to a Banach space E' (densely included in E) in such a manner that the restricted problem (1.1) in E' will be well posed in $L_p(E')$ for all p, $1 \leq p \leq \infty$. The role of E' will be played here by the fractional spaces $E_{\alpha,q} = E_{\alpha,q}(E,A)$, $0 < \alpha < 1$, which consist of all $v \in E$ for which the following norm is finite:

$$\|v\|_{E_{\alpha,q}} = \Big(\int_0^\infty \|\tau^{1-\alpha}A\exp\{-\tau A\}v\|^q_E \frac{d\tau}{\tau}\Big)^{1/q}, \quad \text{if } 1 \leq q < \infty,$$

$$\|v\|_{E_{\alpha,\infty}} = \sup_{\tau>0}\tau^{1-\alpha}\|A\exp\{-\tau A\}v\|_E.$$

First let us consider problem (3.1).

Theorem 4.1. *Let $f(t) \in L_q(E_{\alpha,q})$ for $0 < \alpha < 1$, $1 \leq q \leq \infty$. Then there exists a unique absolutely continuous solution of problem (3.1) such that $Av(t), v'(t) \in$*

$L_q(E_{\alpha,q})$ *and the coercivity inequality holds:*

$$\|v'\|_{L_q(E_{\alpha,q})} + \|Av\|_{L_q(E_{\alpha,q})} \le \frac{M}{\alpha(1-\alpha)}\|f\|_{L_q(E_{\alpha,q})},$$

where M does not depend on α, q and $f(t)$.

Proof. By formula (1.19),

$$v(t) = \int_0^t \exp\{-(t-s)A\}f(s)ds.$$

Let us estimate $Av(t)$ in the norm of $L_q(E_{\alpha,q})$. To this end we extend the function $f(t)$ by zero for $t < 0$ and $t > 1$, denoting the extension by $f^*(t)$. Then

$$v(t) = \int_0^t \exp\{-sA\}f(t-s)ds = \int_0^\infty \exp\{-sA\}f^*(t-s)ds. \qquad (4.1)$$

Next, using the semigroup property we obtain

$$A\exp\{-\tau A\}Av(t) = \int_0^\infty A\exp\left\{-\frac{\tau+s}{2}A\right\}\exp\left\{-\frac{\tau}{2}A\right\}A\exp\left\{-\frac{s}{2}A\right\}f^*(t-s)ds.$$

Using the estimates (1.10) and (1.14) we obtain the inequality

$$\|\tau^{1-\alpha}A\exp\{-\tau A\}Av(t)\|_E \le M\int_0^\infty \frac{\tau^{1-\alpha}s^{\alpha-1}}{\tau+s}\left\|s^{1-\alpha}A\exp\left\{-\frac{s}{2}A\right\}f^*(t-s)\right\|_E ds.$$

From Minkowski's integral inequality it follows that

$$\left(\int_0^1 \|\tau^{1-\alpha}A\exp\{-\tau A\}Av(t)\|_E^q dt\right)^{1/q}$$

$$\le M\int_0^\infty \frac{\tau^{1-\alpha}s^{\alpha-1}}{\tau+s}\left(\int_0^\infty \left\|s^{1-\alpha}A\exp\left\{-\frac{s}{2}A\right\}f^*(t-s)\right\|_E^q dt\right)^{1/q}ds.$$

Since $f^*(z) = 0$ for $z < 0$, the lower limit in the inner integral can be replaced by s. Then, making the substitution $t = s + z$ in the inner integral and extending the domain of integration in that integral, we obtain the inequality

$$\left(\int_0^1 \|\tau^{1-\alpha}A\exp\{-\tau A\}Av(t)\|_E^q dt\right)^{1/q}$$

$$\le M\int_0^\infty \frac{\tau^{1-\alpha}s^{\alpha-1}}{\tau+s}\left(\int_0^1 \left\|s^{1-\alpha}A\exp\left\{-\frac{s}{2}A\right\}f(z)\right\|_E^q dz\right)^{1/q}ds. \qquad (4.2)$$

Let us consider the scalar operator

$$J(\alpha)\varphi(\tau) = \int_0^\infty \frac{\tau^{1-\alpha}s^{\alpha-1}}{\tau+s}\varphi(s)ds = \int_0^\infty (1+z)^{-1}z^{\alpha-1}\varphi(\tau z)dz.$$

From Minkowski's integral inequality it follows that

$$\left(\int_0^\infty |J(\alpha)\varphi(\tau)|^q \frac{d\tau}{\tau}\right)^{1/q} \leq \frac{\pi}{\sin\pi\alpha}\left(\int_0^\infty |\varphi(z)|^q \frac{dz}{z}\right)^{1/q}.$$

Applying this inequality in (4.2) and using Fubini's theorem we conclude (after the substitution $s = 2z$) that

$$\|Av\|_{L_q(E_{\alpha,q})} \leq M_1 \frac{\pi}{\sin\pi\alpha}\|f\|_{L_q(E_{\alpha,q})}.$$

Theorem 4.1 is proved.

Clearly, the conditions of Theorem 4.1 are necessary.

From Theorems 3.1 and 4.1 we derive the following result.

Theorem 4.2. *Let* $1 < p,q < \infty$, $0 < \alpha < 1$, *and* $f(t) \in L_p(E_{\alpha,q})$. *Then formula* (1.19) *gives the unique absolutely continuous solution* $v(t)$ *of problem* (3.1) *such that* $v'(t), Av(t) \in L_p(E_{\alpha,q})$, *and the coercivity inequality holds:*

$$\|v'\|_{L_p(E_{\alpha,q})} + \|Av\|_{L_p(E_{\alpha,q})} \leq \frac{M(q)p^2}{p-1}\frac{1}{\alpha(1-\alpha)}\|f\|_{L_p(E_{\alpha,q})}, \qquad (4.3)$$

where $M(q)$ *does not depend on* α, p, *and* $f(t)$.

Clearly, the condition of Theorem 4.2 is necessary.

Note that what Theorems 4.1 and 4.2 say is that problem (3.1) is well posed in the space $L_p(E_{\alpha,q})$ whenever $\alpha \in (0,1)$ and $p,q \in (1,\infty)$ or $p = q = 1$, or $p = q = \infty$.

Now let us turn to the general problem (1.1). If $f(t) \in L_p(E_{\alpha,q})$ then, by Theorems 4.1 and 4.2, formula (1.19) gives its unique absolutely continuous solution $v(t)$ with the indicated properties if and only if the function $A\exp\{-tA\}v_0$ belongs to $L_p(E_{\alpha,q})$.

The formula (see Section 3)

$$|v_0|_{1+\alpha-\frac{1}{p},q} = \left(\int_0^1 \|A\exp\{-tA\}v_0\|_{E_{\alpha,q}}^p dt\right)^{1/p} \qquad (4.4)$$

defines a norm in the space of initial data $E_{1+\alpha-\frac{1}{p},q}$, which consists of all $v_0 \in E$ for which the quantity (4.4) is finite.

Theorem 4.3. *Suppose $1 < p, q < \infty$ or $p = q = \infty$. Then for problem (1.1) to admit a unique absolutely continuous solution $v(t)$ such that $v'(t), Av(t) \in L_p(E_{\alpha,q})$ and $v(t)$ is continuous as a function with values in $E_{1+\alpha-\frac{1}{p},q}$ it is necessary and sufficient that $f(t) \in L_p(E_{\alpha,q})$ and $v_0 \in E_{1+\alpha-\frac{1}{p},q}$. Such solutions of problem (1.1) obey the coercivity inequality*

$$\|v'\|_{L_p(E_{\alpha,q})} + \|Av\|_{L_p(E_{\alpha,q})} + \max_{0 \le t \le 1} |v(t)|_{1+\alpha-\frac{1}{p},q}$$

$$\le M\Big[|v_0|_{1+\alpha-\frac{1}{p},q} + M_1(p,q)\frac{1}{\alpha(1-\alpha)}\|f\|_{L_p(E_{\alpha,q})}\Big]. \tag{4.5}$$

Here $M_1(p,q) = M(q)p^2/(p-1)$ if $p \ne q$ and $M(p,p) = 1$.

Proof. The necessity of the conditions on the right-hand side $f(t)$ in Theorem 4.3 is obvious. The necessity of the restriction on v_0 was explained above, before we introduced the space of initial data. To verify the necessity of the conditions of Theorem 4.3 we need only to establish that the function $v(t)$ defined by formula (1.19) is continuous in the norm of the space $E_{1+\alpha-\frac{1}{p},q}$. Since $D(A)$ is dense in E, it suffices to obtain an estimate for $v(t)$ in the norm of this space that is uniform in t.

By (1.19),

$$v(t) = \exp\{-tA\}v_0 + \int_0^t \exp\{-(t-s)A\}f(s)ds = w(t) + g(t). \tag{4.6}$$

From the definition of the space $E_{1+\alpha-\frac{1}{p},q}$ it follows that

$$|w(t)|_{1+\alpha-\frac{1}{p},q} = \Big(\int_0^\infty \|A\exp\{-(t+\tau)A\}v_0\|_{E_{\alpha,q}}^p d\tau\Big)^{1/p}$$

$$\le \|\exp\{-tA\}\|_{E \to E}\Big(\int_0^\infty \|A\exp\{-\tau A\}v_0\|_{E_{\alpha,q}}^p d\tau\Big)^{1/p}. \tag{4.7}$$

Inequalities (1.10) and (4.7) yield

$$|w(t)|_{1+\alpha-\frac{1}{p},q} \le M|v_0|_{1+\alpha-\frac{1}{p},q}. \tag{4.8}$$

Now let us consider the solution $g(t)$ of the nonhomogeneous problem (3.1) (with null initial condition).

Since the semigroup $\exp\{-tA\}$ is analytic in E, it is analytic in any space $E_{\alpha,q}$, $1 \leq q \leq \infty$. The estimate (3.7) implies that

$$\|A\exp\{-tA\}\|_{E_{\alpha,q}\to E_{\alpha,q}} \leq Me^{-\delta t}t^{-1}. \tag{4.9}$$

Then from Theorem 3.1 we derive the estimate

$$\|g\|_{C(E_{\alpha,q})} \leq M\frac{\pi}{\sin(\pi/p)}\|f\|_{L_p(E_{\alpha,q})} \tag{4.10}$$

for any $p \in (1,\infty)$.

In the case $p = q = \infty$ the space of traces consists of the elements $z \in D(A)$ with the property that $Az \in E_{\alpha,\infty}$. Therefore, in the present case the estimate in the norm of traces is a corollary of Theorem 4.1. Theorem 4.3 is proved.

Theorem 4.3 does not cover all possible pairs $p, q \in [1,\infty]$. Theorems 4.1 and 4.3 are supplemented by the following result.

Theorem 4.4. a) *Let $p = q = 1$ or $p = q = \infty$. Then for the problem (1.1) to have an absolutely continuous solution $v(t)$ such that $v'(t), Av(t) \in L_p(E_{\alpha,p})$ it is necessary and sufficient that $f(t) \in L_p(E_{\alpha,p})$ and $v_0 \in E_{1+\alpha-\frac{1}{p},p}$. The solution $v(t)$ obeys the coercivity estimate*

$$\|v'\|_{L_p(E_{\alpha,p})} + \|Av\|_{L_p(E_{\alpha,p})} \leq M\Big[|v_0|_{1+\alpha-\frac{1}{p},p} + \frac{1}{\alpha(1-\alpha)}\|f\|_{L_p(E_{\alpha,p})}\Big]. \tag{4.11}$$

b) *Let $1 < p < \infty$ and let $f(t) \in L_p(E_{\alpha,\infty})$, $v_0 \in E_{1+\alpha-\frac{1}{p},\infty}$. Then the function $v(t)$ defined by formula (4.6) obeys the estimate*

$$\max_{0\leq t\leq 1} |v(t)|_{1+\alpha-\frac{1}{p},\infty} \leq M\Big[|v_0|_{1+\alpha-\frac{1}{p},\infty} + \frac{p^2}{p-1}\|f\|_{L_p(E_{\alpha,\infty})}\Big]. \tag{4.12}$$

Proof. a) By Theorem 4.1, to prove inequality (4.11) we need only to justify the estimate of the solution of the homogeneous problem (1.7). Since the semigroup $\exp\{-tA\}$ is bounded in the norm of the space E (see the proof of Theorem 4.3), the desired estimate follows from the definition of the norm in the space of traces.

b) To prove (4.12) we shall use relation (4.6). The needed estimate for $w(t)$ follows from the definition of the norm of traces and the boundedness of the semigroup $\exp\{-tA\}$ in the norm of E. From (4.6) we obtain the identity

$$\varphi(\lambda,\tau;t) = [\lambda^{1-\alpha}A\exp\{-\lambda A\}][A\exp\{-\tau A\}]g(t) =$$

$$\lambda^{1-\alpha} \int_0^t A \exp\left\{-\frac{\lambda+\tau+t-s}{2}A\right\} A \exp\left\{-\frac{\lambda+\tau+t-s}{2}A\right\} f(s)ds, \quad (4.13)$$

which holds for all $\lambda > 0, \tau > 0, t > 0$. Using the estimate (4.8), we obtain from this the bound

$$\|\varphi(\lambda,\tau;t)\|_E \le \lambda^{1-\alpha} M \int_0^t \frac{1}{\lambda+\tau+t-s}\left\|A \exp\left\{-\frac{\lambda+\tau+t-s}{2}A\right\} f(s)\right\|_E ds. \tag{4.14}$$

Now let us transform the integrand in the right-hand side of inequality (4.14) so that it can be estimated by means of the norm in $E_{\alpha,\infty}$. To this end let us rewrite (4.14) as

$$\|\varphi(\lambda,\tau;t)\|_E \le M\lambda^{1-\alpha}2^{1-\alpha} \int_0^t \frac{\left\|(\frac{\lambda+\tau+t-s}{2})A \exp\{-\frac{\lambda+\tau+t-s}{2}A\} f(s)\right\|_E ds}{(\lambda+\tau+t-s)^{2-\alpha}}.$$

This obviously yields the inequality

$$\|\varphi(\lambda,\tau;t)\|_E \le M\lambda^{1-\alpha}2^{1-\alpha} \int_0^t \frac{\sup_{z>0}\|z^{1-\alpha}A \exp\{-zA\} f(s)\|_E ds}{(\lambda+\tau+t-s)^{2-\alpha}},$$

which (by the definition of the norm in $E_{\alpha,\infty}$) can be recast as

$$\|\varphi(\lambda,\tau;t)\|_E \le M\lambda^{1-\alpha}2^{1-\alpha} \int_0^t \frac{\|f(s)\|_{E_{\alpha,\infty}} ds}{(\lambda+\tau+t-s)^{2-\alpha}}.$$

Finally, since $\lambda(\lambda+\tau+t-s) \le 1$, we conclude that

$$\|\varphi(\lambda,\tau;t)\|_E \le M2^{1-\alpha} \int_0^t \frac{1}{\tau+t-s}\|f(s)\|_{E_{\alpha,\infty}} ds. \tag{4.15}$$

From identity (4.13) and the definition of the norm in $E_{\alpha,\infty}$ we obtain

$$\sup_{\lambda>0}\|\varphi(\lambda,\tau;t)\|_E = \|A \exp\{-\tau A\}g(t)\|_{E_{\alpha,\infty}}.$$

From this and (4.15) it follows that

$$\|A \exp\{-\tau A\}g(t)\|_{E_{\alpha,\infty}} \le M2^{1-\alpha} \int_0^t \frac{1}{\tau+t-s}\|f(s)\|_{E_{\alpha,\infty}} ds. \tag{4.16}$$

In the right-hand side of (4.16) there appears an integral operator of the type of Hilbert's operator. Making the substitution $t - s = z$ and extending the function

$f(z)$ by zero for $z \leq 0$ and $z > 1$, we derive the following inequality for the extended function $f^*(z)$:

$$\|A\exp\{-\tau A\}g(t)\|_{E_{\alpha,\infty}} \leq M2^{1-\alpha}\int_0^\infty \frac{1}{\tau+z}\|f^*(t-z)\|_{E_{\alpha,\infty}}dz. \qquad (4.17)$$

Next, putting $z = \tau s$ in the last integral, we get

$$\|A\exp\{-\tau A\}g(t)\|_{E_{\alpha,\infty}} \leq M2^{1-\alpha}\int_0^\infty \frac{1}{1+s}\|f^*(t-s\tau)\|_{E_{\alpha,\infty}}ds. \qquad (4.18)$$

Now we can apply the Minkowski integral inequality (with respect to the variable τ). In this way we conclude that the following inequality holds for all p, $1 < p < \infty$:

$$\left(\int_0^\infty \|A\exp\{-\tau A\}g(t)\|_{E_{\alpha,\infty}}^p d\tau\right)^{1/p}$$

$$\leq M2^{1-\alpha}\int_0^\infty \frac{1}{1+s}\left(\int_0^\infty \|f^*(t-s\tau)\|_{E_{\alpha,\infty}}^p d\tau\right)^{1/p}ds. \qquad (4.19)$$

By the definition of the norm in the space of traces, the expression in the left-hand side of (4.19) coincides with $|g(t)|_{1+\alpha-\frac{1}{p},\infty}$. Making the substitution $t - s\tau = z$ in the inner integral in the right-hand side of (4.19), we get

$$|g(t)|_{1+\alpha-\frac{1}{p},\infty} \leq M2^{1-\alpha}\int_0^\infty \frac{1}{(1+s)s^{1/p}}\left(\int_0^\infty \|f^*(z)\|_{E_{\alpha,\infty}}^p dz\right)^{1/p}ds.$$

Recalling the definition of the function $f^*(z)$, we obtain the final estimate

$$|g(t)|_{1+\alpha-\frac{1}{p},\infty} \leq M2^{1-\alpha}\frac{\pi}{\sin(\pi/p)}\|f\|_{L_p(E_{\alpha,\infty})}. \qquad (4.20)$$

Theorem 4.4 is proved.

Remark. The estimate (4.12) is in fact an estimate of the function $v(t)$ in the norm of the space of traces for the function spaces $L_p(E_{\alpha,\infty})$. Under the assumptions of Theorem 4.4, the continuous embedding $E_{\alpha,\infty} \subset E_{\alpha,p}$, which is a consequence of the exponential decay (for $t \to \infty$) of the semigroup $\exp\{-tA\}$ in E, implies the memberships

$$v_0 \in E_{1+\alpha-\frac{1}{p},p}, \quad f(t) \in L_p(E_{\alpha,p}).$$

Then it follows from Theorem 4.3 that $v(t)$ is an absolutely continuous solution of problem (1.1) with the property that $v'(t), Av(t) \in L_p(E_{\alpha,p})$. We have not succeeded in establishing the memberships

$$v'(t), Av(t) \in L_p(E_{\alpha,\infty}),$$

i.e., the coercivity of problem (1.1) in the space $L_p(E_{\alpha,\infty})$.

5.   WELL-POSEDNESS OF THE CAUCHY PROBLEM IN SPACES OF SMOOTH FUNCTIONS

1.   The space $C_0^{\beta,\gamma}(E)$. The nonhomogeneous problem.

In Sections 2 and 4 the well-posedness of the Cauchy problem (1.1) was established in the spaces $C_0^\alpha(E)$ and $C(E_\alpha) = C(E_{\alpha,\infty})$. The latter are spaces of smooth functions in the time variable or in the variable $v \in E$. This section is devoted to a generalization of the results of Sections 2 and 4.

Let us denote by $C_0^{\beta,\gamma}([0,1], E)$, where $0 \le \gamma \le \beta$, $0 < \beta < 1$, the space obtained by completion of the space of all smooth E-valued functions $f(t)$ on $[0,1]$ in the norm

$$\|f\|_{C_0^{\beta,\gamma}(E)} = \sup_{0 \le t \le 1} \|f(t)\|_E + \sup_{0 \le t < t+\tau \le 1} \frac{(t+\tau)^\gamma \|f(t+\tau) - f(t)\|_E}{\tau^\beta}. \quad (5.1)$$

For $\beta = \gamma = \alpha$ the spaces $C_0^{\beta,\gamma}(E)$ coincide with $C_0^\alpha(E)$. Moreover, the norms of the spaces $C_0^{\alpha,\alpha}(E)$ and $C_0^\alpha(E)$ are equivalent uniformly in $\alpha \in (0,1)$.

First let us consider the special nonhomogeneous Cauchy problem

$$v'(t) + Av(t) = f(t), \quad 0 \le t \le 1, \quad v(0) = 0, f(0) = 0. \quad (5.2)$$

A function $v(t)$ is called a *solution of problem* (5.2) *in* $C_0^{\beta,\gamma}(E)$ if it is a solution of this problem in $C(E)$ and if the functions $v'(t)$ and $Av(t)$ belong to $C_0^{\beta,\gamma}(E)$. If $v(t)$ is a solution of problem (5.2) in $C_0^{\beta,\gamma}(E)$ then clearly $f(t) \in C_0^{\beta,\gamma}(E)$. As in the case of the spaces $C_0^\alpha(E)$, let us give the following

Definition 5.1. Problem (5.2) is said to be *well posed in* $C_0^{\beta,\gamma}(E)$ if the following two conditions are satisfied:

1) For any $f(t) \in C_0^{\beta,\gamma}(E)$ there exists a unique solution $v(t) = v(t; f(t), 0)$ of (5.2).

2) The operator $v(t; f(t), 0)$ is continuous in $C_0^{\beta,\gamma}(E)$.

From the well-posedness of problem (5.2) one derives, as above, the coercivity inequality

$$\|v'\|_{C_0^{\beta,\gamma}(E)} + \|Av\|_{C_0^{\beta,\gamma}(E)} \leq M(\beta,\gamma)\|f\|_{C_0^{\beta,\gamma}(E)}. \tag{5.3}$$

Conversely, if inequality (5.3) can be established for continuously differentiable functions $f(t)$ then, as in the space $C(E)$, one can show that problem (5.2) is well posed in $C_0^{\beta,\gamma}(E)$. As we did in the case of the space $C_0^\alpha(E)$, we can prove that from the well-posedness of problem (5.2) in $C_0^{\beta,\gamma}(E)$ it follows that the semigroup $\exp\{-tA\}$ is analytic in E. Thus, the analyticity of this semigroup in E is a necessary condition for the well-posedness in $C_0^{\beta,\gamma}(E)$ of the problem (5.2). It turns out that this condition is not only necessary but also sufficient.

Theorem 5.1. *Suppose* $-A$ *is the generator of an analytic semigroup in* E *and* $f(t) \in C_0^{\beta,\gamma}(E)$. *Then the Cauchy problem* (5.2) *is well posed in* $C_0^{\beta,\gamma}(E)$, *and the coercivity inequality* (5.3) *holds with* $M(\beta,\gamma) = M/[\beta(1-\beta)]$, *where* M *does not depend on* β, γ, *and* $f(t)$.

Proof. If $v(t)$ is a solution of problem (5.2) in $C_0^{\beta,\gamma}(E)$, then it is also a solution of this problem in $C(E)$. Hence, by (1.19), one has the relation

$$v(t) = \int_0^t \exp\{-(t-s)A\}f(s)ds. \tag{5.4}$$

To prove the theorem we need to verify that $v(t)$ is a solution in $C(E)$, then show that the functions $Av(t)$ and $v'(t)$ belong to $C_0^{\beta,\gamma}(E)$, and finally obtain estimates of the norms of these functions. Here we shall confine ourselves to the last task. From (5.4) we derive the identity

$$v'(t) = \exp\{-tA\}(f(t) - f(0))$$

$$+ \int_0^t A\exp\{-(t-s)A\}(f(t) - f(s))ds \equiv p(t) + g(t). \tag{5.5}$$

Let us estimate the functions $p(t)$ and $g(t)$ in $C_0^{\beta,\gamma}(E)$. By (1.10), we have

$$\|p(t)\|_E \leq \|\exp\{-tA\}\|_{E \to E}\|f(t) - f(0)\|_E \leq Mt^{\beta-\gamma}\|f\|_{C_0^{\beta,\gamma}(E)}. \tag{5.6}$$

This implies that

$$\|p(t)\|_E \le M\|f\|_{C_0^{\beta,\gamma}(E)}$$

for all $t \in [0,1]$, and hence

$$\|p\|_{C(E)} \le M\|f\|_{C_0^{\beta,\gamma}(E)}. \tag{5.7}$$

If $t \le \tau$ then, by (5.6) and the triangle inequality,

$$\|p(t+\tau) - p(t)\|_E \le \|p(t+\tau)\|_E + \|p(t)\|_E \le$$

$$M[(t+\tau)^{\beta-\gamma} + t^{\beta-\gamma}]\|f\|_{C_0^{\beta,\gamma}(E)} \le 2^{1+\beta}M\frac{\tau^\beta}{(t+\tau)^\gamma}\|f\|_{C_0^{\beta,\gamma}(E)}.$$

Thus, for $t \le \tau$ we have the estimate

$$\|p(t+\tau) - p(t)\|_E \le M_1\frac{\tau^\beta}{(t+\tau)^\gamma}\|f\|_{C_0^{\beta,\gamma}(E)}. \tag{5.8}$$

Now let $t > \tau$. Then, by the definition of $p(t)$, we have

$$p(t+\tau) - p(t) = [\exp\{-(t+\tau)A\} - \exp\{-tA\}](f(t+\tau) - f(0))$$

$$+\exp\{-tA\}(f(t+\tau) - f(t)).$$

Applying the estimate (1.10), we obtain

$$\|\exp\{-tA\}(f(t+\tau) - f(t))\|_E \le \|\exp\{-tA\}\|_{E\to E}\|f(t+\tau) - f(t))\|_E$$

$$\le M\frac{\tau^\beta}{(t+\tau)^\gamma}\|f\|_{C_0^{\beta,\gamma}(E)}. \tag{5.9}$$

Further, using the estimate (1.14) we get

$$\|[\exp\{-(t+\tau)A\} - \exp\{-tA\}](f(t+\tau) - f(0))\|_E \le M\frac{\tau}{t}(t+\tau)^{\beta-\gamma}\|f\|_{C_0^{\beta,\gamma}(E)}$$

$$\le 2M\frac{\tau^\beta}{(t+\tau)^\gamma}\|f\|_{C_0^{\beta,\gamma}(E)}.$$

The last inequality in conjunction with (5.9) implies the estimate (5.8) for $t > \tau$. By (5.7) and (5.8), we have

$$\|p\|_{C_0^{\beta,\gamma}(E)} \le M_1\|f\|_{C_0^{\beta,\gamma}(E)}. \tag{5.10}$$

Now let us establish the estimate

$$\|g\|_{C_0^{\beta,\gamma}(E)} \le \frac{M}{\beta(1-\beta)}\|f\|_{C_0^{\beta,\gamma}(E)}. \tag{5.11}$$

To this end it suffices to show that

$$\|g(t)\|_E \le \frac{M}{\beta}\|f\|_{C_0^{\beta,\gamma}(E)}, \quad 0 \le t \le 1, \tag{5.12}$$

and

$$\|g(t+\tau) - g(t)\|_E \le \frac{M}{\beta(1-\beta)}\frac{\tau^\beta}{(t+\tau)^\gamma}\|f\|_{C_0^{\beta,\gamma}(E)}, \quad 0 \le t < t+\tau \le 1. \tag{5.13}$$

First let us prove (5.12). Using the estimate (1.14), we obtain

$$\|g(t)\|_E \le \int_0^t \|A\exp\{-(t-s)A\}\|_{E \to E}\|f(t) - f(s)\|_E ds$$

$$\le M\int_0^t \frac{(t-s)^\beta ds}{(t-s)t^\gamma}\|f\|_{C_0^{\beta,\gamma}(E)} = \frac{M}{t^\gamma}\int_0^t \frac{ds}{(t-s)^{1-\beta}}\|f\|_{C_0^{\beta,\gamma}(E)}$$

$$\le \frac{Mt^\beta}{\beta t^\gamma}\|f\|_{C_0^{\beta,\gamma}(E)} = \frac{Mt^{\beta-\gamma}}{\beta}\|f\|_{C_0^{\beta,\gamma}(E)}.$$

Hence,

$$\|g(t)\|_E \le \frac{M}{\beta}t^{\beta-\gamma}\|f\|_{C_0^{\beta,\gamma}(E)} \tag{5.14}$$

for all $t \in [0,1]$. From the last inequality we obtain, first, (5.12), and second, (5.13) for $t \le \tau$. Indeed, by (5.14) and the triangle inequality we have

$$\|g(t+\tau) - g(t)\|_E \le \|g(t+\tau)\|_E + \|g(t)\|_E \le \frac{M}{\beta}[(t+\tau)^{\beta-\gamma} + t^{\beta-\gamma}]\|f\|_{C_0^{\beta,\gamma}(E)}$$

$$\le \frac{2M}{\beta}(t+\tau)^{\beta-\gamma}\|f\|_{C_0^{\beta,\gamma}(E)} \le \frac{2^{1+\beta}M}{\beta}\frac{\tau^\beta}{(t+\tau)^\gamma}\|f\|_{C_0^{\beta,\gamma}(E)}.$$

Now let $t > \tau$. Let us represent the difference $g(t+\tau) - g(t)$ as the sum of the following integrals:

$$g(t+\tau) - g(t) = \int_{t-\tau}^{t+\tau} A\exp\{-(t+\tau-s)A\}(f(t+\tau) - f(s))ds$$

$$+ \int_{t-\tau}^t A\exp\{-(t-s)A\}(f(s) - f(t))ds+$$

$$\int_0^{t-\tau} A[\exp\{-(t+\tau-s)A\} - \exp\{-(t-s)A\}](f(t) - f(s))ds$$

$$+ \int_0^{t-\tau} A\exp\{-(t+\tau-s)A\}ds(f(t+\tau) - f(t)) = I_1 + I_2 + I_3 + I_4.$$

Let us estimate I_1, I_2, I_3 and I_4 separately. We begin with I_1. Using the estimate (5.14) we obtain

$$\|I_1\|_E \leq \int_{t-\tau}^{t+\tau} \|A\exp\{-(t+\tau-s)A\}\|_{E\to E}\|f(t+\tau) - f(s)\|_E ds$$

$$\leq M \int_{t-\tau}^{t+\tau} \frac{(t+\tau-s)^\beta ds}{(t+\tau-s)(t+\tau)^\gamma}\|f\|_{C_0^{\beta,\gamma}(E)}$$

$$= \frac{M}{(t+\tau)^\gamma} \int_{t-\tau}^{t+\tau} \frac{ds}{(t+\tau-s)^{1-\beta}}\|f\|_{C_0^{\beta,\gamma}(E)} = \frac{M}{\beta}\frac{(2\tau)^\beta}{(t+\tau)^\gamma}\|f\|_{C_0^{\beta,\gamma}(E)}.$$

Therefore,

$$\|I_1\|_E \leq \frac{M_1}{\beta}\frac{\tau^\beta}{(t+\tau)^\gamma}\|f\|_{C_0^{\beta,\gamma}(E)}.$$

In exactly the same manner one establishes the estimate

$$\|I_2\|_E \leq \frac{M}{\beta}\frac{\tau^\beta}{t^\gamma}\|f\|_{C_0^{\beta,\gamma}(E)}.$$

Since $t > \tau$, we have that

$$\|I_2\|_E \leq \frac{2^\gamma M}{\beta}\frac{\tau^\beta}{(t+\tau)^\gamma}\|f\|_{C_0^{\beta,\gamma}(E)}.$$

Next, let us estimate I_4. Since

$$I_4 = [\exp\{-2\tau A\} - \exp\{-(t+\tau)A\}](f(t+\tau) - f(t)),$$

by (1.10) and the triangle inequality we have

$$\|I_4\|_E \leq [\|\exp\{-2\tau A\}\|_{E\to E} + \|\exp\{-(t+\tau)A\}\|_{E\to E}]\|f(t+\tau) - f(t)\|_E$$

$$\leq M\frac{\tau^\beta}{(t+\tau)^\gamma}\|f\|_{C_0^{\beta,\gamma}(E)}.$$

Finally, let us estimate I_3. Using the estimate (2.3) with $\alpha = 1$ we obtain

$$\|I_3\|_E \leq \int_0^{t-\tau} \|A[\exp\{-(t+\tau-s)A\} - \exp\{-(t-s)A\}]\|_{E\to E}\|f(t) - f(s)\|_E ds$$

$$\le M \int_0^{t-\tau} \frac{\tau(t-s)^\beta ds}{(t-s)^2 t^\gamma} \|f\|_{C_0^{\beta,\gamma}(E)} = \frac{M\tau}{t^\gamma} \int_0^{t-\tau} \frac{ds}{(t-s)^{2-\beta}} \|f\|_{C_0^{\beta,\gamma}(E)}$$

$$\le \frac{M\tau}{(1-\beta)t^\gamma \tau^{1-\beta}} \|f\|_{C_0^{\beta,\gamma}(E)} = \frac{M\tau^\beta}{(1-\beta)t^\gamma} \|f\|_{C_0^{\beta,\gamma}(E)}.$$

Since $t > \tau$, we conclude that

$$\|I_3\|_E \le \frac{2^\gamma M}{1-\beta} \frac{t^\beta}{(t+\tau)^\gamma} \|f\|_{C_0^{\beta,\gamma}(E)}.$$

Combining the estimates for I_1, I_2, I_3 and I_4 we obtain (5.13). It remains to use the estimates (5.10) and (5.11) and obtain the inequality

$$\|v'\|_{C_0^{\beta,\gamma}(E)} \le \frac{M}{\beta(1-\beta)} \|f\|_{C_0^{\beta,\gamma}(E)}.$$

The estimate for $Av(t)$ in the norm of $C^{\beta,\gamma}(E)$ follows from the triangle inequality. Theorem 5.1 is proved.

2.   Well-posedness of the general problem.

Let us now consider the Cauchy problem (1.1). From formula (2.10) it follows that the solution of the homogeneous problem (1.1) with $f(t) = 0$ and $w_0 = v_0 - A^{-1}f(0)$ has the form

$$w(t) = \exp\{-tA\}w_0, \quad t \ge 0.$$

Accordingly, for the function $w(t)$ to be a solution of problem (1.1) with $f(t) = 0$ in $C^{\beta,\gamma}(E)$ it is necessary and sufficient that $Aw(t) = A \exp\{-tA\}w_0 \in C_0^{\beta,\gamma}(E)$. By estimate (1.10), the set of elements $w_0 \in E$ with this property is obviously a linear space containing $D(A)$. This space becomes a Banach space $E_1^{\beta,\gamma}$ when endowed with the norm

$$|w_0|_1^{\beta,\gamma} = \max_{0 \le z \le 1} \|A \exp\{-zA\}w_0\|_E$$

$$+ \sup_{0 \le z < z+\tau \le 1} \tau^{-\beta}(z+\tau)^\gamma \|A[\exp\{-(z+\tau)A\} - \exp\{-zA\}]w_0\|_E. \qquad (5.15)$$

Let $v(t)$ be a solution of the general problem (1.1). Then we can write

$$v(t) = g(t) + w(t) + A^{-1}f(0), \quad t \ge 0,$$

where $g(t)$ is the solution of problem (5.2) with $g(0) = 0$ and the right-hand side equal to $f(t) - f(0)$, and $w(t)$ is the solution of the homogeneous problem (1.1) with $w_0 = v_0 - A^{-1}f(0)$. From this and from the solvability in $C_0^{\beta,\gamma}(E)$ of the problem (5.2) and the homogeneous problem (1.1) with $w_0 = v_0 - A^{-1}f(0)$, we deduce the unique solvability of the general Cauchy problem (1.1) for any $v_0 - A^{-1}f(0) \in E$ and the validity of the coercivity inequality for the solutions of this problem:

$$\|v'\|_{C_0^{\beta,\gamma}(E)} + \|Av\|_{C_0^{\beta,\gamma}(E)} \le M(\beta,\gamma)\|f\|_{C_0^{\beta,\gamma}(E)} + M|Av_0 - f(0)|_0^{\beta,\gamma}. \quad (5.16)$$

The general Cauchy problem (1.1) will be said to be *well posed in* $C_0^{\beta,\gamma}(E)$ if for any $v_0 \in D(A)$ such that $v_0 - A^{-1}f(0) \in E_1^{\beta,\gamma}$ and any $f(t) \in C_0^{\beta,\gamma}(E)$ problem (1.1) has a unique solution and the coercivity inequality (5.16) holds.

The foregoing argument shows that the analyticity of the semigroup $\exp\{-tA\}$ in E is not only a necessary but also a sufficient condition for the well-posedness in $C_0^{\beta,\gamma}(E)$ of problem (1.1). The conditions formulated above on the initial data of problem (1.1) are such that the element $v_0 - A^{-1}f(0)$ must belong to a more restricted space than the element $v(t) - A^{-1}f(t)$. Is this really the case? The answer to this question is given by

Theorem 5.2. *For any* $v_0 - A^{-1}f(0) \in E_1^{\beta,\gamma}$ *and* $f(t) \in C_0^{\beta,\gamma}(E)$, *problem* (1.1) *has a unique solution* $v(t)$ *in* $C(E_1^{\beta,\gamma})$, *and*

$$\max_{0 \le t \le 1} |v(t) - A^{-1}f(t)|_1^{\beta,\gamma} \le M\left[|v_0 - A^{-1}f(0)|_1^{\beta,\gamma} + \frac{1}{\beta(1-\beta)}\|f\|_{C_0^{\beta,\gamma}(E)}\right], \quad (5.17)$$

where M *does not depend on* β, γ, v_0 *and* $f(t)$.

Proof. By (1.19), we have the formula

$$v(t) - A^{-1}f(t) = \exp\{-tA\}(v_0 - A^{-1}f(0)) + A^{-1}\exp\{-tA\}(f(0) - f(t))$$

$$+ \int_0^t \exp\{-(t-s)A\}(f(s) - f(t))ds = v_1(t) + v_2(t) + v_3(t). \quad (5.18)$$

From the definition of the space $E_1^{\beta,\gamma}$ and the estimate (1.10) it follows that

$$|v_1(t)|_1^{\beta,\gamma} \le \|\exp\{-tA\}\|_{E \to E}|v_0 - A^{-1}f(0)|_1^{\beta,\gamma} \le M|v_0 - A^{-1}f(0)|_1^{\beta,\gamma}. \quad (5.19)$$

Now let us estimate v_2. Using the estimate (1.10), we obtain

$$\max_{0 \le z \le 1} \|A\exp\{-zA\}A^{-1}\exp\{-tA\}(f(t) - f(0))\|_E$$

$$\leq \|\exp\{-tA\}\|_{E\to E} \max_{0\leq z\leq 1} \|\exp\{-zA\}\|_{E\to E}\|f(t) - f(0)\|_E$$

$$\leq M_1 t^{\beta-\gamma}\|f\|_{C_0^{\beta,\gamma}(E)} \leq M_1\|f\|_{C_0^{\beta,\gamma}(E)}. \tag{5.20}$$

If $z + \tau \leq t$, then by (2.2) with $\alpha = \beta$ we have

$$\frac{(z+\tau)^\gamma}{\tau^\beta}\|A[\exp\{-(z+\tau)A\} - \exp\{-zA\}]A^{-1}\exp\{-tA\}(f(t) - f(0))\|_E$$

$$\leq \frac{(z+\tau)^\gamma}{\tau^\beta}\|\exp\{-(z+\tau+t)A\} - \exp\{-(z+t)A\}\|_{E\to E}\|f(t) - f(0)\|_E$$

$$\leq M\frac{(z+\tau)^\gamma}{\tau^\beta}\frac{\tau^\beta}{(t+z)^\beta}t^{\beta-\gamma}\|f\|_{C_0^{\beta,\gamma}(E)} \leq M\|f\|_{C_0^{\beta,\gamma}(E)}. \tag{5.21}$$

If $z + \tau \geq t$ and $z \leq \tau$, then by (2.2) with $\alpha = 0$ we have

$$\frac{(z+\tau)^\gamma}{\tau^\beta}\|A[\exp\{-(z+\tau)A\} - \exp\{-zA\}]A^{-1}\exp\{-tA\}(f(t) - f(0))\|_E$$

$$\leq \frac{(z+\tau)^\gamma}{\tau^\beta}\|\exp\{-(z+\tau+t)A\} - \exp\{-(z+t)A\}\|_{E\to E}\|f(t) - f(0)\|_E$$

$$\leq M\frac{(z+\tau)^\gamma}{\tau^\beta}t^{\beta-\gamma}\|f\|_{C_0^{\beta,\gamma}(E)} \leq M\left(\frac{z+\tau}{\tau}\right)^{1-\beta}\|f\|_{C_0^{\beta,\gamma}(E)}$$

$$\leq 2^{1-\beta}M\|f\|_{C_0^{\beta,\gamma}(E)}. \tag{5.22}$$

If $z + \tau \geq t$ and $z \geq \tau$, then by (2.2) with $\alpha = 1$ we have

$$\frac{(z+\tau)^\gamma}{\tau^\beta}\|A[\exp\{-(z+\tau)A\} - \exp\{-zA\}]A^{-1}\exp\{-tA\}(f(t) - f(0))\|_E$$

$$\leq \frac{(z+\tau)^\gamma}{\tau^\beta}\|\exp\{-(z+\tau+t)A\} - \exp\{-(z+t)A\}\|_{E\to E}\|f(t) - f(0)\|_E$$

$$\leq M\frac{(z+\tau)^\gamma}{\tau^\beta}\frac{\tau}{z+t}t^{\beta-\gamma}\|f\|_{C_0^{\beta,\gamma}(E)} \leq 2M\left(\frac{\tau}{z+\tau}\right)^{1-\beta}\|f\|_{C_0^{\beta,\gamma}(E)}$$

$$\leq M_1\|f\|_{C_0^{\beta,\gamma}(E)}. \tag{5.23}$$

From the estimates (5.21), (5.22), and (5.23) it follows that

$$\sup_{0\leq z\leq z+\tau\leq 1}\frac{(z+\tau)^\gamma}{\tau^\beta}\|A[\exp\{-(z+\tau)A\} - \exp\{-zA\}]v_2(t)\|_E \leq M_2\|f\|_{C_0^{\beta,\gamma}(E)},$$

which in conjunction with (5.20) gives

$$|v_2(t)|_1^{\beta,\gamma} \leq M\|f\|_{C_0^{\beta,\gamma}(E)}. \tag{5.24}$$

Finally, let us estimate $v_3(t)$. Using the estimate (1.14) and the definition of the space $C_0^{\beta,\gamma}(E)$, we find that

$$\|A\exp\{-zA\}v_3(t)\|_E \leq \int_0^t \|A\exp\{-(z+t-s)A\}\|_{E\to E}\|f(t)-f(s)\|_E ds$$

$$\leq M\int_0^t \frac{(t-s)^\beta ds}{(z+t-s)t^\gamma}\|f\|_{C_0^{\beta,\gamma}(E)} \leq \frac{M}{t^\gamma}\int_0^t \frac{ds}{(t-s)^{1-\beta}}\|f\|_{C_0^{\beta,\gamma}(E)}$$

$$= \frac{M}{\beta}t^{\beta-\gamma}\|f\|_{C_0^{\beta,\gamma}(E)} \leq \frac{M}{\beta}\|f\|_{C_0^{\beta,\gamma}(E)}$$

for all $z \geq 0$. Consequently,

$$\max_{0\leq z\leq 1}\|A\exp\{-zA\}v_3(t)\|_E \leq \frac{M}{\beta}\|f\|_{C_0^{\beta,\gamma}(E)}. \tag{5.25}$$

By the definition of the space $E_0^{\beta,\gamma}$, we have

$$\frac{(z+\tau)^\gamma}{\tau^\beta}\|A[\exp\{-(z+\tau)A\}-\exp\{-zA\}]v_3(t)\|_E$$

$$\leq \frac{(z+\tau)^\gamma}{\tau^\beta t^\gamma}\int_0^t \|A[\exp\{-(z+\tau)A\}-\exp\{-zA\}]\exp\{-(t-s)A\}\|_{E\to E}\times$$

$$\times(t-s)^\beta ds\,\|f\|_{C_0^{\beta,\gamma}(E)} = Q\|f\|_{C_0^{\beta,\gamma}(E)}.$$

Here

$$Q = \frac{(z+\tau)^\gamma}{\tau^\beta t^\gamma}\int_0^t \|A[\exp\{-(z+\tau)A\}-\exp\{-zA\}]\exp\{-(t-s)A\}\|_{E\to E}(t-s)^\beta ds.$$

If $t \leq z$ and $\tau \leq z$, then by (1.10) and (2.3) with $\alpha = \beta$ we have

$$Q \leq \frac{(z+\tau)^\gamma}{\tau^\beta t^\gamma}\int_0^t \|A[\exp\{-(z+\tau)A\}-\exp\{-zA\}]\|_{E\to E}\times$$

$$\times\|\exp\{-(t-s)A\}\|_{E\to E}(t-s)^\beta ds$$

$$\leq M\frac{(z+\tau)^\gamma}{\tau^\beta t^\gamma}\frac{\tau^\beta t^{1+\beta}}{z^{1+\beta}} \leq 2^\gamma M\left(\frac{t}{z}\right)^{1+\beta-\gamma} \leq M_1.$$

If $t \leq z$ and $\tau \geq z$, then by (1.10) and (2.5) with $\alpha = 0$,

$$Q \leq \frac{(z+\tau)^\gamma}{\tau^\beta t^\gamma}\int_0^t \|A[\exp\{-(z+\tau)A\}-\exp\{-zA\}]\|_{E\to E}\times$$

$$\times \|\exp\{-(t-s)A\}\|_{E\to E}(t-s)^{\beta}ds$$

$$\leq M\frac{(z+\tau)^{\gamma}}{\tau^{\beta}t^{\gamma}}\frac{t^{1+\beta}}{z} \leq 2^{\beta}M\frac{(z+\tau)^{\gamma}}{(z+\tau)^{\beta}}\frac{t^{1+\beta-\gamma}}{z} \leq 2^{\beta}M\left(\frac{t}{z}\right)^{1+\beta-\gamma} \leq M_1.$$

Further, if $z \leq t \leq z+\tau$ and $\tau \leq z$, then by (1.14) and (2.2) with $\alpha = \beta$,

$$Q \leq \frac{(z+\tau)^{\gamma}}{\tau^{\beta}t^{\gamma}}\int_0^t \|\exp\{-(z+\tau)A\} - \exp\{-zA\}\|_{E\to E}\times$$

$$\times \|A\exp\{-(t-s)A\}\|_{E\to E}(t-s)^{\beta}ds$$

$$\leq M\frac{(z+\tau)^{\gamma}}{\tau^{\beta}t^{\gamma}}\frac{\tau^{\beta}}{z^{\beta}}\int_0^t \frac{ds}{(t-s)^{1-\beta}} = \frac{M}{\beta}\frac{(z+\tau)^{\gamma}}{z^{\beta}}t^{\beta-\gamma} \leq \frac{2M}{\beta}\left(\frac{t}{z+\tau}\right)^{\beta-\gamma} \leq \frac{M_1}{\beta}.$$

If $z \leq t \leq z+\tau$ and $\tau \geq z$, then by (1.14) and (2.2) with $\alpha = 0$ we have

$$Q \leq \frac{(z+\tau)^{\gamma}}{\tau^{\beta}t^{\gamma}}\int_0^t \|\exp\{-(z+\tau)A\} - \exp\{-zA\}\|_{E\to E}\times$$

$$\times \|A\exp\{-(t-s)A\}\|_{E\to E}(t-s)^{\beta}ds \leq M\frac{(z+\tau)^{\gamma}}{\tau^{\beta}t^{\gamma}}\int_0^t \frac{ds}{(t-s)^{1-\beta}}$$

$$\leq \frac{2^{\beta}M}{\beta}\frac{(z+\tau)^{\gamma}}{(z+\tau)^{\beta}}t^{\beta-\gamma} = \frac{2^{\beta}M}{\beta}\left(\frac{t}{z+\tau}\right)^{\beta-\gamma} \leq \frac{M_1}{\beta}.$$

If $z+\tau \leq t$ and $\tau \leq z$, then by (2.3) with $\alpha = 2$ we have

$$Q \leq \frac{(z+\tau)^{\gamma}}{\tau^{\beta}t^{\gamma}}\int_0^t \|A[\exp\{-(z+\tau+t-s)A\} - \exp\{-(z+t-s)A\}]\|_{E\to E}(t-s)^{\beta}ds$$

$$\leq M\frac{(z+\tau)^{\gamma}}{\tau^{\beta}t^{\gamma}}\int_0^t \frac{\tau(t-s)^{\beta}ds}{(z+t-s)^2} \leq \frac{M\tau}{\tau^{\beta}}\int_0^t \frac{ds}{(z+t-s)^{2-\beta}}$$

$$\leq \frac{M}{1-\beta}\left(\frac{\tau}{z}\right)^{1-\beta} \leq \frac{M_1}{1-\beta}.$$

If $z+\tau \leq t$ and $\tau > z$, then by the triangle inequality and (2.3) with $\alpha = 0$ and $\alpha = 1$ we have

$$Q \leq \frac{(z+\tau)^{\gamma}}{\tau^{\beta}t^{\gamma}}\Big[\int_0^{t-\tau} \|A[\exp\{-(z+\tau+t-s)A\} - \exp\{-(z+t-s)A\}]\|_{E\to E}(t-s)^{\beta}ds$$

$$+ \int_{t-\tau}^t \|A[\exp\{-(z+\tau+t-s)A\} - \exp\{-(z+t-s)A\}]\|_{E\to E}(t-s)^{\beta}ds\Big]$$

$$\leq \frac{M(z+\tau)^{\gamma}}{\tau^{\beta}t^{\gamma}}\Big[\int_0^{t-\tau} \frac{\tau(t-s)^{\beta}ds}{(z+t-s)^2} + \int_{t-\tau}^t \frac{(t-s)^{\beta}ds}{z+t-s}\Big]$$

$$\leq \frac{2^\gamma M}{(z+\tau)^{\beta-\gamma}t^\gamma}\left[\tau\int_0^{t-\tau}\frac{ds}{(t-s)^{2-\beta}}+\int_{t-\tau}^t\frac{ds}{(t-s)^{1-\beta}}\right]$$

$$\leq \frac{2^\gamma M}{\tau^\beta}\left[\frac{\tau}{(1-\beta)\tau^{1-\beta}}+\frac{\tau^\beta}{\beta}\right]=\frac{2^\gamma M}{\beta(1-\beta)}\leq\frac{M_1}{\beta(1-\beta)}.$$

Combining the estimates for Q we obtain

$$\sup_{0\leq z<z+\tau\leq 1}\frac{(z+\tau)^\gamma}{\tau^\beta}\|A[\exp\{-(z+\tau)A\}-\exp\{-zA\}]v_3(t)\|_E$$

$$\leq\frac{M_1}{\beta(1-\beta)}\|f\|_{C_0^{\beta,\gamma}(E)}.$$

From this inequality and (5.25) it follows that

$$|v_3(t)|_1^{\beta,\gamma}\leq\frac{M}{\beta(1-\beta)}\|f\|_{C_0^{\beta,\gamma}(E)}. \tag{5.26}$$

Finally, combining the estimates (5.19), (5.24), and (5.26) we obtain (5.17). Theorem 5.2 is proved.

Note that the spaces of smooth functions $C_0^{\beta,\gamma}(E)$, in which well-posedness has been established, depend on the parameters β and γ. However, the constants in the coercivity inequality (5.16) depend only on β. Hence, γ can be chosen freely in $[0,\beta]$, which increases the number of function spaces in which problem (1.1) is well posed. In particular, problem (1.1) is well posed in the Hölder space without a weight ($\gamma=0$).

In the following subsections we will study the Cauchy problem (1.1) in certain restrictions of an arbitrary space E. For such restrictions we are able to establish the coercivity inequality with constants that are independent not only on γ but also on β.

3.   Semigroup estimates.

Let us establish a number of estimates for the semigroup $\exp\{-tA\}$ in the fractional norms $E_\alpha=E_{\alpha,\infty}$ that will be needed in the sequel.

Lemma 5.1. *For any $t>0$ the following estimates hold:*

$$\|A^n\exp\{-tA\}\|_{E_\alpha\to E_\alpha}\leq\frac{M}{t^n},\quad n=0,1,2,\quad 0<\alpha<1, \tag{5.27}$$

where M does not depend on t and α.

Proof. The estimates (5.27) are consequences of (1.10) and (1.14) because the operator A and the semigroup $\exp\{-tA\}$ commute.

Lemma 5.2. *The following estimates hold:*

$$\|\exp\{-tA\} - \exp\{-(t+\tau)A\}\|_{E_{\alpha-\gamma} \to E_{\alpha-\beta}} \le M(t+\tau)^{\beta-\gamma}, \tag{5.28}$$

$$0 \le t \le t+\tau \le 1, \quad 0 \le \gamma \le \beta < \alpha, \quad \text{or} \quad \alpha = \beta = \gamma > 0, \quad 0 < \alpha < 1,$$

where M does not depend on α, β, γ, t, and τ.

Proof. Let us estimate in the norm of E the expression

$$\Delta(z,t)x = z^{1-(\alpha-\beta)} A \exp\{-zA\}[\exp\{-tA\} - \exp\{-(t+\tau)A\}]x.$$

We have to examine three cases: $z \le t$, $t < z < t+\tau$, and $z \ge t+\tau$. In the first case, using the triangle inequality and the definition of the norm in $E_{\alpha-\gamma}$, we obtain

$$\|\Delta(z,t)x\|_E \le z^{1-(\alpha-\beta)}[\|A\exp\{-(z+t)A\}x\|_E + \|A\exp\{-(z+t+\tau)A\}x\|_E]$$

$$\le \left[\frac{z^{1-(\alpha-\beta)}}{(z+t)^{1-(\alpha-\gamma)}} + \frac{z^{1-(\alpha-\beta)}}{(z+t+\tau)^{1-(\alpha-\gamma)}}\right]\|x\|_{\alpha-\gamma}.$$

Since $z \le t < t+\tau$ and $\tau > 0, t \ge 0$, it follows that

$$\|\Delta(z,t)x\|_E \le 2t^{\beta-\gamma}\|x\|_{\alpha-\gamma} \le 2(t+\tau)^{\beta-\gamma}\|x\|_{\alpha-\gamma}.$$

Thus, if $z \le t$, then

$$\|\Delta(z,t)x\|_E \le 2(t+\tau)^{\beta-\gamma}\|x\|_{\alpha-\gamma}.$$

Now let $z \ge t+\tau$. Then, using the identity

$$\exp\{-tA\}x - Ix = -\int_0^t A\exp\{-sA\}x\,ds, \tag{5.29}$$

we obtain

$$z^{1-(\alpha-\beta)} A\exp\{-zA\}[\exp\{-tA\}x - Ix]$$

$$= z^{1-(\alpha-\beta)} \int_0^t A \exp\left\{-\frac{z+s}{2}A\right\} A \exp\left\{-\frac{z+s}{2}A\right\} x \, ds.$$

From this and the estimate (1.14) it follows that

$$\|z^{1-(\alpha-\beta)} A \exp\{-zA\}[\exp\{-tA\} - I]x\|_E \leq$$

$$z^{1-(\alpha-\beta)} \int_0^t \left\|A \exp\left\{-\frac{z+s}{2}A\right\}\right\|_{E\to E} \left\|A \exp\left\{-\frac{z+s}{2}A\right\}x\right\|_E ds$$

$$\leq M z^{1-(\alpha-\beta)} \int_0^t \frac{ds}{(z+s)^{2-\alpha+\gamma}} \|x\|_{\alpha-\gamma}.$$

Since $z(z+s)^{-1} \leq 1$ and $z \geq t + \tau$, we obtain

$$\|z^{1-(\alpha-\beta)} A \exp\{-zA\}[\exp\{-tA\} - I]x\|_E$$

$$\leq M \int_0^t \frac{ds}{(z+s)^{1-\beta+\gamma}} \|x\|_{\alpha-\gamma} \leq \frac{Mt}{z^{1-\beta+\gamma}} \|x\|_{\alpha-\beta} \leq \frac{Mt}{(t+\tau)^{1-\beta+\gamma}} \|x\|_{\alpha-\gamma},$$

whence

$$\|z^{1-(\alpha-\beta)} A \exp\{-zA\}[\exp\{-tA\} - I]x\|_E \leq M(t+\tau)^{\beta-\gamma} \|x\|_{\alpha-\gamma}.$$

In a similar manner one can establish the estimate

$$\|z^{1-(\alpha-\beta)} A \exp\{-zA\}[I - \exp\{-(t+\tau)A\}]x\|_E \leq M(t+\tau)^{\beta-\gamma} \|x\|_{\alpha-\gamma}.$$

By the triangle inequality and the last two inequalities, we have that

$$\|\Delta(z,t)x\|_E \leq \|z^{1-(\alpha-\beta)} A \exp\{-zA\}[\exp\{-tA\} - I]x\|_E$$

$$+\|z^{1-(\alpha-\beta)} A \exp\{-zA\}[I - \exp\{-(t+\tau)A\}]x\|_E \leq 2M(t+\tau)^{\beta-\gamma} \|x\|_{\alpha-\gamma}.$$

Therefore, if $z \geq t + \tau$ we conclude that

$$\|\Delta(z,t)x\|_E \leq 2M(t+\tau)^{\beta-\gamma} \|x\|_{\alpha-\gamma}.$$

Now let $t < z < t + \tau$. Then, using the definition of the norm in $E_{\alpha-\gamma}$, we obtain

$$\|z^{1-(\alpha-\beta)} A \exp\{-zA\}\exp\{-(t+\tau)A\}x\|_E$$

$$\leq \frac{z^{1-(\alpha-\beta)}}{(z+t+\tau)^{1-(\alpha-\gamma)}} \|x\|_{\alpha-\gamma} \leq z^{\beta-\gamma} \|x\|_{\alpha-\gamma}.$$

This yields

$$\|z^{1-(\alpha-\beta)}A\exp\{-zA\}\exp\{-(t+\tau)A\}x\|_E \leq (t+\tau)^{\beta-\gamma}\|x\|_{\alpha-\gamma}. \tag{5.30}$$

Next, using identity (5.29) we obtain

$$\|z^{1-(\alpha-\beta)}A\exp\{-zA\}[\exp\{-tA\} - I]x\|_E$$

$$\leq z^{1-(\alpha-\beta)}\int_0^t \|A\exp\Big\{-\frac{z+s}{2}A\Big\}\|_{E\to E}\|A\exp\Big\{-\frac{z+s}{2}A\Big\}x\|_E ds.$$

From this and the estimates (1.14) for $t < z$ it follows that

$$\|z^{1-(\alpha-\beta)}A\exp\{-zA\}[\exp\{-tA\} - I]x\|_E$$

$$\leq Mz^{1-(\alpha-\beta)}\int_0^t \frac{ds}{(z+s)^{2-\alpha-\gamma}}\|x\|_{\alpha-\gamma} \leq \frac{Mt}{z^{1-\beta+\gamma}}\|x\|_{\alpha-\gamma} \leq Mz^{\beta-\gamma}\|x\|_{\alpha-\gamma}.$$

Since $z < t+\tau$,

$$\|z^{1-(\alpha-\beta)}A\exp\{-zA\}[\exp\{-tA\} - I]x\|_E \leq M(t+\tau)^{\beta-\gamma}\|x\|_{\alpha-\gamma}. \tag{5.31}$$

Finally, from the definition of the spaces $E_{\alpha-\gamma}$ it follows that

$$\|z^{1-(\alpha-\beta)}A\exp\{-zA\}Ix\|_E \leq \frac{z^{1-(\alpha-\beta)}}{z^{1-(\alpha-\gamma)}}\|x\|_{\alpha-\gamma} = z^{\beta-\gamma}\|x\|_{\alpha-\gamma}.$$

Since $z < t+\tau$, we conclude that

$$\|z^{1-(\alpha-\beta)}A\exp\{-zA\}Ix\|_E \leq (t+\tau)^{\beta-\gamma}\|x\|_{\alpha-\gamma}. \tag{5.32}$$

From the triangle inequality and inequalities (5.30) to (5.32) it follows that

$$\|\Delta(z,t)x\|_E \leq \|z^{1-(\alpha-\beta)}A\exp\{-zA\}\exp\{-(t+\tau)A\}x\|_E +$$

$$\|z^{1-(\alpha-\beta)}A\exp\{-zA\}[\exp\{-tA\} - I]x\|_E + \|z^{1-(\alpha-\beta)}A\exp\{-zA\}Ix\|_E$$

$$\leq (2+M)(t+\tau)^{\beta-\gamma}\|x\|_{\alpha-\gamma}.$$

Hence, in the second case, too, we have

$$\|\Delta(z,t)x\|_E \leq M_1(t+\tau)^{\beta-\gamma}\|x\|_{\alpha-\gamma}.$$

Combining the estimates for $\|\Delta(z,t)x\|_E$, we obtain

$$\|\Delta(z,t)x\|_E \le M_2(t+\tau)^{\beta-\gamma}\|x\|_{\alpha-\gamma}$$

for any $z > 0$ and $0 \le t < t+\tau \le 1$. From the definition of the norm in the spaces $E_{\alpha-\beta}$ it follows that

$$\sup_{z>0}\|\Delta(z,t)x\|_E = \|[\exp\{-tA\} - \exp\{-(t+\tau)A\}]x\|_{\alpha-\beta}.$$

Therefore,

$$\|[\exp\{-tA\} - \exp\{-(t+\tau)A\}]x\|_{\alpha-\beta} \le M_2(t+\tau)^{\beta-\gamma}\|x\|_{\alpha-\gamma}.$$

This yields the estimates (5.28) in the case $0 < \gamma \le \beta < \alpha$. The estimates (5.28) in the cases $\gamma = \beta = \alpha > 0$ or $\beta = \gamma = 0$ are plain. Lemma 5.2 is proved.

Lemma 5.3. *The following estimates hold:*

$$\|\exp\{-tA\} - \exp\{-(t+\tau)A\}\|_{E_{\alpha-\gamma}\to E_{\alpha-\beta}} \le \frac{M\tau}{t^{1+\gamma-\beta}}, \qquad (5.33)$$

$$0 < t < t+\tau \le 1, \quad 0 \le \gamma \le \beta \le \alpha, \quad 0 < \alpha < 1,$$

where M does not depend on α, β, γ, t, and τ.

Proof. Using the identity (5.29) we obtain

$$z^{1-(\alpha-\beta)}A\exp\{-zA\}[\exp\{-(t+\tau)A\}x - \exp\{-tA\}x]$$

$$= -\int_0^\tau z^{1-(\alpha-\beta)}A\exp\left\{-\frac{z+s+t}{2}A\right\}A\exp\left\{-\frac{z+s+t}{2}A\right\}x\,ds.$$

From this relation, (1.14), and the definition of the norm in the spaces $E_{\alpha-\gamma}$ it follows that

$$z^{1-(\alpha-\beta)}\|A\exp\{-zA\}[\exp\{-(t+\tau)A\}x - \exp\{-tA\}x]\|_E$$

$$\le M\int_0^\tau \frac{z^{1-\alpha+\beta}ds}{(s+t+z)^{2-\alpha+\gamma}}\|x\|_{\alpha-\gamma}.$$

Since

$$\int_0^\tau \frac{z^{1-\alpha+\beta}ds}{(s+t+z)^{2-\alpha+\gamma}} \le \int_0^\tau \frac{ds}{(s+t+z)^{1-\beta+\gamma}} \le \frac{\tau}{t^{1-\beta+\gamma}},$$

we have that

$$z^{1-(\alpha-\beta)}\|A\exp\{-zA\}[\exp\{-(t+\tau)A\}x - \exp\{-tA\}x]\|_E \leq \frac{M\tau}{t^{1-\beta+\gamma}}\|x\|_{\alpha-\gamma}.$$

From this inequality and the definition of the norm in the spaces $E_{\alpha-\beta}$ it follows that

$$\|[\exp\{-(t+\tau)A\} - \exp\{-tA\}]x]\|_{\alpha-\beta} \leq \frac{M\tau}{t^{1-\beta+\gamma}}\|x\|_{\alpha-\gamma}.$$

Thus, we have established the estimates (5.33) in the case $0 < \gamma \leq \beta < \alpha$. In the cases $\gamma = \beta = \alpha$, $\beta = \alpha$, and $\gamma = \beta = 0$, the estimates (5.33) are plain. Lemma 5.3 is proved.

4.   The coercivity inequality for the general problem.

Here we study the Cauchy problem (1.1) in the spaces $C_0^{\beta,\gamma}(E_{\alpha-\beta})$ ($0 < \gamma \leq \beta < \alpha, 0 < \alpha < 1$). To these spaces there correspond the spaces of traces $E_{1+\alpha-\beta}^{\beta,\gamma}$, which consist of the elements $w_0 \in E$ for which the following norm is finite:

$$|w_0|_{1+\alpha-\beta}^{\beta,\gamma} = \max_{0 \leq z \leq 1} \|A\exp\{-zA\}w_0\|_{\alpha-\beta}$$

$$+ \sup_{0 \leq z < z+\tau \leq 1} \tau^{-\beta}(z+\tau)^{\gamma}\|A[\exp\{-(z+\tau)A\} - \exp\{-zA\}]w_0\|_{\alpha-\beta}. \qquad (5.34)$$

Theorem 5.3. *Suppose $-A$ is the generator of an analytic semigroup in E and $v_0 - A^{-1}f(0) \in E_{1+\alpha-\beta}^{\beta,\gamma}$, $f(t) \in C_0^{\beta,\gamma}(E_{\alpha-\beta})$. Then there exists a unique solution $v(t)$ of the Cauchy problem (1.1) such that $v'(t), Av(t) \in C_0^{\beta,\gamma}(E_{\alpha-\beta})$, and the coercivity inequality holds:*

$$\|v'\|_{C_0^{\beta,\gamma}(E_{\alpha-\beta})} + \|Av\|_{C_0^{\beta,\gamma}(E_{\alpha-\beta})} + \max_{0 \leq t \leq 1} |v(t) - A^{-1}f(t)|_{1+\alpha-\beta}^{\beta,\gamma}$$

$$\leq M\left[|v_0 - A^{-1}f(0)|_{1+\alpha-\beta}^{\beta,\gamma} + \frac{1}{\alpha(1-\alpha)}\|f\|_{C_0^{\beta,\gamma}(E_{\alpha-\beta})}\right], \qquad (5.35)$$

where M does not depend on $\alpha, \beta, \gamma, v_0$ and $f(t)$.

Proof. The well-posedness of problem (1.1) in $C_0^{\beta,\gamma}(E)$ and the estimate (5.27) imply the inequality (5.35) with the constant $M/[\beta(1 - \beta)]$, $0 < \beta < 1$. Here in the case $E = E_{\alpha-\beta}$, $0 < \beta < \alpha$, $0 < \alpha < 1$, we sharpen the estimates obtained in the first section for the solution of the Cauchy problem (1.1). To this end we shall follow the scheme of the proof of Theorems 5.1 and 5.2. From (5.4) we derive the identity

$$v'(t) = -\exp\{-tA\}(Av_0 - f(0)) + \exp\{-tA\}(f(t) - f(0))$$

$$+ \int_0^t A \exp\{-(t-s)A\}(f(s) - f(t))ds \equiv w(t) + p(t) + g(t). \tag{5.36}$$

First let us estimate the functions $A^{-1}w(t), A^{-1}p(t)$, and $A^{-1}g(t)$ in the norm of $C(E_{1+\alpha-\beta}^{\beta,\gamma})$. From Theorem 5.2 and the estimate (5.27) it follows that

$$|A^{-1}w(t)|_{1+\alpha-\beta}^{\beta,\gamma} \le M|v_0 - A^{-1}f(0)|_{1+\alpha-\beta}^{\beta,\gamma} \tag{5.37}$$

and

$$|A^{-1}p(t)|_{1+\alpha-\beta}^{\beta,\gamma} \le M\|f\|_{C(E_{1+\alpha-\beta}^{\beta,\gamma})} \tag{5.38}$$

for any $t \in [0,1]$. Now let us estimate $A^{-1}g(t)$ in the norm of $C(E_{1+\alpha-\beta}^{\beta,\gamma})$. Using the identity

$$\lambda^{1-(\alpha-\beta)}A\exp\{-\lambda A\}A\exp\{-zA\}A^{-1}g(t)$$

$$= -\int_0^t A\exp\left\{-\frac{z+t-s+\lambda}{2}A\right\}A\exp\left\{-\frac{z+t-s+\lambda}{2}A\right\}(f(t) - f(s))ds,$$

the estimate (1.14), and the definition of the norm in the space $C_0^{\beta,\gamma}(E_{\alpha-\beta})$, we obtain

$$\lambda^{1-(\alpha-\beta)}\|A\exp\{-\lambda A\}A\exp\{-zA\}A^{-1}g(t)\|_E$$

$$\le M\int_0^t \frac{\lambda^{1-\alpha+\beta}\|f(t)-f(s)\|_{\alpha-\beta}ds}{(z+t-s+\lambda)^{2-\alpha+\beta}} \le \frac{M}{t^\gamma}\int_0^t \frac{\lambda^{1-\alpha}(t-s)^\beta ds}{(t-s+\lambda)^{2-\alpha}}\|f\|_{C_0^{\beta,\gamma}(E_{\alpha-\beta})}$$

$$\le Mt^{\beta-\gamma}\int_0^t \frac{\lambda^{1-\alpha}ds}{(t-s+\lambda)^{2-\alpha}}\|f\|_{C_0^{\beta,\gamma}(E_{\alpha-\beta})} \le \frac{M}{1-\alpha}\|f\|_{C_0^{\beta,\gamma}(E_{\alpha-\beta})}$$

for any $\lambda > 0$ and $z \ge 0$. Consequently,

$$\max_{0\le z\le 1}\|A\exp\{-zA\}A^{-1}g(t)\|_{\alpha-\beta} \le \frac{M}{1-\alpha}\|f\|_{C_0^{\beta,\gamma}(E_{\alpha-\beta})}. \tag{5.39}$$

Further, we have

$$\frac{(z+\tau)^\gamma}{\tau^\beta}\|A[\exp\{-(z+\tau)A\} - \exp\{-zA\}]A^{-1}g(t)\|_{\alpha-\beta}$$

$$\le \frac{(z+\tau)^\gamma}{\tau^\beta}\|\exp\{-(z+\tau)A\} - \exp\{-zA\}\|_{E_{\alpha-\gamma}\to E_{\alpha-\beta}}\times$$

$$\times \|A \int_0^t \exp\{-(t-s)A\}(f(t)-f(s))ds\|_{\alpha-\gamma}. \tag{5.40}$$

If $z \le \tau$, then, by Lemma 5.2,

$$\frac{(z+\tau)^\gamma}{\tau^\beta}\|\exp\{-(z+\tau)A\}-\exp\{-zA\}\|_{E_{\alpha-\gamma}\to E_{\alpha-\beta}} \le M\frac{(z+\tau)^\gamma}{\tau^\beta}(z+\tau)^{\beta-\gamma}$$

$$= M\left(\frac{z+\tau}{\tau}\right)^\beta \le 2^\beta M. \tag{5.41}$$

If $z > \tau$, then, by Lemma 5.3,

$$\frac{(z+\tau)^\gamma}{\tau^\beta}\|\exp\{-(z+\tau)A\}-\exp\{-zA\}\|_{E_{\alpha-\gamma}\to E_{\alpha-\beta}}$$

$$\le M\frac{(z+\tau)^\gamma}{\tau^\beta}\frac{\tau}{z^{1-\beta+\gamma}} \le 2^{1-\beta+\gamma}M\frac{\tau^{1-\beta}}{(z+\tau)^{1-\beta}} \le 2^{1-\beta+\gamma}M.$$

From this and the inequality (5.41) it follows that

$$\sup_{0\le z<z+\tau\le1} \tau^{-\beta}(z+\tau)^\gamma\|\exp\{-(z+\tau)A\}-\exp\{-zA\}\|_{E_{\alpha-\gamma}\to E_{\alpha-\beta}} \le M_1. \tag{5.42}$$

Now let us establish the estimate

$$\|A\int_0^t \exp\{-(t-s)A\}(f(t)-f(s))ds\|_{\alpha-\gamma} \le \frac{M}{\alpha(1-\alpha)}\|f\|_{C_0^{\beta,\gamma}(E_{\alpha-\beta})}. \tag{5.43}$$

Using the identity

$$\lambda^{1-(\alpha-\gamma)}A\exp\{-\lambda A\}A\int_0^t \exp\{-(t-s)A\}(f(s)-f(t))ds$$

$$= \lambda^{1-(\alpha-\gamma)}\int_0^t A\exp\left\{-\frac{\lambda+t-s}{2}A\right\}A\exp\left\{-\frac{\lambda+t-s}{2}A\right\}(f(s)-f(t))ds$$

and the estimate (1.14), we obtain

$$\lambda^{1-\alpha+\gamma}\|A\exp\{-\lambda A\}A\int_0^t \exp\{-(t-s)A\}(f(s)-f(t))ds\|_E$$

$$\le M\lambda^{1-\alpha+\gamma}\int_0^t \frac{\|f(s)-f(t)\|_{\alpha-\beta}ds}{(\lambda+t-s)^{2-\alpha+\beta}}$$

$$\le \frac{M\lambda^{1-\alpha+\gamma}}{t^\gamma}\int_0^t \frac{(t-s)^\beta ds}{(\lambda+t-s)^{2-\alpha+\beta}}\|f\|_{C_0^{\beta,\gamma}(E_{\alpha-\beta})}. \tag{5.44}$$

Since

$$\lambda^{1-\alpha+\gamma} \int_0^t \frac{(t-s)^\beta ds}{(\lambda+t-s)^{2-\alpha+\beta}} \le \lambda^{1-\alpha+\gamma} \int_0^t \frac{ds}{(\lambda+t-s)^{2-\alpha}} \le \frac{\lambda^\gamma}{1-\alpha},$$

the last inequality implies that for $\lambda \le t$,

$$\frac{\lambda^{1-\alpha+\gamma}}{t^\gamma} \int_0^t \frac{(t-s)^\beta ds}{(\lambda+t-s)^{2-\alpha+\beta}} \le \frac{1}{1-\alpha}. \tag{5.45}$$

Next, since

$$\lambda^{1-\alpha+\gamma} \int_0^t \frac{(t-s)^\beta ds}{(\lambda+t-s)^{2-\alpha+\beta}} \le \lambda^{-\alpha+\gamma} \int_0^t \frac{ds}{(t-s)^{1-\alpha}} \le \frac{1}{\alpha} \frac{t^\alpha}{\lambda^{\alpha-\gamma}},$$

we see that for $\lambda > t$,

$$\frac{\lambda^{1-\alpha+\gamma}}{t^\gamma} \int_0^t \frac{(t-s)^\beta ds}{(\lambda+t-s)^{2-\alpha+\beta}} \le \frac{1}{\alpha}.$$

From this and inequalities (5.44), (5.45) we derive the estimate (5.43). Combining the estimates (5.40), (5.42), and (5.43) we obtain

$$\sup_{0 \le z < z+\tau \le 1} \frac{(z+\tau)^\gamma}{\tau^\beta} \|A[\exp\{-(z+\tau)A\} - \exp\{-zA\}]A^{-1}g(t)\|_{\alpha-\beta}$$

$$\le \frac{M}{\alpha(1-\alpha)} \|f\|_{C_0^{\beta,\gamma}(E_{\alpha-\beta})}.$$

From this and (5.39) it follows that

$$|A^{-1}g(t)|_{1+\alpha-\gamma}^{\beta,\gamma} \le \frac{M}{\alpha(1-\alpha)} \|f\|_{C_0^{\beta,\gamma}(E_{\alpha-\beta})} \tag{5.46}$$

for any $t \in [0,1]$. Combining the estimates (5.37), (5.38), and (5.46), we conclude that

$$\max_{0 \le t \le 1} |v(t) - A^{-1}f(t)|_{1+\alpha-\beta}^{\beta,\gamma}$$

$$\le M\left[|v_0 - A^{-1}f(0)|_{1+\alpha-\beta}^{\beta,\gamma} + \frac{1}{\alpha(1-\alpha)} \|f\|_{C_0^{\beta,\gamma}(E_{\alpha-\beta})} \right]. \tag{5.47}$$

Now let us estimate the functions $w(t)$, $p(t)$, and $g(t)$ in the norm of the space $C_0^{\beta,\gamma}(E_{\alpha-\beta})$. From the definition of the space $E_{1+\alpha-\beta}^{\beta,\gamma}$ it follows that

$$\|w\|_{C_0^{\beta,\gamma}(E_{\alpha-\beta})} = |v_0 - A^{-1}f(0)|_{1+\alpha-\beta}^{\beta,\gamma}. \tag{5.48}$$

By (5.10) and (5.27), we have

$$\|p\|_{C_0^{\beta,\gamma}(E_{\alpha-\beta})} \le M\|f\|_{C_0^{\beta,\gamma}(E_{\alpha-\beta})}. \qquad (5.49)$$

Finally, let us establish the estimate

$$\|g\|_{C_0^{\beta,\gamma}(E_{\alpha-\beta})} \le \frac{M}{\alpha(1-\alpha)}\|f\|_{C_0^{\beta,\gamma}(E_{\alpha-\beta})}. \qquad (5.50)$$

To this end it suffices to show that

$$\|g(t)\|_{\alpha-\beta} \le \frac{M}{\alpha(1-\alpha)}\|f\|_{C_0^{\beta,\gamma}(E_{\alpha-\beta})}, \quad 0 \le t \le 1, \qquad (5.51)$$

and

$$\|g(t+\tau)-g(t)\|_{\alpha-\beta} \le \frac{M\tau^\beta}{\alpha(1-\alpha)(t+\tau)^\gamma}\|f\|_{C_0^{\beta,\gamma}(E_{\alpha-\beta})}, \quad 0 \le t < t+\tau \le 1, \quad (5.52)$$

First let us prove (5.51). Using the identity

$$z^{1-(\alpha-\beta)}A\exp\{-zA\}g(t)$$

$$= -\int_0^t z^{1-(\alpha-\beta)}A\exp\Big\{-\frac{z+t-s}{2}A\Big\}A\exp\Big\{-\frac{z+t-s}{2}A\Big\}(f(t)-f(s))ds,$$

the estimate (1.14), and the definition of the spaces $E_{\alpha-\beta}$, we obtain

$$z^{1-(\alpha-\beta)}\|A\exp\{-zA\}g(t)\|_E \le \int_0^t z^{1-(\alpha-\beta)}\Big\|A\exp\Big\{-\frac{z+t-s}{2}A\Big\}\Big\|_{E\to E} \times$$

$$\times \Big\|A\exp\Big\{-\frac{z+t-s}{2}A\Big\}(f(t)-f(s))\Big\|_E ds$$

$$\le M\int_0^t \frac{z^{1-(\alpha-\beta)}ds}{(z+t-s)^{2-\alpha+\beta}}\|f(t)-f(s)\|_{\alpha-\beta}ds$$

$$\le \frac{M}{t^\gamma}\int_0^t \frac{z^{1-(\alpha-\beta)}(t-s)^\beta ds}{(z+t-s)^{2-\alpha+\beta}}\|f\|_{C_0^{\beta,\gamma}(E_{\alpha-\beta})}. \qquad (5.53)$$

Since

$$\int_0^t \frac{z^{1-(\alpha-\beta)}(t-s)^\beta ds}{(z+t-s)^{2-\alpha+\beta}} \le \int_0^t \frac{z^{1-\alpha+\beta}ds}{(z+t-s)^{2-\alpha}} \le \frac{z^\beta}{1-\alpha},$$

we have

$$z^{1-(\alpha-\beta)}\|A\exp\{-zA\}g(t)\|_E \le \frac{M}{1-\alpha}\frac{z^\beta}{t^\gamma}\|f\|_{C_0^{\beta,\gamma}(E_{\alpha-\beta})}.$$

If $z \leq t$, then the last inequality yields

$$z^{1-(\alpha-\beta)}\|A\exp\{-zA\}g(t)\|_E \leq \frac{M}{1-\alpha}t^{\beta-\gamma}\|f\|_{C_0^{\beta,\gamma}(E_{\alpha-\beta})}. \qquad (5.54)$$

Further, since

$$\int_0^t \frac{z^{1-(\alpha-\beta)}(t-s)^\beta ds}{(z+t-s)^{2-\alpha+\beta}} \leq \frac{1}{z^{\alpha-\beta}}\int_0^t \frac{ds}{(t-s)^{1-\alpha}} = \frac{t^\alpha}{\alpha z^{\alpha-\beta}},$$

inequality (5.53) yields

$$z^{1-(\alpha-\beta)}\|A\exp\{-zA\}g(t)\|_E \leq \frac{M}{\alpha}\frac{t^{\alpha-\gamma}}{z^{\alpha-\beta}}\|f\|_{C_0^{\beta,\gamma}(E_{\alpha-\beta})}.$$

Now, if $z > t$, then by the last inequality we have that

$$z^{1-(\alpha-\beta)}\|A\exp\{-zA\}g(t)\|_E \leq \frac{M}{\alpha}t^{\beta-\gamma}\|f\|_{C_0^{\beta,\gamma}(E_{\alpha-\beta})},$$

which in conjunction with (5.54) yields

$$\|z^{1-(\alpha-\beta)}A\exp\{-zA\}g(t)\|_E \leq \frac{M}{\alpha(1-\alpha)}t^{\beta-\gamma}\|f\|_{C_0^{\beta,\gamma}(E_{\alpha-\beta})}.$$

By the definition of the space $E_{\alpha-\beta}$, we have

$$\|g(t)\|_{\alpha-\beta} \leq \frac{M}{\alpha(1-\alpha)}t^{\beta-\gamma}\|f\|_{C_0^{\beta,\gamma}(E_{\alpha-\beta})} \qquad (5.55)$$

for any $t \in [0,1]$. This last inequality implies (5.51) as well as the estimate (5.52) for $t \leq \tau$. Indeed, by (5.55) and the triangle inequality,

$$\|g(t+\tau) - g(t)\|_{\alpha-\beta} \leq \|g(t+\tau)\|_{\alpha-\beta} + \|g(t)\|_{\alpha-\beta}$$

$$\leq \frac{M}{\alpha(1-\alpha)}[(t+\tau)^{\beta-\gamma} + t^{\beta-\gamma}]\|f\|_{C_0^{\beta,\gamma}(E_{\alpha-\beta})}$$

$$\leq \frac{2M}{\alpha(1-\alpha)}(t+\tau)^{\beta-\gamma}\|f\|_{C_0^{\beta,\gamma}(E_{\alpha-\beta})} \leq \frac{2^{1+\beta}M}{\alpha(1-\alpha)}\frac{\tau^\beta}{(t+\tau)^\gamma}\|f\|_{C_0^{\beta,\gamma}(E_{\alpha-\beta})}.$$

Now suppose $t > \tau$. Let us represent the difference $g(t+\tau) - g(t)$ as the sum of the following integrals:

$$-(g(t+\tau) - g(t)) = \int_{t-\tau}^{t+\tau} A\exp\{-(t+\tau-s)A\}(f(t+\tau) - f(s))ds$$

$$+ \int_{t-\tau}^{t} A \exp\{-(t-s)A\}(f(s)-f(t))ds+$$

$$\int_{0}^{t-\tau} A[\exp\{-(t+\tau-s)A\} - \exp\{-(t-s)A\}](f(t)-f(s))ds$$

$$+ \int_{0}^{t-\tau} A \exp\{-(t+\tau-s)A\}ds\, (f(t+\tau)-f(t)) = I_1 + I_2 + I_3 + I_4.$$

Let us estimate I_1, I_2, I_3 and I_4 separately. We start with I_1. Using the identity

$$z^{1-(\alpha-\beta)} A \exp\{-zA\}I_1$$

$$= \int_{t-\tau}^{t+\tau} z^{1-(\alpha-\beta)} A \exp\left\{ - \frac{z+t+\tau-s}{2}A \right\}A\times$$

$$\times \exp\left\{ - \frac{z+t+\tau-s}{2}A \right\}(f(t+\tau)-f(s))ds$$

and the estimate (1.14), we obtain

$$z^{1-(\alpha-\beta)} \|A \exp\{-zA\}I_1\|_E$$

$$\leq \int_{t-\tau}^{t+\tau} z^{1-(\alpha-\beta)} \left\|A \exp\left\{ - \frac{z+t+\tau-s}{2}A \right\}\right\|_{E\to E}\times$$

$$\times \left\|A \exp\left\{ - \frac{z+t+\tau-s}{2}A \right\}(f(t+\tau)-f(s))\right\|_E ds$$

$$\leq M \int_{t-\tau}^{t+\tau} \frac{z^{1-(\alpha-\beta)}}{(z+t+\tau-s)^{2-\alpha+\beta}} \|f(t+\tau)-f(s)\|_{\alpha-\beta} ds$$

$$\leq M \int_{t-\tau}^{t+\tau} \frac{z^{1-(\alpha-\beta)}(t+\tau-s)^{\beta}ds}{(z+t+\tau-s)^{2-\alpha+\beta}(t+\tau)^{\gamma}} \|f\|_{C_0^{\beta,\gamma}(E_{\alpha-\beta})}. \tag{5.56}$$

Since

$$\int_{t-\tau}^{t+\tau} \frac{z^{1-(\alpha-\beta)}(t+\tau-s)^{\beta}ds}{(z+t+\tau-s)^{2-\alpha+\beta}} \leq \int_{t-\tau}^{t+\tau} \frac{z^{1-(\alpha-\beta)}ds}{(z+t+\tau-s)^{2-\alpha}} \leq \frac{z^{\beta}}{1-\alpha},$$

it follows that

$$z^{1-(\alpha-\beta)}\|A \exp\{-zA\}I_1\|_E \leq \frac{Mz^{\beta}}{(1-\alpha)(t+\tau)^{\gamma}} \|f\|_{C_0^{\beta,\gamma}(E_{\alpha-\beta})}.$$

If $z \leq \tau$, then the last inequality yields

$$z^{1-(\alpha-\beta)}\|A \exp\{-zA\}I_1\|_E \leq \frac{M\tau^{\beta}}{(1-\alpha)(t+\tau)^{\gamma}} \|f\|_{C_0^{\beta,\gamma}(E_{\alpha-\beta})}. \tag{5.57}$$

Further, since

$$\int_{t-\tau}^{t+\tau} \frac{z^{1-(\alpha-\beta)}(t+\tau-s)^\beta ds}{(z+t+\tau-s)^{2-\alpha+\beta}} \le \int_{t-\tau}^{t+\tau} \frac{z^{-\alpha+\beta} ds}{(z+t+\tau-s)^{1-\alpha}}$$

$$\le \frac{1}{z^{\alpha-\beta}} \int_{t-\tau}^{t+\tau} \frac{ds}{(t+\tau-s)^{1-\alpha}} = \frac{2^\alpha}{\alpha} \frac{\tau^\alpha}{z^{\alpha-\beta}},$$

inequality (5.56) gives

$$z^{1-(\alpha-\beta)}\|A\exp\{-zA\}I_1\|_E \le \frac{2^\alpha M}{\alpha} \frac{\tau^\alpha}{z^{\alpha-\beta}(t+\tau)^\gamma}\|f\|_{C_0^{\beta,\gamma}(E_{\alpha-\beta})}.$$

If $z > \tau$, then the last inequality yields

$$z^{1-(\alpha-\beta)}\|A\exp\{-zA\}I_1\|_E \le \frac{2^\alpha M}{\alpha} \frac{\tau^\beta}{(t+\tau)^\gamma}\|f\|_{C_0^{\beta,\gamma}(E_{\alpha-\beta})}. \tag{5.58}$$

Hence, by the definition of the spaces $E_{\alpha-\beta}$ and by the inequalities (5.57), (5.58), we have the estimate

$$\|I_1\|_{\alpha-\beta} \le \frac{M_1}{\alpha(1-\alpha)} \frac{\tau^\beta}{(t+\tau)^\gamma}\|f\|_{C_0^{\beta,\gamma}(E_{\alpha-\beta})}.$$

In exactly the same manner one establishes the estimate

$$\|I_2\|_{\alpha-\beta} \le \frac{M_1}{\alpha(1-\alpha)} \frac{\tau^\beta}{t^\gamma}\|f\|_{C_0^{\beta,\gamma}(E_{\alpha-\beta})}.$$

Since $t > \tau$, this yields

$$\|I_2\|_{\alpha-\beta} \le \frac{2^\gamma M_1}{\alpha(1-\alpha)} \frac{\tau^\beta}{(t+\tau)^\gamma}\|f\|_{C_0^{\beta,\gamma}(E_{\alpha-\beta})}.$$

Now let us estimate I_4. Since

$$I_4 = [\exp\{-\tau A\} - \exp\{-(t+\tau)A\}][f(t+\tau) - f(t)],$$

by (5.27) and the triangle inequality, we have

$$\|I_4\|_{\alpha-\beta} \le \left(\|\exp\{-\tau A\}\|_{E_{\alpha-\beta}\to E_{\alpha-\beta}} + \|\exp\{-(t+\tau)A\}\|_{E_{\alpha-\beta}\to E_{\alpha-\beta}} \right) \times$$

$$\times \|f(t+\tau) - f(t)\|_{\alpha-\beta} M \frac{\tau^\beta}{(t+\tau)^\gamma}\|f\|_{C_0^{\beta,\gamma}(E_{\alpha-\beta})}.$$

Finally, let us estimate I_3. Using the identity

$$A\exp\{-zA\}I_3 = \int_0^{t-\tau}\int_{t-s}^{t-s+\tau} A^3\exp\{-(z+s_1)A\}ds_1\,[f(t)-f(s)]ds$$

and the estimate (5.27), we obtain

$$\|A\exp\{-zA\}I_3\|_E \le \int_0^{t-\tau}\int_{t-s}^{t-s+\tau}\left\|A^2\exp\left\{-\frac{z+s_1}{2}A\right\}\right\|_{E\to E}\times$$

$$\times\left\|A\exp\left\{-\frac{z+s_1}{2}A\right\}[f(t)-f(s)]\right\|_E ds_1\,ds$$

$$\le M\int_0^{t-\tau}\int_{t-s}^{t-s+\tau}\frac{ds_1}{(z+s_1)^{3-\alpha+\beta}}\|f(t)-f(s)\|_{\alpha-\beta}ds$$

$$\le \frac{M\tau}{t^\gamma}\int_0^{t-\tau}\frac{(t-s)^\beta ds}{(t-s+z)^{3-\alpha+\beta}}\|f\|_{C_0^{\beta,\gamma}(E_{\alpha-\beta})}.$$

Thus, we have established the estimate

$$z^{1-(\alpha-\beta)}\|A\exp\{-zA\}I_3\|_E \le \frac{Mz^{1-\alpha+\beta}\tau}{t^\gamma}\int_0^{t-\tau}\frac{(t-s)^\beta ds}{(t-s+z)^{3-\alpha+\beta}}\|f\|_{C_0^{\beta,\gamma}(E_{\alpha-\beta})}.$$

$$\tag{5.59}$$

Since

$$z^{1-(\alpha-\beta)}\tau\int_0^{t-\tau}\frac{(t-s)^\beta ds}{(t-s+z)^{3-\alpha+\beta}} \le z^{1-(\alpha-\beta)}\tau\int_0^{t-\tau}\frac{ds}{(z+t-s)^{3-\alpha}}$$

$$\le \frac{z^{1-\alpha+\beta}\tau}{(2-\alpha)(z+\tau)^{2-\alpha}} \le \frac{z^{1-\alpha+\beta}\tau^{1-\beta}}{(z+\tau)^{2-\alpha}}\tau^\beta \le \tau^\beta,$$

from (5.59) it follows that

$$z^{1-(\alpha-\beta)}\|A\exp\{-zA\}I_3\|_E \le M_1\frac{\tau^\beta}{(t+\tau)^\gamma}\|f\|_{C_0^{\beta,\gamma}(E_{\alpha-\beta})}.$$

Hence, by the definition of the spaces $E_{\alpha-\beta}$, we have

$$\|I_3\|_{\alpha-\beta} \le M_1\frac{\tau^\beta}{(t+\tau)^\gamma}\|f\|_{C_0^{\beta,\gamma}(E_{\alpha-\beta})}.$$

Combining the estimates for I_1, I_2, I_3 and I_4, we obtain (5.52). Now from (5.49) and (5.50) we obtain the estimate

$$\|v'\|_{C_0^{\beta,\gamma}(E_{\alpha-\beta})} \le M\left[|v_0 - A^{-1}f(0)|_{1+\alpha-\beta}^{\beta,\gamma} + \frac{1}{\alpha(1-\alpha)}\|f\|_{C_0^{\beta,\gamma}(E_{\alpha-\beta})}\right].$$

The estimate for $Av(t)$ in the norm of $C_0^{\beta,\gamma}(E_{\alpha-\beta})$ follows from the triangle inequality. Theorem 5.3 is proved.

By Lemmas 5.3 and 5.2, we have

$$|w_0|_{1-\alpha+\beta}^{\beta,\gamma} \leq M\|Aw_0\|_{\alpha-\gamma} \quad (Aw_0 \in E_{\alpha-\gamma}). \tag{5.60}$$

We have not been able to establish the opposite inequality necessary for the equivalence of norms. Nevertheless, we have the following result.

Theorem 5.4. *Let $v_0' = f(0) - Av_0 \in E_{\alpha-\gamma}$, $f(t) \in C_0^{\beta,\gamma}(E_{\alpha-\beta})$ for some $0 < \gamma \leq \beta < \alpha, 0 < \alpha < 1$. Then there exists a unique solution of the Cauchy problem (1.1) such that*

$$v'(t), Av(t) \in C_0^{\beta,\gamma}(E_{\alpha-\beta}), \quad v'(t) \in C(E_{\alpha-\gamma}),$$

and the coercivity inequality holds:

$$\|v'\|_{C_0^{\beta,\gamma}(E_{\alpha-\beta})} + \|Av\|_{C_0^{\beta,\gamma}(E_{\alpha-\beta})} + \|v'\|_{C(E_{\alpha-\gamma})}$$

$$\leq M\left[\|v_0'\|_{\alpha-\gamma} + \frac{1}{\alpha(1-\alpha)}\|f\|_{C_0^{\beta,\gamma}(E_{\alpha-\beta})}\right],$$

where M does not depend on $\alpha, \beta, \gamma, v_0'$, and $f(t)$.

Proof. By (1.19), we have

$$v'(t) = \exp\{-tA\}v_0' + \exp\{-tA\}(f(t) - f(0)) +$$

$$\int_0^t A\exp\{-(t-s)A\}(f(t) - f(s))ds = w(t) + p(t) + g(t).$$

To prove the theorem it suffices to establish the needed estimates for $w(t), p(t)$, and $g(t)$ in $C(E_{\alpha-\gamma})$ and $w(t)$ in $C_0^{\beta,\gamma}(E_{\alpha-\beta})$. From (5.43) it follows that

$$\|g\|_{C(E_{\alpha-\gamma})} \leq \frac{M}{\alpha(1-\alpha)}\|f\|_{C_0^{\beta,\gamma}(E_{\alpha-\beta})}.$$

Using estimate (5.27) with $n = 0$, we obtain

$$\|w(t)\|_{\alpha-\gamma} \leq \|\exp\{-tA\}\|_{E_{\alpha-\gamma} \to E_{\alpha-\gamma}}\|v_0'\|_{\alpha-\gamma} \leq M\|v_0'\|_{\alpha-\gamma}, \quad t \geq 0,$$

whence

$$\|w\|_{C(E_{\alpha-\gamma})} \leq M\|v_0'\|_{\alpha-\gamma}.$$

Next, using the identity

$$z^{1-\alpha+\gamma}A\exp\{-zA\}p(t) = z^{1-\alpha+\gamma}A\exp\{-(z+t)A\}[f(t)-f(0)]$$

and the definition of the space $E_{\alpha-\beta}$, we obtain

$$z^{1-\alpha+\gamma}\|A\exp\{-zA\}p(t)\|_E \leq \frac{z^{1-\alpha+\gamma}}{(z+t)^{1-\alpha+\beta}}\|f(t)-f(0)\|_{\alpha-\beta}$$

$$\leq \frac{t^{\beta-\gamma}}{(z+t)^{\beta-\gamma}}\|f\|_{C_0^{\beta,\gamma}(E_{\alpha-\beta})} \leq \|f\|_{C_0^{\beta,\gamma}(E_{\alpha-\beta})}.$$

From this it follows that, for any $t \in [0,1]$,

$$\|p(t)\|_{\alpha-\gamma} \leq \|f\|_{C_0^{\beta,\gamma}(E_{\alpha-\beta})}.$$

Consequently,

$$\|p\|_{C(E_{\alpha-\gamma})} \leq \|f\|_{C_0^{\beta,\gamma}(E_{\alpha-\beta})}.$$

Now let us estimate $w(t)$ in $C_0^{\beta,\gamma}(E_{\alpha-\beta})$. To this end it suffices to establish the estimates

$$\|w(t)\|_{\alpha-\beta} \leq M\|v_0'\|_{\alpha-\gamma}, \quad 0 \leq t \leq 1, \tag{5.61}$$

and

$$\frac{(t+\tau)^\gamma}{\tau^\beta}\|w(t+\tau)-w(t)\|_{\alpha-\beta} \leq M\|v_0'\|_{\alpha-\gamma}, \quad 0 \leq t < t+\tau \leq 1. \tag{5.62}$$

Since $\alpha - \beta \leq \alpha - \gamma$, we have

$$\|w(t)\|_{\alpha-\beta} \leq M\|w(t)\|_{\alpha-\gamma}.$$

Therefore, the estimate (5.61) follows from the estimate (5.27) for $n = 0$. The estimate (5.62) follows from (5.23) and (5.33). Theorem 5.4 is proved.

Let us note the the spaces of smooth functions in which well-posedness has been established depend on the parameters α, β, γ. However, the constants in the coercivity inequality depend only on α. Hence, we can choose the parameters β, γ

freely, which increases considerably the number of function spaces in which problem (1.1) is well posed. In particular, Theorem 5.4 implies the well-posedness theorem established in Sections 2 and 4.

One can enrich the family of spaces of smooth functions in which the Cauchy problem (1.1) is well posed by introducing the spaces $C_{\alpha\beta\gamma}$ $(0 < \gamma \leq \beta \leq \alpha,$ $0 < \alpha < 1)$, obtained by completion of the set of all smooth E-valued functions $f(t)$ on $[0,1]$ with respect to the norm

$$\|f\|_{C_{\alpha\beta\gamma}} = \max_{0 \leq t \leq 1} \|f(t)\|_\gamma t^{\alpha-\gamma} + \max_{0 \leq t < t+\tau \leq 1} \frac{\|f(t+\tau) - f(t)\|_E}{\tau^\beta} t^{\alpha-\gamma}.$$

Let us give, without proof, the following result.

Theorem 5.5. *Let $Av_0 \in E_{\alpha-\beta}$, $f(t) \in C_{\alpha\beta\gamma}$ for some $0 < \gamma \leq \beta \leq \alpha, 0 < \alpha < 1$. Then there exists a unique solution of the Cauchy problem (1.1) such that*

$$v'(t), Av(t) \in C_{\alpha\beta\gamma},$$

and the coercivity inequality holds:

$$\|v'\|_{C_{\alpha\beta\gamma}} + \|Av\|_{C_{\alpha\beta\gamma}} \leq M\left[\|Av_0\|_{\alpha-\beta} + \frac{1}{\alpha(1-\alpha)}\|f\|_{C_{\alpha\beta\gamma}}\right],$$

where M does not depend on $\alpha, \beta, \gamma, v_0$, and $f(t)$.

CHAPTER 2

THE ROTHE DIFFERENCE SCHEME

0. STABILITY OF THE DIFFERENCE PROBLEM

1. The difference problem.

Let us associate to the Cauchy problem (1.1) of Chapter 1 the corresponding difference problem

$$\mathcal{D}u_k + Au_k = \varphi_k, \quad 1 \le k \le N, \ u_0 = u_0(\tau). \tag{0.1}$$

Here N is a fixed positive integer, $\tau = 1/N$, $\mathcal{D}u_k = (u_k - u_{k-1})/\tau$; $u^\tau = \{u_k\}_1^N$, $\varphi^\tau = \{\varphi_k\}_1^N$ are the unknown and the given grid functions with values in the Banach space E. It is assumed that the function φ^τ and the elements $u_0(\tau)$ approximate $f(t)$ and v_0, respectively, in a specified way.

For example, if $f(t) \in C(E)$, then one usually sets $\varphi_k = f(t_k)$, $t_k = k\tau$, $k = 1, \cdots, N$, $u_0 = v_0$. If $f(t) \in L_1(E)$, then one sets

$$\varphi_k = \frac{1}{\tau} \int_{t_{k-1}}^{t_k} f(s) ds.$$

The difference problem (0.1) is called *the Rothe difference scheme* for the approximate solution of the Cauchy problem (1.1) of Chapter 1.

Concerning the operator A we will assume that the necessary conditions for the well-posedness of the differential problem are satisfied. As we have shown in Chapter 1, the analyticity in E of the semigroup $\exp\{-tA\}$, $t > 0$ is such a necessary condition in various function spaces. We shall assume that A is a strongly positive operator, i.e., the norm of the semigroup admits an exponentially decaying estimate.

Concerning the initial element v_0 we shall also assume that the necessary conditions for the well-posedness of the differential problem are satisfied. Recall that for the spaces $C(E)$ and $C_0^\alpha(E)$ it is required that $v_0 \in D(A)$. For the space $L_p(E)$ the necessary condition is different.

The grid function u^τ with τ fixed will be said to be a *solution of the difference problem* (0.1) if its components $u_k \in D(A)$, $k = 1, \cdots, N$, satisfy the system of equations (0.1).

From the strong positivity of the operator A it follows that there exists the bounded operator $(I + \tau A)^{-1}$, defined on the whole space E. Hence, for any φ^τ and u_0 the solution of problem (0.1) exists, and the following formula holds

$$u_k = R^k(\tau A)u_0 + \sum_{r=1}^{k} R^{k-r+1}(\tau A)\varphi_r \tau = w_k + g_k, \quad k = 1, \cdots, N. \qquad (0.2)$$

Here $R(\tau A) = (I + \tau A)^{-1}$, $k = 1, \cdots, N$, $w^\tau = \{w_k\}_1^N$ is the solution of the homogeneous problem

$$\mathcal{D}w_k + Aw_k = 0, \quad k = 1, \cdots, N, w_0 = u_0, \qquad (0.3)$$

and $g^\tau = \{g\}_1^N$ is the solution of the nonhomogeneous problem

$$\mathcal{D}g_k + Ag_k = \varphi_k, \quad k = 1, \cdots, N, \ g_0 = 0. \qquad (0.4)$$

2.   Banach spaces of grid functions.

Let us denote by $E(\tau) = E(\tau, E)$ the space of grid functions φ^τ for fixed $\tau = 1/N$. Thus, $E(\tau)$ is the vector space whose elements are ordered N-tuples of elements of E. The space $E(\tau)$ can be equipped with various norms and thus become a normed space. Thus, for instance, the vector space $E(\tau)$ generates the normed space $C(\tau, E) = C(E(\tau))$ with the norm

$$\|\varphi^\tau\|_{C(\tau,E)} = \max_{1 \leq k \leq N} \|\varphi_k\|_E,$$

the space $C_0^\alpha(\tau, E)$, $0 < \alpha < 1$, with the norm

$$\|\varphi^\tau\|_{C_0^\alpha(\tau,E)} = \|\varphi^\tau\|_{C(\tau,E)} + \max_{1 \le k < k+r \le N} \|\varphi_{k+r} - \varphi_k\|_E k^\alpha r^{-\alpha},$$

and the space $L_p(\tau, E)$, $1 \le p < \infty$, with the norm

$$\|\varphi^\tau\|_{L_p(\tau,E)} = \Big(\sum_{k=1}^N \|\varphi_k\|_E^p \tau \Big)^{1/p}.$$

The fact that the last three formulas define norms in the space $E(\tau)$ is an obvious consequence of the properties of the norm in E and the finiteness of N.

Furthermore, from the completeness of E it follows that the spaces $C(\tau, E)$, $C_0^\alpha(\tau, E)$, $L_p(\tau, E)$ are not only normed but also Banach spaces.

Suppose that on some vector space of elements φ there are defined two norms, $\|\cdot\|_1$ and $\|\cdot\|_2$. These two norms are said to be *equivalent* if the following inequalities hold:

$$0 < \delta_1 \le \|\varphi\|_1 / \|\varphi\|_2 \le \delta_2 < \infty, \quad \varphi \ne 0.$$

Suppose the vector space in question is a Banach space with respect to both norms. Then from Banach's theorem on the inverse operator it follows that the two norms are equivalent. In particular, the norms of the spaces $C(\tau, E)$, $C_0^\alpha(\tau, E)$, $L_p(\tau, E)$ will be equivalent for any fixed τ; however, Banach's theorem does not allow us to establish how the quantities $\delta_1 = \delta_1(N)$ and $\delta_2 = \delta_2(N)$ depend on N.

3. The operator equation in $\mathcal{E}(E)$. Definition of the stability of the difference scheme.

Let us consider again the difference problem (0.1). We shall reduce it to an operator problem in the space $E(\tau)$. First we will define an operator $\mathcal{D}$ acting from the space $E \times E(\tau)$ of vectors $u^\tau = (u_0, u_1, \cdots, u_N)$ into the space $E(\tau)$ of vectors $v^\tau = (v_1, \cdots, v_N)$ according to the rule

$$v^\tau = \mathcal{D}u^\tau, \quad v_k = (u_k - u_{k-1})/\tau, \quad k = 1, \cdots, N.$$

Next, let us introduce the continuation operator $\Pi(u_0)$, which acts from $E \times E(\tau)$ to $E(\tau)$ according to the rule

$$\Pi(u_0)(u_1, \cdots, u_N) = (u_0, u_1, \cdots, u_N).$$

Then, clearly, the difference problem (0.1) is equivalent to the operator problem

$$\mathcal{D}\Pi(u_0)u^\tau + Au^\tau = \varphi^\tau. \tag{0.5}$$

Here Au^τ and φ^τ are defined by the formula

$$Au^\tau = \{Au_1, \cdots, Au_N\}, \quad \varphi^\tau = \{\varphi_1, \cdots, \varphi_N\}.$$

The operator problem (0.5) will be considered in the space $E(\tau)$. From its unique solvability for any $u_0 \in E$ and $\varphi^\tau \in E(\tau)$ it follows that its solution u^τ defines an additive and homogeneous operator $u^\tau(\varphi^\tau, u_0)$, acting from $E(\tau) \times E$ to $E(\tau)$. To define the well-posedness in this space we have to treat $E(\tau) \times E$ and $E(\tau)$ as normed spaces. The problem (0.1) is said to be *well-posed in the space $E(\tau)$* if the operator $u^\tau(\varphi^\tau, u_0)$ is continuous. Since $u^\tau(\varphi^\tau, u_0)$ is additive and homogeneous, the problem (0.1) is well posed if and only if the inequality

$$\|u^\tau(\varphi^\tau, u_0)\|_{E(\tau)} \le M[\|u_0\|_E + \|\varphi^\tau\|_{E(\tau)}]$$

holds, where M does not depend on u_0 and φ^τ, but, generally speaking, depends on τ. If we take for $E(\tau)$ one of the spaces $C(\tau, E)$, $C_0^\alpha(\tau, E)$, $L_p(\tau, E)$, then from the boundedness of the operator $(I + \tau A)^{-1}$ and formulas (0.2) it follows that problem (0.1) is well posed in these spaces.

Now let us consider the spaces $E(\tau)$, $0 < \tau \le \tau_0$, and associate to them the space $\mathcal{E}(E)$ of sequences of grid functions $\varphi = \{\varphi^\tau\}$, $0 < \tau \le \tau_0$. Obviously, $\mathcal{E}(E)$ is a vector space but is already infinite-dimensional. Suppose now that $E(\tau)$ is a Banach space for all $0 < \tau \le \tau_0$. Assign to the sequence $\varphi = \{\varphi^\tau\}$ the sequence of numbers

$$|\varphi| = \{\|\varphi^\tau\|_{E(\tau)}\}.$$

Any norm in the space of sequences of numbers defines a norm in the space $\mathcal{E}(E)$. Below we shall use only the norm of the space of bounded sequences of numbers.

Denote by $\mathcal{C}(E) = C(\mathcal{E}(E))$ the space of elements $\varphi = \{\varphi^\tau\}$ for which the norm

$$\|\varphi\|_{\mathcal{C}(E)} = \sup_{0 < \tau \le \tau_0} \|\varphi^\tau\|_{C(\tau, E)}$$

is finite. Denote by $C_0^\alpha(E) = C_0^\alpha(\mathcal{E}(E))$ the space of elements $\varphi = \{\varphi^\tau\}$ for which the norm

$$\|\varphi\|_{C_0^\alpha(E)} = \sup_{0 < \tau \le \tau_0} \|\varphi^\tau\|_{C_0^\alpha(\tau, E)}$$

is finite. Finally, denote by $\mathcal{L}_p(E) = L_p(\mathcal{E}(E))$ the space of elements $\varphi = \{\varphi^\tau\}$ for which the norm

$$\|\varphi\|_{\mathcal{L}_p(E)} = \sup_{0 < \tau \leq \tau_0} \|\varphi^\tau\|_{\mathcal{L}_p(\tau, E)}$$

is finite.

In contrast to the case where τ is fixed, these spaces are not identical. Specifically, we have the embeddings

$$\mathcal{C}_0^\alpha(E) \subset \mathcal{C}(E) \subset \mathcal{L}_p(E),$$

and the embedding operators are continuous.

We shall consider problem (0.1) as an operator problem in $\mathcal{E}(E)$. To this end let us introduce operators $\overline{\mathcal{D}}$, $\overline{\Pi}$, and $\overline{A}$ which act in $\mathcal{E}(E)$ componentwise as $\mathcal{D}$, Π, and A, respectively. This leads to the operator equation

$$\overline{\mathcal{D}}\,\overline{\Pi}(u_0)u + \overline{A}u = \varphi \qquad (0.6)$$

in the vector space $\mathcal{E}(E)$.

The operator equation (0.6) is obviously uniquely solvable for any $u_0 \in E$ and $\varphi \in \mathcal{E}(E)$. This solvability is equivalent to the unique solvability in $E(\tau)$ of problem (0.5) for all $0 < \tau \leq \tau_0$. The solution of problem (0.6) in the form (0.2) defines an additive and homogeneous operator $u(\varphi, u_0)$ acting from $\mathcal{E}(E) \times E$ to $\mathcal{E}$.

Now let us assume that $\mathcal{E}(E)$ is one of the spaces $\mathcal{C}(E)$, $\mathcal{C}_0^\alpha(E)$, or $\mathcal{L}_p(E)$.

Definition 0.1. The problem (0.6) will be said to be *stable* in the Banach space $\mathcal{E}(E)$ if the operator $u(\varphi, u_0)$ is continuous from $\mathcal{E}(E) \times E$ to $\mathcal{E}$.

Since $u(\varphi, u_0)$ is additive and homogeneous, problem (0.6) is stable in $\mathcal{E}(E)$ if and only if the following inequality holds:

$$\|u^\tau\|_{E(\tau)} \leq M[\|u_0\|_E + \|\varphi^\tau\|_{E(\tau)}], \qquad (0.7)$$

where M is not only independent of u_0 and φ^τ, but also of τ.

4. Stability of the difference scheme.

From formula (0.2) one derives the following result.

Theorem 0.1. *For the stability of the difference problem* (0.6) *in* $C(E)$ *it is necessary and sufficient that the following estimate hold for all* $k = 1, \cdots, N$ *and all* $\tau > 0$:

$$\|R^k(\tau A)\|_{E \to E} \leq M, \tag{0.8}$$

where M *does not depend on* k *and* τ.

The estimate (0.8) is the criterion for the operator $-A$ to generate a strongly continuous semigroup $\exp\{-tA\}$, $t \geq 0$ (the Hille-Yosida-Phillips-Miyadera theorem; see Chapter 1, Section 1).

Now let us consider problem (0.6) in the space $C_0^\alpha(E)$.

Theorem 0.2. *For the stability of the difference problem* (0.6) *in* $C_0^\alpha(E)$ *it is necessary and sufficient that the estimate* (0.8) *hold and that*

$$\|R^k(\tau A) - R^{k+r}(\tau A)\|_{E \to E} \leq M \frac{r^\alpha}{k^\alpha} \quad (1 \leq k < k + r \leq N, \ 0 \leq \alpha \leq 1), \tag{0.9}$$

where M *does not depend on* τ, α, k, *and* τ.

Proof. Let the difference problem (0.6) be stable in the space $C_0^\alpha(E)$. Since $w^\tau = \{R^k u_0\}_1^N$ is a solution of the homogeneous problem (0.4), the estimates (0.8) and (0.9) follow from the definition of the norm in $C_0^\alpha(E)$. Now suppose that the estimates (0.8) and (0.9) are valid. By formula (0.2), the solution of the difference problem (0.6) has the form

$$u^\tau = w^\tau + g^\tau. \tag{0.10}$$

Let us estimate the norm of each term. Using (0.8), we obtain

$$\|w_k\|_E \leq \|R^k(\tau A)\|_{E \to E} \|u_0\|_E \leq M \|u_0\|_E.$$

Next, using the estimate (0.9) one can show that, for $1 \leq k < k + r \leq N$,

$$\|w_{k+r} - w_k\|_E \leq \|R^{k+r}(\tau A) - R^k(\tau A)\|_{E \to E} \|u_0\|_E \leq M \frac{r^\alpha}{k^\alpha} \|u_0\|_E.$$

Thus, we have shown that

$$\|w^\tau\|_{C_0^\alpha(\tau, E)} \leq M \|u_0\|_E.$$

Now let us estimate g^τ in $C_0^\alpha(\tau, E)$. Using (0.8), we obtain

$$\|g_k\|_E \leq \sum_{i=1}^k \|R^{k-i+1}(\tau A)\|_{E\to E} \|\varphi_i\|_E \cdot \tau \leq Mk\tau \|\varphi^\tau\|_{C(\tau,E)} \leq M\|\varphi^\tau\|_{C_0^\alpha(\tau,E)},$$

whence

$$\|g^\tau\|_{C(\tau,E)} \leq M\|\varphi^\tau\|_{C_0^\alpha(\tau,E)}. \tag{0.11}$$

Now let us estimate the difference $g_{k+r} - g_k$ for $1 \leq k < k + r \leq N$. We shall consider two separate cases: $k \leq 2r$ and $k > 2r$. If $k \leq 2r$, then from (0.11) and the triangle inequality it follows that

$$\|g_{k+r} - g_k\|_E \leq \|g_{k+r}\|_E + \|g_k\|_E \leq 2M\|\varphi^\tau\|_{C_0^\alpha(\tau,E)} r^\alpha r^{-\alpha}$$

$$\leq 2^{1+\alpha} M\|\varphi^\tau\|_{C_0^\alpha(\tau,E)} r^\alpha k^{-\alpha}.$$

Hence, for $k \leq 2r$ we have the estimate

$$\|g_{k+r} - g_k\|_E \leq M_1 \|\varphi^\tau\|_{C_0^\alpha(\tau,E)} r^\alpha k^{-\alpha}. \tag{0.12}$$

Now let $k > 2r$. Using formula (0.2), we obtain the identity

$$g_k = A^{-1}(I - R^k(\tau A))\varphi_k + \sum_{i=1}^k R^{k-i+1}(\tau A)[\varphi_i - \varphi_k]\tau, \quad k = 2, \cdots, N. \tag{0.13}$$

Since $r > 0$, we have $k + r > 1$, and identity (0.13) yields

$$g_{k+r} - g_k = A^{-1}\big[\varphi_{k+r} - \varphi_k + R^k(\tau A)\varphi_k - R^{k+r}(\tau A)\varphi_{k+r}\big]$$

$$+ \sum_{i=k-r+1}^{k+r} R^{k+r+1-i}(\tau A)(\varphi_i - \varphi_{k+r})\tau + \sum_{i=k-r+1}^k R^{k+1-i}(\tau A)(\varphi_k - \varphi_i)\tau$$

$$+ \sum_{i=1}^{k-r} R^{k+r+1-i}(\tau A)(\varphi_k - \varphi_{k+r})\tau + \sum_{i=1}^{k-r} [R^{k+r+1-i}(\tau A) - R^{k+1-i}(\tau A)](\varphi_i - \varphi_k)\tau$$

$$= P_1 + P_2 + P_3 + P_4 + P_5.$$

Using the fact that the operator A^{-1} is bounded in the norm of E, the triangle inequality, and the estimates (0.8) and (0.9), we obtain the inequality

$$\|P_1\|_E \leq \|A^{-1}\|_{E\to E}[\|\varphi_{k+r} - \varphi_k\|_E + \|R^k(\tau A)\|_{E\to E}\|\varphi_k - \varphi_{k+r}\|_E$$

$$+\|R^k(\tau A) - R^{k+r}(\tau A)\|_{E\to E}\|\varphi_{k+r}\|_E] \le M\|\varphi^\tau\|_{C_0^\alpha(\tau,E)}r^\alpha k^{-\alpha}.$$

Further, by (0.8), we have

$$\|P_2\|_E \le \sum_{i=k-r+1}^{k+r} \|R^{k+r+1-i}(\tau A)\|_{E\to E}\|\varphi_i - \varphi_{k+r}\|_E\tau$$

$$\le M \sum_{i=k-r+1}^{k+r} \frac{(k+r-i)^\alpha}{i^\alpha}\tau\|\varphi^\tau\|_{C_0^\alpha(\tau,E)}.$$

Since

$$\sum_{i=k-r+1}^{k+r} \frac{(k+r-i)^\alpha\tau}{i^\alpha} \le \frac{(2r-1)^\alpha(2r-1)\tau}{(k-r+1)^\alpha} \le \frac{(2r)^\alpha 2r\tau}{(\frac{k}{2}+\frac{k}{2}-r+1)^\alpha} \le 2^{2\alpha+1}\frac{r^\alpha}{k^\alpha},$$

we have

$$\|P_2\|_E \le M_1\|\varphi^\tau\|_{C_0^\alpha(\tau,E)}r^\alpha k^{-\alpha}.$$

In a similar way one shows that

$$\|P_3\|_E \le M_1\|\varphi^\tau\|_{C_0^\alpha(\tau,E)}r^\alpha k^{-\alpha}.$$

Further, using the identity

$$P_4 = [R^k(\tau A) - R^{2r}(\tau A)]A^{-1}(\varphi_{k+r} - \varphi_k),$$

the boundedness of the operator A^{-1} in the norm of E, the triangle inequality, and estimate (0.8), we obtain

$$\|P_4\|_E \le [\|R^k(\tau A)\|_{E\to E} + \|R^{2r}(\tau A)\|_{E\to E}]\|A^{-1}\|_{E\to E}\|\varphi_{k+r} - \varphi_k\|_E$$

$$\le M\|\varphi^\tau\|_{C_0^\alpha(\tau,E)}r^\alpha k^{-\alpha}.$$

Finally, let us estimate P_5. To this end we write P_5 as the sum

$$\left(\sum_{i=1}^{[\frac{k-r}{2}]} + \sum_{i=[\frac{k-r}{2}]+1}^{k-r}\right)(R^{k+r+1-i}(\tau A) - R^{k+1-i}(\tau A))(\varphi_i - \varphi_k)\tau = P_{5,1} + P_{5,2},$$

where $[x]$ denotes the integer part of $x > 0$. Using the triangle inequality and estimate (0.10), we obtain

$$\|P_{5,1}\|_E \le \sum_{i=1}^{[\frac{k-r}{2}]} \|R^{k+r+1-i}(\tau A) - R^{k+1-i}(\tau A)\|_{E\to E}[\|\varphi_i\|_E + \|\varphi_k\|_E]\tau$$

$$\leq M \sum_{i=1}^{[\frac{k-r}{2}]} \frac{r^\alpha \tau}{(k+1-i)^\alpha} \|\varphi^\tau\|_{C_0^\alpha(\tau,E)}.$$

Since

$$\sum_{i=1}^{[\frac{k-r}{2}]} \frac{r^\alpha \tau}{(k+1-i)^\alpha} \leq \frac{2^\alpha r^\alpha \frac{k-r}{2} \tau}{k^\alpha} \leq 2^{\alpha-1} r^\alpha k^{-\alpha},$$

we have

$$\|P_{5,1}\|_E \leq M_1 \|\varphi^\tau\|_{C_0^\alpha(\tau,E)} r^\alpha k^{-\alpha}. \tag{0.14}$$

Now, using estimate (0.10) we obtain

$$\|P_{5,2}\|_E \leq \sum_{i=[\frac{k-r}{2}]+1}^{k-r} \|R^{k+r+1-i}(\tau A) - R^{k+1-i}(\tau A)\|_{E \to E} \|\varphi_i - \varphi_k\|_E \tau$$

$$\leq M \sum_{i=[\frac{k-r}{2}]+1}^{k-r} \frac{r^\alpha \tau}{i^\alpha} \|\varphi^\tau\|_{C_0^\alpha(\tau,E)}.$$

Since

$$\sum_{i=[\frac{k-r}{2}]+1}^{k-r} \frac{r^\alpha \tau}{i^\alpha} \leq \frac{2^\alpha r^\alpha (k-r)\tau}{k-r} \leq 2^{2\alpha} \frac{r^\alpha}{k^\alpha},$$

it follows that

$$\|P_{5,2}\|_E \leq M_1 \|\varphi^\tau\|_{C_0^\alpha(\tau,E)} r^\alpha k^{-\alpha}. \tag{0.15}$$

Estimates (0.14) and (0.15) yield

$$\|P_5\|_E \leq M_1 \|\varphi^\tau\|_{C_0^\alpha(\tau,E)} r^\alpha k^{-\alpha}.$$

Combining the estimates in E for P_1, P_2, P_3, P_4, and P_5, we obtain (0.12) for $k > 2r$.

Finally, from (0.11) and and (0.12) it follows that

$$\|g^\tau\|_{C_0^\alpha(\tau,E)} \leq M_2 \|\varphi^\tau\|_{C_0^\alpha(\tau,E)}.$$

Theorem 0.2 is proved.

Later it will be shown that the estimates (0.8) and (0.9) hold if $-A$ is the generator of an analytic semigroup $\exp\{-tA\}$, $t \geq 0$. The necessity of this condition can be established if the difference problem (0.6) is not only stable in $\mathcal{C}_0^\alpha(E)$ but also well posed.

Now let us consider problem (0.6) in the space $\mathcal{L}_p(E)$.

Theorem 0.3. *For the stability of the difference problem (0.6) it is necessary and sufficient that the following estimate hold:*

$$\left(\sum_{k=1}^N \|R^k(\tau A)u_0\|_{E \to E}^p \tau\right)^{1/p} \le M\|u_0\|_E, \tag{0.16}$$

where M does not depend on τ, p and $u_0 \in E$.

Proof. Suppose problem (0.6) is stable in $\mathcal{L}_p(E)$. Then, in particular, we have the estimate

$$\left(\sum_{k=1}^N \|R^k(\tau A)u_0\|_E^p \tau\right)^{1/p} \le M\|u_0\|_E.$$

The necessity of the condition (0.16) is thus established. Now suppose (0.16) holds for any $u_0 \in E$. Let us show that problem (0.6) is stable in $\mathcal{L}_p(E)$. Clearly, it suffices to estimate the grid function g^τ. Let us use the identity

$$g_k = \sum_{i=1}^k R^{k-i+1}(\tau A)\varphi_i\tau = \sum_{i=1}^N \mathcal{G}(k-i+1)\varphi_i\tau.$$

Here $\mathcal{G}(s) = R^s(\tau A)$ if $s \ge 1$ and $\mathcal{G}(s) = 0$ if $s < 1$. This yields the inequality

$$\|g_k\|_E \le \sum_{i=1}^N \|\mathcal{G}(k-i+1)\varphi_i\|_E\tau.$$

Using the Minkowski inequality for sums, we obtain

$$\left(\sum_{k=1}^N \|g_k\|_E^p\tau\right)^{1/p} \le \sum_{i=1}^N \left(\sum_{k=1}^N \|\mathcal{G}(k-i+1)\varphi_i\|_E^p\tau\right)^{1/p}\tau.$$

From the definition of the operator $\mathcal{G}$ it follows that

$$\left(\sum_{k=1}^N \|g_k\|_E^p\tau\right)^{1/p} \le \sum_{i=1}^N \left(\sum_{k=i}^N \|R^{k-i+1}(\tau A)\varphi_i\|_E^p\tau\right)^{1/p}\tau.$$

Changing the variable in the inner sum, we get

$$\Big(\sum_{k=1}^{N}\|g_k\|_E^p\tau\Big)^{1/p} \le \sum_{k=1}^{N}\Big(\sum_{s=1}^{N-i+1}\|R^s(\tau A)\varphi_i\|_E^p\tau\Big)^{1/p}\tau$$

$$\le \sum_{i=1}^{N}\Big(\sum_{s=1}^{N}\|R^s(\tau A)\varphi_i\|_E^p\tau\Big)^{1/p}\tau.$$

Using condition (0.16), this implies that

$$\Big(\sum_{k=1}^{N}\|g_k\|_E^p\tau\Big)^{1/p} \le M\sum_{i=1}^{N}\|\varphi_i\|_E\tau.$$

To complete the proof of Theorem 0.3, it remains to use Hölder's inequality.

Condition (0.16) is obviously satisfied if condition (0.8) is, i.e., if $-A$ is the generator of a strongly continuous semigroup. The necessity of this condition can be established if the difference scheme (0.6) is not only stable in $\mathcal{L}_p(E)$ but also well posed.

1.   WELL-POSEDNESS OF THE DIFFERENCE PROBLEM IN $\mathcal{C}(E)$

1.   The homogeneous difference problem.

We shall now consider equation (0.6) in the Banach space $\mathcal{C}(E)$. An element $u \in \mathcal{C}(E)$ will be called a *solution of* (0.6) if, in addition, the elements $\overline{\mathcal{D}\Pi}(u_0)u$ and $\overline{A}u$ belong to $\mathcal{C}(E)$. If problem (0.6) is solvable in $\mathcal{C}(E)$, then obviously $\varphi \in \mathcal{C}(E)$. It is harder to find the necessary condition on the initial data u_0, since we are not assuming that the difference equation (0.1) is satisfied at the initial point. As we mentioned above, we shall assume that $u_0 \in D(A)$.

Definition 1.1. The problem (0.6) is said to be *well posed in* $\mathcal{C}(E)$ if the following conditions are satisfied:

   1) For any $u_0 \in D(A)$ and $\varphi \in \mathcal{C}(E)$ it has a unique solution in $\mathcal{C}(E)$.
   2) Problem (0.6) is stable.

This notion of well-posedness is obviously equivalent to that of the well-posedness of problem (0.5) in $C(\tau, E)$ uniformly in τ, $0 < \tau \leq \tau_0$.

Let problem (0.6) be well posed in $\mathcal{C}(E)$. This means, in particular, that the corresponding homogeneous problem ($\varphi = 0$) is well posed. From condition 2) it follows that the homogeneous problem is stable in $\mathcal{C}(E)$. As we established in Section 0, this is the case if and only if $-A$ is the generator of a strongly continuous semigroup $\exp\{-tA\}$.

Thus, we have shown that for the stability of problem (0.6) it is necessary that $-A$ be the generator of a strongly continuous semigroup $\exp\{-tA\}$. If we consider only the homogeneous problem, then this condition is also sufficient for its well-posedness, and the solution $u(0, u_0)$ of this problem obeys the coercivity inequality

$$\|\overline{\mathcal{D}}\,\overline{\Pi}(u_0)u\|_{C(E)} + \|\overline{A}u\|_{C(E)} \leq M_0\|Au_0\|_E. \tag{1.1}$$

Before we turn to the investigation of the well-posedness of the nonhomogeneous difference problem, let us discuss the connections between the solutions of the homogeneous difference and differential problems.

In Chapter 1 we have shown that the solution of the homogeneous differential problem (1.1) therein is given by $\exp\{-tA\}v_0$. Further, the solution of the homogeneous difference problem has the form $\{(I + \tau A)^{-k}u_0\}_k^N$ (see formula (0.2)). The character of the convergence of the solution of the difference problem to the solution of the differential problem is described by the following assertion.

Theorem 1.1. *For any $0 \leq t \leq 1$ and any $x \in E$ one has the limit relation*

$$\lim_{\substack{\tau \to 0 \\ k\tau \to t \\ 0 \leq k \leq N}} \|(I + \tau A)^{-k}x - \exp\{-tA\}x\|_E = 0, \tag{1.2}$$

and the convergence in (1.2) is uniform in $t \in [0, 1]$.

Proof. Since we have the estimate

$$\|(I + \tau A)^{-k}\|_{E \to E} \leq M, \|\exp\{-tA\}\|_{E \to E} \leq M \tag{1.3}$$

(see the estimates (0.8) and (1.10) in Chapter 1), it suffices to verify relation (1.2) for elements x in a dense subset of E. We take $D(A^2)$ as such a subset. Consider the following E-valued function on $[0, 1]$:

$$\varphi(s) = (1 + s\tau A)^{-k}\exp\{-(1 - s)tA\}x.$$

Since $x \in D(A^2)$, $\varphi(s)$ is continuously differentiable (with respect to the norm in E) and

$$\varphi'(s) = (I + s\tau A)^{-k-1} \exp\{-(1-s)tA\}[(t - k\tau)Ax + st\tau A^2 x]. \qquad (1.4)$$

Further, from the definition of $\varphi(s)$ it follows that

$$(I + \tau A)^{-k}x - \exp\{-tA\}x = \varphi(1) - \varphi(0) = \int_0^1 \varphi'(s)ds.$$

Using formula (1.4) and the estimate (1.3), we obtain the inequality

$$\|(I + \tau A)^{-k}x - \exp\{-tA\}x\|_E \leq M^2[|t - k\tau|\|Ax\|_E + \tau\|A^2x\|_E],$$

which implies the limit relation (1.2) for $x \in D(A^2)$. Theorem 1.1 is proved.

2. The nonhomogeneous problem. A real-field criterion for analyticity.

Now let us consider the nonhomogeneous problem with null initial condition ($u_0 = 0$). Its solution is provided by the operator $u(\varphi, 0)$. From condition 2) in the definition of well-posedness it follows that the operator $u(\varphi, 0)$ is continuous in $\mathcal{C}(E)$. Condition 1) means that $\overline{A}u(\varphi, 0)$ is an additive, homogeneous, and everywhere defined operator in $\mathcal{C}(E)$. From the boundedness of the operator A^{-1} in E follows the boundedness of the operator $(\overline{A})^{-1}$ in $\mathcal{C}(E)$, and hence the closedness of $\overline{A}$ in this space. Here one can apply Banach's theorem asserting that a closed linear operator defined everywhere in a Banach space is bounded. Consequently, the solution $u(\varphi, 0)$ obeys the coercivity estimate

$$\|\overline{\mathcal{D}}\,\overline{\Pi}(u_0)u\|_{\mathcal{C}(E)} + \|\overline{A}u\|_{\mathcal{C}(E)} \leq M_c\|\varphi\|_{\mathcal{C}(E)}. \qquad (1.5)$$

Note that the constant M_0 (see (1.1) is readily determined from the estimate satisfied by the semigroup $\exp\{-tA\}$, whereas the constant M_c cannot be determined effectively (the existence of M_c follows from Banach's theorem).

From inequalities (1.1) and (1.5) we derive the coercivity inequality

$$\|\overline{\mathcal{D}}\,\overline{\Pi}(u_0)u\|_{\mathcal{C}(E)} + \|\overline{A}u\|_{\mathcal{C}(E)} \leq M(\|\varphi\|_{\mathcal{C}(E)} + \|Au_0\|_E) \qquad (1.6)$$

for the solutions of the general problem (0.6).

Now let the data φ and u_0 be sufficiently smooth. Here and in what follows it suffices to assume that $\overline{A}\varphi \in \mathcal{C}(E)$ and $u_0 \in D(A^2)$. Then there exists a unique solution of problem (0.6) in $\mathcal{C}(E)$. If inequality (1.6) holds for all such solutions, then by passing to the limit one can establish the well-posedness in $\mathcal{C}(E)$ of problem (0.6). Hence, the coercivity inequality is a necessary and sufficient condition for well-posedness.

At the end of the preceding section we showed that a necessary condition for stability in $\mathcal{C}(E)$ of the difference problem (0.6) is the strong continuity of the semigroup $\exp\{-tA\}$ $(t \geq 0)$. Let us show that a necessary condition for the stronger property of well-posedness in $\mathcal{C}(E)$ of problem (0.6) is the analyticity of this semigroup.

Let us consider problem (0.6) for $u_0 = 0$ and $\varphi_j = R^{j-1}(\tau A)\psi$, $j = 1, \cdots, N$, $\psi \in E$. The components u_k of the solution of this problem have the form

$$u_k = \sum_{j=1}^{k} R^{k+1-j}(\tau A)R^{j-1}(\tau A)\psi\tau = R^k(\tau A)\psi k\tau.$$

Since problem (0.6) is well posed in $\mathcal{C}(E)$ by hypothesis, the estimate

$$\|k\tau A R^k(\tau A)\psi\|_E \leq M_1 \|\varphi^\tau\|_{C(\tau, E)}$$

holds, where M_1 does not depend on τ (and ψ). Since problem (0.6) is stable in $\mathcal{C}(E)$,

$$\|\varphi^\tau\|_{C(\tau, E)} \leq M_2 \|\psi\|_E,$$

where M_2 also does not depend on τ (and ψ). Hence, we have the estimate

$$\|k\tau A R^k(\tau A)\|_{E \to E} \leq M \quad (k = 1, 2, \cdots, N). \tag{1.7}$$

Now let ψ be an arbitrary element in $D(A)$. Then (1.7) is equivalent to the estimate

$$\|k\tau(\overline{I} + \tau A)^{-k} A\psi\|_E \leq M \|\psi\|_E. \tag{1.8}$$

Now let $N \to \infty$ and $k \to \infty$ in such a way that $k\tau = t$, where $t > 0$ is a fixed number. From Theorem 1.1 it follows that $k\tau(I + \tau A)^{-k} A\psi \to t \exp\{-tA\}A\psi$ in the norm of E. Hence, by (1.5), we have

$$\|t \exp\{-tA\}A\psi\|_E \leq M \|\psi\|_E.$$

Since $\psi \in D(A)$ is arbitrary and $D(A)$ is dense in E, this implies that the operator $A \exp\{-tA\}$ is bounded for $0 < t \leq 1$ and obeys the estimate

$$\|tA \exp\{-tA\}\|_{E \to E} \leq M. \tag{1.9}$$

This estimate implies (see Section 1, Chapter 1) that the semigroup $\exp\{-tA\}$ admits an analytic continuation to the sector $|\arg z| < \arcsin(1/M)$.

There arises the question whether the analyticity of the semigroup $\exp\{-tA\}$ is sufficient for the well-posedness of problem (0.6) in $\mathcal{C}(E)$. By passing to the limit the coercivity inequality (2.1) yields the coercivity inequality (1.1) of Chapter 1 for the solutions of the corresponding differential problem. However, as shown in Section 1 of Chapter 1, not every such differential problem is well-posed in $\mathcal{C}(E)$. Hence, the analyticity of the semigroup $\exp\{-tA\}$ cannot guarantee the well-posedness in $\mathcal{C}(E)$ of every problem of the form (0.6).

We can nevertheless end this section with a positive assertion. Recall that above we have shown that the estimates (0.8) and (1.7) are sufficient for the analyticity of the semigroup $\exp\{-tA\}$, $t \leq 0$. Now let us show that these estimates are also necessary. The necessity of the estimate (0.8) is provided by the Hille-Yosida-Phillips-Miyadera theorem even for the wider class of strongly continuous semigroups. We shall now establish the necessity of (1.7) under the assumption that the estimate (1.9) holds for all $t > 0$. From the formula connecting the resolvent of the generator of a semigroup with the semigroup (see [31]) it follows that, for $k \geq 2$,

$$(I + \tau A)^{-k} = \frac{1}{(k-1)!} \int_0^\infty t^{k-1} e^{-t} \exp\{-\tau t A\} dt. \tag{1.10}$$

Using (1.9) and the fact that the operator A is closed, this yields the estimate

$$\|A(I + \tau A)^{-k}\|_{E \to E} \leq \frac{M}{\tau(k-1)!} \int_0^\infty t^{k-2} e^{-t} dt = \frac{M}{\tau(k-1)}.$$

This clearly implies (1.7) for $k \geq 2$. For $k = 1$ (1.7) is obvious.

Thus, we established the following

Real-field criterion for analyticity. *For the operator $-A$ with dense domain $D(A)$ to be the generator of an analytic semigroup it is necessary and sufficient that the estimates (0.8) and (1.7) hold.*

We call this criterion a *real-field criterion* to distinguish it from the analyticity criterion given in Chapter 1, which is based on the estimate of the resolvent of the operator $-A$ in some complex right half-plane.

3.   An almost coercive inequality in $C(E)$.

In the preceding subsections we have seen that the coercivity inequality holds uniformly in τ in $C(\tau, E)$ only for solutions of the homogeneous difference problem. It follows that the solutions of the general difference problem (0.5) obey the inequality

$$\|\mathcal{D}\Pi(u_0)u^\tau\|_{C(\tau,E)} + \|Au^\tau\|_{C(\tau,E)} \le M_1\|Au_0\|_E + M_2(\tau)\|\varphi^\tau\|_{C(\tau,E)}, \quad (1.11)$$

where M_1 is some positive constant and $M_2(\tau) \to +\infty$ as $\tau \to +0$. This follows from the investigation of the well-posedness of the limiting differential problem in $C(E)$. The study of the difference problem (0.5) in the space $C(\tau, E)$ allows us to obtain more information on how fast $M_2(\tau)$ converges to $+\infty$.

As we mentioned in the Introduction, the Cauchy problem and other initial-boundary value problems for parabolic partial differential equations reduce to the abstract Cauchy problem (1.1) of Chapter 1. In such cases the operator A is given by an elliptic differential operator. The difference scheme for the parabolic equations in question with partial discretization with respect to the time variable only reduces to the difference scheme (0.5). A full discretization (with respect to the time and space variables) leads to a series of difference problems of the form (0.5) with bounded operators A_h acting in Banach spaces E_h. Here h is the step in the difference grid for the space variables. Of course, the bounded operators $-A_h$ generate analytic semigroups $\exp\{-tA_h\}$, $t \ge 0$, in the spaces E_h. We shall assume that these operators obey the estimates (0.8) and (1.7) with constants which are independent of h. Since the operator A_h is bounded in E_h for fixed $h > 0$, the difference problem (0.5) corresponding to A_h is well posed in $C(\tau, E_h)$ uniformly in τ, and the following inequality holds:

$$\|\mathcal{D}\Pi(u_0)u^{\tau,h}\|_{C(\tau,E_h)} + \|A_hu^{\tau,h}\|_{C(\tau,E_h)}$$

$$\le \tilde{M}_1\|A_hu_0\|_{E_h} + \tilde{M}_2(\|A_h\|_{E_h\to E_h})\|\varphi^{\tau,h}\|_{C(\tau,E_h)}, \quad (1.12)$$

in which $\tilde{M}_2(\lambda) \to \infty$ as $\lambda \to \infty$.

Since the bounded operators A_h approximate the unbounded operator A (the difference operators A_h approximate the differential operator A with respect to the space variables), $\|A_h\|_{E_h \to E_h} \to \infty$ as $h \to 0+$. Consequently, $M_2(\|A_h\|_{E_h \to E_h}) \to \infty$ as $h \to 0+$. This follows from an analysis of the differential problem. The study of the difference problem allows one to estimate the rate at which the quantity $M_2(\|A_h\|_{E_h \to E_h})$ tends to infinity.

Below we shall consider a problem of the form (0.5) in Banach spaces E_h with bounded operators A_h, but to simplify the notation the index h will be omitted.

Theorem 1.2. *Let $-A$ be the generator of an analytic semigroup. Then the solutions of the difference problem (0.5) in $C(\tau, E)$ obey the almost coercive inequality*

$$\|\mathcal{D}\Pi(u_0)u^\tau\|_{C(\tau,E)} + \|Au^\tau\|_{C(\tau,E)}$$

$$\leq M\left[\|Au_0\|_E + \min\left\{\ln(1/\tau), |\ln\|A\|_{E\to E}| + 1\right\}\|\varphi^\tau\|_{C(\tau,E)}\right], \tag{1.13}$$

where M does not depend on τ, u_0, and φ^τ.

Proof. From the fact that the homogeneous equation in well posed in $C(\tau, E)$ uniformly in τ it follows that to prove the theorem it suffices to consider the case $u_0 = 0$.

From (0.2) and (0.9) we obtain

$$\|Au_k\|_E \leq \sum_{i=1}^{k} \|AR^{k+1-i}(\tau A)\|_{E\to E}\, \tau \|\varphi_i\|_E \leq M\sum_{i=1}^{k} \frac{1}{k+1-i}\|\varphi^\tau\|_{C(\tau,E)}.$$

Since

$$\sum_{i=1}^{k} \frac{1}{k+1-i} \leq \int_1^k \frac{ds}{k+1-s} = \ln k,$$

we have

$$\|Au_k\|_E \leq M\ln k\|\varphi^\tau\|_{C(\tau,E)},$$

whence

$$\|Au^\tau\|_{C(\tau,E)} \leq M\ln(\frac{1}{\tau})\|\varphi^\tau\|_{C(\tau,E)}. \tag{1.14}$$

Further, using identity (0.2) we obtain

$$\|Au_k\|_E \leq \sum_{i=1}^{k} \|AR^{k+1-i}(\tau A)\|_{E\to E}\|\varphi_i\|_E\, \tau$$

$$\leq \sum_{i=1}^{k} \|AR^{k+1-i}(\tau A)\|_{E \to E} \tau \|\varphi^{\tau}\|_{C(\tau,E)}.$$

It remains to estimate the quantity

$$J_k = \sum_{i=1}^{k} \|AR^{k+1-i}(\tau A)\|_{E \to E} \tau = \sum_{s=1}^{k} \|AR^s(\tau A)\|_{E \to E} \tau.$$

From the last identity it is clear that it suffices to estimate J_N. Using the estimates (0.8) and (1.7), we obtain

$$\|AR^s(\tau A)\|_{E \to E} \leq M \min \left\{ \frac{1}{s\tau}, \|A\|_{E \to E} \right\}.$$

If $\|A\|_{E \to E} > N$, then

$$J_N \leq M \sum_{s=1}^{N} \frac{\tau}{s\tau} \leq M \int_{1}^{\|A\|_{E \to E}} \frac{ds}{s} \leq M |\ln \|A\|_{E \to E}|.$$

If $\|A\|_{E \to E} \leq 1$, then

$$J_N \leq M \sum_{s=1}^{N} \|A\|_{E \to E} \tau \leq M \|A\|_{E \to E} \leq M.$$

Finally, if $1 \leq \|A\|_{E \to E} \leq N$, then

$$J_N \leq M \left\{ \sum_{s=1}^{[N\|A\|_{E \to E}^{-1}]} \|A\|_{E \to E} \tau + \sum_{[N\|A\|_{E \to E}^{-1}]+1}^{N} \frac{\tau}{s\tau} \right\}$$

$$\leq M \left(1 + \int_{\|A\|_{E \to E}^{-1}}^{1} \frac{ds}{s} \right) = M(1 + \ln \|A\|_{E \to E}).$$

Thus, in all three cases we have the estimate

$$I_N \leq M(1 + |\ln \|A\|_{E \to E}|),$$

which yields

$$\|Au^{\tau}\|_{C(\tau,E)} \leq M \left[1 + |\ln \|A\|_{E \to E}| \right] \|\varphi^{\tau}\|_{C(\tau,E)}. \tag{1.15}$$

From the estimates (1.14) and (1.15) (and from the uniform well-posedness of the homogeneous difference problem) we obtain the estimate (1.13). Theorem 1.2 is proved.

For most difference approximations A_h one has the estimate

$$\ln \|A_h\|_{E_h \to E_h} \leq M \left| \ln \frac{1}{h} \right|.$$

Consequently, for small τ and h (1.13) implies the estimate

$$\|\mathcal{D}\Pi(u_0)u^{\tau,h}\|_{C(\tau,E_h)} + \|A_h u^{\tau,h}\|_{C(\tau,E_h)}$$

$$\leq M \left[\|A_h u_0\|_{E_h} + \ln \left(\frac{1}{\tau + h} \right) \|\varphi^{\tau,h}\|_{C(\tau,E_h)} \right]. \tag{1.16}$$

Further, if one studies only time discretizations, then from the foregoing arguments (here $\|A\|_{E \to E} = \infty$) it follows that the solutions of problem (0.5) obey the estimate

$$\|\mathcal{D}\Pi(u_0)u^{\tau}\|_{C(\tau,E)} + \|Au^{\tau}\|_{C(\tau,E)} \leq M \left[\|Au_0\|_E + \ln \frac{1}{\tau} \|\varphi^{\tau}\|_{C(\tau,E)} \right]. \tag{1.17}$$

Let us show how the almost coercive inequalities established above apply in the estimation of accuracy of difference schemes. We will examine the last inequality, (1.17), for the solutions of the difference problem (0.5).

Let $v(t)$ be a solution in $C(E)$ of problem (1.1), Chapter 1, such that $v''(t) \in C(E)$ and $v''(t) \not\equiv 0$. Then

$$\frac{1}{\tau}(v(t_k) - v(t_{k-1})) + Av(t_k)$$

$$= f(t_k) + \frac{1}{\tau}(v(t_k) - v(t_{k-1})) - v'(t_k), \quad k = 1, \cdots, N. \tag{1.18}$$

Set $\varphi_k = f(t_k)$ and $u_0 = v_0$ in (0.5). The error vector z^{τ} with components $z_k = v(t_k) - u_k$ satisfies the identity

$$\frac{1}{\tau}(z_k - z_{k-1}) + Az_k = a_k, \quad k = 1, \cdots, N, \quad z_0 = 0, \tag{1.19}$$

and the component a_k of the approximation vector a^{τ} is given by the formula

$$a_k = \frac{1}{\tau}(v(t_k) - v(t_{k-1})) - v'(t_k) = \frac{\tau}{2}v''(t_k) + \frac{1}{\tau}\int_{t_{k-1}}^{t_k} (z - t_{k-1})[v''(s) - v''(t_k)]ds.$$

$$\tag{1.20}$$

For the error vector z^τ we shall use the coercive norm

$$\|z^\tau\|_{C^\tau} = \|\mathcal{D}\Pi(0)z^\tau\|_{C(\tau,E)} + \|Az^\tau\|_{C(\tau,E)}. \tag{1.21}$$

Then from (1.17) and the triangle inequality we obtain the inequalities

$$\|a^\tau\|_{C(\tau,E)} \leq \|z^\tau\|_{C^\tau} \leq M \ln \frac{1}{\tau} \|a^\tau\|_{C(\tau,E)}, \tag{1.22}$$

i.e., the coercive norm of the error vector z^τ tends to zero almost in the same way as the $C(\tau, E)$-norm of the approximation vector a^τ. Further, from formula (1.20) it follows that for sufficiently small τ the inequalities

$$\delta\|v''\|_{C(E)}\,\tau \leq \|a^\tau\|_{C(\tau,E)} \leq \delta^{-1}\|v''\|_{C(E)}\,\tau$$

hold for some $\delta > 0$. From this and (1.22) we obtain the following almost sharp estimates of the rate of convergence to zero of the coercive norm of the error of the solution:

$$\delta\|v''\|_{C(E)}\,\tau \leq \|z^\tau\|_{C^\tau} \leq M\delta^{-1}\|v''\|_{C(E)}\,\tau \ln \frac{1}{\tau}. \tag{1.23}$$

We have thus demonstrated the application of coercive inequalities to the determination of the accuracy of difference problems.

2.   WELL-POSEDNESS OF THE DIFFERENCE PROBLEM IN $\mathcal{C}_0^\alpha(E)$

The analyticity of the semigroup $\exp\{-tA\}$ remains a necessary condition for the well-posedness of the difference problem (0.6) when the space $\mathcal{C}(E)$ is restricted or extended in certain ways. In doing so, as in the differential case, one succeeds in introducing restrictions and extensions for which the analyticity of the semigroup is not only necessary but also sufficient for well-posedness.

A solution u of problem (0.6) is called a *solution in* $\mathcal{C}_0^\alpha(E)$ if $\overline{\mathcal{D}}\,\overline{\Pi}(u_0)u \in \mathcal{C}_0^\alpha(E)$ and $\overline{A}u \in \mathcal{C}_0^\alpha(E)$.

As in the case of the space $\mathcal{C}(E)$ one can show that for the solvability in $\mathcal{C}_0^\alpha(E)$ of problem (0.6) it is necessary that $\varphi \in \mathcal{C}_0^\alpha(E)$.

We shall assume that $u_0 \in D(A)$.

Definition 2.1. Problem (0.6) will be said to be *well-posed in the space* $\mathcal{C}_0^\alpha(E)$ if the following conditions are satisfied:

1) For any $\varphi \in \mathcal{C}_0^\alpha(E)$ and $u_0 \in D(A)$ there exists a unique solution $u = u(\varphi, u_0)$ of problem (0.6).

2) Problem (0.6) is stable in $\mathcal{C}_0^\alpha(E)$.

From the well-posedness in $\mathcal{C}_0^\alpha(E)$ of the nonhomogeneous problem (0.6) with $u_0 = 0$ it follows that the operator $\overline{A}u(\varphi, 0)$ is bounded in $\mathcal{C}_0^\alpha(E)$. Furthermore, from the solvability in $\mathcal{C}_0^\alpha(E)$ of problem (0.6) with $u_0 = 0$ one infers, via the substitution $u = w + u_0$, that the general problem (0.6) is unique solvable and that the following coercivity inequality holds:

$$\|\overline{\mathcal{D}}\,\overline{\Pi}(u_0)u\|_{\mathcal{C}_0^\alpha(E)} + \|\overline{A}u\|_{\mathcal{C}_0^\alpha(E)} \leq M(\alpha)[\|\varphi\|_{\mathcal{C}_0^\alpha(E)} + \|Au_0\|_E]. \qquad (2.1)$$

Now let the data φ and u_0 be sufficiently smooth. Assume that $\overline{A}\varphi \in \mathcal{C}_0^\alpha(E)$ and $u_0 \in D(\overline{A}^2)$. Then there exists a unique solution of problem in $\mathcal{C}_0^\alpha(E)$. If all such solutions obey inequality (2.1), then by passage to the limit one can establish the well-posedness in $\mathcal{C}_0^\alpha(E)$ of problem (0.6). Thus, the coercivity inequality (2.1) is a necessary and sufficient condition for well-posedness.

By letting $\tau \to 0$ in (2.1), we obtain the corresponding inequality for the differential problem. Therefore, A must be the generator of an analytic semigroup. Let us show how this fact can be derived directly from the well-posedness of the difference problem. Since the well-posedness of the homogeneous problem ($\varphi = 0$) in $\mathcal{C}_0^\alpha(E)$ implies its well-posedness in $\mathcal{C}(E)$, $-A$ is the generator of a strongly continuous semigroup. To prove that this semigroup is analytic it suffices to establish the inequality

$$|\lambda|\,\|w\|_E \leq M\|\psi\|_E \qquad (2.2)$$

for the solutions of the equation

$$\lambda w + Aw = \psi \qquad (2.3)$$

for sufficiently large $\operatorname{Re}\lambda > 0$. The grid function $u_k = (I - \tau\lambda)^{-k}w$, $k = 1, \cdots, N$, $\tau < |\lambda|^{-1}$, is a solution of (0.1) with $\varphi_k = (1 - \tau\lambda)^{-k}\psi$ and $u_0 = w$. Using inequality (2.1), we obtain

$$|\lambda|\|w\|_E + \|Aw\|_E \leq M(\alpha)\Big[\|\psi\|_E + \|\{(1 - \tau\lambda)^{-k}\}\|_{\mathcal{C}_0^\alpha(\mathbf{R})}^{-1}\|Aw\|_E\Big].$$

Here in the right-hand side we have the Hölder norm of the scalar grid function $\{(1 - \tau\lambda)^{-k}\}_1^N$. This norm is not smaller than the maximum of the grid function. Hence, passing to the limit $\tau \to 0$, we obtain the inequality

$$|\lambda|\|w\|_E + \|Aw\|_E \leq M(\alpha)\big(\|\psi\|_E + e^{-\operatorname{Re}\lambda}\|Aw\|_E\big),$$

which for sufficiently large $\operatorname{Re}\lambda$ yields (2.2).

Thus, the analyticity of the semigroup $\exp\{-tA\}$ is a necessary condition for the well-posedness in $\mathcal{C}_0^\alpha(E)$ of problem (0.6). It turns out that this condition is also sufficient for well-posedness. First, let us prove a lemma that will be needed in the sequel.

Lemma 2.1. *Let* $-A$ *be the generator of an analytic semigroup. Then for any* $1 \leq k < k + r \leq N$ *and* $0 \leq \alpha \leq 1$ *one has the estimates*

$$\|R^k(\tau A) - R^{k+r}(\tau A)\|_{E \to E} \leq M \frac{r^\alpha}{k^\alpha}, \tag{2.4}$$

$$\|\tau A[R^k(\tau A) - R^{k+r}(\tau A)]\|_{E \to E} \leq M \frac{r^\alpha}{k^{1+\alpha}}, \tag{2.5}$$

where M *does not depend on* τ, α, k *and* r.

Proof. We shall use the formula

$$R^k(\tau A) - R^{k+r}(\tau A) = \tau A \sum_{i=1}^{r} R^{k+i}(\tau A). \tag{2.6}$$

By the estimate (1.7), we have

$$\|R^k(\tau A) - R^{k+r}(\tau A)\|_{E \to E} \leq M \sum_{i=1}^{r} \frac{1}{k+i} \leq M \frac{r}{k}. \tag{2.7}$$

Clearly, the estimate

$$\|R^k(\tau A) - R^{k+r}(\tau A)\|_{E \to E} \leq M \tag{2.8}$$

follows from (0.8) via the triangle inequality. From (2.7) and (2.8) we obtain the estimate (2.4). Next, using formula (2.6), we get

$$\|\tau A[R^k(\tau A) - R^{k+r}(\tau A)]\|_{E \to E} \leq \sum_{i=1}^{r} \|(\tau A)^2 R^{k+i}(\tau A)\|_{E \to E}$$

$$\leq \sum_{i=2}^{r} \|\tau A R^{[\frac{k+1}{2}]}\|_{E\to E} \|\tau A R^{k+i-[\frac{k+1}{2}]}\|_{E\to E} + \|(\tau A)^2 R^{k+1}(\tau A)\|_{E\to E},$$

where $[\cdot]$ stands for integer part. Using the estimate (1.7), we obtain

$$\|\tau A[R^k(\tau A) - R^{k+r}(\tau A)]\|_{E\to E} \leq M \sum_{i=1}^{r} \frac{1}{(k+i)^2} \leq M \frac{r}{k^2}.$$

Thus,

$$\|\tau A[R^k(\tau A) - R^{k+r}(\tau A)]\|_{E\to E} \leq M \frac{r}{k^2}. \tag{2.9}$$

Clearly, the estimate

$$\|\tau A[R^k(\tau A) - R^{k+r}(\tau A)]\|_{E\to E} \leq M \frac{r}{k} \tag{2.10}$$

follows from (1.7) via the triangle inequality. Inequalities (2.9) and (2.10) yield (2.5). Lemma 2.1 is proved.

By Theorem 0.2, the estimate (2.4) yields the following result.

Theorem 2.1. *Let* $-A$ *be the generator of an analytic semigroup. Then the difference problem* (0.6) *is stable in* $\mathcal{C}_0^\alpha(E)$.

Now let us show that the analyticity of the semigroup $\exp\{-tA\}$ is a sufficient condition for the well-posedness in $\mathcal{C}_0^\alpha(E)$ of the difference problem (0.6).

Theorem 2.2. *Let* $-A$ *be the generator of an analytic semigroup. Then the difference problem* (0.6) *is well posed in* $\mathcal{C}_0^\alpha(E)$.

Proof. To prove the theorem it suffices to establish the coercivity inequality

$$\|\mathcal{D}\Pi(u_0)u^\tau\|_{\mathcal{C}_0^\alpha(\tau,E)} + \|Au^\tau\|_{\mathcal{C}_0^\alpha(\tau,E)} \leq M(\alpha)\|\varphi^\tau\|_{\mathcal{C}_0^\alpha(\tau,E)} + M\|Au_0\|_E, \tag{2.11}$$

where $M(\alpha)$ and M do not depend on τ. It obviously suffices to estimate the norm $\|Au^\tau\|_{\mathcal{C}_0^\alpha(\tau,E)}$. By formula (0.2), for the solution of problem (0.5) we can write

$$Au^\tau = Aw^\tau + Ag^\tau.$$

Hence, it suffices to estimate the norm of each of the two terms. Since

$$Aw_k = R^k(\tau A)u_0, \quad 1 \leq k \leq N,$$

the estimate

$$\|Aw^\tau\|_{C_0^\alpha(\tau,E)} \le M\|Au_0\|_E \tag{2.12}$$

is a consequence of Theorem 2.1.

The estimation of the term Ag^τ is more difficult. Using formula (0.13) and the estimate (1.7), we obtain

$$\|Ag_k\|_E \le M\|\varphi_k\|_E + \sum_{i=1}^{k-1} M\|\varphi_k - \varphi_i\|_E (k-i)^{-1}\tau$$

$$\le M\|\varphi^\tau\|_{C_0^\alpha(\tau,E)} \Big[1 + \sum_{i=1}^{k-1}(k-i)^{\alpha-1}i^{-\alpha}\tau\Big].$$

The sum enclosed in the right-hand side square brackets is the lower Darboux integral sum for the integral

$$\int_0^1 (1-s)^{\alpha-1}s^{-\alpha}ds = \frac{\pi}{\sin \pi\alpha}.$$

Thus, for $k > 1$ we established the bound

$$\|Ag_k\|_E \le \frac{M}{\alpha(1-\alpha)}\|\varphi^\tau\|_{C_0^\alpha(\tau,E)}. \tag{2.13}$$

It obviously holds also for $k = 1$.

Now let us estimate the difference $Ag_{k+r} - Ag_k$ for $1 \le k < k+r \le N$. To this end we shall examine separately the two cases $k \le 2r$ and $k > 2r$. If $k \le 2r$, then (2.13) implies

$$\|Ag_{k+r} - Ag_k\|_E \le \|Ag_{k+r}\|_E + \|Ag_k\|_E$$

$$\le \frac{2M}{\alpha(1-\alpha)}\|\varphi^\tau\|_{C_0^\alpha(\tau,E)}r^\alpha r^{-\alpha} \le \frac{2^{1+\alpha}M}{\alpha(1-\alpha)}\|\varphi^\tau\|_{C_0^\alpha(\tau,E)}r^\alpha k^{-\alpha}.$$

Now let $k > 2r$. Since $k+r > k$, we have $k+r > 1$, and then identity (0.13) implies

$$Ag_{k+r} - Ag_k = \{\varphi_{k+r} - \varphi_k + R^k(\tau A)\varphi_k - R^{k+r}(\tau A)\varphi_{k+r}\}$$

$$+ \sum_{i=k-r+1}^{k+r} AR^{k+r+1-i}(\tau A)(\varphi_i - \varphi_{k+r})\tau + \sum_{i=k-r+1}^{k} AR^{k+1-i}(\tau A)(\varphi_i - \varphi_k)\tau$$

$$+\sum_{i=1}^{k-r} AR^{k+r+1-i}(\tau A)(\varphi_k-\varphi_{k+r})\tau+\sum_{i=1}^{k-r} A[R^{k+1-i}(\tau A)-R^{k+r+1-i}(\tau A)](\varphi_i-\varphi_k)\tau$$

$$= I_1 + I_2 + I_3 + I_4 + I_5.$$

Using the estimates (0.8) and (2.4), we obtain

$$\|I_1\|_E \le M\|\varphi^\tau\|_{C_0^\alpha(\tau,E)} r^\alpha k^{-\alpha}.$$

Next, by (1.7), we have

$$\|I_2\|_E \le M \sum_{i=k-r+1}^{k+r-1} (k+r-i)^{-1}\|\varphi_i - \varphi_{k+r}\|_E$$

$$\le M\|\varphi^\tau\|_{C_0^\alpha(\tau,E)} \int_{(k-r+1)\tau}^{(k+r)\tau} \frac{ds}{((k+r)\tau - s)^{1-\alpha}s^\alpha}$$

$$\le M\|\varphi^\tau\|_{C_0^\alpha(\tau,E)} \frac{1}{((k-r+1)\tau)^\alpha} \int_{(k-r+1)\tau}^{(k+r)\tau} \frac{ds}{((k+r)\tau - s)^{1-\alpha}}$$

$$\le \frac{M}{\alpha}\|\varphi^\tau\|_{C_0^\alpha(\tau,E)} \frac{r^\alpha}{(k-r+1)^\alpha}.$$

Since $k - r = k/2 + k/2 - 2 \ge k/2$, it follows that

$$\|I_2\|_E \le \frac{2^\alpha M}{\alpha}\|\varphi^\tau\|_{C_0^\alpha(\tau,E)} r^\alpha k^{-\alpha}.$$

In a similar manner one can show that

$$\|I_3\|_E \le \frac{2^\alpha M}{\alpha}\|\varphi^\tau\|_{C_0^\alpha(\tau,E)} r^\alpha k^{-\alpha}.$$

Further, using the identity

$$I_4 = [R^k(\tau A) - R^{2r}(\tau A)](\varphi_{k+r} - \varphi_k)$$

and the estimate (0.8), we obtain the estimate

$$\|I_4\|_E \le M\|\varphi^\tau\|_{C_0^\alpha(\tau,E)} r^\alpha k^{-\alpha}.$$

Finally, using (2.5) for $\alpha = 1$, we obtain the estimate

$$\|I_5\|_E \le M\|\varphi^\tau\|_{C_0^\alpha(\tau,E)} \int_0^{(k-r)\tau} r\tau(k\tau - s)^{\alpha-2}s^{-\alpha}ds.$$

Since for $t > 2\tau > 0$ we have the inequality

$$\int_0^{t-\tau} \tau(t-s)^{\alpha-2} s^{-\alpha} ds \leq \frac{M}{1-\alpha} \tau^\alpha t^{-\alpha},$$

we conclude that

$$\|I_5\|_E \leq \frac{M}{1-\alpha} \|\varphi^\tau\|_{C_0^\alpha(\tau,E)} r^\alpha k^{-\alpha}.$$

Thus, we have shown that the inequality

$$\|Ag_{k+r} - Ag_k\|_E \leq \frac{M}{\alpha(1-\alpha)} \|\varphi^\tau\|_{C_0^\alpha(\tau,E)} r^\alpha k^{-\alpha}$$

holds for all $1 \leq k < k+r \leq N$. From this and (2.13) we obtain

$$\|Ag^\tau\|_{C_0^\alpha(\tau,E)} \leq \frac{M}{\alpha(1-\alpha)} \|\varphi^\tau\|_{C_0^\alpha(\tau,E)},$$

which in conjunction with the estimate (2.12) yields the desired inequality

$$\|Au^\tau\|_{C_0^\alpha(\tau,E)} \leq \frac{M}{\alpha(1-\alpha)} \|\varphi^\tau\|_{C_0^\alpha(\tau,E)} + M\|Au_0\|_E.$$

Theorem 2.2 is proved.

Let us remark that from the coercivity inequality (2.11) and the identity (1.18) it follows that

$$\left\|\left\{\frac{v(t_k) - v(t_{k-1})}{\tau}\right\}_1^N\right\|_{C_0^\alpha(\tau,E)} + \left\|\{Av(t_k)\}_1^N\right\|_{C_0^\alpha(\tau,E)} \leq$$

$$M(\alpha)\left[\left\|\{f(t_k)\}_1^N\right\|_{C_0^\alpha(\tau,E)} + \left\|\left\{\frac{v(t_k) - v(t_{k-1})}{\tau} - v'(t_k)\right\}\right\|_{C_0^\alpha(\tau,E)}\right].$$

If $v''(t) \in C_0^\alpha(E)$, then by passing to the limit $\tau \to 0$ in the last inequality one can obtain Theorem 2.1, Chapter 1 on the well-posedness of the differential Cauchy problem in the space $C_0^\alpha(E)$.

From the proofs of Theorem 2.1 of Chapter 1 and Theorem 2.2 it follows that the quantity $M(\alpha)$ in the coercivity inequalities for the difference and differential problem is subject to the bound

$$M(\alpha) \leq \frac{M}{\alpha(1-\alpha)}. \qquad (2.14)$$

In contrast to the differential problem, the estimate (2.11) established for the difference problem allows us to prove Theorem 1.2. Indeed, by the definition of the norm in $C_0^\alpha(\tau, E)$,

$$\|\varphi^\tau\|_{C(\tau,E)} \le \|\varphi^\tau\|_{C_0^\alpha(\tau,E)} \le \frac{3}{\tau^\alpha}\|\varphi^\tau\|_{C(\tau,E)}.$$

Then, by (2.14), Theorem 2.2 implies that

$$\|\mathcal{D}\Pi(u_0)u^\tau\|_{C(\tau,E)} + \|Au^\tau\|_{C(\tau,E)} \le M\left[\frac{1}{\alpha\tau^\alpha}\|\phi^\tau\|_{C(\tau,E)} + \|Au_0\|_E\right]$$

for all sufficiently small $\alpha > 0$. Choosing α in the best way, we obtain inequality (1.17) for the solutions of the difference problem (0.5) in the case of discretization only with respect to time. The estimate (2.14) for the difference problem (0.5) in the space $C_0^\alpha(\tau, E)$ is an upper estimate for

$$M(\alpha) = \left\|\left\{\sum_{i=1}^{k} AR^{k+1-i}(\tau A)\varphi_i \tau\right\}_1^N\right\|_{C_0^\alpha(\tau,E)\to C_0^\alpha(\tau,E)}.$$

We were not able to obtain a sharp estimate for $M(\alpha)$.

3.   WELL-POSEDNESS OF THE DIFFERENCE PROBLEM IN $\mathcal{L}_p(E)$

1.   Definition of the well-posedness of the difference problem in $\mathcal{L}_p(E)$.

Let us study problem (0.6) in the Lebesgue (Bochner) space $\mathcal{L}_p(E)$, which is larger than $\mathcal{C}(E)$. A solution of problem (0.6) is called *a solution of this problem in $\mathcal{L}_p(E)$* if $\overline{\mathcal{D}}\,\overline{\Pi}(u_0)u$ and $\overline{A}u$ belong to $\mathcal{L}_p(E)$. For the solvability in $\mathcal{L}_p(E)$ of problem (0.6) it is necessary that $\varphi \in \mathcal{L}_p(E)$. The necessary condition that u_0 must satisfy is more complicated. For this reason we shall proceed as for the differential case and begin our analysis with the nonhomogeneous problem with null initial condition $u_0 = 0$.

Definition 3.1. We say that problem (0.6) with $u_0 = 0$ is *well posed in $\mathcal{L}_p(E)$* if the following conditions are satisfied:

1) For any $\varphi \in \mathcal{L}_p(E)$ there exists a unique solution $u = u(\varphi, 0)$ of (0.6). The unique solvability of problem (0.6) implies the linearity of the operator $u(\varphi, 0)$.

2) The operator $u(\varphi, 0)$ is continuous in $\mathcal{L}_p(E)$.

From the continuity of the linear operator $u(\varphi, 0)$ it follows that this operator is bounded. As in Chapter 1, one shows that the operator $\overline{A}u(\varphi, 0)$ is continuous in $\mathcal{L}_p(E)$. Further, from the solvability in $\mathcal{L}_p(E)$ of problem (0.6) with $u_0 = 0$ it follows, via the substitution $u = w + u_0$, that the general problem (0.6) with $u_0 \in D(A)$ is uniquely solvable and that the coercivity inequality holds:

$$\|\overline{D}\,\overline{\Pi}(u_0)u\|_{\mathcal{L}_p(E)} + \|\overline{A}u\|_{\mathcal{L}_p(E)} \leq M(p)\Big[\|\varphi\|_{\mathcal{L}_p(E)} + \|Au_0\|_E\Big]. \qquad (3.1)$$

From this one infers that the semigroup $\exp\{-tA\}$ is analytic. This can be done either by passage to the limit to the differential case or by the method used for the spaces $C_0^{\alpha}(E)$.

2.   Spaces of initial data.

It follows from formula (0.3) that the solution of the homogeneous problem (0.5) with $\varphi^{\tau} = 0$ has the form

$$w_k = R^k(\tau A)u_0, \quad 1 \leq k \leq N.$$

Hence, for w to be a solution of problem (0.6) in $\mathcal{L}_p(E)$ it is necessary and sufficient that

$$\sup_{0 < \tau \leq \tau_0} \sum_{k=1}^{N} \|AR^k(\tau A)u_0\|_E^p \tau < \infty.$$

By estimate (0.8), the collection of all $u_0 \in E$ with this last property is a linear set that contains $D(A)$. It becomes a Banach space $E'_{1-\frac{1}{p}}$ when equipped with the norm

$$\langle u_0 \rangle_{1-\frac{1}{p}} = \sup_{0 < \tau \leq \tau_0} \left(\sum_{k=1}^{N} \|AR^k(\tau A)u_0\|_E^p \tau \right)^{1/p} + \|u_0\|_E. \qquad (3.2)$$

Let u^{τ} be a solution of problem (0.5). Since problem (0.4) has a solution g^{τ} for any $\varphi^{\tau} \in \mathcal{L}_p(E)$, the grid function $w^{\tau} = u^{\tau} - g^{\tau}$ will be a solution of problem (0.3) in $\mathcal{L}_p(E)$, and the function w will be a solution of problem (0.6) with $\varphi = 0$. Consequently, $u_0 \in E'_{1-\frac{1}{p}}$. Thus, for the solvability in $\mathcal{L}_p(E)$ of the general problem (0.6) (under the assumption that problem (0.6) with $u_0 = 0$ is well posed) it is necessary and sufficient that $u_0 \in E'_{1-\frac{1}{p}}$. Below it will be shown that the spaces E'_α coincide with the spaces E_α introduced in Chapter 1, Section 3.

Now let us assume that $-A$ is the generator of an analytic semigroup. The fact that this is a natural assumption was discussed in Subsection 1. Since problem (1.1) of Chapter 1 and problem (0.6) are problems on a bounded interval, with no loss of generality we can assume that the norm of $\exp\{-tA\}$ decays exponentially (to do this we must substitute in equations $v(t) = e^{kt}w(t)$, with $k > 0$ large enough). Then we have the estimates

$$\|\exp\{-tA\}\|_{E \to E} \leq M e^{-\delta t}, \quad \|A \exp\{-tA\}\|_{E \to E} \leq M t^{-1} e^{-\delta t}, \quad t > 0, \quad \delta > 0, \tag{3.3}$$

or the equivalent estimate for the resolvent of the operator $-A$,

$$\|(\lambda I + A)^{-1}\|_{E \to E} \leq \frac{M}{1 + |\lambda|}, \quad \operatorname{Re} \lambda \geq 0.$$

By the assumption that the semigroup $\exp\{-tA\}$ is analytic, to (3.3) there now correspond the estimates

$$\left.\begin{array}{l} \|(I + \tau A)^{-k}\|_{E \to E} \leq M(1 + \delta\tau)^{-k}, \\[2mm] \|A(I + \tau A)^{-(k+1)}\|_{E \to E} \leq M(1 + \delta\tau)^{-k} \dfrac{1}{k\tau}, \quad k = 1, 2, \cdots. \end{array}\right\} \tag{3.4}$$

We shall need the following estimates, which are more general than (3.3) and (3.4):

$$\|A^n \exp\{-tA\}\|_{E \to E} \leq M^n n^n e^{-\delta t} t^{-n}, \quad t > 0, \quad n = 1, 2, \cdots \tag{3.5}$$

and

$$\|A^n R^{k+1}(\tau A)\|_{E \to E} \leq \frac{M^n n^n (k - n)!}{k! \tau^n} (1 + \delta\tau)^{-(k+1-n)}, \quad k + 1 \geq n. \tag{3.6}$$

Estimate (3.6) follows from the integral representation (1.10) of the powers of the resolvent of the operator $-A$. Estimate (3.5) allows us to obtain the following smoothness estimate for the semigroup $\exp\{-tA\}$, which will be nedeed in the sequel:

$$\|A[\exp\{-tA\} - \exp\{-k\tau A\}]\|_{E \to E} \leq 4M^2 (k\tau)^{-2} \tau e^{-\delta k\tau}, \tag{3.7}$$

where $k\tau \leq t \leq (k+1)\tau, \ k = 1, 2, \cdots$.

By Theorem 1.1, a semigroup can be obtained as a limit of powers of the resolvent of its generator. We shall need an estimate of the rate of this convergence.

Lemma 3.1. *Let $-A$ be the generator of an analytic semigroup. Then the following estimate holds for any $k \geq n+2, \ n = 1, 2, \cdots$:*

$$\|A^n R^k(\tau A) - A^n \exp\{-k\tau A\}\|_{E \to E} \leq M(n)[(k - n - 1)\tau]^{-n-1}\tau. \qquad (3.8)$$

Proof. Using the identity

$$P \equiv A^n[R^k(\tau A) - \exp\{-k\tau A\}] = A^n \int_0^1 \frac{d}{ds}[(I + \tau sA)^{-k}\exp\{-k\tau(1-s)A\}]ds$$

and the estimates (3.5) and (3.6), we obtain

$$\|P\|_{E \to E} = \| \int_0^1 ks\tau^2 A^{n+2}(I + \tau sA)^{-k-1}\exp\{-k\tau(1-s)A\}ds\|_{E \to E}$$

$$\leq \int_0^{1/2} ks\tau^2\|A(I + \tau sA)^{-k-1}\|_{E \to E}\|A^{n+1}\exp\{-k\tau(1-s)A\}\|_{E \to E}ds$$

$$+ \int_{1/2}^1 ks\tau^2\|A^{n+2}(I + \tau sA)^{-k-1}\|_{E \to E}\|\exp\{-k\tau(1-s)A\}\|_{E \to E}ds$$

$$\leq M_1(n)(k\tau)^{-n-1}\tau + M_2(n)[(k-n-1)\tau]^{-n-1}\tau \leq M(n)[(k-n-1)\tau]^{-n-1}\tau.$$

Lemma 3.1 is proved.

Lemma 3.1 and the estimates for the semigroup and the powers of the resolvent given above allow us to investigate the spaces E_α and E'_α of initial data for the differential and difference problem, respectively.

Theorem 3.1. *The spaces E_α and E'_α coincide for all $0 < \alpha < 1$.*

Proof. Clearly, it suffices to establish the equivalence of the norms in E_α and E'_α. From (3.3) it follows that the norm in the spaces E_α of initial data for the differential problem (1.1), introduced in Section 3 of Chapter 1, is equivalent to the norm

$$|x|_\alpha = \left(\int_0^\infty \|A \exp\{-tA\}x\|_E^{1/1-\alpha}dt \right)^{1-\alpha}.$$

To show this it suffices to establish the bounds

$$\left(\int_0^1 \|A\exp\{-tA\}x\|_E^{1/1-\alpha}dt\right)^{1-\alpha} \leq \left(\int_0^\infty \|A\exp\{-tA\}x\|_E^{1/1-\alpha}\right)^{1-\alpha}$$

$$\leq M\left(\int_0^1 \|A\exp\{-tA\}x\|_E^{1/1-\alpha}\right)^{1-\alpha}.$$

The lower bound is obvious. To prove the upper one we observe that, by the triangle inequality,

$$\left(\int_0^\infty \|A\exp\{-tA\}x\|_E^{1/1-\alpha}\right)^{1-\alpha}$$

$$\leq \left(\int_0^1 \|A\exp\{-tA\}x\|_E^{1/1-\alpha}dt\right)^{1-\alpha} + \left(\int_1^\infty \|A\exp\{-tA\}x\|_E^{1/1-\alpha}dt\right)^{1-\alpha}.$$

From (3.3) it follows that

$$\left(\int_1^\infty \|A\exp\{-tA\}x\|_E^{1/(1-\alpha)}dt\right)^{1-\alpha} \leq M_1\left(\int_1^\infty e^{-\frac{\delta t}{1-\alpha}}t^{-\frac{1}{1-\alpha}}dt\right)^{1-\alpha}\|x\|_E$$

$$\leq M_1\left(\frac{1-\alpha}{\delta}e^{-\frac{\delta}{1-\alpha}}\right)^{1-\alpha}\|x\|_E = M_1\frac{(1-\alpha)^{1-\alpha}}{\delta^{1-\alpha}}e^{-\delta}\|x\|_E.$$

The estimate

$$\|x\|_E \leq M_2\left(\int_0^1 \|A\exp\{-tA\}x\|_E^{1/(1-\alpha)}dt\right)^{1-\alpha}$$

was established in Chapter 1, Section 3. It follows that

$$\left(\int_0^\infty \|A\exp\{-tA\}x\|_E^{1/(1-\alpha)}dt\right)^{1-\alpha}$$

$$\leq \left(1 + M_1 M_2\frac{(1-\alpha)^{1-\alpha}}{\delta^{1-\alpha}}e^{-\delta}\right)\left(\int_0^1 \|A\exp\{-tA\}x\|_E^{1/(1-\alpha)}dt\right)^{1-\alpha},$$

i.e., the upper bound is also established.

Let $x \in E_\alpha$. Using representation (1.10) for powers of the resolvent, we obtain

$$AR^k(\tau A)x = \tau^{-k}\int_0^\infty \left[\frac{t^{k-1}}{(k-1)!}\exp\left(-\frac{t}{\tau}\right)\right]^\alpha \times$$

$$\times \left[\frac{t^{k-1}}{(k-1)!}\exp\left(-\frac{t}{\tau}\right)\right]^{1-\alpha} A\exp\{-tA\}xdt.$$

Further, applying Hölder's inequality, we have

$$\|AR^k(\tau A)x\|_E^{1/(1-\alpha)} \le \tau^{-\frac{k}{1-\alpha}} \left(\int_0^\infty \frac{t^{k-1}}{(k-1)!} \exp\left(-\frac{t}{\tau}\right) dt \right)^{\frac{\alpha}{1-\alpha}} \times$$

$$\times \int_0^\infty \frac{t^{k-1}}{(k-1)!} \exp\left(-\frac{t}{\tau}\right) \|A\exp\{-tA\}x\|_E^{1/(1-\alpha)} dt$$

$$= \tau^{-1} \int_0^\infty \frac{1}{(k-1)!} \left(\frac{t}{\tau}\right)^{k-1} \exp\left(-\frac{t}{\tau}\right) \|A\exp\{-tA\}x\|_E^{1/(1-\alpha)} dt.$$

The estimate obtained and the fact that the operator A is closed allow us to justify the above calculations for any $x \in E_\alpha$. Since

$$|x|_\alpha^{1/(1-\alpha)} = \sum_{k=1}^\infty \int_0^\infty \frac{1}{(k-1)!} \left(\frac{t}{\tau}\right)^{k-1} \exp\left(-\frac{t}{\tau}\right) \|A\exp\{-tA\}x\|_E^{1/(1-\alpha)} dt,$$

we have

$$\sum_{k=1}^N \|AR^k(\tau A)x\|_E^{1/(1-\alpha)} \tau$$

$$\le \sum_{k=1}^N \int_0^\infty \frac{1}{(k-1)!} \left(\frac{t}{\tau}\right)^{k-1} \exp\left(-\frac{t}{\tau}\right) \|A\exp\{-tA\}x\|_E^{1/(1-\alpha)} dt \le |x|_\alpha^{1/(1-\alpha)}.$$

Consequently,

$$\sum_{k=1}^N \|AR^k(\tau A)x\|_E^{1/(1-\alpha)} \tau \le |x|_\alpha^{1/(1-\alpha)},$$

i.e., $\langle x \rangle_\alpha \le |x|_\alpha$. Now let $x \in E_\alpha'$. Let us consider the quantity

$$Q(\eta) = \left(\int_\eta^1 \|A\exp\{-tA\}x\|_E^{1/(1-\alpha)} \right)^{1-\alpha},$$

where $\eta = ([\mu/\tau] + 2)\tau$ for some $0 < \mu < 1$; here $[a]$ denotes the integer part of $a > 0$, and $\tau = 1/N$. By (3.3), $Q(\eta) < \infty$. Let us show that $Q(\eta)$ can be estimated by $\langle x \rangle_\alpha$ with small corrections. To this end let us represent $Q(\eta)$ as

$$Q(\eta) = \left(\sum_{k\tau \ge \eta}^{N-1} \int_{k\tau}^{(k+1)\tau} \|A\exp\{-tA\}x\|_E^{1/(1-\alpha)} dt \right)^{1-\alpha}.$$

Using the triangle inequality, we obtain

$$Q(\eta) \le \left(\sum_{k\tau \ge \eta}^{N-1} \int_{k\tau}^{(k+1)\tau} \|A[\exp\{-tA\} - \exp\{-k\tau A\}]x\|_E^{1/(1-\alpha)} dt \right)^{1-\alpha}$$

$$+ \left(\sum_{\substack{k\tau \geq \eta}}^{N-1} \|A[\exp\{-k\tau A\} - R^k(\tau A)]x\|_E^{1/(1-\alpha)} \tau \right)^{1-\alpha}$$

$$+ \left(\sum_{\substack{k\tau \geq \eta}}^{N-1} \|AR^k(\tau A)x\|_E^{1/(1-\alpha)} \tau \right)^{1-\alpha} = b_1 + b_2 + b_3.$$

Using (3.7), we obtain for b_1 the estimate

$$b_1 \leq \left(\sum_{\substack{k\tau \geq \eta}}^{N-1} \left(\frac{4M^2\tau}{(k\tau)^2} \right)^{1/(1-\alpha)} \tau \right)^{1-\alpha} \leq 4M^2 \eta^{-2} \tau^\alpha.$$

To estimate b_2 we use Lemma 3.1 for $n = 1$:

$$b_2 \leq \left(\sum_{\substack{k\tau \geq \eta}}^{N-1} \left[\frac{M(1)\tau}{((k-2)\tau)^2} \right]^{1/(1-\alpha)} \tau \right)^{1-\alpha} \leq \tilde{M} \mu^{-2} \tau^\alpha.$$

The third term b_3 is obviously bounded by the norm $\langle x \rangle_\alpha$. Hence, $Q(\eta)$ obeys the estimate

$$Q(\eta) \leq 4M^2 \eta^{-2} \tau^\alpha + \tilde{M} \mu^{-2} \tau^\alpha + \langle x \rangle_\alpha.$$

Letting $\tau \to 0$, we conclude that $\limsup_{\tau \to 0} Q(\eta) \leq \langle x \rangle_\alpha$. This means that $x \in E_\alpha$ and $|x|_\alpha \leq \langle x \rangle_\alpha$. Theorem 3.1 is proved.

Now let us consider problem (0.6) in $\mathcal{L}_p(E)$, $1 < p < \infty$. Clearly, the operator $u(0, u_0)$ corresponding to the solutions of the homogeneous equation (0.6), with $\varphi = 0$, is continuous as an operator from $E_{1-\frac{1}{p}}$ into $\mathcal{C}(E_{1-\frac{1}{p}})$. It turns out that a similar property is enjoyed by the operator $u(\varphi, u_0)$ corresponding to the general problem (0.6). To show this it obviously suffices to consider the case $u_0 = 0$.

Theorem 3.2. *The solutions $g^\tau = \{g_k\}_1^N$ of problem (0.5) obey the estimate*

$$\sup_{0 \leq \tau \leq \tau_0} \max_{1 \leq k \leq N} |g_k|_{1-\frac{1}{p}} \leq M \frac{p^2}{p-1} \|\varphi\|_{\mathcal{L}_p(E)}, \quad 1 < p < \infty. \tag{3.9}$$

Proof. Using formula (0.2) for the solutions of the problem (0.5) and the integral representation (1.10), we obtain

$$\|A\exp\{-\tau s A\}g_k\|_E \leq \int_0^\infty \tau \sum_{j=1}^k \frac{t^{k-j}}{(k-j)!} \exp(-t) \|A\exp\{-\tau(t+s)\varphi_j\|_E dt.$$

Next, using the estimate (3.3), we find that

$$\|A\exp\{-\tau sA\}g_k\|_E \le M \int_0^\infty \frac{1}{t+s} \sum_{j=1}^k \frac{t^{k-j}}{(k-j)!} \exp(-t)\|\varphi_j\|_E dt,$$

where M does not depend on k and τ. Further, using the fact that the Hilbert operator is bounded in $L_p(0,\infty)$ if $1 < p < \infty$, we obtain

$$|g_k|_{1-\frac{1}{p}} = \left(\int_0^\infty \|A\exp\{-\tau sA\}g_k\|_E^p \, \tau ds \right)^{1/p}$$

$$\le \frac{Mp^2}{p-1} \left(\int_0^\infty \left(\sum_{j=1}^k \frac{t^{k-j}}{(k-j)!} \exp(-t)\|\varphi_j\|_E \right)^p \tau dt \right)^{1/p}.$$

As in Theorem 3.1, we derive the estimate

$$\sum_{j=1}^k \left(\frac{t^{k-j}}{(k-j)!} \exp(-t) \right)^{1-1/p} \left(\frac{t^{k-j}}{(k-j)!} \exp(-t) \right)^{1/p} \|\varphi_j\|_E$$

$$\le \left(\sum_{j=1}^k \frac{t^{k-j}}{(k-j)!} \exp(-t) \right)^{\frac{p-1}{p}} \left(\sum_{j=1}^k \frac{t^{k-j}}{(k-j)!} \exp(-t)\|\varphi_j\|_E^p \right)^{1/p}$$

$$\le \left(\sum_{j=1}^k \frac{t^{k-j}}{(k-j)!} \exp(-t)\|\varphi_j\|_E^p \right)^{1/p}.$$

Finally, since $(k-j)! = \int_0^\infty t^{k-j}e^{-t}dt$, we conclude that

$$|g_k|_{1-\frac{1}{p}} \le \frac{Mp^2}{p-1} \|\varphi^\tau\|_{L_p(\tau,E)}, \ 1 \le k \le N, \tag{3.10}$$

where M does not depend on k and τ. Theorem 3.2 is proved.

3.   The coercivity inequality for the solutions in $\mathcal{L}_p(E)$ of the general problem (0.6).

The fact that the operator $u(\varphi, u_0)$ is bounded (see Theorem 3.1) allows us to sharpen inequality (3.1) and to establish the coercivity inequality for the $\mathcal{L}_p(E)$-solutions of problem (0.6):

$$\|\overline{\mathcal{D}}\,\overline{\Pi}(u_0)u\|_{\mathcal{L}_p(E)} + \|\overline{A}u\|_{\mathcal{L}_p(E)} + \|u\|_{C(E_{1-\frac{1}{p}})} \le M(p)\left[\|\varphi\|_{\mathcal{L}_p(E)} + |u_0|_{1-\frac{1}{p}}\right]. \tag{3.11}$$

The assertions made above concerning the solvability in $\mathcal{L}_p(E)$ of the general problem (0.6) were derived from the well-posedness in $\mathcal{L}_p(E)$ of problem (0.6) with $u_0 = 0$. As we observed earlier, a necessary condition for the latter is the analyticity of the semigroup $\exp\{-tA\}$. One can ask whether this analyticity is a sufficient condition for the well-posedness of problem (0.6) with $u_0 = 0$. It turns out that, as in the differential case, the following extrapolation result holds true.

Theorem 3.3. *Suppose problem* (0.6) *with* $u_0 = 0$ *is well posed in* $\mathcal{L}_{p_0}(E)$ *for some* p_0, $1 < p_0 < \infty$. *Then it is well posed in* $\mathcal{L}_p(E)$ *for any* p, $1 < p < \infty$, *and* $M(p) = Mp^2/(p-1)$ *in* (3.11).

The **proof** of this theorem is carried out according to the following scheme. The solution of problem (0.4) defines a difference convolution operator, acting as

$$g_k = \sum_{j=1}^{k} R^{k+1-j}(\tau A)\varphi_j \tau.$$

This operator extends to a difference operator on the full real line, to which there corresponds an operator B in a function space, such that the boundedness of B in the function space is equivalent to the boundedness of the original difference operator. Then to the investigation of B one applies the extrapolation Theorem 3.3 of Chapter 1.

First let us introduce the necessary concepts and notations. For nonnegative real numbers x let $[x]$ denote the smallest integer $\leq x$. Extend the function $[x]$ to negative x by the rule $[x] = -[|x|]$, and for an arbitrary real x put $\{x\} = x - [x]$. Then for any x one has the representation

$$x = [x] + \{x\}. \tag{3.12}$$

In the space $L_p^\infty(E)$ we will distinguish the set of step functions. To each positive integer N we associate the step function $\tilde{\psi}^\tau(t)$ with step $\tau = 1/N$, defined as

$$\tilde{\psi}^\tau(t) = (\psi_k, \ (k-1)\tau < t \leq k\tau, \ k = 0, \pm 1, \pm 2, \cdots),$$

where ψ_k are elements of the Banach space E. The function $\tilde{\psi}^\tau(t)$ belongs to $L_p^\infty(E)$ if and only if

$$\|\tilde{\psi}^\tau\|_{L_p^\infty(E)}^p = \int_{-\infty}^{\infty} \|\tilde{\psi}^\tau(t)\|_E^p dt = \sum_{-\infty}^{\infty} \|\psi_k\|_E^p \tau < \infty.$$

For each τ define the averaging operator $W_\tau f(t)$ in $L_p^\infty(E)$, acting according to the rule

$$W_\tau f(t) = \tilde{f}^\tau(t) = \left((W_\tau f)_k = \tau^{-1} \int_{t_{k-1}}^{t_k} f(\tau)dt, \ t_{k-1} < t \le t_k, \right.$$

$$\left. k = \left[\frac{t}{\tau}\right], \ t_k = k\tau, \ k = 0, \pm 1, \pm 2, \cdots \right).$$

Clearly, the operator $W_\tau f(t)$, which maps $L_p^\infty(E)$ into the set of step functions with step τ, acts as the identity operator on the set of step functions with step τ and maps the space $L_\infty^0(E)$ into itself. Furthermore, W_τ is a linear (additive, homogeneous, and continuous) operator in $L_p^\infty(E)$ for any p, $1 < p < \infty$, and $\|W_\tau\|_{L_p^\infty(E) \to L_p^\infty(E)} = 1$.

For fixed τ we define an operator B on functions $f(t) \in L_\infty^0(E)$ as follows. First, we define the kernel of B as

$$\tilde{B}\left(\left[\frac{t}{\tau}\right]\right) = \begin{cases} AR^{[\frac{t}{\tau}]}(\tau A), & t \ge \tau, \\ 0, & t < \tau. \end{cases} \tag{3.13}$$

Then B acts as

$$Bf(t) = \int_{-\infty}^\infty \tilde{B}\left(\left[\frac{t}{\tau}\right] - \left[\frac{s}{\tau}\right]\right) W_\tau f(s)ds. \tag{3.14}$$

Clearly, B maps the function $f(t) \in L_\infty^0$ into the step function $(\widetilde{Bf})^\tau(t)$. When we speak about the boundedness of the operator B we mean boundedness uniform in τ. If B is bounded in $L_p^\infty(E) \cap L_\infty^0(E)$, then it admits a natural extension by continuity to a bounded operator on $L_p^\infty(E)$.

Theorem 3.4. *If the operator B is bounded on $L_{p_0}^\infty(E)$ for some p_0, $1 < p_0 < \infty$, then it is bounded on $L_p^\infty(E)$ for any p, $1 < p < \infty$.*

The **proof** of the theorem consists of two steps. In the first, main step, one shows that the operator B satisfies the conditions of Theorem 3.3 of Chapter 1 with constants that are independent of τ. This will imply that B is bounded in $L_p^\infty(E)$ for $p \in (1, p_0)$. In the second part one proves that B is bounded in $L_p^\infty(E)$ for all $p \in (1, \infty)$.

Let $f(t)$ be such that $\operatorname{supp} f(t) \subset \{t : \ |t - t_0| < \rho\}$ and $\int_{-\infty}^\infty f(t)dt = 0$. Then we have the equality

$$\int_{-\infty}^\infty W_\tau f(t)dt = \sum_{k=-\infty}^\infty \left(\tau^{-1} \int_{t_{k-1}}^{t_k} f(t)dt \right)\tau = \int_{-\infty}^\infty f(t)dt = 0.$$

Using it, we obtain

$$\int_{|t-t_0|>M_1\rho} \|Bf(t)\|_E dt = \int_{|z|>M_1\rho} \left\| \int_{-\infty}^{\infty} \tilde{B}\left(\left[\frac{z+t_0}{\tau}\right] - \left[\frac{s}{\tau}\right]\right) W_\tau f(s) ds \right\|_E dz$$

$$= \int_{|z|>M_1\rho} \left\| \int_{-\infty}^{\infty} \left(\tilde{B}\left(\left[\frac{z+t_0}{\tau}\right] - \left[\frac{s}{\tau}\right]\right) \right. \right.$$

$$\left. \left. - \tilde{B}\left(\left[\frac{z}{\tau}\right] + \left[\left\{\frac{z}{\tau}\right\} + \left\{\frac{t_0}{\tau}\right\}\right]\right) \right) W_\tau f(s) ds \right\|_E dz$$

$$\leq \int_{-\infty}^{\infty} \|W_\tau f(u+t_0)\|_E \times$$

$$\times \int_{|z|>M_1\rho} \left\| \tilde{B}\left(\left[\frac{z+t_0}{\tau}\right] - \left[\frac{u+t_0}{\tau}\right]\right) - \tilde{B}\left(\left[\frac{z}{\tau}\right] + \left[\left\{\frac{z}{\tau}\right\} + \left\{\frac{t_0}{\tau}\right\}\right]\right) \right\|_{E\to E} dz\, du.$$

Thus, to show that B satisfies the conditions of Theorem 3.3, Chapter 1, it suffices to show that for some M_1 the following quantity is bounded for any $t_0 \in (-\infty, \infty)$ and $|u| < \rho + \tau$:

$$Q = \int_{|z|>M_1\rho} \left\| \tilde{B}\left(\left[\frac{z+t_0}{\tau}\right] - \left[\frac{u+t_0}{\tau}\right]\right) \right.$$

$$\left. - \tilde{B}\left(\left[\frac{z}{\tau}\right] + \left[\left\{\frac{z}{\tau}\right\} + \left\{\frac{t_0}{\tau}\right\}\right]\right) \right\|_{E\to E} dz. \qquad (3.15)$$

To this end we use (3.12) to write

$$\left[\frac{z+t_0}{\tau}\right] - \left[\frac{u+t_0}{\tau}\right] = \left[\frac{z}{\tau}\right] + \left[\left\{\frac{z}{\tau}\right\} + \left\{\frac{t_0}{\tau}\right\}\right] - \left[\frac{u}{\tau}\right] - \left[\left\{\frac{u}{\tau}\right\} + \left\{\frac{t_0}{\tau}\right\}\right].$$

Let

$$k(z) = \left[\frac{z}{\tau}\right], \quad n(z) = \left[\left\{\frac{z}{\tau}\right\} + \left\{\frac{t_0}{\tau}\right\}\right], \quad j = \left[\frac{u}{\tau}\right], \quad m = \left[\left\{\frac{u}{\tau}\right\} + \left\{\frac{t_0}{\tau}\right\}\right].$$

If $\tau > \rho$, we can write

$$Q \leq \left(\int_{|z|>M_1(\rho+\tau)/2} + \int_{|z|\leq M_1\tau} \right) \|\tilde{B}(k(z)+n(z)-j-m) - \tilde{B}(k(z)+n(z))\|_{E\to E} dz$$

$$= I_1 + I_2.$$

From the estimates (3.4) and the inequality $|z| \leq M_1\tau$ it follows that I_2 is bounded. Now, if $\tau \leq \rho$, then $Q \leq I_1$. Let $M_1 \geq 4$. Then from the inequality $|z| > M_1(p + \tau)/2$, relation (3.12), and the condition $|z| \leq \rho + \tau$ ($\text{supp}\, W_\tau f \subset \{t : |t - t_0| < \rho + \tau\}$) it follows that $|[z/\tau]| \geq M_1|[u/\tau]|/4$. Hence, it suffices to estimate the quantity

$$P = \int_{k(z)>M_1|j|/4} \|\tilde{B}(k(z) + n(z) - j - m) - \tilde{B}(k(z) + n(z))\|_{E\to E}dz. \quad (3.16)$$

Since m does not depend on z and $|m| \leq 1$, it suffices to examine three cases: 1) $j = m = 0$; 2) $j = 0$, $m = \pm1$; and 3) $|j| \geq 1$, $m = 0, \pm1$.

In the case 1) $P = 0$. Let us consider the case 2). First let $m = 1$. By the definition (3.13) of the kernel $\tilde{B}$, the quantity P given by (3.16) admits the bound

$$P \leq \int_{(k(z)+n(z))\tau \geq \tau} \|AR^{k(z)+n(z)-1}(\tau A) - AR^{k(z)+n(z)}(\tau A)\|_{E\to E}dz$$

$$\leq \| - AR(\tau A)\|_{E\to E}\tau + \|AR(\tau A) - AR^2(\tau A)\|_{E\to E}\tau$$

$$+ \int_{(k(z)+n(z))\tau \geq 3\tau} \|AR^{k(z)+n(z)-1}(\tau A) - AR^{k(z)+n(z)}(\tau A)\|_{E\to E}dz = T_1 + T_2 + T_3.$$

Using (3.4), we obtain for T_1 the bound

$$T_1 = \|I - R(\tau A)\|_{E\to E} \leq M.$$

To estimate T_2 and T_3 we shall use the equality

$$AR^{k(z)+n(z)-1}(\tau A) - AR^{k(z)+n(z)}(\tau A) = \tau A^2 R^{k(z)+n(z)}(\tau A). \quad (3.17)$$

Using (3.4), we obtain for T_2 the bound

$$T_2 = \tau^2\|AR(\tau A)AR(\tau A)\|_{E\to E} \leq M^2.$$

To estimate T_3, we use (3.6). We get

$$T_3 = \int_{(k(z)+n(z))\tau \geq 3\tau} \tau\|A^2 R^{k(z)+n(z)}(\tau A)\|_{E\to E}dz \leq \sum_{r\geq3} \frac{M^2}{(r - 2)^2} \leq \frac{3}{2}M^2.$$

Thus, Q admits a bound that does not depend on τ

The case $m = -1$ is dealt with in a similar manner.

In the case 3) let us put $M_1 \geq 16$. Since $|j| \geq 1$, proceeding as in case 2) we infer from the definition of the kernel $\tilde{B}$ that

$$P \leq \int_{k(z)\tau \geq 4|j|\tau} \|AR^{k(z)+n(z)-j-m} - AR^{k(z)+n(z)}(\tau A)\|_{E \to E} dz.$$

Let us estimate each term of this sum using (3.6) and (3.17). We have

$$\|AR^{k(z)+n(z)-j-m}(\tau A) - AR^{k(z)+n(z)}(\tau A)\|_{E \to E}$$

$$= \left\| \sum_{s=k(z)+n(z)-j-m}^{k(z)+n(z)} \frac{AR^{s+1}(\tau A) - AR^s(\tau A)}{\tau} \right\|_{E \to E}$$

$$= \left\| \sum_{s=k(z)+n(z)-j-m}^{k(z)+n(z)} A^2 R^{s+1}(\tau A)\tau \right\|_{E \to E}$$

$$\leq M^2 \sum_{s=k(z)+n(z)-j-m}^{k(z)+n(z)} \frac{\tau}{[(s-1)\tau]^2} \leq M^2 \frac{(|j|+|m|)\tau\tau^{-2}}{(k(z)+n(z)-|j|-|m|-1)^2}.$$

Therefore Q admits the bound

$$Q \leq M^2 \sum_{s>3|j|-2} \frac{(|j|+|m|)\tau}{(s\tau)^2}\tau \leq M^2 \int_{3|j|-2)\tau}^{\infty} \frac{(|j|+|m|)\tau}{s^2} ds$$

$$= M^2 \frac{(|j|+|m|)\tau}{(3|j|-2)\tau} = \frac{M^2}{3}\left(1 + \frac{3}{3|j|-2}\right) \leq \frac{4}{3}M^2.$$

Thus, all the conditions of Theorem 3.3 of Chapter 1 are satisfied, and so the operator B is bounded in $L_p^\infty(E)$ for $1 < p < p_0$.

Let us extend this result for arbitrary $p \in (1, \infty)$. Since B is bounded in $L_p^\infty(E)$, the conjugate operator B^* exists and acts in $(L_p^\infty(E))^* = L_q^\infty(E^*)$, $p^{-1} + q^{-1} = 1$. Let us find the explicit form of B^*. We denote by $\{x, y\}$ the value of the functional $y \in E^*$ on the element $x \in E$, and by $\langle f, g \rangle$ the value of the functional $g \in L_q^\infty(E^*)$ on the element $f \in L_p^\infty(E)$. we have

$$\langle Bf, g \rangle = \int_{-\infty}^{\infty} \int_{-\infty}^{\infty} \left\{ \tilde{B}\left(\left[\frac{t}{\tau}\right] - \left[\frac{s}{\tau}\right]\right) \frac{1}{\tau} \int_{s_{k-1}}^{s_k} f(z)dz, g(t) \right\} dsdt$$

$$= \sum_{n=-\infty}^{\infty} \int_{t_{n-1}}^{t_n} \left\{ \sum_{k=-\infty}^{\infty} \int_{z_{k-1}}^{z_k} (f(z), (\tilde{B}(n(t) - k(s)))^* g(t) \right\} dzdt$$

$$= \int_{-\infty}^{\infty} \int_{-\infty}^{\infty} \left\{ f(z), \left(\tilde{B} \left(\left[\frac{t}{\tau} \right] - \left[\frac{s}{\tau} \right] \right) \right)^* \frac{1}{\tau} \int_{s_{n-1}}^{s_n} g(t)dt \right\} dsdz = \langle f, B^*g \rangle.$$

Consequently,

$$B^*g(s) = \int_{-\infty}^{\infty} \tilde{B} \left(\left[\frac{t}{\tau} \right] - \left[\frac{s}{\tau} \right] \right)^* W_\tau g(t)dt.$$

Since the operator $\tilde{B}^*$ obviously admits the same estimates as $\tilde{B}$, it follows that B^* is bounded in $L_{q_0}^\infty(E^*)$, where $q_0 = p_0/(p_0 - 1)$. Hence, it is bounded in $L_q^\infty(E^*)$ for any $q \in (1, p_0/(p_0 - 1))$, which in turn implies that B is bounded in $L_p^\infty(E)$ for any $p = q/(q - 1) \in (p_0, \infty)$ Therefore, B is bounded in $L_p^\infty(E)$ for any $p \in (1, \infty)$.

Thus, we have shown that

$$\|B\|_{L_p^\infty(E) \to L_p^\infty(E)} \leq M(p),$$

where $M(p) = \frac{M_1(p_0)}{(p-1)(p_0-p)}$ if $1 < p < p_0$, and $M(p) = \frac{M_2(p_0)p^2}{(p-p_0)}$ if $p_0 < p < \infty$. From this it is readily seen that $M(p)$ has the form $M(p) = \frac{M(p_0)p^2}{(p-1)}$. The proof of Theorem 3.4 is complete.

Let us consider the operator G generated by the solution of the difference problem (0.4) with null initial condition, acting on the space $L_p(\tau, E)$ of grid functions as

$$G\varphi^\tau = \left\{ \tau \sum_{j=1}^k AR^{k+1-j}(\tau A)\varphi_j \right\}_1^N. \tag{3.18}$$

We will show next that the boundedness (uniform in τ) of G is equivalent to the boundedness of the operator B. This will allow us to establish the well-posedness of the general difference problem (0.6).

Theorem 3.5. *The operator G is bounded in $L_p(\tau, E)$ if and only if the operator B is bounded in $L_p^\infty(E)$.*

Proof. Using the definition (3.13) of the kernel $\tilde{B}$, we represent B as the sum

$$Bf(t) = \int_{-\infty}^{[\frac{t}{\tau}]\tau} \tilde{B} \left(\left[\frac{t}{\tau} \right] - \left[\frac{s}{\tau} \right] \right) W_\tau f(s)ds$$

$$= \int_{[\frac{t}{\tau}]\tau-1}^{[\frac{t}{\tau}]\tau} \tilde{B} \left(\left[\frac{t}{\tau} \right] - \left[\frac{s}{\tau} \right] \right) W_\tau f(s)ds + \int_{-\infty}^{[\frac{t}{\tau}]\tau-1} \tilde{B} \left(\left[\frac{t}{\tau} \right] - \left[\frac{s}{\tau} \right] \right) W_\tau f(s)ds$$

$$= B_1 f(t) + B_2 f(t).$$

The operator B_2 is bounded in $L_p^\infty(E)$ for any $1 < p < \infty$. Indeed, by the definition of the kernel $\tilde{B}$, we have

$$B_2 f(t) = \sum_{j=-\infty}^{k(t)-1-N} A R^{k(t)-j}(\tau A)(W_\tau f)_j \tau,$$

where, as before, we let $k(t) = [t/\tau]$. Applying inequality (3.4), we obtain

$$\|B_2 f(t)\|_E \leq \sum_{j=-\infty}^{k(t)-1-N} \frac{M\|(W_\tau f)_j\|_E \tau}{(1+\delta\tau)^{k(t)-1-j}(k(t)-1-j)\tau}$$

$$\leq M\tau \sum_{j=-\infty}^{k(t)-1-N} \frac{\|(W_\tau f)_j\|_E}{(1+\delta\tau)^{k(t)-1-j}} = M\tau \sum_{z=\infty}^{N} \frac{\|(W_\tau f)_{k(t)-1-z}\|_E}{(1+\delta\tau)^z}.$$

Next, using the Minkowski inequality and the boundedness of the operator W_τ, we obtain for $\|B_2 f\|_{L_p^\infty(E))}$ the bound

$$\|B_2 f\|_{L_p^\infty(E)} \leq \left(\sum_{k(t)=-\infty}^{\infty} \left(M\tau \sum_{z=\infty}^{N} \frac{\|(W_\tau f)_{k(z)-1-z}\|_E}{(1+\delta\tau)^z} \right)^p \tau \right)^{1/p}$$

$$\leq M\tau \sum_{z=\infty}^{N} \frac{1}{(1+\delta\tau)^z} \left(\sum_{k(t)=-\infty}^{\infty} \|(W_\tau f)_{k(t)-z-1}\|_E^p \tau \right)^{1/p}$$

$$= M\tau \frac{1}{\delta\tau}(1+\delta\tau)^{-N+1}\|W_\tau f\|_{L_p^\infty(E)} \leq \frac{M}{\delta}\|f\|_{L_p^\infty(E)}.$$

Thus, the problem of the boundedness of the operator B reduces to that of the boundedness of the operator B_1.

Suppose the operator G is bounded. Since B_1 maps $L_p^\infty(E)$ into the set of step functions, we can write

$$\|B_1 f\|_{L_p^\infty(E)} = \left(\sum_{m=-\infty}^{\infty} \sum_{k(t)=m+1}^{m+N} \|(B_1 f)_{k(t)}\|_E^p \tau \right)^{1/p}.$$

Using the definition (3.13) of the kernel $\tilde{B}$ and the triangle inequality, we obtain first the estimate

$$\left(\sum_{m=-\infty}^{\infty} \sum_{k(t)=m+1}^{m+N} \|(B_1 f)_{k(t)}\|_E^p \tau \right)^{1/p}$$

$$= \left(\sum_{m=-\infty}^{\infty} \sum_{s=1}^{N} \| \sum_{z=s-1-N}^{s-1} AR^{s-z}(\tau A)(W_\tau f)_{z+m}\tau \|_E^p \tau \right)^{1/p}$$

$$\leq \left(\sum_{m=-\infty}^{\infty} \sum_{s=1}^{N} \| \sum_{z=0}^{s-1} AR^{s-z}(\tau A)(W_\tau f)_{z+m}\tau \|_E^p \tau \right)^{1/p}$$

$$+ \left(\sum_{m=-\infty}^{\infty} \sum_{s=1}^{N} \| \sum_{z=s-1-N}^{-1} AR^{s-z}(\tau A)(W_\tau f)_{z+m}\tau \|_E^p \tau \right)^{1/p} = J_1 + J_2.$$

Since by assumption the operator G is bounded and its norm does not depend on τ,

$$J_1 \leq M_1(p) \left(\sum_{m=-\infty}^{\infty} \sum_{k(z)=m}^{m+N-1} \|(W_\tau f)_{k(t)}\|_E^p \tau \right)^{1/p},$$

where $M_1(p)$ does not depend on τ.

To estimate J_2 we use (3.4), which yields

$$\| \sum_{z=t-1-N}^{-1} AR^{t-z}(\tau A)(W_\tau f)_{z+m}\tau \|_E$$

$$\leq \sum_{z=t-1-N}^{-1} \frac{M\|(W_\tau f)_{z+m}\|_E}{(1+\delta\tau)^{t-z-1}(t-z-1)} \leq \sum_{s=0}^{N} \frac{M\|(W_\tau f)_{m-s-1}\|_E}{t+s}.$$

The last expression is a difference analogue of Hilbert's operator. Using its boundedness in the spaces $L_p(\tau, E)$, $1 < p < \infty$, of grid functions, we deduce that

$$J_2 \leq \left(M^p \sum_{m=-\infty}^{\infty} \sum_{t=1}^{N} \left(\sum_{s=0}^{N} \frac{\|(W_\tau f)_{m-1-s}\|}{t+s} \right)^p \tau \right)^{1/p}$$

$$\leq \frac{Mp^2}{p-1} \left(\sum_{m=-\infty}^{\infty} \sum_{k(t)=m}^{m-1-N} \|(W_\tau f)_{k(t)}\|_E^p \tau \right)^{1/p}.$$

Using the boundedness of the operator W_τ we obtain the estimate

$$\|B_1 f\|_{L_p^\infty(E)} \leq M(p)\|W_\tau f\|_{L_p^\infty(E)} \leq M(p)\|f\|_{L_p^\infty(E)},$$

where $M(p) = M_1(p) + Mp^2/(p-1)$, i.e., the operator B_1, as well as the operator B, is bounded in $L_p^\infty(E)$.

Now assume that B is bounded. Pick grid functions $\varphi^\tau = \{\varphi_k\}_1^N$, and extend them by zero to grid functions defined on the entire real line. From these grid functions construct step functions $\tilde{\varphi}^\tau$, defined on the real axis, by the rule

$$\tilde{\varphi}^\tau(t) = (\varphi_j, \ t_{j-1} < t \le t_j, \ t_j = j\tau, \ j = 0, \pm 1, \pm 2, \cdots).$$

Clearly, we have

$$\|\varphi^\tau\|_{L_p(\tau, E)} = \|\tilde{\varphi}^\tau\|_{L_p^\infty(E)}.$$

Moreover, for such functions we have the inequality

$$\|G\varphi^\tau\|_{L_p(\tau, E)} \le \|B\tilde{\varphi}^\tau\|_{L_p^\infty(E)}.$$

Therefore,

$$\|G\varphi^\tau\|_{L_p(\tau, E)} \le M\|\tilde{\varphi}^\tau\|_{L_p^\infty(E)} = M\|\varphi^\tau\|_{L_p(\tau, E)},$$

where $M = \|B\|_{L_p(\tau, E) \to L_p(\tau, E)}$. This means that the operator G is bounded in $L_p(\tau, E)$, and its norm does not exceed the norm of the operator B in $L_p^\infty(E)$. Theorem 3.5 is proved.

We have the following consequence.

Theorem 3.6. *If the operator $\overline{A}u(\varphi, 0)$ is bounded in $\mathcal{L}_{p_0}(E)$ for some p_0, $1 < p_0 < \infty$, then it is bounded in $\mathcal{L}_p(E)$ for any p, $1 < p < \infty$.*

Proof. From the boundedness of the operator $\overline{A}u(\varphi, 0)$ in $\mathcal{L}_{p_0}(E)$ and formula (0.2) it follows that the operator G, defined by (3.15), is bounded in $L_{p_0}(\tau, E)$, uniformly in $\tau \in (0, \tau_0]$. Then, by Theorem 3.5, the operator B is bounded in $L_{p_0}^\infty(E)$. Hence, by Theorem 3.4, B is bounded in $L_p^\infty(E)$ for any p, $1 < p < \infty$. Again using Theorem 3.5, we conclude that the operator G is bounded in $L_p(\tau, E)$, uniformly in τ, and hence $\overline{A}u(\varphi, 0)$ is bounded in $\mathcal{L}_p(E)$ for any p, $1 < p < \infty$. Theorem 3.6 is proved.

Theorem 3.7. *Let the difference problem (0.6) be well posed in $\mathcal{L}_{p_0}(E)$ for some p_0, $1 < p_0 < \infty$. Then it is well posed in $\mathcal{L}_p(E)$ for any p, $1 < p < \infty$.*

Proof. As we remarked earlier, the difference problem (0.6) is well posed in $\mathcal{L}_p(E)$ if and only if its solutions satisfy the coercivity inequality (3.11). From the explicit form (0.2) of the solutions of the difference problems and Theorem 3.3 it follows that the coercivity inequality (3.11) holds if and only the operator $\overline{A}u(\varphi, 0)$ is bounded in $\mathcal{L}_p(E)$. Now the assertion of the theorem follows from Theorem 3.6.

As we have shown in Subsection 3, the estimate of the norm

$$\sup_{0<\tau\leq\tau_0} \max_{1\leq k\leq N} |u_k|_{1-\frac{1}{p}}$$

of a solution of problem (0.6) in terms of the right-hand side and the initial condition can be established for arbitrary operators that generate analytic semigroups. In contrast, the full inequality (3.11) in the space $\mathcal{L}_p(E)$ can be established under the assumption that it holds in $\mathcal{L}_{p_0}(E)$ for some $p_0, 1 < p_0 < \infty$. One is therefore led to asking when this last condition is satisfied. This can be answered in the case in which $E = H$ is a Hilbert space.

Theorem 3.8. *The difference problem* (0.6) *is well posed in* $\mathcal{L}_p(H)$.

Proof. As shown in Theorems 3.7 and 3.6, to prove this assertion it suffices to establish the boundedness in the space $L_2^\infty(H)$ of the operator B introduced in Subsection 3. To this end we shall regard the set of values of a grid function from $L_2^\infty(H)$, $\tilde{\varphi}^\tau = (\varphi_k, (k-1)\tau < t \leq k\tau, k = 0, \pm 1, \cdots)$, as the set of Fourier coefficients of some function $\hat{\varphi}(t) \in L_2(H)$, which by the Parseval equality can be represented by a Fourier series

$$\hat{\varphi}(t) \sim \sum_{k=-\infty}^{\infty} \varphi_k \exp(2\pi i k t).$$

From the definition (3.14) of the operator B it follows that its value on a function $f(t) \in L_2^\infty(H)$ is a step function of the form

$$Bf(t) = (\widetilde{Bf})^\tau(t) = ((Bf)_k, (k-1)\tau < t \leq k\tau, k = 0, \pm 1, \cdots). \qquad (3.19)$$

Let us denote the corresponding function from $L_2(H)$ by $(\widehat{Bf})(t)$. Using the explicit form of the operator B, we obtain

$$(\widehat{Bf})(t) = \sum_{k=-\infty}^{\infty} (Bf)_k \exp(2\pi i k t) = \sum_{k=0}^{\infty} \sum_{j=-\infty}^{\infty} \tilde{B}(k-j)(W_\tau f)_j \tau \exp(2\pi i k t)$$

$$= \left(\sum_{s=-\infty}^{\infty} \tilde{B}(s) \exp(2\pi i s t) \right) \left(\sum_{j=-\infty}^{\infty} (W_\tau f)_j \exp(2\pi i j t) \tau \right).$$

Let

$$\hat{B}(t) = \sum_{s=-\infty}^{\infty} \tilde{B}(s) \exp(2\pi i s t), \quad (\widehat{W_\tau f})(t) = \sum_{j=-\infty}^{\infty} (W_\tau f)_j \exp(2\pi i j t).$$

Then, using the definition (3.13) of the kernel $\tilde{B}$, we deduce that

$$\hat{B}(t) = \sum_{s=1}^{\infty} A R^s(\tau A) \exp(2\pi i t) = A \sum_{s=1}^{\infty} [(I + \tau A)^{-1} \exp(2\pi i t)]^s$$

$$= A \exp(2\pi i t)[(1 - \exp(2\pi i t)) + \tau A]^{-1}$$

$$= \tau^{-1} A \exp(2\pi i t)[N(1 - \exp(2\pi i t)) + A]^{-1}. \tag{3.20}$$

Applying (3.19), (3.20), and the Parseval equality, we first obtain for $\|Bf\|_{L_2^\infty(H)}$ the estimate

$$\|Bf\|_{L_2^\infty(H)}^2 = \|(\widehat{Bf})^\tau\|_{L_2^\infty(H)}^2 = 2\pi\tau \|\widehat{Bf}\|_{L_2(H)}^2 \leq$$

$$2\pi\tau \max_{0 \leq t \leq 1} \|\tau^{-1} A \exp(2\pi i t)[N(1 - \exp(2\pi i t)) + A]^{-1}\|_{H \to H}^2 \|\hat{W}_\tau f\|^2 \|_{L_2(H)}. \tag{3.21}$$

Further, since the operator $-A$ is the generator of an analytic semigroup, we have that

$$\tau^{-1} \max_{0 \leq t \leq 1} \|A \exp(2\pi i t)[N(1 - \exp(2\pi i t)) + A]^{-1}\|_{H \to H} \leq M_0 < \infty,$$

where M_0 does not depend on τ. Again using the Parseval equality, and the boundedness of the operator W_τ, we derive from (3.21) the estimate

$$\|Bf\|_{L_2^\infty(H)}^2 \leq 2\pi\tau M_0 \|\widehat{W_\tau f}\|_{L_2(H)}^2 = M_0 \|W_\tau f\|_{L_2^\infty(H)}^2 \leq M_0 \|f\|_{L_2^\infty(H)}^2.$$

Theorem 3.8 is proved.

Let us note that by passing to the limit from Theorem 3.7 we can derive Theorem 3.6 of Chapter 1 on the well-posedness of the differential Cauchy problem in $L_p(E)$.

From Theorem 3.7 one can also obtain Theorem 1.2. Indeed, by the definition of the norm in $L_p(\tau, E)$,

$$\tau^{1/p} \|\varphi^\tau\|_{C(\tau,E)} \leq \|\varphi^\tau\|_{L_p(\tau,E)} \leq \|\varphi^\tau\|_{C(\tau,E)}. \tag{3.22}$$

Consequently, Theorem 3.7 yields the inequality

$$\|\mathcal{D}u^\tau\|_{C(\tau,E)} + \|Au^\tau\|_{C(\tau,E)} \leq \frac{M}{\tau^{1/p}}\left[\frac{p^2}{p-1}\|\varphi^\tau\|_{C(\tau,E)} + \|Au_0\|_E\right]$$

for all $p \in (1,\infty)$. Taking the minimum of the right-hand side over all p, we obtain inequality (1.10) for the solutions of the difference problem (0.6) in the case of discretization with respect to time only.

Note that the constant $M(p)$ figuring in the coercivity inequality (3.11) admits the estimate

$$M(p) = \left\|\left\{\sum_{j=1}^{k} AR^{k-j+1}(\tau A)\varphi_j\tau\right\}_1^N\right\|_{L_p(\tau,E)\to L_p(\tau,E)} \leq \frac{M(p_0)p^2}{p-1}.$$

We have not been able to find a sharp estimate for $M(p)$.

4. WELL-POSEDNESS OF THE DIFFERENCE PROBLEM IN $\mathcal{L}_p(E_{\alpha,q})$

1. Strongly positive operators and fractional spaces.

Let E be an arbitrary Banach space and A be a linear operator in E with dense domain $D(A)$.

Definition 4.1. The operator A is said to be *strongly positive* if its spectrum $\sigma(A)$ lies in the interior of the sector of angle ϕ, $0 < 2\phi < \pi$, symmetric with respect to the real axis, and if on the edges of this sector, $S_1(\phi) = \{\rho e^{i\phi} : 0 \leq \rho < \infty\}$ and $S_2(\phi) = \{\rho e^{-i\phi} : 0 \leq \rho < \infty\}$, and outside it the resolvent $(\lambda - A)^{-1}$ is subject to the bound

$$\|(\lambda - A)^{-1}\|_{E\to E} \leq \frac{M(\phi)}{1 + |\lambda|}. \tag{4.1}$$

The infimum of all such angles ϕ is called the *spectral angle* of the strongly positive operator A and is denoted by $\phi(A) = \phi(A, E)$. Since the spectrum $\sigma(A)$ is a closed set, it lies inside the sector formed by the rays $S_1(\phi(A))$ and $S_2(\phi(A))$, and some neighborhood of the apex of this sector does not intersect $\sigma(A)$. We shall consider contours $\Gamma = \Gamma(\phi, r)$ composed by the rays $S_1(\phi)$, $S_2(\phi)$ and an arc of circle of radius r centered at the origin; ϕ and r will be chosen so that $\phi(A) < |\phi| < \pi/2$ and the arc of circle of radius r lies in the resolvent set $\rho(A)$ of the operator A.

Let $f(z)$ be an analytic function on the set bounded by such a contour Γ, and suppose that f satisfies the estimate

$$|f(z)| \leq M|z|^{-\varepsilon}$$

for some $\varepsilon > 0$. Then the operator Cauchy-Riesz integral

$$f(A) = \frac{1}{2\pi i} \int_\Gamma f(z)(z - A)^{-1} dz \qquad (4.2)$$

converges in the operator norm and defines a bounded linear operator $f(A)$, a function of the strongly positive operator A. If $f(z)$ is continuous in a neighborhood of the origin, then in (4.2) we shall consider that $r = 0$, i.e., $\Gamma = S_1(\phi) \cup S_2(\phi)$.

As in the case of a bounded operator A one shows that $f(A)$ does not depend on the choice of the contour Γ in the domain of analyticity of the function $f(z)$, and that the correspondence between the function $f(z)$ and the operator $f(A)$ is linear and multiplicative.

The function $f(z) = z^{-\alpha}$ defines a bounded operator $A^{-\alpha}$ whenever $\alpha > 0$. Here the contour Γ is chosen with $r > 0$. By the multiplicativity property, $A^{-(\alpha+\beta)} = A^{-\alpha}A^{-\beta} = A^{-\beta}A^{-\alpha}$ for any powers of the strongly positive operator A, and not only for negative integer ones. From this identity it follows (when $\alpha + \beta$ is an integer) that the equation $A^{-\alpha}x = 0$ has the unique solution $x = 0$. Hence, the positive powers $A^{\alpha} = (A^{-\alpha})^{-1}$ of the strongly positive operator are defined. The operators A^{α} ($\alpha > 0$) are unbounded if A is unbounded; they have dense domains $D(A^{\alpha})$ and one has the continuous embeddings $D(A^{\alpha}) \subset D(A^{\beta})$ if $\beta < \alpha$.

The theory of fractional powers of operators can be constructed for a wider class of positive operators. For such operators the estimate 4.1 is required to hold for some ϕ and not only from the interval $[0, \pi/2]$, but from the larger interval $[0, \pi)$.

Now let us consider the function $f(z) = e^{-tz}$. For any $t > 0$ this function tends to zero faster than any power $z^{-\alpha}$ as $|z| \to \infty$ and its values lie inside any sector bounded by a contour Γ. Therefore, formula (4.2) can be used to define the function $\exp\{-tA\}$ of the strongly positive operator A. By multiplicativity, the semigroup property holds:

$$\exp\{-(t_1 + t_2)A\} = \exp\{-t_1 A\}\exp\{-t_2 A\}, \quad t_1, t_2 > 0.$$

Consider the function $\Psi(z) = z^\alpha e^{-tz}$ for some $\alpha > 0$ and $t > 0$. Since, obviously, $\Psi(z) \to 0$ faster than any negative power of z as $|z| \to \infty$, $\Psi(z)$ defines the operator function

$$\Psi(A) = \frac{1}{2\pi i}\int_\Gamma z^\alpha e^{-tz}(z - A)^{-1}dz. \tag{4.3}$$

Let us show that the operator $\exp\{-tA\}$ maps E into $D(A^\alpha)$ and $A^\alpha \exp\{-tA\} = \Psi(A)$. Let x be an arbitrary element of E. By the multiplicativity property, (4.3) implies that

$$A^{-\alpha}\Psi(A)x = \frac{1}{2\pi i}\int_\Gamma e^{-tz}(z - A)^{-1}xdz = \exp\{-tA\}x,$$

which proves our assertion. Thus, we have the formula

$$A^\alpha \exp\{-tA\} = \frac{1}{2\pi i}\int_\Gamma z^\alpha e^{-tz}(z - A)^{-1}dz. \tag{4.4}$$

In the above argument we must assume that the contour Γ contains an arc of radius r, since we applied the operator $A^{-\alpha}$, which corresponds to the function $z^{-\alpha}$. The final formula (4.4) is valid for any (small) $r > 0$. Since the integrand in (4.4) is continuous at the point $z = 0$, letting $z \to 0$ we obtain the formula

$$A^\alpha \exp\{-tA\} = \frac{1}{2\pi i}\left[\int_\infty^0 \rho^\alpha e^{i\alpha\phi}e^{-t\rho e^{i\phi}}(\rho e^{i\phi} - A)^{-1}d\rho\right.$$

$$\left. + \int_0^\infty \rho^\alpha e^{-i\alpha\phi}e^{-t\rho e^{-i\phi}}(\rho e^{-i\phi} - A)^{-1}d\rho\right]$$

for some $0 < \phi < \pi/2$. From this and the estimate (4.1) it follows that

$$\|A^\alpha \exp\{-tA\}\|_{E\to E} \leq \frac{M(\phi)}{\pi}\int_0^\infty \rho^{\alpha-1}e^{-t\rho\cos\phi}d\rho = \frac{M(\phi)\Gamma(\alpha)}{\pi(\cos\phi)^\alpha}t^{-\alpha}. \tag{4.5}$$

In particular, we have the estimate

$$\|\exp\{-tA\}\|_{E\to E} \leq \frac{M(\phi)}{\pi}. \tag{4.6}$$

Let us show that the estimate (4.5) can be sharpened by a factor that decays exponentially when $t \to +\infty$.

Let A be a strongly positive operator. We claim that for sufficiently small $\delta > 0$ the operator $A - \delta$ is also strongly positive, and $\phi(A - \delta) = \phi(A)$. Indeed, let $\lambda \in \Gamma(\phi)$. Consider the equation $\lambda x - (A - \delta)x = y$ for an arbitrary $y \in E$. The substitution $\lambda x - Ax = z$ yields the equation $z + \delta(\lambda - A)^{-1}z = y$. Since $\|\delta(\lambda - A)^{-1}\|_{E \to E} \leq \delta M(\phi)$ if $\lambda \in \Gamma(\phi)$, we see that for $\delta \leq [2M(\phi)]^{-1}$ the equation for z has a unique solution, and $\|z\| \leq 2\|y\|$. Consequently, the equation for x has a unique solution, and $\|x\| \leq M(\phi)[|\lambda| + 1]^{-1}\|z\| \leq 2M(\phi)[|\lambda| + 1]^{-1}\|y\|$. This means that the operator $\lambda - (A + \delta)$ has a bounded inverse for $0 < \delta \leq [2M(\phi)]^{-1}$ and

$$\|[\lambda - (A - \delta)]^{-1}\|_{E \to E} \leq 2M(\phi)[|\lambda| + 1]^{-1}.$$

Thus, we have shown that $A - \delta$ is a strongly positive operator. Hence, by (4.6), we have the estimate

$$\|\exp\{-(A - \delta)t\}\|_{E \to E} \leq \frac{2M(\phi)}{\pi}.$$

This obviously yields

$$\|\exp\{-tA\}\|_{E \to E} \leq \frac{2M(\phi)}{\pi}e^{-\delta t}, \tag{4.7}$$

where we can put $\delta = [2M(\phi)]^{-1}$.

Let $t > 1$. Then, using the semigroup property, we can write

$$\exp\{-tA\} = \exp\{-A\}\exp\{-(t - 1)A\}.$$

Next, applying the estimates (4.5) with $t = 1$ and (4.7), we obtain

$$\|A^\alpha \exp\{-tA\}\|_{E \to E} \leq \frac{M(\phi)}{\pi(\cos\phi)^\alpha} \frac{2M(\phi)}{\pi} e^{-\delta(t-1)}.$$

Hence, the following estimate holds for $t > 1$:

$$\|A^\alpha \exp\{-tA\}\|_{E \to E} \leq M_1(\phi)e^{-\delta t}.$$

If $0 < t \leq 1$, then estimate (4.5) prevails. Combining these two estimates, we conclude that

$$\|A^\alpha \exp\{-tA\}\|_{E \to E} \leq \tilde{M}(\phi)e^{-\delta t}t^{-\alpha} \tag{4.8}$$

for some $\tilde{M}(\phi) > 0$ and $\delta > 0$.

Further, formula (4.2) allows us to establish that the operator-valued function $\exp\{-tA\}$ is differentiable in the operator norm for $t > 0$ and

$$\frac{d}{dt}\exp\{-tA\} = -A\exp\{-tA\}. \tag{4.9}$$

In particular, this implies that $\exp\{-tA\}$ is continuous in the operator norm. Using the semigroup property we deduce that the derivative of $\exp\{-tA\}$ is also continuous in the operator norm for $t > 0$. Finally, formula (4.9) shows that the operator-valued function $\exp\{-tA\}$ has derivatives of arbitrary order in the operator norm for $t > 0$.

Now let $x \in D(A)$. Then the (E-valued) function $\exp\{-tA\}x$ has a derivative for $t > 0$ and, by (4.9),

$$\frac{d}{dt}\exp\{-tA\}x = -\exp\{-tA\}Ax. \tag{4.10}$$

Next, for x as above we can write

$$(z - A)^{-1}x = z^{-1}x + z^{-1}(z - A)^{-1}Ax.$$

Using formula (4.2), we obtain

$$\exp\{-tA\}x = \frac{1}{2\pi i}\int_\Gamma e^{-tz}[z^{-1}x + z^{-1}(z - A)^{-1}Ax]dz.$$

Here the contour Γ has the form

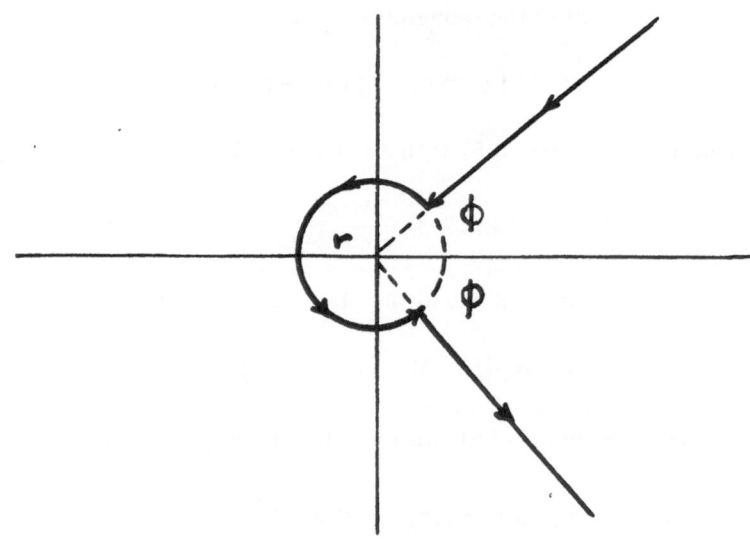

Using the Cauchy theorem, we get

$$\exp\{-tA\}x = \frac{1}{2\pi i}\int_\Gamma e^{-tz}z^{-1}(z-A)^{-1}Axdz + x.$$

The estimate (4.1) shows that in the last equality one can pass to the limit under the integral sign when $t \to +0$. Hence, the limit

$$\lim_{t\to+0}\exp\{-tA\}x = x + \frac{1}{2\pi i}\int_\Gamma z^{-1}(z-A)^{-1}Axdz$$

exists (in the norm of E). By Cauchy's theorem, the integral

$$\mathcal{J} = \frac{1}{2\pi i}\int_\Gamma z^{-1}(z-A)^{-1}Axdz = \frac{1}{2\pi i}\int_{-\sigma-i\infty}^{-\sigma+i\infty} z^{-1}(z-A)^{-1}Axdz$$

for some $\sigma > 0$. Hence, by (4.1),

$$\|\mathcal{J}\|_E \leq \frac{M}{2\pi}\int_{-\infty}^{\infty}\frac{dt}{\sigma^2+t^2}\|Ax\|_E.$$

Since $\mathcal{J}$ does not depend on σ, it follows that $\mathcal{J} \equiv 0$. Hence, we proved that

$$\lim_{t\to+0}\exp\{-tA\}x = x \tag{4.11}$$

for any $x \in D(A)$. Since the norm $\|\exp\{-tA\}\|_{E\to E}$ is uniformly bounded for $t > 0$ (see (4.7)), the limit relation (4.11) holds for any $x \in E$.

Thus, if we extend the operator-valued function $U(t) = \exp\{-tA\}$, $t > 0$, at $t = 0$ by $U(0) = I$, we obtain a strongly continuous semigroup. From the estimate (4.8) (with $\alpha = 0$) it follows that this semigroup is analytic. Finally, let us show that its generator is $U'(0) = -A$. From (4.10) and the estimate (4.7) we derive the identity

$$U(t)x - x = -\int_0^t U(s)Axds$$

for $x \in D(A)$. Since $U(t)$ is strongly continuous to the left at the point $t = 0$, this implies that $x \in D(U'(0))$ and $U'(0)x = -Ax$. Hence, $U'(0)$ is an extension of the operator $-A$. By the estimate (4.7), the operator $U'(0) + \lambda$ and $-A + \lambda$ have bounded inverses for any $\lambda < 0$. Therefore, $U'(0) = -A$.

We have shown that the operator-valued function $\exp\{-tA\}$ is an analytic semigroup with generator $-A$ and with an exponentially decaying norm. In Chapter 1 operators $-A$ that generate such semigroups were called strongly positive operators. In this subsection we gave a sufficient condition for the strong positivity of A in terms of the resolvent of $-A$. One can show that this condition is also necessary.

Let A be a strongly positive operator. With the help of A we introduce the fractional spaces $E'_{\alpha,q}(E, A)$, $0 < \alpha < 1$, consisting of all $v \in E$ for which the following norms are finite:

$$\|v\|'_{\alpha,q} = \left(\int_0^\infty \|\lambda^\alpha A(\lambda + A)^{-1} v\|_E^q \frac{d\lambda}{\lambda} \right)^{1/q}, \quad 1 \leq q < \infty,$$

$$\|v\|'_\alpha = \|v\|'_{\alpha,\infty} = \sup_{\lambda > 0} \|\lambda^\alpha A(\lambda + A)^{-1} v\|_E.$$

Recall that in Section 4 of Chapter 1 we introduced the fractional spaces $E_{\alpha,q}$ in which the norms are defined with the help of the semigroup $\exp\{-tA\}$ generated by a strongly positive operator A.

Theorem 4.1. $E'_{\alpha,q} = E_{\alpha,q}$ for all $0 < \alpha < 1$ and $1 \leq q \leq \infty$.

Proof. From (1.10) it follows that

$$\lambda^\alpha A(\lambda + A)^{-1} = \int_0^\infty \lambda^\alpha e^{-\lambda t} A \exp\{-tA\} dt.$$

Let $v \in E_{\alpha,\infty}$. Then

$$\|\lambda^\alpha A(\lambda + A)^{-1} v\|_E \leq \int_0^\infty \frac{\lambda^\alpha e^{-\lambda t}}{t^{1-\alpha}} dt \|v\|_{\alpha,\infty}.$$

Since

$$\int_0^\infty \frac{\lambda^\alpha e^{-\lambda t}}{t^{1-\alpha}} dt = \int_0^\infty \frac{e^{-\tau} dt}{\tau^{1-\alpha}} = \Gamma(\alpha) \leq \frac{M}{\alpha},$$

we have

$$\|\lambda^\alpha A(\lambda + A)^{-1} v\|_E \leq \frac{M}{\alpha} \|v\|_{\alpha,\infty}$$

for all $\lambda > 0$. This shows that $v \in E'_{\alpha,\infty}$ and

$$\|v\|'_{\alpha,\infty} \leq \frac{M}{\alpha} \|v\|_{\alpha,\infty}.$$

Next, using the Cauchy-Riesz representation formula (4.2) for the operator-valued function $\exp\{-tA\}$, we obtain

$$t^{1-\alpha} A \exp\{-tA\} = \frac{1}{2\pi i} \int_{S_1 \cup S_2} t^{1-\alpha} e^{-tz} A(z - A)^{-1} dz.$$

Now let $v \in E'_{\alpha,\infty}$. Using the fact that $z = \rho e^{\pm i\phi}$, with $|\phi| \leq \pi/2$, and the estimate (4.1), we deduce that

$$\|t^{1-\alpha} A \exp\{-tA\} v\|_E \leq M \int_0^\infty t^{1-\alpha} e^{-tp\cos\phi} \|A(p+A)^{-1} v\|_E dp$$

$$\leq M \int_0^\infty t^{1-\alpha} e^{-tp\cos\phi} p^{-\alpha} dp \|v\|'_{\alpha,\infty}.$$

Since

$$\int_0^\infty t^{1-\alpha} e^{-tp\cos\phi} p^{-\alpha} dp = \int_0^\infty \frac{e^{-\tau\cos\phi} dt}{\tau^\alpha} = \frac{1}{(\cos\phi)^{1-\alpha}} \int_0^\infty s^{(1-\alpha)-1} e^{-s} ds$$

$$= \frac{\Gamma(1-\alpha)}{(\cos\phi)^{1-\alpha}} \leq \frac{M(\phi)}{1-\alpha},$$

it follows that

$$\|t^{1-\alpha} A \exp\{-tA\} v\|_E \leq \frac{M(\phi)}{1-\alpha} \|v\|'_{\alpha,\infty}$$

for all $t > 0$. This shows that $v \in E_{\alpha,\infty}$ and

$$\|v\|_{\alpha,\infty} \leq \frac{M(\phi)}{1-\alpha} \|v\|'_{\alpha,\infty}.$$

Thus, we have established the inequalities

$$(1-\alpha)m(\phi)\|v\|_{\alpha,\infty} \leq \|v\|'_{\alpha,\infty} \leq \frac{M(\phi)}{\alpha} \|v\|_{\alpha,\infty}, \quad 0 < \alpha < 1, \tag{4.12}$$

which completes the proof of the theorem for the case $q = \infty$. In the case $q \neq \infty$ the proof is based on Minkowski's inequality and follows the same scheme. Theorem 4.1 is proved.

Let us note that the fractional spaces $E'_{\alpha,q}$ can be defined for a larger class of positive operators, since the definition involves the resolvent $(\lambda + A)^{-1}$ only for $\lambda \geq 0$.

2.   Well-posedness of the difference problem in $\mathcal{L}_p(E'_{\alpha,q})$.

In Section 3 we have shown that well-posedness in $\mathcal{L}_p(E)$ implies the strong analyticity of the semigroup $\exp\{-tA\}$ in E. At this time it is not known whether the analyticity of this semigroup is sufficient for the well-posedness of the problem (0.6) in $\mathcal{L}_p(E)$ for arbitrary E and A, as is the case for the differential problem.

In this subsection we restrict the Banach space E to the Banach space $E'_{\alpha,q}$, $0 < \alpha < 1$, $1 \le q \le \infty$, and we study the well-posedness of the difference problem (0.6) in $\mathcal{L}_q(E'_{\alpha,q})$. First we will examine the nonhomogeneous problem (0.6) with $u_0 = 0$.

Theorem 4.2. Let $\varphi \in \mathcal{L}_p(E'_{\alpha,q})$, where $0 < \alpha < 1$ and $1 \le q \le \infty$. Then the coercivity inequality holds:

$$\|\overline{\mathcal{D}}\,\overline{\Pi}(u_0)u\|_{\mathcal{L}_q(E'_{\alpha,q})} + \|\overline{A}u\|_{\mathcal{L}_q(E'_{\alpha,q})} \le \frac{M}{\alpha(1-\alpha)}\|\varphi\|_{\mathcal{L}_q(E'_{\alpha,q})}, \qquad (4.13)$$

where M does not depend on α, q, and φ.

Proof. First let us consider the case $q = \infty$. By formula (0.2),

$$u_k = \sum_{j=1}^{k} R^{k-j+1}(\tau A)\varphi_j\tau.$$

To estimate Au^τ in the norm of $L_q(\tau, E_{\alpha,q})$ we use the Cauchy-Riesz representation formula (4.2) for the operator $A(\lambda + A)^{-1}R^k(\tau A)$. We obtain

$$\lambda^\alpha A(\lambda + A)^{-1}Au_k = \frac{1}{2\pi i}\int_{S_1 \cup S_2}\sum_{j=1}^{k}\frac{z}{(1+z)^{k-j+1}}\frac{\lambda^\alpha}{\lambda + z\tau^{-1}}A(z - \tau A)^{-1}\varphi_j dz$$

$$= \frac{1}{2\pi i}\int_{S_1 \cup S_2}\sum_{j=1}^{k}\frac{\left(\frac{z}{\tau}\right)^{-\alpha}}{(1+z)^{k-j+1}}\frac{\lambda^\alpha}{\lambda\tau + z}\left(\frac{z}{\tau}\right)^\alpha A\left(\frac{z}{\tau} - A\right)^{-1}\varphi_j dz. \qquad (4.14)$$

Since $z = \rho e^{\pm i\phi}$, with $|\phi| \le \pi/2$, the estimate (4.1) yields

$$\left\|\left(\frac{z}{\tau}\right)^\alpha A\left(\frac{z}{\tau} - A\right)^{-1}\varphi_j\right\|_E \le M\left\|\left(\frac{\rho}{\tau}\right)^\alpha A\left(\frac{\rho}{\tau} + A\right)^{-1}\varphi_j\right\|_E, \quad \frac{1}{|\lambda\tau + z|} \le \frac{M}{\lambda\tau + \rho}. \qquad (4.15)$$

Hence,

$$\|\lambda^\alpha A(\lambda + A)^{-1}Au_k\|_E$$

$$\le M\int_0^\infty\sum_{j=1}^{k}\frac{\rho^{1-\alpha}}{[1 + 2\rho\cos\phi + \rho^2]^{\frac{k-j+1}{2}}}\frac{(\lambda\tau)^\alpha d\rho}{\lambda\tau + \rho}\|\varphi^\tau\|_{L_\infty(\tau, E'_{\alpha,\infty})}.$$

Summing the geometric progression, we get

$$\|\lambda^\alpha A(\lambda + A)^{-1} Au_k\|_E \leq M \int_0^\infty \frac{\rho^{1-\alpha}}{\sqrt{1 + 2\rho \cos\phi + \rho^2}} \times$$

$$\times \left[1 - \frac{1}{\sqrt{1 + 2\rho \cos\phi + \rho^2}} \right]^{-1} \frac{(\lambda\tau)^\alpha d\rho}{\lambda\tau + \rho} \|\varphi^\tau\|_{L_\infty(\tau, E'_{\alpha,\infty})}$$

$$\leq M \int_0^\infty \frac{(\tau\lambda)^\alpha \Psi(\rho) d\rho}{(\tau\lambda + \rho)\rho^\alpha} \|\varphi^\tau\|_{L_\infty(\tau, E'_\alpha)}.$$

Since the function

$$\Psi(\rho) = \frac{1 + \sqrt{1 + 2\rho \cos\phi + \rho^2}}{2 \cos\phi + \rho}$$

does not increase for $\rho \geq 0$, we have $\Psi(0) = 1/\cos\phi \geq \Psi(\rho)$ for all $\rho > 0$. Consequently,

$$\|\lambda^\alpha A(\lambda + A)^{-1} Au_k\|_E \leq \frac{M}{\cos\phi} \int_0^\infty \frac{(\tau\lambda)^\alpha d\rho}{\rho^\alpha(\tau\lambda + \rho)} \|\varphi^\tau\|_{L_\infty(\tau, E'_{\alpha,\infty})}$$

$$\leq \frac{M(\phi)}{\alpha(1 - \alpha)} \|\varphi^\tau\|_{L_\infty(\tau, E'_{\alpha,\infty})}$$

for any $k = 1, \cdots, N$. Therefore,

$$\|Au^\tau\|_{L_\infty(\tau, E'_{\alpha,\infty})} \leq \frac{M}{\alpha(1 - \alpha)} \|\varphi^\tau\|_{L_\infty(\tau, E'_{\alpha,\infty})}. \tag{4.16}$$

Since M does not depend on τ and $\varphi \in \mathcal{L}_\infty(E'_{\alpha,\infty})$, we have $\overline{\mathcal{D}}\,\overline{\Pi}(u_0)u, \overline{A}u \in \mathcal{L}_\infty(E'_{\alpha,\infty})$ and inequality (4.13) holds.

Now let $q \neq \infty$. Set $\varphi^*_j = \varphi_j$ if $j = 1, \cdots, N$, and $\varphi^*_j = 0$ otherwise. From (0.2) it follows that

$$Au_k = A \sum_{j=1}^k R^j(\tau A)\varphi_{k-j+1}\tau = A \sum_{j=1}^N R^j(\tau A)\varphi^*_{k-j+1}\tau. \tag{4.17}$$

Using this formula and the estimate (4.15), we obtain

$$\|\lambda^\alpha A(\lambda + A)^{-1} Au_k\|_E \leq M \int_0^\infty \sum_{j=1}^N \frac{\rho^{1-\alpha}}{[1 + 2\rho \cos\phi + \rho]^{j/2}} \frac{(\tau\lambda)^\alpha}{\tau\lambda + \rho} \times$$

$$\times \left\| \left(\frac{\rho}{\tau}\right)^\alpha A \left(\frac{\rho}{\tau} + A\right)^{-1} \varphi^*_{k-j+1} \right\|_E d\rho.$$

From the Minkowski sum inequality (with respect to k) it follows that

$$\left(\sum_{k=1}^{N}\|\lambda^\alpha A(\lambda+A)^{-1}Au_k\|_E^q\tau\right)^{1/q} \leq M\int_0^\infty\sum_{j=1}^{N}\frac{\rho^{1-\alpha}}{(1+2\rho\cos\phi+\rho^2)^{j/2}}\times$$

$$\times\frac{(\tau\lambda)^\alpha}{\tau\lambda+\rho}\left(\sum_{k=1}^{N}\tau\left\|\left(\frac{\rho}{\tau}\right)^\alpha A\left(\frac{\rho}{\tau}+A\right)^{-1}\varphi^*_{k-j+1}\right\|_E^q\right)^{1/q}d\rho. \qquad (4.18)$$

By the definition of the grid function φ_k^*,

$$\sum_{k=1}^{N}\left\|\left(\frac{\rho}{\tau}\right)^\alpha A\left(\frac{\rho}{\tau}+A\right)^{-1}\varphi^*_{k-j+1}\right\|_E^q\tau \leq \sum_{l=1}^{N}\left\|\left(\frac{\rho}{\tau}\right)^\alpha A\left(\frac{\rho}{\tau}+A\right)^{-1}\varphi_l\right\|_E^q\tau,$$

which in conjunction with (4.18) yields

$$\left(\sum_{k=1}^{N}\tau\|\lambda^\alpha A(\lambda+A)^{-1}Au_k\|_E^q\right)^{1/q} \leq M\int_0^\infty\sum_{j=1}^{N}\frac{\rho^{1-\alpha}}{(1+2\rho\cos\phi+\rho^2)^{j/2}}\frac{(\lambda\tau)^\alpha}{\tau\lambda+\rho}\times$$

$$\times\left(\sum_{l=1}^{N}\tau\left\|\left(\frac{\rho}{\tau}\right)^\alpha A\left(\frac{\rho}{\tau}+A\right)^{-1}\varphi_l\right\|_E^q\right)^{1/q}d\rho.$$

Summing the geometric progression (with respect to j), we obtain

$$\left(\sum_{k=1}^{N}\tau\|\lambda^\alpha A(\lambda+A)^{-1}Au_k\|_E^q\right)^{1/q}$$

$$\leq\frac{M}{\cos\phi}\int_0^\infty\frac{(\tau\lambda)^\alpha}{(\tau\lambda+\rho)\rho^\alpha}\left(\sum_{l=1}^{N}\tau\left\|\left(\frac{\rho}{\tau}\right)^\alpha A\left(\frac{\rho}{\tau}+A\right)^{-1}\varphi_l\right\|_E^q\right)^{1/q}d\rho.$$

The substitution $\lambda\tau r=\rho$ yields

$$\left(\sum_{k=1}^{N}\tau\|\lambda^\alpha A(\lambda+A)^{-1}Au_k\|_E^q\right)^{1/q}$$

$$\leq\frac{M}{\cos\phi}\int_0^\infty\frac{1}{(1+r)r^\alpha}\left(\sum_{l=1}^{N}\tau\|(r\tau)^\alpha A(r\tau+A)^{-1}\varphi_l\|_E^q\right)^{1/q}dr.$$

Applying the Minkowski integral inequality, we obtain

$$\left(\int_0^\infty \sum_{k=1}^N \tau \| \lambda^\alpha A(\lambda + A)^{-1} A u_k \|_E^q \frac{d\lambda}{\lambda} \right)^{1/q}$$

$$\leq \frac{M}{\cos\phi} \int_0^\infty \frac{dr}{(1+r)r^\alpha} \left(\sum_{l=1}^N \tau \int_0^\infty \| (r\lambda)^\alpha A(r\lambda + A)^{-1} \varphi_l \|_E^q \frac{d\lambda}{\lambda} \right)^{1/q}$$

$$\leq \frac{M}{\cos\phi} \frac{\pi}{\sin\pi\alpha} \| \varphi^\tau \|_{L_q(\tau, E'_{\alpha,q})}.$$

Therefore,

$$\| A u^\tau \|_{L_q,(\tau, E'_{\alpha,q})} \leq \frac{M(\phi)}{\alpha(1-\alpha)} \| \varphi^\tau \|_{L_q(\tau, E'_{\alpha,q})}.$$

Since here $M(\phi)$ does not depend on τ, we obtain the assertion of the theorem for $q \neq \infty$.

Note that since $\frac{M(\phi)}{\alpha(1-\alpha)}$ does not depend on q, the inequality for $q = \infty$ can be obtained from inequality (4.11) by letting $q \to \infty$. Theorem 4.2 is proved.

Theorems 3.3 and 4.2 admit the following corollary.

Theorem 4.3. *Let* $1 < p, q < \infty$, $0 < \alpha < 1$ *and* $\varphi \in \mathcal{L}_p(E'_{\alpha,q})$. *Then the coercivity inequality holds:*

$$\| \overline{\mathcal{D}} \, \overline{\Pi}(0) u \|_{\mathcal{L}_p(E'_{\alpha,q})} + \| \overline{A} u \|_{\mathcal{L}_p(E'_{\alpha,q})} \leq \frac{Mp^2}{(p-1)\alpha(1-\alpha)} \| \varphi \|_{\mathcal{L}_p(E'_{\alpha,q})},$$

where M *does not depend on* α, p, q, *and* φ.

Note that Theorems 4.2 and 4.3 actually assert that the difference problem (0.6) with $u_0 = 0$ is well posed in the space $\mathcal{L}_p(E'_{\alpha,q})$ whenever $p, q \in (1,\infty)$, $0 < \alpha < 1$, or $p = q = 1$, or $p = q = \infty$.

Now let us turn to the general problem (0.6). By (0.2), $u = w + g$. In order for w to be a solution of (0.6) with $\varphi = 0$ in $\mathcal{L}_p(E'_{\alpha,q})$ it is necessary and sufficient that

$$\sup_{0 < \tau \leq \tau_0} \sum_{k=1}^N \| A R^k(\tau A) u_0 \|_{\alpha,q}^p \tau < \infty.$$

The quantity (see Section 3)

$$\langle u_0 \rangle'_{1+\alpha-\frac{1}{p},q} = \sum_{0 < \tau \leq \tau_0} \left(\sum_{k=1}^N \| A R^k(\tau A) u_0 \|_{\alpha,q}^p \tau \right)^{1/p}$$

defines a norm in the space of initial data $E'_{1+\alpha-\frac{1}{p},q}$, which consists of all elements $u_0 \in E'_{\alpha,q}$ for which this quantity is finite.

Theorem 4.4. *Let* $1 < p,q < \infty$ *or* $p = q = \infty$, *and let* $u_0 \in E'_{1+\alpha-\frac{1}{p},q}$, $\varphi \in \mathcal{L}_p(E'_{\alpha,q})$. *Then the coercivity inequality holds:*

$$\|\overline{\mathcal{D}}\,\overline{\Pi}(u_0)u\|_{\mathcal{L}_p(E'_{\alpha,q})} + \|\overline{A}u\|_{\mathcal{L}_p(E'_{\alpha,q})} + \max_{1 \le k \le N}\langle u_k\rangle'_{1+\alpha-\frac{1}{p},q}$$

$$\le M\left[\langle u_0\rangle'_{1+\alpha-\frac{1}{p},q} + \frac{M_1(p,q)}{\alpha(1-\alpha)}\|\varphi\|_{\mathcal{L}_p(E'_{\alpha,q})}\right], \tag{4.19}$$

where $M_1(p,q) = \frac{p^2}{p-1}$ *if* $p \ne q$, $M(p,p) = 1$.

Proof. By formula (0.2),

$$u_k = w_k + g_k = R^k(\tau A)u_0 + \sum_{j=1}^{k} R^{k-j+1}(\tau A)\varphi_j\tau. \tag{4.20}$$

For the solution w of the homogeneous equation with nonzero initial condition, the estimate of $\overline{A}w$, and hence that of $\overline{\mathcal{D}}\,\overline{\Pi}(u_0)w$ in $\mathcal{L}_p(E'_{\alpha,q})$, follows from the definition of the norm in the space $E'_{1+\alpha-\frac{1}{p},q}$. The estimate of w in the norm of $\mathcal{C}(E'_{1+\alpha-\frac{1}{p},q})$ follows from the uniform boundedness of the powers $(I + \tau A)^{-k}$. For the solution g of the nonhomogeneous problem, the estimate of $\overline{A}g$, and hence that of $\overline{\mathcal{D}}\,\overline{\Pi}(0)g$ in $\mathcal{L}_p(E'_{\alpha,q})$, was obtained in Theorem 4.3. Hence, it remains to estimate g in the norm of $\mathcal{C}(E'_{1+\alpha-\frac{1}{p},q})$.

Now consider g_k. From estimate (3.3) it follows that

$$\|AR^k(\tau A)\|_{E'_{\alpha,q} \to E'_{\alpha,q}} \le \frac{M}{k\tau}(1 + \delta\tau)^{-k}. \tag{4.21}$$

From this inequality and Theorem 3.2 we obtain the estimate

$$\langle g_k\rangle'_{1+\alpha-\frac{1}{p},q} \le M\frac{\pi}{\sin(\pi/p)}\|\varphi\|_{\mathcal{L}_p(E'_{\alpha,q})}$$

for any $p \in (1,\infty)$.

In the case $p = q = \infty$ the space of traces consists of the elements $z \in D(A)$ such that $Az \in E_{\alpha,\infty}$. Hence, here the estimate in the trace norm is a corollary of Theorem 4.2. Theorem 4.4 is proved.

Theorem 4.4 does not cover all cases $p, q \in [1, \infty]$. Theorems 4.2 and 4.4 are supplemented by the following result.

Theorem 4.5. *Let* $p = q = 1$ *or* $p = q = \infty$, *and let* $u_0 \in E'_{1+\alpha-\frac{1}{p},q}$, $\varphi \in \mathcal{L}_p(E'_{\alpha,q})$. *Then the solution of the difference problem* (0.6) *obeys the coercivity inequality*

$$\|\overline{\mathcal{D}}\,\overline{\Pi}(u_0)u\|_{\mathcal{L}_p(E'_{\alpha,p})} + \|\overline{A}u\|_{\mathcal{L}_p(E'_{\alpha,p})}$$

$$\leq M \left[\langle u_0 \rangle'_{1+\alpha-\frac{1}{p},q} + \frac{1}{\alpha(1-\alpha)} \|\varphi\|_{\mathcal{L}_p(E'_{\alpha,p})} \right]. \tag{4.22}$$

Let $1 < p < \infty$, $u_0 \in E'_{1+\alpha-\frac{1}{p},\infty}$, *and* $\varphi \in \mathcal{L}_p(E'_{\alpha,\infty})$. *Then the solution of the difference problem* (0.6) *obeys the inequality*

$$\max_{1 \leq k \leq N} \langle u_k \rangle'_{1+\alpha-\frac{1}{p},\infty} \leq M \left[\langle u_0 \rangle'_{1+\alpha-\frac{1}{p},\infty} + \frac{p^2}{p-1} \|\varphi\|_{\mathcal{L}_p(E'_{\alpha,\infty})} \right]. \tag{4.23}$$

Proof. By Theorem 4.2, in inequality (4.22) we only need to justify the estimate for the solution of the homogeneous problem (0.4). Since $R^k(\tau A)$ is bounded in the norm of the space E, the needed estimate follows from the definition of the norm in the space of traces.

Now let us prove inequality (4.23). To this end we use (4.20). The needed estimate for w_k follows from the definition of the norm in the space of traces and the boundedness of $R^k(\tau A)$ in the norm of E. Now consider g_k. From (4.20) we obtain the identity

$$\Psi(\lambda, i, k, \tau) \equiv \lambda^\alpha A(\lambda + A)^{-1} A R^i(\tau A) g_k = \lambda^\alpha \sum_{j=1}^{k} A R^{i+k-j+1}(\tau A) A(\lambda + A)^{-1} \varphi_j \tau$$

$$\tag{4.24}$$

for any $\lambda > 0$ and $k \geq 1$, $i \geq 1$. Using the estimate (3.3), we obtain

$$\|\Psi(\lambda, i, k, \tau)\|_E \leq M \sum_{j=1}^{k} \tau \frac{\lambda^\alpha \|A(\lambda + A)^{-1}\varphi_j\|_E}{(i+k-j+1)\tau}$$

$$\leq M \sum_{j=1}^{k} \frac{\tau}{(i+k-j+1)\tau} \|\varphi_j\|'_{\alpha,\infty}. \tag{4.25}$$

From the identity (4.24) and the definition of the norm in the space $E'_{\alpha,\infty}$ it follows that

$$\|A R^i(\tau A) g_k\|'_{\alpha,\infty} = \sup_{\lambda > 0} \|\Psi(\lambda, i, k, \tau)\|_E,$$

which in conjunction with (4.25) yields the inequality

$$\|AR^i(\tau A)g_k\|'_{\alpha,\infty} \leq M \sum_{j=1}^{k} \frac{\tau}{(i+k-j+1)\tau}\|\varphi_j\|'_{\alpha,\infty}.$$

Making the substitution $k - j + 1 = l$ and setting $\varphi_j^* = \varphi_j$ if $j = 1, \cdots, N$, and $\varphi_j^* = 0$ otherwise, we obtain the inequality

$$\|AR^i(\tau A)g_k\|'_{\alpha,\infty} \leq M \sum_{l=1}^{N} \frac{\tau}{(i+l)\tau}\|\varphi_{k-l+1}^*\|'_{\alpha,\infty}. \tag{4.26}$$

In the left-hand side of (4.26) we recognize the difference analogue of Hilbert's integral operator. Using the fact that this operator is bounded in the space $L_p(\tau, E)$ $(1 < p < \infty)$ of grid functions, we see that

$$\left(\sum_{i=1}^{N} \|AR^i(\tau A)g_k\|'^p_{\alpha,\infty}\tau\right)^{1/p} \leq M_1 \frac{\pi}{\sin(\pi/p)}\|\varphi\|_{\mathcal{L}_p(E'_{\alpha,\infty})}.$$

By the definition of the norm in the space of traces, the expression in the left-hand side of this last inequality is identical to $\langle g_k\rangle_{1+\alpha-\frac{1}{p},\infty}$. Hence,

$$\langle g_k\rangle'_{1+\alpha-\frac{1}{p},\infty} \leq \frac{M_2 p^2}{p-1}\|\varphi\|_{\mathcal{L}_p(E'_{\alpha,\infty})}.$$

Theorem 4.5 is proved.

5. WELL-POSEDNESS OF THE DIFFERENCE PROBLEM IN DIFFERENCE ANALOGUES OF SPACES OF SMOOTH FUNCTIONS

1. The space $\mathcal{C}_0^{\beta,\gamma}(E)$. The nonhomogeneous difference problem.

In Sections 2 and 4 the well-posedness of problem (0.6) was established in the spaces $\mathcal{C}_0^\alpha(E)$ and $\mathcal{C}(E_\alpha) = \mathcal{L}_\infty(E_{\alpha,\infty})$. These are difference analogues of spaces of smooth functions with respect to time or the space E.

This section is devoted to the generalization of the results of Sections 2 and 4 of the present chapter and of Section 5 of Chapter 1.

Let us introduce the Banach space $C_0^{\beta,\gamma}(\tau, E)$ $(0 \le \gamma \le \beta < 1)$ of grid functions φ^τ with the norm

$$\|\varphi^\tau\|_{C_0^{\beta,\gamma}(\tau,E)} = \|\varphi^\tau\|_{C(\tau,E)} + \max_{1 \le k < k+r \le N} \frac{((k+r)\tau)^\gamma \|\varphi_{k+r} - \varphi_k\|_E}{(r\tau)^\beta}. \tag{5.1}$$

Next, let us denote by $C_0^{\beta,\gamma}(E) = C_0^{\beta,\gamma}(\mathcal{E}(E))$ the set of collections $\{\varphi^\tau\} = \varphi$ for which the norm

$$\|\varphi\|_{C_0^{\beta,\gamma}(E)} = \sup_{0 < \tau \le \tau_0} \|\varphi^\tau\|_{C_0^{\beta,\gamma}(\tau,E)}$$

is finite. The norm (5.1) in the space $C_0^{\alpha,\alpha}(\tau, E)$ and the norm in the space $C_0^\alpha(\tau, E)$ are equivalent uniformly in τ and α. Hence, the spaces $C_C^{\alpha,\alpha}(E)$ and $\mathcal{C}_0^\alpha(E)$ coincide.

First let us consider the special nonhomogeneous difference problem (0.6) with $u_0 = 0$ and $\varphi_1 = 0$. A solution u of this problem is said to be *a solution in* $C_0^{\beta,\gamma}(E)$ if $\overline{\mathcal{D}}\,\overline{\Pi}(0)u$, $\overline{A}u \in C_0^{\beta,\gamma}(E)$. If u is a solution of the difference problem (0.6) with $u_0 = 0$ and $\varphi_1 = 0$, then obviously $\varphi \in C_0^{\beta,\gamma}(E)$. As above, let us give

Definition 5.1. The difference problem (0.6) with $u_0 = 0$ and $\varphi_1 = 0$ is said to be *well posed in* $C_0^{\beta,\gamma}(E)$ if the following two conditions are satisfied:

1) For any $\varphi \in C_0^{\beta,\gamma}(E)$ there exists a unique solution $u(\varphi, 0)$ of this problem.
2) The operator $u(\varphi, 0)$ is continuous in $C_0^{\beta,\gamma}(E)$.

As in the case of the space $\mathcal{C}(E)$ one shows that for the well-posedness in $C_0^{\beta,\gamma}(E)$ of this difference problem it is necessary and sufficient that the following coercivity inequality hold:

$$\|\overline{\mathcal{D}}\,\overline{\Pi}(0)u\|_{C_0^{\beta,\gamma}(E)} + \|\overline{A}u\|_{C_0^{\beta,\gamma}(E)} \le M(\beta,\gamma)\|\varphi\|_{C_0^{\beta,\gamma}(E)}. \tag{5.2}$$

By passing to the limit $\tau \to 0$ we deduce from this the corresponding inequality for the differential problem. Hence, $-A$ must be the generator of an analytic semigroup in E. It turns out that this last condition is not only necessary but also sufficient for the well-posedness of the difference problem (0.6) with $u_0 = 0$ and $\varphi_1 = 0$.

Theorem 5.1. *Suppose $-A$ is the generator of an analytic semigroup in E and let $\varphi \in C_0^{\beta,\gamma}(E)$. Then the coercivity inequality (5.2) holds with $M(\beta,\gamma) = M/\{\beta(1-\beta)\}$, where M does not depend on β, γ, and φ.*

Proof. From (0.2) we derive the identity

$$\mathcal{D}\Pi(0)u_k = R^k(\tau A)(\varphi_k - \varphi_1) + \sum_{j=1}^{k} AR^{k-j+1}(\tau A)(\varphi_k - \varphi_j)\tau$$

$$\equiv p_k + g_k, \quad k = 1, \cdots, N. \tag{5.3}$$

Let us estimate the grid functions $p^\tau = \{p_k\}_1^N$ and $g^\tau = \{g_k\}_1^N$ in the norm of $C_0^{\beta,\gamma}(\tau, E)$. By (0.8), we have

$$\|p_k\|_E \le \|R^k(\tau A)\|_{E\to E}\|\varphi_k - \varphi_1\|_E \le M(k\tau)^{\beta-\gamma}\|\varphi^\tau\|_{C_0^{\beta,\gamma}(\tau,E)}, \tag{5.4}$$

whence

$$\|p_k\|_E \le M\|\varphi^\tau\|_{C_0^{\beta,\gamma}(\tau,E)}$$

for any $k \ge 1$. Consequently,

$$\|p^\tau\|_{C(\tau,E)} \le M\|\varphi^\tau\|_{C_0^{\beta,\gamma}(\tau,E)}. \tag{5.5}$$

First let $k \le r$. Then, by (5.4) and the triangle inequality,

$$\|p_{k+r} - p_k\|_E \le \|p_{k+r}\|_E + \|p_k\|_E$$

$$\le M[((k+r)\tau)^{\beta-\gamma} + (k\tau)^{\beta-\gamma}]\|\varphi^\tau\|_{C_0^{\beta,\gamma}(\tau,E)} \le 2^{1+\beta}M\frac{(r\tau)^\beta}{((k+r)\tau)^\gamma}\|\varphi^\tau\|_{C_0^{\beta,\gamma}(\tau,E)}.$$

Thus, for $k \le r$ we have the estimate

$$\|p_{k+r} - p_k\|_E \le M_1\frac{(r\tau)^\beta}{((k+r)\tau)^\gamma}\|\varphi^\tau\|_{C_0^{\beta,\gamma}(\tau,E)}. \tag{5.6}$$

Now let $k > r$. Then

$$p_{k+r} - p_k = [R^{k+r}(\tau A) - R^k(\tau A)][\varphi_{k+r} - \varphi_1] + R^k(\tau A)(\varphi_{k+r} - \varphi_k).$$

By (0.8) and (2.4) with $\alpha = 1$, we have

$$\|p_{k+r} - p_k\|_E \le \|R^{k+r}(\tau A) - R^k(\tau A)\|_{E\to E}\|\varphi_{k+r} - \varphi_1\|_E$$

$$+ \|R^k(\tau A)\|_{E\to E}\|\varphi_{k+r} - \varphi_k\|_E$$

$$\le M\left[\frac{r\tau}{k\tau}((k+r)\tau)^{\beta-\gamma} + \frac{(r\tau)^\beta}{((k+r)\tau)^\gamma}\right]\|\varphi^\tau\|_{C_0^{\beta,\gamma}(\tau,E)}.$$

Since

$$\frac{r\tau}{k\tau}((k+r)\tau)^{\beta-\gamma} \le 2\frac{(r\tau)^\beta}{((k+r)\tau)^\gamma},$$

it follows that for $k > r$

$$\|p_{k+r} - p_k\|_E \le M_1 \frac{(r\tau)^\beta}{((k+r)\tau)^\gamma}\|\varphi^\tau\|_{C_0^{\beta,\gamma}(\tau,E)}.$$

By (5.5) and (5.6),

$$\|p^\tau\|_{C_0^{\beta,\gamma}(\tau,E)} \le M\|\varphi^\tau\|_{C_0^{\beta,\gamma}(\tau,E)}. \tag{5.7}$$

Now let us establish the estimate

$$\|g^\tau\|_{C_0^{\beta,\gamma}(\tau,E)} \le \frac{M}{\beta(1-\beta)}\|\varphi^\tau\|_{C_0^{\beta,\gamma}(\tau,E)}. \tag{5.8}$$

To this end it suffices to show that

$$\|g_k\|_E \le \frac{M}{\beta}\|\varphi^\tau\|_{C_0^{\beta,\gamma}(\tau,E)}, \quad 1 \le k \le N, \tag{5.9}$$

$$\|g_{k+r} - g_k\|_E \le \frac{M}{\beta(1-\beta)} \frac{(r\tau)^\beta}{((k+r)\tau)^\gamma}\|\varphi^\tau\|_{C_0^{\beta,\gamma}(\tau,E),1\le k<k+r\le N}. \tag{5.10}$$

First we deal with (5.9). Using the estimate (1.7), we obtain

$$\|g_k\|_E \le \sum_{j=1}^{k} \|AR^{k+1-j}(\tau A)\|_{E\to E}\|\varphi_k - \varphi_j\|_E \tau$$

$$\le M \sum_{j=1}^{k-1} \frac{((k-j)\tau)^\beta}{(k+1-j)\tau} \frac{\tau}{(k\tau)^\gamma}\|\varphi^\tau\|_{C_0^{\beta,\gamma}(\tau,E)}.$$

Since

$$\sum_{j=1}^{k-1} \frac{((k-j)\tau)^\beta\tau}{(k+1-j)\tau} \le \int_0^{k\tau} \frac{ds}{(k\tau-s)^{1-\beta}} = \frac{(k\tau)^\beta}{\beta},$$

it follows that

$$\|g_k\|_E \le \frac{M(k\tau)^{\beta-\gamma}}{\beta}\|\varphi^\tau\|_{C_0^{\beta,\gamma}(\tau,E)} \tag{5.11}$$

for any $k = 1, \cdots, N$. The last inequality yields the estimate (5.9) as well as the estimate (5.10) for $k \le r$. Indeed, by (5.11) and the triangle inequality,

$$\|g_{k+r} - g_k\|_E \le \|g_{k+r}\|_E + \|g_k\|_E \le \frac{M}{\beta}[((k+r)\tau)^{\beta-\gamma} + (k\tau)^{\beta-\gamma}]\|\varphi^\tau\|_{C_0^{\beta,\gamma}(\tau,E)}$$

$$\leq \frac{2M}{\beta}((k+r)\tau)^{\beta-\gamma}\|\varphi^\tau\|_{C_0^{\beta,\gamma}(\tau,E)} \leq \frac{2^{1+\beta}M}{\beta}\frac{(r\tau)^\beta}{((k+r)\tau)^\gamma}\|\varphi^\tau\|_{C_0^{\beta,\gamma}(\tau,E)}.$$

Now suppose that $k > r$. Let us write the difference $g_{k+r} - g_k$ as a sum of four terms:

$$g_{k+r} - g_k = \sum_{j=k-r+1}^{k+r} AR^{k+r+1-j}(\tau A)(\varphi_{k+r} - \varphi_j)\tau$$

$$+ \sum_{j=k-r+1}^{k} AR^{k+1-j}(\tau A)(\varphi_j - \varphi_k)\tau$$

$$+ \sum_{j=1}^{k-r} A[R^{k+r+1-j}(\tau A) - R^{k+1-j}(\tau A)](\varphi_k - \varphi_j)\tau$$

$$+ \sum_{j=1}^{k-r} AR^{k+r+1-j}(\tau A)\tau(\varphi_{k+r} - \varphi_k) = \mathcal{J}_1 + \mathcal{J}_2 + \mathcal{J}_3 + \mathcal{J}_4.$$

We shall estimate these four terms separately. First let us consider $\mathcal{J}_1$. Using the estimate (1.7), we obtain

$$\|\mathcal{J}_1\|_E \leq \sum_{j=k-r+1}^{k+r} \|AR^{k+r+1-j}(\tau A)\|_{E\to E}\|\varphi_{k+r} - \varphi_j\|_E\tau$$

$$\leq M \sum_{j=k-r+1}^{k+r-1} \frac{((k+r-j)\tau)^\beta \tau}{(k+r+1-j)\tau((k+r)\tau)^\gamma}\|\varphi^\tau\|_{C_0^{\beta,\gamma}(\tau,E)}.$$

Since

$$\sum_{j=k-r+1}^{k+r-1} \frac{((k+r-j)\tau)^\beta}{(k+r+1-j)\tau} \leq \int_{(k-r)\tau}^{(k+r)\tau} \frac{ds}{((k+r)\tau - s)^{1-\beta}} = \frac{(2r\tau)^\beta}{\beta},$$

we conclude that

$$\|\mathcal{J}_1\|_E \leq \frac{2^\beta M}{\beta}\frac{(r\tau)^\beta}{((k+r)\tau)^\gamma}\|\varphi^\tau\|_{C_0^{\beta,\gamma}(\tau,E)}.$$

In exactly the same way one establishes the estimate

$$\|\mathcal{J}_2\|_E \leq \frac{M}{\beta}\frac{(r\tau)^\beta}{(k\tau)^\gamma}\|\varphi^\tau\|_{C_0^{\beta,\gamma}(\tau,E)}.$$

Since $k > r$, we obtain

$$\|\mathcal{J}_2\|_E \leq \frac{2^\gamma M}{\beta} \frac{(r\tau)^\beta}{((k+r)\tau)^\gamma} \|\varphi^\tau\|_{C_0^{\beta,\gamma}(\tau,E)}.$$

Now let us estimate $\mathcal{J}_4$. Since

$$\mathcal{J}_4 = [R^{2r}(\tau A) - R^{k+r}(\tau A)](\varphi_{k+r} - \varphi_k),$$

the bound (0.8) yields

$$\|\mathcal{J}_4\|_E \leq \|R^{2r}(\tau A) - R^{k+r}(\tau A)\|_{E \to E} \|\varphi_{k+r} - \varphi_k\|_E \leq M \frac{(r\tau)^\beta}{((k+r)\tau)^\gamma} \|\varphi^\tau\|_{C_0^{\beta,\gamma}(\tau,E)}.$$

Finally, let us estimate $\mathcal{J}_3$. Using the estimate (2.5) with $\alpha = 1$, we obtain

$$\|\mathcal{J}_3\|_E \leq \sum_{j=1}^{k-r} \|A[R^{k+r+1-j}(\tau A) - R^{k+1-j}(\tau A)]\|_{E \to E} \|\varphi_k - \varphi_j\|_E \tau$$

$$\leq M \sum_{j=1}^{k-r} \frac{r\tau((k-j)\tau)^\beta \tau}{((k+1-j)\tau)^2 (k\tau)^\gamma} \|\varphi^\tau\|_{C_0^{\beta,\gamma}(\tau,E)}.$$

Since

$$\sum_{j=1}^{k-r} \frac{((k-j)\tau)^\beta \tau}{((k+1-j)\tau)^2} \leq \int_0^{(k-r)\tau} \frac{ds}{(k\tau - s)^{2-\beta}} \leq \frac{1}{1-\beta} \frac{1}{(r\tau)^{1-\beta}},$$

it follows that

$$\|\mathcal{J}_3\|_E = \frac{M}{1-\beta} \frac{r\tau}{(r\tau)^{1-\beta}(k\tau)^\gamma} \|\varphi^\tau\|_{C_0^{\beta,\gamma}(\tau,E)} = \frac{M}{1-\beta} \frac{(r\tau)^\beta}{(k\tau)^\gamma} \|\varphi^\tau\|_{C_0^{\beta,\gamma}(\tau,E)}.$$

Hence, since $k > r$, we have

$$\|\mathcal{J}_3\|_E \leq \frac{2^\gamma M}{1-\beta} \frac{(r\tau)^\beta}{((k+r)\tau)^\gamma} \|\varphi^\tau\|_{C_0^{\beta,\gamma}(\tau,E)}.$$

Combining the estimates for $\mathcal{J}_i$, $i = 1, \cdots, 4$, we obtain (5.10). From (5.7) and (5.8) we derive the inequality

$$\|\mathcal{D}\Pi(0)u^\tau\|_{C_0^{\beta,\gamma}(\tau,E)} \leq \frac{M}{\beta(1-\beta)} \|\varphi^\tau\|_{C_0^{\beta,\gamma}(\tau,E)}.$$

The estimate for Au^τ in the norm of $C_0^{\beta,\gamma}(\tau, E)$ is obtained by means of the triangle inequality.

Since M does not depend on τ and $\varphi \in \mathcal{C}_0^{\beta,\gamma}(E)$, it follows that $\overline{\mathcal{D}}\,\overline{\Pi}(0)u$ and $\overline{A}u$ belong to $\mathcal{C}_0^{\beta,\gamma}(E)$, and the coercivity inequality (5.2) holds. Theorem 5.1 is proved.

2. Well-posedness of the general difference problem.

Now let us consider the difference problem (0.6). Formula (0.3) shows that the solution of the homogeneous problem (0.5) with $\varphi^\tau = 0$ and $w_0 = u_0 - A^{-1}\varphi_1$ has the form

$$w_k = R^k(\tau A)(u_0 - A^{-1}\varphi_1), \quad 1 \le k \le N.$$

Hence, for w to be a solution in $\mathcal{C}_0^{\beta,\gamma}(E)$ of problem (0.6) with $\varphi = 0$ it is necessary and sufficient that

$$\sup_{0<\tau\le\tau_0} \|\{AR^i(\tau A)w_0\}_1^N\|_{C_0^{\beta,\gamma}(\tau,E)} < \infty.$$

By the estimate (0.8), the collection of all $w_0 \in E$ with this property is clearly a linear set that contains $D(A)$. It becomes a Banach space $E_1^{\beta,\gamma}$ when endowed with the norm

$$\langle w_0 \rangle_1^{\beta,\gamma} = \sup_{0<\tau\le\tau_0} \left[\max_{1\le i\le N} \|AR^i(\tau A)w_0\|_E \right.$$

$$\left. + \sup_{1\le i<i+r\le N} \frac{((i+r)\tau)^\gamma \|A[R^{i+r}(\tau A) - R^i(\tau A)]w_0\|_E}{(r\tau)^\beta} \right]. \tag{5.12}$$

Let $u^\tau = (u_1, \cdots, u_N)$ be a solution of the general problem (0.5). Then we can write $u_k = g_k + w_k + A^{-1}\varphi_1$, $1 \le k \le N$, where g^τ is a solution of problem (0.5) with $g_0 = 0$ and right-hand side $\varphi_k - \varphi_1$, and w^τ is a solution of problem (0.5) with $w_0 = u_0 - A^{-1}\varphi_1$. From this representation and the solvability in $\mathcal{C}_0^{\beta,\gamma}(E)$ of these two problems it follows that the general problem (0.6) is uniquely solvable for any $u_0 - A^{-1}\varphi_1 \in E_1^{\beta,\gamma}$ and its solutions obey the coercivity inequality

$$\|\overline{\mathcal{D}}\,\overline{\Pi}(u_0)u\|_{\mathcal{C}_0^{\beta,\gamma}(E)} + \|\overline{A}u\|_{\mathcal{C}_0^{\beta,\gamma}(E)}$$

$$\le M(\beta,\gamma)\|\varphi\|_{\mathcal{C}_0^{\beta,\gamma}(E)} + M\langle u_0 - A^{-1}\varphi_1 \rangle_1^{\beta,\gamma}. \tag{5.13}$$

We say that the general difference problem (0.6) is *well-posed in* $\mathcal{C}_0^{\beta,\gamma}(E)$ if for any $u_0 \in D(A)$, $u_0 - A^{-1}\varphi_1 \in E_1^{\beta,\gamma}$ and $\varphi \in \mathcal{C}_0^{\beta,\gamma}(E)$ this problem has a unique solution and the coercivity inequality (5.13) holds. The foregoing arguments show that the analyticity of the semigroup $\exp\{-tA\}$ is not only necessary but also sufficient for the well-posedness in $\mathcal{C}_0^{\beta,\gamma}(E)$ of the difference problem (0.6). The conditions on the data of the problem (0.6) formulated above are such that the element $u_0 - A^{-1}\varphi_1$ should belong to a smaller space than the elements $u_k - A^{-1}\varphi_k$, $k = 1, \cdots, N$. Is this really the case? The answer is provided by

Theorem 5.2. *Suppose* $u_0 - A^{-1}\varphi_1 \in E_1^{\beta,\gamma}$ *and* $\varphi \in \mathcal{C}_0^{\beta,\gamma}(E)$. *Then the solutions of the difference problem (0.6) obey the inequality*

$$\max_{1 \le k \le N} \langle u_k - A^{-1}\varphi_k \rangle_1^{\beta,\gamma} \le M \left[\langle u_0 - A^{-1}\varphi_1 \rangle_1^{\beta,\gamma} + \frac{1}{\beta(1-\beta)} \|\varphi\|_{\mathcal{C}_0^{\beta,\gamma}(E)} \right], \quad (5.14)$$

where M does not depend on β, γ, u_0, and φ.

Proof. By (0.2), we have

$$u_k - A^{-1}\varphi_k = R^k(\tau A)(u_0 - A^{-1}\varphi_1) + A^{-1}R^k(\tau A)(\varphi_1 - \varphi_k)$$

$$+ \sum_{j=1}^{k} R^{k+1-j}(\tau A)(\varphi_j - \varphi_k)\tau = V_1 + V_2 + V_3.$$

From the definition of the space $E_1^{\beta,\gamma}$ and the estimate (0.8) it follows that

$$\langle V_1 \rangle_1^{\beta,\gamma} \le \|R^k(\tau A)\|_{E \to E} \langle u_0 - A^{-1}\varphi_1 \rangle_1^{\beta,\gamma} \le M \langle u_0 - A^{-1}\varphi_1 \rangle_1^{\beta,\gamma}. \quad (5.15)$$

Now let us estimate V_2. Applying (0.8), we obtain

$$\max_{1 \le i \le N} \|AR^i(\tau A)A^{-1}R^k(\tau A)(\varphi_k - \varphi_1)\|_E \le \max_{1 \le i \le N} \|R^{k+i}(\tau A)\|_{E \to E} \|\varphi_k - \varphi_1\|_E$$

$$\le M(k\tau)^{\beta-\gamma} \|\varphi^\tau\|_{\mathcal{C}_0^{\beta,\gamma}(\tau,E)} \le M_1 \|\varphi^\tau\|_{\mathcal{C}_0^{\beta,\gamma}(\tau,E)}. \quad (5.16)$$

If $i + r \le k$, then by (2.4) with $\alpha = \beta$ we have

$$\frac{((i+r)\tau)^\gamma}{(r\tau)^\beta} \|A[R^{i+r}(\tau A) - R^i(\tau A)]A^{-1}R^k(\tau A)(\varphi_k - \varphi_1)\|_E$$

$$\le \frac{((i+r)\tau)^\gamma}{(r\tau)^\beta} \|R^{i+r+k}(\tau A) - R^{i+k}(\tau A)\|_{E \to E} \|\varphi_k - \varphi_1\|_E$$

$$\le M \frac{((i+r)\tau)^\gamma}{(r\tau)^\beta} \frac{(r\tau)^\beta (k\tau)^{\beta-\gamma}}{((i+k)\tau)^\beta} \|\varphi^\tau\|_{\mathcal{C}_0^{\beta,\gamma}(\tau,E)} \le M \|\varphi^\tau\|_{\mathcal{C}_0^{\beta,\gamma}(\tau,E)}.$$

If $i + r > k$ and $i \leq r$, then by (2.4) with $\alpha = 0$ we have

$$\frac{((i+r)\tau)^\gamma}{(r\tau)^\beta} \| A[R^{i+r}(\tau A) - R^i(\tau A)]A^{-1}R^k(\tau A)(\varphi_k - \varphi_1)\|_E$$

$$\leq \frac{((i+r)\tau)^\gamma}{(r\tau)^\beta} \| R^{i+r+k}(\tau A) - R^{i+k}(\tau A)\|_{E \to E} \|\varphi_k - \varphi_1\|_E$$

$$\leq M \frac{((i+r)\tau)^\gamma}{(r\tau)^\beta} (k\tau)^{\beta-\gamma} \|\varphi^\tau\|_{C_0^{\beta,\gamma}(\tau,E)}$$

$$\leq M \left(\frac{i+r}{r}\right)^\beta \|\varphi^\tau\|_{C_0^{\beta,\gamma}(\tau,E)} \leq 2^\beta M \|\varphi^\tau\|_{C_0^{\beta,\gamma}(\tau,E)}.$$

Finally, if $i + r > k$ and $i > r$, then by (2.4) with $\alpha = 1$ we have

$$\frac{((i+r)\tau)^\gamma}{(r\tau)^\beta} \| A[R^{i+r}(\tau A) - R^i(\tau A)]A^{-1}R^k(\tau A)(\varphi_k - \varphi_1)\|_E$$

$$\leq \frac{((i+r)\tau)^\gamma}{(r\tau)^\beta} \| R^{i+r+k}(\tau A) - R^{i+k}(\tau A)\|_{E \to E} \|\varphi_k - \varphi_1\|_E$$

$$\leq M \frac{((i+r)\tau)^\gamma}{(r\tau)^\beta} \frac{r\tau(k\tau)^{\beta-\gamma}}{(i+k)\tau} \|\varphi^\tau\|_{C_0^{\beta,\gamma}(\tau,E)}$$

$$\leq 2M((i+r)\tau)^\gamma \frac{(r\tau)^{1-\beta}((i+r)\tau)^{\beta-\gamma}}{(i+r)\tau} \|\varphi^\tau\|_{C_0^{\beta,\gamma}(\tau,E)}$$

$$= 2M \left(\frac{r}{i+r}\right)^{1-\beta} \|\varphi^\tau\|_{C_0^{\beta,\gamma}(\tau,E)} \leq M_1 \|\varphi^\tau\|_{C_0^{\beta,\gamma}(\tau,E)}.$$

Therefore,

$$\sup_{1 \leq i < i+r \leq N} \frac{((i+r)\tau)^\gamma \|A[R^{i+r}(\tau A) - R^i(\tau A)]V_2\|_E}{(r\tau)^\beta} \leq M \|\varphi^\tau\|_{C_0^{\beta,\gamma}(\tau,E)},$$

which in conjunction with (5.15) yields

$$\langle V_2 \rangle_1^{\beta,\gamma} \leq M \|\varphi^\tau\|_{C_0^{\beta,\gamma}(\tau,E)}. \tag{5.17}$$

Next let us estimate V_3. Using (1.7), we obtain

$$\|AR^i(\tau A)V_3\|_E \leq \sum_{j=1}^{k-1} \|AR^{k+1-j+i}(\tau A)\|_{E \to E} \|\varphi_j - \varphi_k\|_E \tau$$

$$\leq M \sum_{j=1}^{k-1} \frac{((k-j)\tau)^{\beta}\tau}{(k+1-j+i)\tau(k\tau)^{\gamma}} \|\varphi^{\tau}\|_{C_0^{\beta,\gamma}(\tau,E)}.$$

Since

$$\sum_{j=1}^{k-1} \frac{((k-j)\tau)^{\beta}\tau}{(k+1-j+i)\tau} \leq \int_0^{k\tau} \frac{ds}{(k\tau-s)^{1-\beta}} = \frac{(k\tau)^{\beta}}{\beta}$$

for any $i = 1, \cdots, N$, it follows that

$$\max_{1 \leq i \leq N} \|AR^i(\tau A)V_3\|_E \leq \frac{M}{\beta}(k\tau)^{\beta-\gamma}\|\varphi^{\tau}\|_{C_0^{\beta,\gamma}(\tau,E)} \leq \frac{M}{\beta}\|\varphi^{\tau}\|_{C_0^{\beta,\gamma}(\tau,E)}. \qquad (5.18)$$

By the definition of the space $C_0^{\beta,\gamma}(\tau, E)$,

$$\frac{((i+r)\tau)^{\gamma}}{(r\tau)^{\beta}}\|A[R^{i+r}(\tau A) - R^i(\tau A)]V_3\|_E$$

$$\leq \frac{((i+r)\tau)^{\gamma}}{(r\tau)^{\beta}(k\tau)^{\gamma}} \sum_{j=1}^{k-1} \|A[R^{i+r}(\tau A) - R^i(\tau A)]R^{k+1-j}(\tau A)\|_{E\to E} \times$$

$$\times ((k-j)\tau)^{\beta}\tau\|\varphi^{\tau}\|_{C_0^{\beta,\gamma}(\tau,E)} = Q\|\varphi^{\tau}\|_{C_0^{\beta,\gamma}(\tau,E)},$$

where Q denotes the quantity multiplying $\|\varphi^{\tau}\|_{C_0^{\beta,\gamma}(\tau,E)}$ in the last inequality. If $k \leq i$ and $r \leq i$, then by (0.8) and (2.5) with $\alpha = \beta$ we have

$$Q \leq \frac{((i+r)\tau)^{\gamma}}{(r\tau)^{\beta}(k\tau)^{\gamma}} \sum_{j=1}^{k-1} \|A[R^{i+r}(\tau A) - R^i(\tau A)]\|_{E\to E} \times \|R^{k+1-j}(\tau A)\|_{E\to E}((k-j)\tau)^{\beta}\tau$$

$$\leq M\frac{2^{\gamma}(i\tau)^{\gamma}}{(r\tau)^{\beta}(k\tau)^{\gamma}} \frac{(r\tau)^{\beta}(k\tau)^{\beta}k\tau}{(i\tau)^{1+\beta}} = M2^{\gamma}\left(\frac{k}{i}\right)^{1+\beta-\gamma} \leq M_1.$$

If $k \leq i$ and $r > i$, then by (0.8) and (2.5) with $\alpha = 0$ we have

$$Q \leq \frac{((i+r)\tau)^{\gamma}}{(r\tau)^{\beta}(k\tau)^{\gamma}} \sum_{j=1}^{k-1} \|A[R^{i+r}(\tau A) - R^i(\tau A)]\|_{E\to E}\|R^{k+1-j}(\tau A)\|_{E\to E}((k-j)\tau)^{\beta}\tau$$

$$\leq M\frac{2^{\beta}((i+r)\tau)^{\gamma}}{((i+r)\tau)^{\beta}(k\tau)^{\gamma}} \frac{k\tau(k\tau)^{\beta}}{i\tau} \leq M2^{\beta}\left(\frac{k}{i}\right)^{1+\beta-\gamma} \leq M_1.$$

Next, if $i < k \leq i+r$ and $r \leq i$, then by (1.7) and (2.4) with $\alpha = \beta$ we have that

$$Q \leq \frac{((i+r)\tau)^{\gamma}}{(r\tau)^{\beta}(k\tau)^{\gamma}} \sum_{j=1}^{k-1} \|R^{i+r}(\tau A) - R^i(\tau A)\|_{E\to E}\|AR^{k+1-j}\|_{E\to E}((k-j)\tau)^{\beta}\tau$$

$$\leq M \frac{((i+r)\tau)^\gamma}{(r\tau)^\beta (k\tau)^\gamma} \frac{(r\tau)^\beta}{(i\tau)^\beta} \sum_{j=1}^{k-1} \frac{\tau}{((k+1-j)\tau)^{1-\beta}}$$

$$\leq M \frac{2^\beta ((i+r)\tau)^\gamma}{((i+r)\tau)^\beta (k\tau)^\gamma} \int_0^{k\tau} \frac{ds}{(k\tau - s)^{1-\beta}}$$

$$= M \frac{2^\beta}{\beta} \frac{(k\tau)^{\beta-\gamma}}{((i+r)\tau)^{\beta-\gamma}} = M \frac{2^\beta}{\beta} \left(\frac{k}{i+r} \right)^{\beta-\gamma} \leq \frac{M_1}{\beta}.$$

If $i < k \leq i+r$ and $r > i$, then by (1.7) and (2.4) with $\alpha = 0$ we have

$$Q \leq \frac{((i+r)\tau)^\gamma}{(r\tau)^\beta (k\tau)^\gamma} \sum_{j=1}^{k-1} \|R^{i+r}(\tau A) - R^i(\tau A)\|_{E \to E} \|AR^{k+1-j}(\tau A)\|_{E \to E} ((k-j)\tau)^\beta \tau$$

$$\leq M \frac{2^\beta ((i+r)\tau)^\gamma}{((i+r)\tau)^\beta (k\tau)^\gamma} \sum_{j=1}^{k-1} \frac{\tau}{((k+1-j)\tau)^{1-\beta}}$$

$$\leq \frac{M 2^\beta}{((i+r)\tau)^{\beta-\gamma} (k\tau)^\gamma} \int_0^{k\tau} \frac{ds}{(k\tau - s)^{1-\beta}}$$

$$= \frac{M 2^\beta}{\beta} \frac{(k\tau)^{\beta-\gamma}}{((i+r)\tau)^{\beta-\gamma}} = \frac{M 2^\beta}{\beta} \left(\frac{k}{i+r} \right)^{\beta-\gamma} \leq \frac{M_1}{\beta}.$$

If $i+r < k$ and $r \leq i$, then by (2.5) with $\alpha = 2$ we have

$$Q \leq \frac{((i+r)\tau)^\gamma}{(r\tau)^\beta (k\tau)^\gamma} \sum_{j=1}^{k-1} \|A[R^{i+r+k-j+1}(\tau A) - R^{i+k-j+1}(\tau A)]\|_{E \to E} ((k-j)\tau)^\beta \tau$$

$$\leq M \frac{((i+r)\tau)^\gamma}{(r\tau)^\beta (k\tau)^\gamma} \sum_{j=1}^{k-1} \frac{r\tau((k-j)\tau)^\beta \tau}{((i+k-j+1)\tau)^2}$$

$$\leq M \left(\frac{i+r}{k} \right)^\gamma (r\tau)^{1-\beta} \int_0^{k\tau} \frac{ds}{(k\tau + i\tau - s)^{2-\beta}}$$

$$\leq \frac{M}{1-\beta} \frac{(r\tau)^{1-\beta}}{(i\tau)^{1-\beta}} = \frac{M}{1-\beta} \left(\frac{r}{i} \right)^{1-\beta} \leq \frac{M}{1-\beta}.$$

Finally, if $i+r < k$ and $r > i$, then using the triangle inequality and (2.5) with $\alpha = 0$ and $\alpha = 1$ we obtain

$$Q \leq \frac{((i+r)\tau)^\gamma}{(r\tau)^\beta (k\tau)^\gamma} \left[\sum_{j=1}^{k-r} \|A[R^{i+r+k-j+1}(\tau A) - R^{i+k-j+1}(\tau A)]\|_{E \to E} ((k-j)\tau)^\beta \tau \right.$$

$$+ \sum_{j=k-r+1}^{k-1} \|A[R^{i+r+k-j+1}(\tau A) - R^{i+k-j+1}(\tau A)]\|_{E \to E}((k-j)\tau)^{\beta}\tau\Big]$$

$$\le M \frac{2^{\gamma}}{(r\tau)^{\beta-\gamma}((i+r)\tau)^{\gamma}} \times$$

$$\times \left[\sum_{j=1}^{k-r} \frac{r\tau\,\tau}{((i+k-j+1)\tau)^{2-\beta}} + \sum_{j=k-r+1}^{k-1} \frac{\tau}{((i+k-j+1)\tau)^{1-\beta}}\right]$$

$$\le \frac{M2^{\gamma}}{(r\tau)^{\beta}} \left[\int_0^{(k-r)\tau} \frac{ds\,r\tau}{(k\tau - s)^{2-\beta}} + \int_{(k-r)\tau}^{k\tau} \frac{ds}{(k\tau - s)^{1-\beta}}\right]$$

$$\le \frac{M2^{\gamma}}{(r\tau)^{\beta}} \left[\frac{r\tau}{(1-\beta)(r\tau)^{1-\beta}} + \frac{(r\tau)^{\beta}}{\beta}\right] = \frac{M2^{\gamma}}{(1-\beta)\beta} = \frac{M_1}{\beta(1-\beta)}.$$

Combining the estimates for Q we obtain

$$\sup_{1 \le i < i+r \le N} \frac{((i+r)\tau)^{\gamma}}{(r\tau)^{\beta}} \|A[R^{i+r}(\tau A) - R^i(\tau A)]V_3\|_E \le \frac{M}{\beta(1-\beta)} \|\varphi^{\tau}\|_{C_0^{\beta,\gamma}(\tau,E)}.$$

This inequality and (5.18) yield

$$\langle V_3 \rangle_1^{\beta,\gamma} \le \frac{M}{\beta(1-\beta)} \|\varphi^{\tau}\|_{C_0^{\beta,\gamma}(\tau,E)}.$$

Combining the estimates for V_1, V_2, and V_3 we obtain

$$\langle u_k - A^{-1}\varphi_k \rangle_1^{\beta,\gamma} \le M\left[\langle u_0 - A^{-1}\varphi_1 \rangle_1^{\beta,\gamma} + \frac{1}{\beta(1-\beta)} \|\varphi^{\tau}\|_{C_0^{\beta,\gamma}(\tau,E)}\right].$$

Since M does not depend on τ and $\varphi \in C_0^{\beta,\gamma}(E)$ and $u_0 - A^{-1} \in E^{\beta,\gamma}$, inequality (5.14) holds. Theorem 5.2 is proved.

Let us comment on the results of the first two subsections of this section. The estimates obtained show that the degree γ of the weight factor may be an arbitrary number in the interval $[0, \beta]$. In particular, this means that the problem (0.6) is well posed in spaces of Hölder functions (with respect to time). The Hölder degree β must belong to $(0, 1)$, which does not allow us to establish the well-posedness of problem (0.6) in spaces of continuous functions. These conclusions refer to the case of an arbitrary Banach space E. In subsequent subsections we will study problem (0.6) in certain restrictions of the arbitrary space E. For such restrictions we are able to establish the well-posedness of problem (0.6) in the case of continuous functions. The first result of this type will be given in Section 4.

3.   Estimates for powers of the resolvent.

Let us establish some estimates for powers of the resolvent in the fractional norms E'_α that will be needed in the sequel. We take $E'_0 = E$.

Lemma 5.1. *The following estimates hold for any* $k = n + 1, \cdots, N$:

$$\|A^n R^k(\tau A)\|_{E'_\alpha \to E'_\alpha} \leq \frac{M}{(k\tau)^n}, \quad n = 0, 1, 2, \ 0 < \alpha < 1, \tag{5.19}$$

where M *does not depend on* τ, α, *and* k.

Proof. By the semigroup property of $R^k(\tau A)$, (1.7) and (0.8) imply

$$\|A^n R^k(\tau A)\|_{E \to E} \leq \frac{M}{(k\tau)^n}, \tag{5.20}$$

which in turn implies (5.19) thanks to the fact that the operators $A^n R^k(\tau A)$ and $(\lambda + A)^{-1}$ commute. Lemma 5.1 is proved.

Further, we shall need the inequality

$$|z^{-\delta}[(1 + \tau z)^{-(k+r)} - (1 + \tau z)^{-k}]| \leq M \min \left\{ ((k + r)\tau)^\delta, \frac{r\tau}{(k\tau)^{1-\delta}} \right\},$$

$$z = \rho e^{\pm i\phi}, \ 0 \leq \rho < \infty, \ 0 \leq \phi < \frac{\pi}{2}, \ 1 \leq k < k + r \leq N. \tag{5.21}$$

To prove it, note first that since

$$|1 + \tau z|^{-k} \leq 1, \quad |z(1 + \tau z)^{-k}| \leq \frac{M(\phi)}{k\tau},$$

we have

$$|z^\alpha (1 + \tau z)^{-k}| \leq \frac{M(\phi)}{(k\tau)^\alpha}, \quad \alpha \geq 0. \tag{5.22}$$

Using the elementary identity

$$|z^{-\delta}[(1 + \tau z)^{-(k+r)} - (1 + \tau z)^{-k}]| = \left| z^{1-\delta} \sum_{l=0}^{r-1} (1 + \tau z)^{-(k+l+1)} \tau \right| \tag{5.23}$$

and the estimate (5.22), we obtain

$$|z^{-\delta}[(1+\tau z)^{-(k+r)} - (1+\tau z)^{-k}]| \leq M\frac{r\tau}{(k\tau)^{1-\delta}}. \tag{5.24}$$

Now let us establish the inequality

$$|z^{-\delta}[(1+\tau z)^{-(k+r)} - (1+\tau z)^{-k}]| \leq M((k+r)\tau)^{\delta}. \tag{5.25}$$

We need to examine two cases: $\rho \leq \frac{1}{(k+r)\tau}$ and $\rho > \frac{1}{(k+r)\tau}$. In the first case, (5.22) and (5.23) yield

$$|z^{-\delta}[(1+\tau z)^{-(k+r)} - (1+\tau z)^{-k}]| \leq \frac{Mr\tau}{((k+r)\tau)^{1-\delta}} \leq M((k+r)\tau)^{\delta}.$$

In the second case, the triangle inequality and (5.22) yield

$$|z^{-\delta}[(1+\tau z)^{-(k+r)} - (1+\tau z)^{-k}]| \leq 2((k+r)\tau)^{\delta}.$$

This completes the proof of (5.25). Inequality (5.21) follows from (5.24) and (5.26).

Lemma 5.2. *The following estimates hold for any $1 \leq k < k + r \leq N$:*

$$\|R^k(\tau A) - R^{k+r}(\tau A)\|_{E'_{\alpha-\gamma} \to E'_{\alpha-\beta}} \leq M(\alpha,\gamma)((k+r)\tau)^{\beta-\gamma}, \tag{5.26}$$

in which

$$M(\alpha,\gamma) = \begin{cases} \frac{M}{(\alpha-\gamma)(1-\alpha)}, & 0 \leq \gamma \leq \beta \leq \alpha, \ 0 < \alpha < 1, \ \alpha \neq \gamma, \\ M, & \alpha = \gamma = \beta, \end{cases}$$

where M does not depend on τ, α, β, γ, k, and r.

Proof. By (4.2), we have

$$\lambda^{\alpha-\beta}A(\lambda+A)^{-1}[R^k(\tau A) - R^{k+r}(\tau A)]x$$

$$= \frac{\lambda^{\alpha-\beta}}{2\pi i}\int_{S_1\cup S_2} \frac{1}{\lambda+z}[(1+\tau z)^{-k} - (1+\tau z)^{-(k+r)}]A(z-A)^{-1}xdz. \tag{5.27}$$

Using this identity and the estimate (4.1) we obtain

$$\lambda^{\alpha-\beta}\|A(\lambda+A)^{-1}[R^k(\tau A) - R^{k+r}(\tau A)]x\|_E$$

$$\leq M \int_0^\infty \frac{\lambda^{\alpha-\beta}}{(\lambda+\rho)\rho^{\alpha-\beta}} \rho^{-(\beta-\gamma)} \times$$

$$\times |(1+\tau\rho e^{i\phi})^{-k} - (1+\tau\rho e^{i\phi})^{-(k+r)}| \rho^{\alpha-\gamma} \|A(p+A)^{-1}x\|_E d\rho$$

$$\leq M \int_0^\infty \frac{\lambda^{\alpha-\beta}}{(\lambda+\rho)\rho^{\alpha-\beta}} \rho^{-(\beta-\gamma)} |(1+\tau\rho e^{i\phi})^{-k} - (1+\tau\rho e^{i\phi})^{-(k+r)}| d\rho \|x\|'_{\alpha-\gamma}.$$

From this and the estimate (5.25) with $\delta = \beta - \gamma$ it follows that

$$\lambda^{\alpha-\beta} \|A(\lambda+A)^{-1}[R^k(\tau A) - R^{k+r}(\tau A)]x\|_E$$

$$\leq M \int_0^\infty \frac{\lambda^{\alpha-\beta} d\rho}{(\lambda+\rho)\rho^{\alpha-\beta}} ((k+r)\tau)^{\beta-\gamma} \|x\|'_{\alpha-\gamma} \leq \frac{M}{(\alpha-\beta)(1-\alpha+\beta)} \|x\|'_{\alpha-\gamma} \quad (5.28)$$

for any $\lambda > 0$, whence

$$\|[R^k(\tau A) - R^{k+r}(\tau A)]x\|'_{\alpha-\beta} \leq \frac{M}{(\alpha-\beta)(1-\alpha+\beta)} \|x\|'_{\alpha-\gamma}. \quad (5.29)$$

On the other hand, using (5.22), (5.25), and the inequality

$$\frac{\lambda^{\alpha-\beta}\rho^{1-\alpha+\beta}}{\lambda+\rho} \leq 1, \quad (5.30)$$

we deduce that

$$\lambda^{\alpha-\beta} \|A(\lambda+A)^{-1}[R^k(\tau A) - R^{k+r}(\tau A)]x\|_E$$

$$\leq M \int_0^\infty \frac{1}{\rho^{1+\beta-\gamma}} |(1+\tau\rho e^{i\phi})^{-k} - (1-\tau\rho e^{i\phi})^{-(k+r)}| d\rho \|x\|'_{\alpha-\gamma} =$$

$$M \left(\int_0^{\frac{1}{(k+r)\tau}} + \int_{\frac{1}{(k+r)\tau}}^{\frac{1}{k\tau}} + \int_{\frac{1}{k\tau}}^\infty \right) \frac{1}{\rho^{1+\beta-\gamma}} |(1+\tau\rho e^{i\phi})^{-k} - (1+\tau\rho e^{i\phi})^{-(k+r)}| d\rho \|x\|'_{\alpha-\gamma}$$

$$\leq M_1 \left[\int_0^{\frac{1}{(k+r)\tau}} d\rho ((k+r)\tau)^{1+\beta-\gamma} + \int_{\frac{1}{(k+r)\tau}}^{\frac{1}{k\tau}} \frac{d\rho}{\rho^{1+\beta-\gamma}} \right.$$

$$\left. + \int_{\frac{1}{k\tau}}^\infty \frac{d\rho}{\rho^2} \frac{1}{(k\tau)^{1-\beta+\gamma}} \right] \|x\|'_{\alpha-\gamma} \leq \frac{M_2}{\beta-\gamma} ((k+r)\tau)^{\beta-\gamma} \|x\|'_{\alpha-\gamma}.$$

Thus, we have shown that the inequality

$$\lambda^{\alpha-\beta} \|A(\lambda+A)^{-1}[R^k(\tau A) - R^{k+r}(\tau A)]x\|_E \leq \frac{M_2}{\beta-\gamma} ((k+r)\tau)^{\beta-\gamma} \|x\|'_{\alpha-\gamma}$$

holds for all $\lambda > 0$. Consequently,

$$\|[R^k(\tau A) - R^{k+r}(\tau A)]x\|_{\alpha-\beta}' \leq \frac{M_2}{\beta-\gamma}((k+r)\tau)^{\beta-\gamma}\|x\|_{\alpha-\gamma}',$$

which in conjunction with (5.29) yields the estimates (5.26) for $0 \leq \gamma \leq \beta \leq \alpha$, $\alpha \neq \gamma$. The estimate (5.26) for $\alpha = \gamma$ follows from the triangle inequality. Lemma 5.2 is proved.

Lemma 5.3. *The following estimates hold for any $1 \leq k < k + r \leq N$:*

$$\|R^k(\tau A) - R^{k+r}(\tau A)\|_{E_{\alpha-\gamma}' \to E_{\alpha-\beta}'}$$

$$\leq M\frac{r\tau}{(k\tau)^{1+\gamma-\beta}}, \quad 0 \leq \gamma \leq \beta \leq \alpha, \ 0 < \alpha < 1, \qquad (5.31)$$

where M does not depend on τ, α, β, γ, k, and r.

Proof. Using the inequalities (5.28), (5.30) and the estimates (5.22), (5.24), we obtain

$$\lambda^{\alpha-\beta}\|A(\lambda+A)^{-1}[R^k(\tau A) - R^{k+r}(\tau A)]x\|_E$$

$$\leq M\Bigg[\int_0^{\frac{1}{k\tau}} \frac{1}{\rho^{1-\beta+\gamma}}|(1+\tau e^{i\phi})^{-k} - (1+\tau e^{i\phi})^{-(k+r)}|d\rho$$

$$+ \int_{\frac{1}{k\tau}}^{\infty} \frac{1}{\rho^2}\rho^{1+\beta-\gamma}|(1+\tau e^{i\phi})^{-k} - (1+\tau e^{i\phi})^{-(k+r)}|d\rho\Bigg]\|x\|_{\alpha-\gamma}'$$

$$\leq M\Bigg[\int_0^{\frac{1}{k\tau}} d\rho\frac{r\tau}{(k\tau)^{-\beta+\gamma}} + \int_{\frac{1}{k\tau}}^{\infty} \frac{d\rho}{\rho^2}\frac{r\tau}{(k\tau)^{2+\beta-\gamma}}\Bigg]\|x\|_{\alpha-\gamma}' \leq \frac{M_1 r\tau}{(k\tau)^{1-\beta+\gamma}}\|x\|_{\alpha-\gamma}'.$$

Thus, we have established the inequality

$$\lambda^{\alpha-\beta}\|A(\lambda+A)^{-1}[R^k(\tau A) - R^{k+r}(\tau A)]x\|_E \leq \frac{M_1 r\tau}{(k\tau)^{1-\beta+\gamma}}\|x\|_{\alpha-\gamma}'$$

for all $\lambda > 0$. This yields (5.31). Lemma 5.3 is proved.

4. The coercivity inequality for the general problem.

Let us study now problem (0.6) in the spaces $C_0^{\beta,\gamma}(E_{\alpha-\beta})$ ($0 \leq \gamma \leq \beta \leq \alpha$, $0 < \alpha < 1$). To these there correspond the spaces of traces $E_{1+\alpha-\beta}^{\beta,\gamma}$, which consist of the elements $w_0 \in E$ for which the norm

$$\langle w_0\rangle_{1+\alpha-\beta}^{\beta,\gamma} = \sup_{0 < \tau \leq \tau_0}\Bigg[\max_{1 \leq l \leq N}\|AR^l(\tau A)w_0\|_{\alpha-\beta}'$$

$$+ \sup_{1 \le l < l+r \le N} \frac{((l+r)\tau)^\gamma \|A[R^{l+r}(\tau A) - R^l(\tau A)]w_0\|_{\alpha - \beta}}{(r\tau)^\beta} \right] \tag{5.32}$$

is finite.

Theorem 5.3. *Let* $-A$ *be the generator of an analytic semigroup in* E *and* $u_0 - A^{-1}\varphi_1 \in E_{1+\alpha-\beta}^{\beta,\gamma}$, $\varphi \in C_0^{\beta,\gamma}(E_{\alpha-\beta})$. *The solution of the difference problem* (0.6) *obeys the coercivity inequality*

$$\|\overline{\mathcal{D}}\,\overline{\Pi}(u_0)u\|_{C_0^{\beta,\gamma}(E_{\alpha-\beta})} + \|\overline{A}u\|_{C_0^{\beta,\gamma}(E_{\alpha-\beta})} + \max_{1 \le k \le N} \langle u_k - A^{-1}\varphi_k \rangle_{1+\alpha-\beta}^{\beta,\gamma}$$

$$\le M \left[\langle u_0 - A^{-1}\varphi_1 \rangle_{1+\alpha-\beta}^{\beta,\gamma} + \frac{1}{\alpha(1-\alpha)}\|\varphi\|_{C_0^{\beta,\gamma}(E_{\alpha-\beta})} \right], \tag{5.33}$$

where M *does not depend on* α, β, γ, u_0, *and* φ.

Proof. The well-posedness of problem (0.6) in $C_0^{\beta,\gamma}(E)$ and the estimate (5.19) imply the inequality (5.33) with the constant $\frac{M}{\beta(1-\beta)}$, $0 < \beta < 1$. Here, in the case $E = E_{\alpha-\beta}'$, $0 \le \beta \le \alpha$, $0 < \alpha < 1$, we sharpen the estimates obtained in the first subsection for the solutions of the difference problem (0.6). To this end we shall use the scheme of the proof of Theorems 5.1 and 5.2.

From (0.2) we obtain the identity

$$-\tau^{-1}(u_k - u_{k-1}) = -R^k(\tau A)[Au_0 - \varphi_1] + R^k(\tau A)(\varphi_k - \varphi_1)$$

$$+ \sum_{j=1}^{k-1} AR^{k+1-j}(\tau A)(\varphi_j - \varphi_k)\tau \equiv w_k + p_k + g_k, \quad 1 \le k \le N. \tag{5.34}$$

First let us estimate the grid functions

$$A^{-1}w^\tau = \{A^{-1}w_k\}_{k=1}^N, \quad A^{-1}p^\tau = \{A^{-1}p_k\}_{k=1}^N, \quad A^{-1}g^\tau = \{A^{-1}g_k\}_{k=1}^N$$

in the norm of the space $C(\tau, E_{1+\alpha-\beta}^{\beta,\gamma})$. By Theorem 5.2 and the estimate (5.19),

$$\langle A^{-1}w_k \rangle_{1+\alpha-\beta}^{\beta,\gamma} \le M \langle u_0 - A^{-1}\varphi_1 \rangle_{1+\alpha-\beta}^{\beta,\gamma}, \quad 1 \le k \le N, \tag{5.35}$$

and

$$\langle A^{-1}p_k \rangle_{1+\alpha-\beta}^{\beta,\gamma} \le M \|\varphi^\tau\|_{C_0^{\beta,\gamma}(\tau, E_{\alpha-\beta}')}, \quad 1 \le k \le N, \tag{5.36}$$

where M does not depend on τ, α, β, and γ (indeed, in the estimates (5.19), (5.16), and (5.17) the constants do not depend on the choice of the space E and the parameters β and γ). Now let us estimate $A^{-1}g^\tau$ in the norm of $C(\tau, E_{1+\alpha-\beta}^{\beta,\gamma})$. From the estimates (5.19) an (5.14) it follows that

$$\langle A^{-1}g_k\rangle_{1+\alpha-\beta}^{\beta,\gamma} \le \frac{M}{\beta(1-\beta)}\|\varphi^\tau\|_{C_0^{\beta,\gamma}(\tau,E_{\alpha-\beta}')}. \tag{5.37}$$

On the other hand, using the identity

$$\lambda^{\alpha-\beta}A(\lambda+A)^{-1}AR^l(\tau A)A^{-1}g_k = \sum_{j=1}^{k-1}\lambda^{\alpha-\beta}A^2(\lambda+A)^{-1}R^{l+k+1-j}(\tau A)(\varphi_j-\varphi_k)\tau$$

$$= \sum_{j=1}^{k-1}\frac{\lambda^{\alpha-\beta}}{2\pi i}\int_{S_1\cup S_2}\frac{z}{\lambda+z}(1+\tau z)^{-(l+k+1-j)}A(z-A)^{-1}(\varphi_j-\varphi_k)\tau dz,$$

the estimate (4.1), and the definition of the spaces $C_0^{\beta,\gamma}(\tau, E_{\alpha-\beta}')$, we obtain

$$\lambda^{\alpha-\beta}\|A(\lambda+A)^{-1}AR^l(\tau A)A^{-1}g_k\|_E$$

$$\le M\int_0^\infty \frac{\lambda^{\alpha-\beta}}{(\lambda+\rho)\rho^{\alpha-\beta}}\sum_{j=1}^{k-1}\rho|1+\tau\rho e^{i\phi}|^{-(l+k+1-j)}\times$$

$$\times((k-j)\tau)^{\beta-\gamma}\tau d\rho\,\|\varphi^\tau\|_{C_0^{\beta,\gamma}(\tau,E_{\alpha-\beta}')}.$$

Since

$$\rho\sum_{j=1}^{k-1}|1+\tau\rho e^{i\phi}|^{-(l+k+1-j)}\tau \le M, \tag{5.38}$$

it follows that

$$\lambda^{\alpha-\beta}\|A(\lambda+A)^{-1}AR^l(\tau A)A^{-1}g_k\|_E$$

$$\le M(k\tau)^{\beta-\gamma}\int_0^\infty \frac{\lambda^{\alpha-\beta}d\rho}{(\lambda+\rho)\rho^{\alpha-\beta}}\,\|\varphi^\tau\|_{C_0^{\beta,\gamma}(\tau,E_{\alpha-\beta}')}$$

$$\le \frac{M}{(\alpha-\beta)(1-\alpha+\beta)}\|\varphi^\tau\|_{C_0^{\beta,\gamma}(\tau,E_{\alpha-\beta}')}$$

for all $\lambda \ge 0$ and $l = 1, 2\cdots, N$. Consequently,

$$\max_{1\le l\le N}\|AR^l(\tau A)A^{-1}g_k\|_{\alpha-\beta}' \le \frac{M}{(\alpha-\beta)(1-\alpha+\beta)}\|\varphi^\tau\|_{C_0^{\beta,\gamma}(\tau,E_{\alpha-\beta}')}. \tag{5.39}$$

Next, using the identity

$$\lambda^{\alpha-\beta}A(\lambda+A)^{-1}A[R^{l+r}(\tau A)-R^l(\tau A)]A^{-1}g_k$$

$$= \sum_{j=1}^{k-1} \lambda^{\alpha-\beta} A^2 (\lambda + A)^{-1} [R^{l+r}(\tau A) - R^l(\tau A)] R^{k+1-j}(\tau A)(\varphi_j - \varphi_k)\tau$$

$$= \frac{1}{2\pi i} \sum_{j=1}^{k-1} \int_{S_1 \cup S_2} \frac{\lambda^{\alpha-\beta} z}{\lambda + z} [(1+\tau z)^{-(l+r)} - (1+\tau z)^{-l}] \times$$

$$\times (1 + \tau z)^{-(k+1-j)} A(z - A)^{-1}(\varphi_j - \varphi_k)\tau dz,$$

the estimate (4.1), and the definition of the spaces $C_0^{\beta,\gamma}(\tau, E'_{\alpha-\beta})$, we obtain

$$\lambda^{\alpha-\beta} \| A(\lambda + A)^{-1} A[R^{l+r}(\tau A) - R^l(\tau A)] A^{-1} g_k \|_E$$

$$\leq M \int_0^\infty \frac{\lambda^{\alpha-\beta}}{(\lambda + \rho)\rho^{\alpha-\beta}} \sum_{j=1}^{k-1} \rho^{1-\gamma+\beta} \rho^{-(\beta-\gamma)} |(1 + \tau\rho e^{i\phi})^{-(l+r)} - (1 + \tau\rho e^{i\phi})^{-l}| \times$$

$$\times |1 + \tau\rho e^{i\phi}|^{-(k+1-j)} ((k-j)\tau)^{\beta-\gamma} \tau d\rho \, \|\varphi^\tau\|_{C_0^{\beta,\gamma}(\tau, E'_{\alpha-\beta})}$$

$$\leq M_1 \min \left\{ \frac{r\tau}{(l\tau)^{1+\gamma-\beta}}, ((l+r)\tau)^{\beta-\gamma} \right\} \int_0^\infty \frac{\lambda^{\alpha-\beta}}{(\lambda + \rho)\rho^{\alpha-\beta}} \sum_{j=1}^{k-1} \rho^{1-\gamma+\beta} \times$$

$$\times |1 + \tau\rho e^{i\phi}|^{-(k+1-j)} ((k-j)\tau)^{\beta-\gamma} \tau d\rho \, \|\varphi^\tau\|_{C_0^{\beta,\gamma}(\tau, E'_{\alpha-\beta})}.$$

By (5.22),

$$\sum_{j=1}^{k-1} \rho^{1-\gamma+\beta} |1 + \tau\rho e^{i\phi}|^{-(k+1-j)} ((k-j)\tau)^{\beta-\gamma} \tau \leq M \sum_{j=1}^{k-1} \rho |1 + \tau\rho e^{i\phi}|^{-\left[\frac{k+1-2}{2}\right]} \tau \leq M_1. \tag{5.40}$$

Hence, we have

$$\lambda^{\alpha-\beta} \| A(\lambda + A)^{-1} A[R^{l+r}(\tau A) - R^l(\tau A)] A^{-1} g_k \|_E$$

$$\leq M \int_0^\infty \frac{\lambda^{\alpha-\beta} d\rho}{(\lambda + \rho)\rho^{\alpha-\beta}} \|\varphi^\tau\|_{C_0^{\beta,\gamma}(\tau, E'_{\alpha-\beta})} \min \left\{ \frac{r\tau}{(l\tau)^{1+\gamma-\beta}}, ((l+r)\tau)^{\beta-\gamma} \right\}$$

$$\leq \frac{M}{(\alpha - \beta)(1 - \alpha + \beta)} \min \left\{ \frac{r\tau}{(l\tau)^{1+\gamma-\beta}}, ((l+r)\tau)^{\beta-\gamma} \right\} \|\varphi^\tau\|_{C_0^{\beta,\gamma}(\tau, E'_{\alpha-\beta})}$$

for any $\lambda \geq 0$. Consequently,

$$\| A[R^{l+r}(\tau A) - R^l(\tau A)] A^{-1} g_k \|'_{\alpha-\beta}$$

$$\leq \frac{M}{(\alpha - \beta)(1 - \alpha + \beta)} \min \left\{ \frac{r\tau}{(l\tau)^{1+\gamma-\beta}}, ((l+r)\tau)^{\beta-\gamma} \right\} \|\varphi^\tau\|_{C_0^{\beta,\gamma}(\tau, E'_{\alpha-\beta})}. \tag{5.41}$$

If $l \geq r$, this yields

$$\frac{((l+r)\tau)^\gamma}{(r\tau)^\beta} \|A[R^{l+r}(\tau A) - R^l(\tau A)]A^{-1}g_k\|'_{\alpha-\beta}$$

$$\leq \frac{M((l+r)\tau)^\gamma}{(\alpha-\beta)(1-\alpha+\beta)(r\tau)^\beta} \frac{r\tau}{(l\tau)^{1+\gamma-\beta}} \|\varphi^\tau\|_{C_0^{\beta,\gamma}(\tau, E'_{\alpha-\beta})}$$

$$\leq \frac{M2^\gamma}{(\alpha-\beta)(1-\alpha+\beta)} \left(\frac{r}{l}\right)^{1-\beta} \|\varphi^\tau\|_{C_0^{\beta,\gamma}(\tau, E'_{\alpha-\beta})}$$

$$\leq \frac{M2^\gamma}{(\alpha-\beta)(1-\alpha+\beta)} \|\varphi^\tau\|_{C_0^{\beta,\gamma}(\tau, E'_{\alpha-\beta})}.$$

If $l \leq r$, (5.39) yields

$$\frac{((l+r)\tau)^\gamma}{(r\tau)^\beta} \|A[R^{l+r}(\tau A) - R^l(\tau A)]A^{-1}g_k\|'_{\alpha-\beta}$$

$$\leq \frac{M((l+r)\tau)^\gamma}{(\alpha-\beta)(1-\alpha+\beta)(r\tau)^\beta} ((l+r)\tau)^{\beta-\gamma} \|\varphi^\tau\|_{C_0^{\beta,\gamma}(\tau, E'_{\alpha-\beta})}$$

$$\leq \frac{M2^\beta}{(\alpha-\beta)(1-\alpha+\beta)} \|\varphi^\tau\|_{C_0^{\beta,\gamma}(\tau, E'_{\alpha-\beta})}.$$

Thus, we have established the inequality

$$\sup_{1 \leq l < l+r \leq N} \frac{((l+r)\tau)^\gamma}{(r\tau)^\beta} \|A[R^{l+r}(\tau A) - R^l(\tau A)]A^{-1}g_k\|'_{\alpha-\beta}$$

$$\leq \frac{M_1}{(\alpha-\beta)(1-\alpha+\beta)} \|\varphi^\tau\|_{C_0^{\beta,\gamma}(\tau, E'_{\alpha-\beta})},$$

which in conjunction with (5.39) gives

$$\langle A^{-1}g_k\rangle_{1+\alpha-\beta}^{\beta,\gamma} \leq \frac{M}{(\alpha-\beta)(1-\alpha+\beta)} \|\varphi^\tau\|_{C_0^{\beta,\gamma}(\tau, E'_{\alpha-\beta})}. \tag{5.42}$$

Further, by (5.37) and (5.42), we have

$$\langle A^{-1}g_k\rangle_{1+\alpha-\beta}^{\beta,\gamma} \leq \frac{M}{1-\alpha} \min\left\{\frac{1}{\beta}, \frac{1}{\alpha-\beta}\right\} \|\varphi^\tau\|_{C_0^{\beta,\gamma}(\tau, E'_{\alpha-\beta})}.$$

Therefore,

$$\langle A^{-1}g_k\rangle_{1+\alpha-\beta}^{\beta,\gamma} \leq \frac{M_1}{\alpha(1-\alpha)} \|\varphi^\tau\|_{C_0^{\beta,\gamma}(\tau, E'_{\alpha-\beta})}. \tag{5.43}$$

From (5.35), (5.36), and (5.43) it follows that

$$\max_{1\leq k\leq N} \langle u_k - A^{-1}\varphi_k\rangle^{\beta,\gamma}_{1+\alpha-\beta}$$

$$\leq M\left[\langle u_0 - A^{-1}\varphi_1\rangle^{\beta,\gamma}_{1+\alpha-\beta} + \frac{1}{\alpha(1-\alpha)}\|\varphi^\tau\|_{C_0^{\beta,\gamma}(\tau,E'_{\alpha-\beta})}\right], \tag{5.44}$$

where M does not depend on τ, α, β, γ, u_0, and φ^τ.

Now let us estimate the grid functions w^τ, p^τ, and g^τ in the norms of $C_0^{\beta,\gamma}(\tau, E'_{\alpha-\beta})$. From the definition of the spaces $E_{1+\alpha-\beta}^{\beta,\gamma}$ it follows that

$$\|w^\tau\|_{C_0^{\beta,\gamma}(\tau,E'_{\alpha-\beta})} = \langle u_0 - A^{-1}\varphi_1\rangle^{\beta,\gamma}_{1+\alpha-\beta}. \tag{5.45}$$

By (5.19) and (5.7), we have

$$\|p^\tau\|_{C_0^{\beta,\gamma}(\tau,E'_{\alpha-\beta})} \leq M\|\varphi^\tau\|_{C_0^{\beta,\gamma}(\tau,E'_{\alpha-\beta})}. \tag{5.46}$$

Finally, let us estimate g^τ in the norm of $C_0^{\beta,\gamma}(\tau, E'_{\alpha-\beta})$. By (5.19) and (5.8),

$$\|g^\tau\|_{C_0^{\beta,\gamma}(\tau,E'_{\alpha-\beta})} \leq \frac{M}{\beta(1-\beta)}\|\varphi^\tau\|_{C_0^{\beta,\gamma}(\tau,E'_{\alpha-\beta})}. \tag{5.47}$$

On the other hand, we can show that

$$\|g^\tau\|_{C_0^{\beta,\gamma}(\tau,E'_{\alpha-\beta})} \leq \frac{M}{(\alpha-\beta)(1-\alpha+\beta)}\|\varphi^\tau\|_{C_0^{\beta,\gamma}(\tau,E'_{\alpha-\beta})}. \tag{5.48}$$

To this end it suffices to show that

$$\|g_k\|'_{\alpha-\beta} \leq \frac{M}{(\alpha-\beta)(1-\alpha+\beta)}\|\varphi^\tau\|_{C_0^{\beta,\gamma}(\tau,E'_{\alpha-\beta})}, \quad 1\leq k\leq N, \tag{5.49}$$

and

$$\|g_{k+r} - g_k\|'_{\alpha-\beta} \leq \frac{M(r\tau)^\beta}{(\alpha-\beta)(1-\alpha+\beta)((k+r)\tau)^\gamma}\|\varphi^\tau\|_{C_0^{\beta,\gamma}(\tau,E'_{\alpha-\beta})}, \tag{5.50}$$

$$1 \leq k < k+r \leq N.$$

First let us prove (5.49). By (4.2) and (5.34), we have

$$\lambda^{\alpha-\beta}A(\lambda+A)^{-1}g_k = \sum_{j=1}^{k-1}\lambda^{\alpha-\beta}A^2(\lambda+A)^{-1}R^{k+1-j}(\tau A)(\varphi_k - \varphi_j)\tau$$

$$= \sum_{j=1}^{k-1} \frac{1}{2\pi i} \int_{S_1 \cup S_2} \lambda^{\alpha-\beta} \frac{z}{\lambda+z} (1+\tau z)^{-(k+1-j)} A(z-A)^{-1} (\varphi_k - \varphi_j) \tau dz.$$

Using the estimates (4.1), (5.38) and the definiton of the spaces $C_0^{\beta,\gamma}(\tau, E'_{\alpha-\beta})$, we obtain

$$\lambda^{\alpha-\beta} \|A(\lambda+A)^{-1} g_k\|_E$$

$$\leq M \int_0^\infty \frac{\lambda^{\alpha-\beta}}{(\lambda+\rho)\rho^{\alpha-\beta}} \sum_{j=1}^{k-1} \rho |1 + \rho\tau e^{i\phi}|^{-(k+1-j)} ((k-j)\tau)^{\beta-\gamma} \|\varphi^\tau\|_{C_0^{\beta,\gamma}(\tau, E'_{\alpha-\beta})}$$

$$\leq M_1 (k\tau)^{\beta-\gamma} \int_0^\infty \frac{\lambda^{\alpha-\beta} d\rho}{(\lambda+\rho)\rho^{\alpha-\beta}} \|\varphi^\tau\|_{C_0^{\beta,\gamma}(\tau, E'_{\alpha-\beta})}$$

$$\leq \frac{M_1 (k\tau)^{\beta-\gamma}}{(\alpha-\beta)(1-\alpha+\beta)} \|\varphi^\tau\|_{C_0^{\beta,\gamma}(\tau, E'_{\alpha-\beta})}$$

for all $\lambda > 0$. Consequently,

$$\|g_k\|'_{\alpha-\beta} \leq M \frac{(k\tau)^{\beta-\gamma}}{(\alpha-\beta)(1-\alpha+\beta)} \|\varphi^\tau\|_{C_0^{\beta,\gamma}(\tau, E'_{\alpha-\beta})} \tag{5.51}$$

for all $k = 1, \cdots, N$. From the last inequality we obtain (5.49), and also the estimate (5.50) if $k \leq r$. Indeed, by (5.51) and the triangle inequality,

$$\|g_{k+r} - g_k\|'_{\alpha-\beta} \leq \|g_{k+r}\|'_{\alpha-\beta} + \|g_k\|'_{\alpha-\beta}$$

$$\leq \frac{M}{(\alpha-\beta)(1-\alpha+\beta)} [((k+r)\tau)^{\beta-\gamma} + (k\tau)^{\beta-\gamma}] \|\varphi^\tau\|_{C_0^{\beta,\gamma}(\tau, E'_{\alpha-\beta})}$$

$$\leq \frac{2^{1+\beta} M}{(\alpha-\beta)(1-\alpha+\beta)} \frac{(r\tau)^\beta}{((k+r)\tau)^\gamma} \|\varphi^\tau\|_{C_0^{\beta,\gamma}(\tau, E'_{\alpha-\beta})}.$$

Now suppose that $k > r$, and let us write the difference $g_{k+r} - g_k$ as a sum of four terms:

$$g_{k+r} - g_k = \sum_{j=k-r+1}^{k+r-1} AR^{k+r+1-j}(\tau A)(\varphi_{k+r} - \varphi_j)\tau$$

$$+ \sum_{j=k-r+1}^{k-1} AR^{k+1-j}(\tau A)(\varphi_j - \varphi_k)\tau$$

$$+ \sum_{j=1}^{k-r} A[R^{k+r+1-j}(\tau A) - R^{k+1-j}(\tau A)](\varphi_k - \varphi_j)\tau$$

$$+\sum_{j=1}^{k-r} AR^{k+r+1-j}(\tau A)\tau(\varphi_{k+r} - \varphi_k) = \mathcal{J}_1 + \mathcal{J}_2 + \mathcal{J}_3 + \mathcal{J}_4.$$

We will estimate the terms $\mathcal{J}_i$, $i = 1, \cdots, 4$, separately. First we deal with $\mathcal{J}_1$. Using the identity

$$\lambda^{\alpha-\beta} A(\lambda + A)^{-1}\mathcal{J}_1 =$$

$$\frac{1}{2\pi i}\sum_{j=k-r+1}^{k+r-1}\lambda^{\alpha-\beta}\int_{S_1\cup S_2}(\lambda + z)^{-1}(1 + \tau z)^{-(k+r+1-j)}A(z - A)^{-1}(\varphi_{k+r} - \varphi_j)\tau dz,$$

the estimates (4.1), (5.22), (5.38), and the definition of the spaces $C_0^{\beta,\gamma}(\tau, E'_{\alpha-\beta})$, we obtain

$$\lambda^{\alpha-\beta}\|A(\lambda + A)^{-1}\mathcal{J}_1\|_E \le$$

$$M\int_0^\infty\sum_{j=k-r+1}^{k+r-1}\frac{\lambda^{\alpha-\beta}\rho|1 + \tau\rho e^{i\phi}|^{-(k+r+1-j)}d\rho((k+r-j)\tau)^\beta\tau}{(\lambda+\rho)\rho^{\alpha-\beta}((k+r)\tau)^\gamma}\|\varphi^\tau\|_{C_0^{\beta,\gamma}(\tau,E'_{\alpha-\beta})}$$

$$\le\frac{M(r\tau)^\beta}{((k+r)\tau)^\gamma}\int_0^\infty\frac{\lambda^{\alpha-\beta}}{(\lambda+\rho)\rho^{\alpha-\beta}}\times$$

$$\times\sum_{j=k-r+1}^{k+r-1}\rho|1 + \tau\rho e^{i\phi}|^{-(k+r+1-j)}\tau d\rho\,\|\varphi^\tau\|_{C_0^{\beta,\gamma}(\tau,E'_{\alpha-\beta})}$$

$$\le\frac{M_1(r\tau)^\beta}{((k+r)\tau)^\gamma}\int_0^\infty\frac{\lambda^{\alpha-\beta}d\rho}{(\lambda+\rho)\rho^{\alpha-\beta}}\|\varphi^\tau\|_{C_0^{\beta,\gamma}(\tau,E'_{\alpha-\beta})}$$

$$\le\frac{M_1}{(\alpha-\beta)(1-\alpha+\beta)}\frac{(r\tau)^\beta}{((k+r)\tau)^\gamma}\|\varphi^\tau\|_{C_0^{\beta,\gamma}(\tau,E'_{\alpha-\beta})}$$

for all $\lambda > 0$. This yields the estimate

$$\|\mathcal{J}_1\|'_{\alpha-\beta} \le \frac{M_1}{(\alpha-\beta)(1-\alpha+\beta)}\frac{(r\tau)^\beta}{((k+r)\tau)^\gamma}\|\varphi^\tau\|_{C_0^{\beta,\gamma}(\tau,E'_{\alpha-\beta})}.$$

In exactly the same manner one obtains the estimate

$$\|\mathcal{J}_2\|'_{\alpha-\beta} \le \frac{M}{(\alpha-\beta)(1-\alpha+\beta)}\frac{(r\tau)^\beta}{(k\tau)^\gamma}\|\varphi^\tau\|_{C_0^{\beta,\gamma}(\tau,E'_{\alpha-\beta})}.$$

Since $k > r$, this yields

$$\|\mathcal{J}_2\|'_{\alpha-\beta} \le \frac{M2^\gamma}{(\alpha-\beta)(1-\alpha+\beta)}\frac{(r\tau)^\beta}{((k+r)\tau)^\gamma}\|\varphi^\tau\|_{C_0^{\beta,\gamma}(\tau,E'_{\alpha-\beta})}.$$

Next let us estimate $\mathcal{J}_4$. Since

$$\mathcal{J}_4 = [R^{2r}(\tau A) - R^{k+r}(\tau A)] \, (\varphi_{k+r} - \varphi_k),$$

(5.19) and the triangle inequality give

$$\|\mathcal{J}_4\|'_{\alpha-\beta} \leq \left[\|R^{2r}(\tau A)\|_{E'_{\alpha-\beta} \to E'_{\alpha-\beta}} + \|R^{k+r}(\tau A)\|_{E'_{\alpha-\beta} \to E'_{\alpha-\beta}} \right] \|\varphi_{k+r} - \varphi_k\|'_{\alpha-\beta}$$

$$\leq \frac{2M(r\tau)^\beta}{((k+r)\tau)^\gamma} \|\varphi^\tau\|_{C_0^{\beta,\gamma}(\tau, E'_{\alpha-\beta})}.$$

Finally, let us estimate $\mathcal{J}_3$. From the definition of the spaces $E'_{\alpha-\beta}$ it follows that

$$\|\mathcal{J}_3\|'_{\alpha-\beta} \leq \sum_{j=1}^{k-r} \|A[R^{k+r+1-j}(\tau A) - R^{k+1-j}(\tau A)]\|_{E \to E} \|\varphi_k - \varphi_j\|'_{\alpha-\beta}\tau,$$

which in conjunction with the estimate (2.4) yields

$$\|\mathcal{J}_3\|'_{\alpha-\beta} \leq M \sum_{j=1}^{k-r} \frac{r\tau((k+1-j)\tau)^\beta \tau}{((k+1-j)\tau)^2 (k\tau)^\gamma} \|\varphi^\tau\|_{C_0^{\beta,\gamma}(\tau, E'_{\alpha-\beta})}.$$

Since

$$\sum_{j=1}^{k-r} \frac{((k+1-j)\tau)^\beta \tau}{((k+1-j)\tau)^2} \leq \int_0^{(k-r)\tau} \frac{ds}{(\tau-s)^{2-\beta}} \leq \frac{1}{1-\beta} \frac{1}{(r\tau)^{1-\beta}} \leq \frac{1}{1-\alpha} \frac{1}{(r\tau)^{1-\beta}},$$

it follows that

$$\|\mathcal{J}_3\|'_{\alpha-\beta} \leq \frac{M}{1-\alpha} \frac{(r\tau)^\beta}{(k\tau)^\gamma} \|\varphi^\tau\|_{C_0^{\beta,\gamma}(\tau, E'_{\alpha-\beta})}.$$

Hence, since $k > r$, we have

$$\|\mathcal{J}_3\|'_{\alpha-\beta} \leq \frac{M 2^\gamma}{1-\alpha} \frac{(r\tau)^\beta}{((k+r)\tau)^\gamma} \|\varphi^\tau\|_{C_0^{\beta,\gamma}(\tau, E'_{\alpha-\beta})}.$$

Combining the estimates for $\mathcal{J}_i$, $i = 1, \cdots, 4$, we obtain (5.50).

Further, (5.47) and (5.48) imply

$$\|g^\tau\|_{C_0^{\beta,\gamma}(\tau, E'_{\alpha-\beta})} \leq \frac{M}{\alpha(1-\alpha)} \|\varphi^\tau\|_{C_0^{\beta,\gamma}(\tau, E'_{\alpha-\beta})}. \tag{5.52}$$

Now, from (5.45), (5.46), and (5.52) we derive the inequality

$$\left\| \left\{ \frac{1}{\tau}(u_k - u_{k-1}) \right\} \right\|_{C_0^{\beta,\gamma}(\tau, E'_{\alpha-\beta})}$$

$$\leq M\left[\langle u_0 - A^{-1}\varphi_1\rangle^{\beta,\gamma}_{1+\alpha-\beta} + \frac{1}{\alpha(1-\alpha)}\|\varphi^\tau\|_{C_0^{\beta,\gamma}(\tau,E'_{\alpha-\beta})}\right].$$

The estimate for Au^τ in the norm of $C_0^{\beta,\gamma}(\tau, E'_{\alpha-\beta})$ follows via the triangle inequality. In the estimates obtained above the constants do not depend on τ and $u_0 - A^{-1}\varphi_1 \in E^{\beta,\gamma}_{1+\alpha-\beta}$, $\varphi \in C_0^{\beta,\gamma}(E'_{\alpha-\beta})$. Hence, $\overline{\mathcal{D}}\,\overline{\Pi}(u_0)u$, $\overline{A}u \in C_0^{\beta,\gamma}(E'_{\alpha-\beta})$, $u - A^{-1}\varphi \in \mathcal{C}(E^{\beta,\gamma}_{1+\alpha-\beta})$, and the coercivity inequality (5.39) holds. Theorem 5.3 is proved.

By (5.26) and (5.31), we have

$$\langle w_0\rangle^{\beta,\gamma}_{1+\alpha-\beta} \leq M(\alpha,\gamma)\|Aw_0\|'_{\alpha-\gamma} \quad \text{(for all } Aw_0 \in E'_{\alpha-\gamma}\text{)}, \tag{5.53}$$

where

$$M(\alpha,\gamma) = \begin{cases} \frac{M}{(\alpha-\gamma)(1-\alpha)}, & 0 \leq \gamma \leq \beta \leq \alpha, \ 0 < \alpha < 1, \ \alpha \neq \gamma, \\ \frac{M}{\alpha(1-\alpha)}, & \alpha = \beta = \gamma. \end{cases}$$

We were not able to establish the opposite inequality expressing the equivalence of these two norms. Nevertheless, the following result holds true.

Theorem 5.4. *Let $-A$ be the generator of an analytic semigroup in E and $Au_0 - \varphi_1 \in E'_{\alpha-\gamma}$, $\varphi \in C_0^{\beta,\gamma}(E'_{\alpha-\beta})$, $0 \leq \gamma \leq \beta \leq \alpha$, $0 < \alpha < 1$. Then the solutions of the difference problem (0.6) obey the coercivity inequality*

$$\|\overline{\mathcal{D}}\,\overline{\Pi}(u_0)u\|_{C_0^{\beta,\gamma}(E'_{\alpha-\beta})} + \|\overline{A}u\|_{C_0^{\beta,\gamma}(E'_{\alpha-\beta})} + \max_{1\leq k\leq N}\|Au_k - \varphi_k\|'_{\alpha-\gamma}$$

$$\leq M(\alpha,\gamma)[\|Au_0 - \varphi_1\|'_{\alpha-\gamma} + \|\varphi\|_{C_0^{\beta,\gamma}(E'_{\alpha-\beta})}], \tag{5.54}$$

in which

$$M(\alpha,\gamma) = \begin{cases} \frac{M}{(\alpha-\gamma)(1-\alpha)}, & 0 \leq \gamma \leq \beta \leq \alpha, \ 0 < \alpha < 1, \ \alpha \neq \gamma, \\ \frac{M}{\alpha(1-\alpha)}, & \alpha = \beta = \gamma, \end{cases}$$

where M does not depend on α, β, γ, u_0, and φ.

Proof. It suffices to establish the needed estimates for w^τ, p^τ, and g^τ in the norm of $C(\tau, E'_{\alpha-\gamma})$ and for w^τ in the norm of $C_0^{\beta,\gamma}(E'_{\alpha-\beta})$. Using the estimate (5.3) with $n = 0$, we obtain

$$\|w_k\|'_{\alpha-\gamma} \leq \|R^k(\tau A)\|_{E'_{\alpha-\gamma}\to E'_{\alpha-\gamma}}\|Au_0 - \varphi_1\|'_{\alpha-\gamma} \leq M\|Au_0 - \varphi_1\|'_{\alpha-\gamma}$$

for all $k = 1, \cdots, N$. Hence

$$\|w^\tau\|_{C(\tau, E'_{\alpha-\gamma})} \le M\|Au_0 - \varphi_1\|_{\alpha-\gamma}. \tag{5.55}$$

Next, using the identity

$$\lambda^{\alpha-\gamma} A(\lambda + A)^{-1} p_k = -\frac{\lambda^{\alpha-\gamma}}{2\pi i} \int_{S_1 \cup S_2} (\lambda + z)^{-1}(1 + \tau z)^{-k} A(z - A)^{-1}(\varphi_k - \varphi_1)dz,$$

the definition of the spaces $E'_{\alpha-\beta}$, and the estimate (5.22), we obtain

$$\lambda^{\alpha-\gamma}\|A(\lambda + A)^{-1}p_k\|_E \le M \int_0^\infty \frac{\lambda^{\alpha-\gamma}|1 + \tau\rho e^{i\phi}|^{-k}}{\lambda + \rho}\|A(\rho + A)^{-1}(\varphi_k - \varphi_1)\|_E d\rho$$

$$\le M \int_0^\infty \frac{\lambda^{\alpha-\gamma}d\rho\,(k\tau)^{\beta-\gamma}}{(\lambda + \rho)\rho^{\beta-\gamma}(k\tau)^{\beta-\gamma}\rho^{\alpha-\beta}}\|\varphi^\tau\|_{C_0^{\beta,\gamma}(\tau, E'_{\alpha-\beta})}$$

$$\le M \int_0^\infty \frac{dr}{(1 + r)r^{\alpha-\gamma}}\|\varphi^\tau\|_{C_0^{\beta,\gamma}(\tau, E'_{\alpha-\beta})} \le \frac{M_1}{(\alpha - \gamma)(1 - \alpha)}\|\varphi^\tau\|_{C_0^{\beta,\gamma}(\tau, E'_{\alpha-\beta})}.$$

This shows that

$$\|p_k\|'_{\alpha-\gamma} \le \frac{M_1}{(\alpha - \gamma)(1 - \alpha)}\|\varphi^\tau\|_{C_0^{\beta,\gamma}(\tau, E'_{\alpha-\beta})}$$

for any $k = 1, \cdots, N$. Consequently,

$$\|p^\tau\|_{C(\tau, E'_{\alpha-\gamma})} \le \frac{M_1}{(\alpha - \gamma)(1 - \alpha)}\|\varphi^\tau\|_{C_0^{\beta,\gamma}(\tau, E'_{\alpha-\beta})}. \tag{5.56}$$

Now let us estimate g^τ in $C(\tau, E'_{\alpha-\gamma})$. Using the identity

$$\lambda^{\alpha-\gamma} A(\lambda + A)^{-1} g_k$$

$$= \frac{\lambda^{\alpha-\gamma}}{2\pi i} \int_{S_1 \cup S_2} \sum_{j=1}^{k-1} \frac{z}{\lambda + z}(1 + \tau z)^{-(k+1-j)} A(z - A)^{-1}(\varphi_j - \varphi_k)dz\tau,$$

and the estimates (4.1), (5.22), and (5.40), we obtain

$$\lambda^{\alpha-\gamma}\|A(\lambda + A)^{-1}g_k\|_E$$

$$\le M \int_0^\infty \frac{\lambda^{\alpha-\gamma}\rho}{\lambda + \rho} \sum_{j=1}^{k-1} |1 + \tau\rho e^{i\phi}|^{-(k+1-j)} \|A(p + A)^{-1}(\varphi_j - \varphi_k)\|_E d\rho\,\tau$$

$$\leq M \int_0^\infty \frac{\lambda^{\alpha-\gamma}\rho^{1-\alpha+\beta}}{\lambda+\rho} \sum_{j=1}^{k-1} |1 + \tau\rho e^{i\phi}|^{-(k+1-j)} d\rho ((k-j)\tau)^{\beta-\gamma}\tau \, \|\varphi^\tau\|_{C_0^{\beta,\gamma}(\tau,E'_{\alpha-\beta})}$$

$$\leq M \int_0^\infty \frac{\lambda^{\alpha-\gamma}}{(\lambda+\rho)\rho^{\alpha-\gamma}} \sum_{j=1}^{k-1} \rho|1 + \tau\rho e^{i\phi}|^{-[\frac{k+1-j}{2}]} d\rho \, \tau \, \|\varphi^\tau\|_{C_0^{\beta,\gamma}(\tau,E'_{\alpha-\beta})}$$

$$\leq M_1 \int_0^\infty \frac{\lambda^{\alpha-\gamma} d\rho}{(\lambda+\rho)\rho^{\alpha-\gamma}} \, \|\varphi^\tau\|_{C_0^{\beta,\gamma}(\tau,E'_{\alpha-\beta})} \leq \frac{M_2}{(\alpha-\gamma)(1-\alpha+\gamma)} \|\varphi^\tau_{C_0^{\beta,\gamma}(\tau,E'_{\alpha-\beta})}.$$

This shows that

$$\|g_k\|'_{\alpha-\gamma} \leq \frac{M_2}{(\alpha-\gamma)(1-\alpha+\gamma)} \|\varphi^\tau\|_{C_0^{\beta,\gamma}(\tau,E'_{\alpha-\beta})}$$

for any $k = 1, \cdots, N$, whence

$$\|g^\tau\|_{C(\tau,E'_{\alpha-\gamma})} \leq \frac{M_2}{(\alpha-\gamma)(1-\alpha+\gamma)} \|\varphi^\tau\|_{C_0^{\beta,\gamma}(\tau,E'_{\alpha-\beta})}. \tag{5.57}$$

Finally, let us estimate w^τ in $C_0^{\beta,\gamma}(\tau, E'_{\alpha-\beta})$. To this end it suffices to establish the estimates

$$\|w_k\|'_{\alpha-\beta} \leq M\|Au_0 - \varphi_1\|'_{\alpha-\gamma}, \quad 1 \leq k \leq N, \tag{5.58}$$

and

$$\|w_{k+r} - w_k\|'_{\alpha-\beta} \leq \frac{M(r\tau)^\beta}{((k+r)\tau)^\gamma(\alpha-\gamma)(1-\alpha+\gamma)} \|Au_0 - \varphi_1\|'_{\alpha-\gamma}, \tag{5.59}$$

$$1 \leq k < k+r \leq N.$$

Since $\alpha - \beta \leq \alpha - \gamma$, we have $\|w_k\|'_{\alpha-\beta} \leq M\|w_k\|'_{\alpha-\gamma}$. Hence, (5.58) follows from (5.55). The estimate (5.59) is a consequence of (5.26) and (5.31). Since in the estimates established above the constants do not depend on τ and $Au_0 - \varphi_1 \in E'_{\alpha-\gamma}$, $\varphi \in C_0^{\beta,\gamma}(E'_{\alpha-\beta})$, we have $\overline{\mathcal{D}}\,\overline{\Pi}(u_0)u$, $\overline{A}u \in C_0^{\beta,\gamma}(E'_{\alpha-\beta})$, $Au - \varphi \in C(E'_{\alpha-\gamma})$, and the coercivity inequality (5.54) holds for $\alpha \neq \gamma$. The estimate (5.54) in the case $\alpha = \gamma$ was established in Subsection 1. Theorem 4.5 is proved.

Theorem 5.4 admits as consequences the theorems on the well-posedness of the difference problem (0.6) proved in Sections 2 and 4.

Let us add that by passing to the limit for $\tau \to 0$ one can recover Theorems 5.1–5.4 of Chapter 1.

CHAPTER 3

PADÉ DIFFERENCE SCHEMES

0. STABILITY OF THE DIFFERENCE PROBLEM

1. Padé approximants of the function e^{-z}.

Let us consider the problem of approximating the function e^{-z} near $z = 0$ by rational functions

$$R_{j,l}(z) = \frac{P_{j,l}(z)}{Q_{j,l}(z)} = \frac{a_0 + a_1 z + \cdots + a_j z^j}{b_0 + b_1 z + \cdots + b_l z^l},$$

$$a_r = a_r(j,l), \ r = 1, \cdots, j, \ b_r = b_r(j,l), \ r = 1, \cdots, l, \ a_j \neq 0, \ b_l \neq 0, \ b_0 \neq 0.$$

With no loss of generality we may assume that $b_0 = 1$. Thus, $R_{j,l}(z)$ contains $j + l + 1$ independent parameters. These parameters can be determined from the condition that the Taylor expansions of the functions e^{-z} and $R_{j,l}(z)$ should coincide up to terms of order $j + l$ as $|z| \to 0$. Notice that $Q_{j,l}(z) \neq 0$ near $z = 0$, since $Q_{j,l}(0) = 1$, and consequently the fraction $R_{j,l}(z)$ is infinitely differentiable near $z = 0$. The coincidence of the indicated Taylor coefficients yields the relation

$$e^{-z} - R_{j,l}(z) = O(|z|^{j+l+1}), \quad |z| \to 0. \tag{0.1}$$

However, this path leads to a nonlinear system of equations for the determination of the coefficients of the polynomials $P_{j,l}(z)$ and $Q_{j,l}(z)$. Since $Q_{j,l}(z) \neq 0$ near $z = 0$, relation (0.1) is equivalent to

$$M(z) = Q_{j,l}(z)e^{-z} - P_{j,l}(z) = O(|z|^{j+l+1}), \quad |z| \to 0. \tag{0.2}$$

This already leads to a linear system of algebraic equations for determining the coefficients of the polynomials in question. To verify relation (0.2), we use the representations

$$Q_{j,l}(z) = \int_0^\infty (z+y)^l y^j e^{-y} dy \frac{1}{(j+l)!}, \quad P_{j,l}(z) = \int_0^\infty (y-z)^j y^l e^{-y} dy \frac{1}{(j+l)!}, \tag{0.3}$$

which follow from Newton's binomial formula and the formula for Euler's gamma function. Since

$$e^{-z} Q_{j,l}(z) = \frac{1}{(j+l)!} \int_0^\infty (z+y)^l y^j e^{-(y+z)} dy = \frac{1}{(j+l)!} \int_z^\infty (y-z)^j y^l e^{-y} dy,$$

we have

$$e^{-z} Q_{j,l}(z) - P_{j,l}(z) = -\frac{1}{(j+l)!} \int_0^z (y-z)^j y^l e^{-y} dy$$

$$= z^{j+l+1} \frac{(-1)^{j+1}}{(j+l)!} \int_0^1 (1-t)^j t^l e^{-tz} dt.$$

Hence, for $z \geq 0$,

$$|e^{-z} Q_{j,l}(z) - P_{j,l}(z)| \leq z^{j+l+1} \frac{1}{(j+l)!} \int_0^1 (1-t)^j t^l dt$$

$$= \frac{z^{j+l+1}}{(j+l)!} B(j+1, l+1) = \frac{z^{j+l+1}}{(j+l)!} \frac{(j+l+1)!}{j! l!} = \frac{z^{j+l+1}(j+l+1)}{j! l!},$$

by the formula for Euler's beta function. On the other hand,

$$|e^{-z} Q_{j,l}(z) - P_{j,l}(z)| \geq z^{j+l+1} e^{-z} \frac{(j+l+1)}{j! l!}.$$

Therefore, the estimate (0.2) is sharp near $z = 0$.

Further, using formulas (0.3), let us find the coefficients of the polynomials $Q_{j,l}(z)$ and $P_{j,l}(z)$. They are given by the formulas

$$Q_{j,l}(z) = \sum_{r=0}^j \frac{(j+l-r)! j!}{(j+l)! r! (j-r)!} (-z)^r, \quad P_{j,l}(z) = \sum_{r=0}^l \frac{(j+l-r)! l!}{(j+l)! r! (l-r)!} z^r. \tag{0.4}$$

Finally, let us give examples of the simplest Padé fractions for the function e^{-z}. From formulas (0.4) it follows that

$$R_{0,1}(z) = \frac{1}{1+z}, \quad R_{1,1}(z) = \frac{1-z/2}{1+z/2}, \quad R_{0,2}(z) = \frac{1}{1+z+z^2/2},$$

$$R_{1,2}(z) = \frac{1 - z/3}{1 + 2z/3 + z^2/6}, \quad R_{2,2}(z) = \frac{1 - z/2 + z^2/8}{1 + z/2 + z^2/8}.$$

The first fraction corresponds to an implicit difference scheme of first order of accuracy for a parabolic equation, the second to a Crank-Nicolson scheme of second order of accuracy, the third to an implicit difference scheme of second order of accuracy, the fourth to an implicit difference scheme of third order of accuracy, and the fifth to a generalized Crank-Nicolson difference scheme of fourth order of accuracy. Here we gave examples of fractions $R_{j,l}(z)$ with $l - 2 \le j \le l$, which generate stable difference schemes.

The investigation of the stability of difference schemes relies in an essential manner on information about the disposition of the roots of the polynomial $Q_{j,l}(z)$. It is known that for $l - 4 \le j \le l$ the roots of this polynomial lie in the open half-plane $\mathbf{C}^- = \{z \in \mathbf{C}, \operatorname{Re} z < 0\}$. For the examples of Padé fractions given above this can be verified directly. The investigation of the stability and well-posedness of Padé difference schemes relies on a number of properties of the rational functions $R_{j,l}(z)$ that generate them.

Lemma 0.1. *The following relation holds:*

$$R_{j,l}(z) + R'_{j,l}(z) = a_j b_l z^{j+l} (Q_{j,l}(z))^{-2}, \tag{0.5}$$

where $a_j = (-1)^j l!/(j + l)!$ and $b_l = j!/(j + l)!$ are the leading coefficients of the polynomials $P_{j,l}(z)$ and $Q_{j,l}(z)$, respectively.

Proof. Since the function $R_{j,l}(z)$ satisfies relation (0.1), we have

$$e^{-z} - R_{j,l}(z) = c_{j+l+1} z^{j+l+1} + c_{j+l+2} z^{j+l+2} + \cdots.$$

This implies that

$$|R_{j,l}(z) + R'_{j,l}(z)| = |R_{j,l}(z) - e^{-z} + R'_{j,l}(z) + e^{-z}|$$

$$\le |R_{j,l}(z) - e^{-z}| + |[R_{j,l}(z) - e^{-z}]'| \le M|z|^{j+l}, \tag{0.6}$$

where M does not depend on z.

On the other hand, we have

$$R_{j,l}(z) + R'_{j,l}(z) = \frac{Q_{j,l}(z)P_{j,l}(z) - P_{j,l}(z)Q'_{j,l}(z) + P'_{j,l}(z)Q_{j,l}(z)}{Q^2_{j,l}(z)}.$$

The numerator of this fraction is a polynomial of degree $j+l$ with leading coefficient $a_j b_l$. Since $Q_{j,l}(0) = 1$, from (0.6) it follows that the numerator of this fraction is equal to $a_j b_l z^{j+l}$. Lemma 0.1 is proved.

Lemma 0.2. *The following inequality holds:*

$$|R_{l,l}(z)| \leq |(z + z_1)/(z - \bar{z}_1)|, \quad z \in S_1 \cup S_2, \tag{0.7}$$

where z_1 is some root of the polynomial $Q_{l,l}(z)$.

Proof. Let $Q_{l,l}(z) = b_l \prod_{r=1}^{l}(z - z_r)$, where z_r, $r = 1, \cdots, l$, are the roots of the polynomial $Q_{l,l}(z)$. Since the coefficients of this polynomial are real, we have

$$R_{l,l}(z) = \frac{(-1)^l \prod_{r=1}^{l_1}(z + a_r) \prod_{r=1}^{l_2}(z + \bar{z}_r)(z + z_r)}{\prod_{r=1}^{l_1}(z - a_r) \prod_{r=1}^{l_2}(z - z_r)(z - \bar{z}_r)}. \tag{0.8}$$

Further, since $a_r < 0$, $r = 1, \cdots, l_1$, and $\mathrm{Re}\, z_r < 0$, $r = 1, \cdots, l_2$, it follows that

$$\left|\frac{z + a_r}{z - a_r}\right| \leq 1,\ r = 1, \cdots, l_1, \quad \left|\frac{z + \bar{z}_r}{z - z_r}\right| \leq 1, \quad \left|\frac{z + z_r}{z - \bar{z}_r}\right| \leq 1,\ r = 1, \cdots, l_2. \tag{0.9}$$

Relations (0.8) and (0.9) imply inequality (0.7). Lemma 0.2 is proved.

Lemma 0.3. *The following inequality holds for all z with $\mathrm{Re}\, z \geq 0$:*

$$|R_{l-1,l}(z)| \leq |Q_{l,l-1}(-z)/Q_{l,l-1}(z)|. \tag{0.10}$$

Proof. Consider the function

$$f(z) = Q_{l,l-1}(z)/Q_{l-1,l}(z),$$

which is analytic for $z \in \mathbf{C}^+ = \mathbf{C} \setminus \mathbf{C}^-$ because the roots of the polynomial $Q_{l,l-1}(z)$ lie in $\mathbf{C}^-$. Since $P_{l-1,l}(z) = Q_{l,l-1}(-z)$, to prove (0.10) it obviously suffices to show that $|f(z)| \leq 1$. But $f(z)$ is a proper fraction, and so $\lim_{|z|\to\infty}|f(z)| = 0$. Hence, by the maximum principle, it suffices to establish the inequality

$$|f(i\lambda)| \leq 1, \quad -\infty < \lambda < \infty. \tag{0.11}$$

To this end we shall use the identity

$$f(i\lambda) = \frac{\int_0^\infty t^{l-1}(t + i\lambda)^{l-1}(t + \frac{i\lambda}{2})e^{-t}dt - \frac{i\lambda}{2}\int_0^\infty t^{l-1}(t + i\lambda)^{l-1}e^{-t}dt}{\int_0^\infty t^{l-1}(t + i\lambda)^{l-1}(t + \frac{i\lambda}{2})e^{-t}dt + \frac{i\lambda}{2}\int_0^\infty t^{l-1}(t + i\lambda)^{l-1}e^{-t}dt}$$

$$= \frac{u_{l-1}(i\lambda) - \frac{i\lambda}{2}}{u_{l-1}(i\lambda) + \frac{i\lambda}{2}}, \tag{0.12}$$

with

$$u_l(i\lambda) \equiv \frac{\int_0^\infty t^l(t+i\lambda)^l(t+\frac{i\lambda}{2})e^{-t}dt}{\int_0^\infty t^l(t+i\lambda)^l e^{-t}dt} \equiv \frac{v_l(i\lambda)}{\left|\int_0^\infty t^l(t+i\lambda)^l e^{-t}dt\right|^2},$$

where

$$v_l(i\lambda) \equiv \int_0^\infty t^l(t+i\lambda)^l(t+\frac{i\lambda}{2})e^{-t}dt \int_0^\infty t^l(t-i\lambda)^l e^{-t}dt. \tag{0.13}$$

It follows from (0.12) that (0.11) holds provided that

$$\frac{\lambda}{2}\operatorname{Im} u_l(i\lambda) \geq 0, \quad l = 0, 1, \cdots, \tag{0.14}$$

which in turn holds if

$$\frac{\lambda}{2}\operatorname{Im} v_l(i\lambda) \geq 0, \quad l = 0, 1, \cdots. \tag{0.15}$$

To prove (0.15), let us establish a recursion relation (in l) for the quantity $\operatorname{Im} v_l(i\lambda)$. To this end we will make the substitutions $s = t + \frac{i\lambda}{2}$, $\tau = t - \frac{i\lambda}{2}$ in the first and second integrals appearing in the right-hand side of (0.13). We obtain

$$v_l(i\lambda) = \int_{0+i\lambda/2}^{\infty+i\lambda/2} \left(s^2 + \frac{\lambda^2}{4}\right)^l s e^{-s} ds \int_{0-i\lambda/2}^{\infty-i\lambda/2} \left(\tau^2 + \frac{\lambda^2}{4}\right)^l e^{-\tau} d\tau. \tag{0.16}$$

Integrating the first integral by parts, we obtain

$$v_l(i\lambda) = \int_{0+i\lambda/2}^{\infty+i\lambda/2} \left(s^2 + \frac{\lambda^2}{4}\right)^l e^{-s} ds \int_{0-i\lambda/2}^{\infty-i\lambda/2} \left(\tau^2 + \frac{\lambda^2}{4}\right)^l e^{-\tau} d\tau$$

$$+ 2l \int_{0+i\lambda/2}^{\infty+i\lambda/2} \left(s^2 + \frac{\lambda^2}{4}\right)^{l-1} s^2 e^{-s} ds \int_{0-i\lambda/2}^{\infty-i\lambda/2} \left(\tau^2 + \frac{\lambda^2}{4}\right)^l e^{-\tau} d\tau.$$

Since the expression

$$\int_{0+i\lambda/2}^{\infty+i\lambda/2} \left(s^2 + \frac{\lambda^2}{4}\right)^l e^{-s} ds \int_{0-i\lambda/2}^{\infty-i\lambda/2} \left(\tau^2 + \frac{\lambda^2}{4}\right)^l e^{-\tau} d\tau$$

is real, we have

$$\operatorname{Im} v_l(i\lambda) = 2l \operatorname{Im}\left[\int_{0+i\lambda/2}^{\infty+i\lambda/2} \left(s^2 + \frac{\lambda^2}{4}\right)^{l-1} s^2 e^{-s} ds \int_{0-i\lambda/2}^{\infty-i\lambda/2} \left(\tau^2 + \frac{\lambda^2}{4}\right)^l e^{-\tau} d\tau\right]$$

$$= 2l \operatorname{Im}\Bigg[\int_{0+i\lambda/2}^{\infty+i\lambda/2} \left(s^2 + \frac{\lambda^2}{4}\right)^l e^{-s} ds \int_{0-i\lambda/2}^{\infty-i\lambda/2} \left(\tau^2 + \frac{\lambda^2}{4}\right)^l e^{-\tau} d\tau$$

$$-\frac{\lambda^2}{4} \int_{0+i\lambda/2}^{\infty+i\lambda/2} \left(s^2 + \frac{\lambda^2}{4}\right)^{l-1} e^{-s} ds \int_{0-i\lambda/2}^{\infty-i\lambda/2} \left(\tau^2 + \frac{\lambda^2}{4}\right)^l e^{-\tau} d\tau\Bigg]$$

$$= -l\frac{\lambda^2}{2} \operatorname{Im}\Bigg[\int_{0+i\lambda/2}^{\infty+i\lambda/2} \left(s^2 + \frac{\lambda^2}{4}\right)^{l-1} e^{-s} ds \int_{0-i\lambda/2}^{\infty-i\lambda/2} \left(\tau^2 + \frac{\lambda^2}{4}\right)^l e^{-\tau} d\tau\Bigg]. \quad (0.17)$$

Integrating the second integral in (0.17) by parts, we obtain

$$\operatorname{Im} v_l(i\lambda)$$

$$= -l^2\lambda^2 \operatorname{Im}\Bigg[\int_{0+i\lambda/2}^{\infty+i\lambda/2} \left(s^2 + \frac{\lambda^2}{4}\right)^{l-1} e^{-s} ds \int_{0-i\lambda/2}^{\infty-i\lambda/2} \left(\tau^2 + \frac{\lambda^2}{4}\right)^{l-1} \tau e^{-\tau} d\tau\Bigg]$$

$$= -l^2\lambda^2 \operatorname{Im} v_{l-1}(-i\lambda) = -l^2\lambda^2 \operatorname{Im} \overline{v_{l-1}(i\lambda)} = l^2\lambda^2 \operatorname{Im} v_{l-1}(i\lambda).$$

Since $v_0(i\lambda) = 1 + i\lambda/2$, it follows that $\operatorname{Im} v_l(i\lambda) = (l!)^2 \lambda^{2m+1}/2$. Hence,

$$\frac{\lambda}{2}\operatorname{Im} v_l(i\lambda) = (l!\lambda^{l+1})^2/4 \geq 0.$$

Lemma 0.3 is proved.

Lemma 0.4. *The following inequality holds for all z, $\operatorname{Re} z \geq 0$:*

$$|R_{l-2,l}(z)| \leq |Q_{l,l-2}(-z)/Q_{l,l-2}(z)|. \qquad (0.18)$$

Proof. We shall follow the scheme of the proof of Lemma 0.3. Let us consider the function

$$f(z) = Q_{l,l-2}(z)/Q_{l-2,l}(z),$$

which is analytic for $z \in \mathbf{C}^+$ because the roots of the polynomial $Q_{l-2,l}(z)$ lie in $\mathbf{C}^-$. Since $P_{l-2,l}(z) = Q_{l,l-2}(-z)$, to prove (0.18) it obviously suffices to verify that $|f(z)| \leq 1$. Since $f(z)$ is a proper fraction, $\lim_{|z|\to\infty} |f(z)| = 0$. Therefore, by the maximum principle for analytic functions, is suffices to show that

$$|f(i\lambda)| \leq 1, \quad -\infty < \lambda < \infty.$$

To this end we shall use the identity

$$f(i\lambda) = \frac{u_{l-1}(i\lambda) - i\lambda}{u_{l-1}(i\lambda) + i\lambda},$$

in which

$$u_l(i\lambda) = \frac{\int_0^\infty t^l(t+i\lambda)^l\left(t+\frac{i\lambda}{2}\right)^2 e^{-t}dt - \frac{\lambda^2}{4}\int_0^\infty t^l(t+i\lambda)^l e^{-t}dt}{\int_0^\infty t^l(t+i\lambda)^l\left(t+\frac{i\lambda}{2}\right)e^{-t}dt}$$

$$= \frac{v_l(i\lambda)}{\left|\int_0^\infty t^l(t+i\lambda)^l\left(t+\frac{i\lambda}{2}\right)e^{-t}dt\right|^2}, \qquad (0.19)$$

where

$$v_l(i\lambda) = \left[\int_0^\infty t^l(t+i\lambda)^l\left(t+\frac{i\lambda}{2}\right)^2 e^{-t}dt - \frac{\lambda^2}{4}\int_0^\infty t^l(t+i\lambda)^l e^{-t}dt\right] \times$$

$$\times \int_0^\infty t^l(t-i\lambda)^l\left(t-\frac{i\lambda}{2}\right)e^{-t}dt. \qquad (0.20)$$

From (0.19) it follows that (0.18) holds if

$$\lambda\,\mathrm{Im}\,u_l(i\lambda) \geq 0, \quad l = 0,1,\cdots, \qquad (0.21)$$

which in turn holds if

$$\lambda\,\mathrm{Im}\,v_l(i\lambda) \geq 0, \quad l = 0,1,\cdots. \qquad (0.22)$$

To prove (0.22) let us calculate $\mathrm{Im}\,v_l(i\lambda)$. Making the substitutions $s = t + i\lambda/2$, $\tau = t - i\lambda/2$ in the first and respectively the second integral in the right-hand side of (0.20), we obtain

$$v_l(i\lambda) = \left[\int_{0+i\lambda/2}^{\infty+i\lambda/2}\left(s^2+\frac{\lambda^2}{4}\right)^l s^2 e^{-s}ds - \frac{\lambda^2}{4}\int_{0+i\lambda/2}^{\infty+i\lambda/2}\left(s^2+\frac{\lambda^2}{4}\right)^l e^{-s}ds\right] \times$$

$$\times \int_{0-i\lambda/2}^{\infty-i\lambda/2}\left(\tau^2+\frac{\lambda^2}{4}\right)^l \tau e^{-\tau}d\tau.$$

Since

$$\int_{0-i\lambda/2}^{\infty-i\lambda/2}\left(\tau^2+\frac{\lambda^2}{4}\right)^l \tau e^{-\tau}d\tau = \frac{1}{2(l+1)}\int_{0-i\lambda/2}^{\infty-i\lambda/2}\left(\tau^2+\frac{\lambda^2}{4}\right)^{l+1} e^{-\tau}d\tau,$$

we have

$$v_l(i\lambda) = \left[\int_{0+i\lambda/2}^{\infty+i\lambda/2}\left(s^2+\frac{\lambda^2}{4}\right)^{l+1} e^{-s}ds - \frac{\lambda^2}{2}\int_{0+i\lambda/2}^{\infty+i\lambda/2}\left(s^2+\frac{\lambda^2}{4}\right)^l e^{-s}ds\right] \times$$

$$\times \frac{1}{2(l+1)} \int_{0-i\lambda/2}^{\infty-i\lambda/2} \left(\tau^2 + \frac{\lambda^2}{4}\right)^{l+1} e^{-\tau} d\tau.$$

Next, since the expression

$$\int_{0+i\lambda/2}^{\infty+i\lambda/2} \left(s^2 + \frac{\lambda^2}{4}\right)^{l+1} e^{-s} ds \int_{0-i\lambda/2}^{\infty-i\lambda/2} \left(\tau^2 + \frac{\lambda^2}{4}\right)^{l+1} e^{-\tau} d\tau \qquad (0.23)$$

is real, it follows that

$$\operatorname{Im} v_l(i\lambda) = -\frac{\lambda^2}{4(l+1)} \operatorname{Im} \int_{0+i\lambda/2}^{\infty+i\lambda/2} \left(s^2 + \frac{\lambda^2}{4}\right)^{l} e^{-s} ds \int_{0-i\lambda/2}^{\infty-i\lambda/2} \left(\tau^2 + \frac{\lambda^2}{4}\right)^{l+1} e^{-\tau} d\tau.$$

$$(0.24)$$

Let us establish a recursion relation for the imaginary part of the quantity

$$w_l(i\lambda) = \int_{0+i\lambda/2}^{\infty+i\lambda/2} \left(s^2 + \frac{\lambda^2}{4}\right)^{l} e^{-s} ds \int_{0-i\lambda/2}^{\infty-i\lambda/2} \left(\tau^2 + \frac{\lambda^2}{4}\right)^{l+1} e^{-\tau} d\tau.$$

Integrating here the second integral by parts, we obtain

$$w_l(i\lambda) = \int_{0+i\lambda/2}^{\infty+i\lambda/2} \left(s^2 + \frac{\lambda^2}{4}\right)^{l} e^{-s} ds \cdot 2(l+1) \int_{0-i\lambda/2}^{\infty-i\lambda/2} \left(\tau^2 + \frac{\lambda^2}{4}\right)^{l} \tau e^{-\tau} d\tau$$

$$= 2(l+1) \int_{0+i\lambda/2}^{\infty+i\lambda/2} \left(s^2 + \frac{\lambda^2}{4}\right)^{l} e^{-s} ds \times$$

$$\times \left[2l \int_{0-i\lambda/2}^{\infty-i\lambda/2} \left(\tau^2 + \frac{\lambda^2}{4}\right)^{l-1} \tau^2 e^{-\tau} d\tau + \int_{0-i\lambda/2}^{\infty-i\lambda/2} \left(\tau^2 + \frac{\lambda^2}{4}\right)^{l} e^{-\tau} d\tau \right]$$

$$= 2(l+1) \int_{0+i\lambda/2}^{\infty+i\lambda/2} \left(s^2 + \frac{\lambda^2}{4}\right)^{l} e^{-s} ds \times$$

$$\times \left[(2l+1) \int_{0-i\lambda/2}^{\infty-i\lambda/2} \left(\tau^2 + \frac{\lambda^2}{4}\right)^{l} e^{-\tau} d\tau - \frac{l\lambda^2}{2} \int_{0-i\lambda/2}^{\infty-i\lambda/2} \left(\tau^2 + \frac{\lambda^2}{4}\right)^{l-1} e^{-\tau} d\tau \right].$$

By (0.23), we have that

$$\operatorname{Im} w_l(i\lambda) = -(l+1)l\lambda^2 \operatorname{Im} \int_{0+i\lambda/2}^{\infty+i\lambda/2} \left(s^2 + \frac{\lambda^2}{4}\right)^{l} e^{-s} ds \int_{0-i\lambda/2}^{\infty-i\lambda/2} \left(\tau^2 + \frac{\lambda^2}{4}\right)^{l-1} e^{-\tau} d\tau$$

$$= -(l+1)l\lambda^2 \operatorname{Im} w_{l-1}(-i\lambda) = -(l+1)l\lambda^2 \operatorname{Im} \overline{w_{l-1}(i\lambda)} = (l+1)l\lambda^2 \operatorname{Im} w_{l-1}(i\lambda).$$

Consequently,

$$\operatorname{Im} w_l(i\lambda) = \lambda^{2l}(l+1)(l!)^2 \operatorname{Im} w_0(i\lambda).$$

Since

$$w_0(i\lambda) = \int_{0+i\lambda/2}^{\infty+i\lambda/2} e^{-s}\,ds \int_{0-i\lambda/2}^{\infty-i\lambda/2} \left(\tau^2 + \frac{\lambda^2}{4}\right) e^{-\tau}\,d\tau$$

$$= 2e^{-i\lambda/2} \int_{0-i\lambda/2}^{\infty-i\lambda/2} \tau e^{-\tau}\,d\tau = 2e^{-i\lambda/2}\left(-\frac{i\lambda}{2}+1\right)e^{i\lambda/2} = -i\lambda + 2,$$

we have that

$$\operatorname{Im} w_l(i\lambda) = -\lambda^{2l+1}(l+1)(l!)^2.$$

From this and (0.24) it follows that $\operatorname{Im} v_l(i\lambda) = \lambda^{2l+3}(l!)^2/4$. Thus, $\lambda\operatorname{Im} v_l(i\lambda) = \lambda^{2(l+2)}(l!)^2/4 \geq 0$. Lemma 0.4 is proved.

Lemma 0.5. *The inequality*

$$|R_{l-1,l}(z)(1+z/l)| \leq 1 \tag{0.25}$$

holds for all z, $z = \rho e^{\pm i\phi}$, $0 \leq \rho < \infty$, $0 \leq \phi \leq \pi/2l$.

Proof. By (0.4),

$$P_{l-1,l}(z)(1+z/l)$$

$$= \sum_{r=0}^{l-1} \frac{(2l-1-r)!(l-1)!}{(2l-1)!r!(l-1-r)!}(-z)^r + \sum_{r=0}^{l-1} \frac{(2l-1-r)!(l-1)!(-1)^r}{(2l-1)!r!(l-1-r)!l}z^{r+1}$$

$$= 1 + \sum_{r=1}^{l-1} \frac{(2l-1-r)!l!(-1)^{r-1}\psi(r,l)}{(2l-1)!r!(l-r)!}z^r + \frac{l!(-1)^{l-1}}{(2l)!}z^l,$$

where $\psi(r,l) = 1 - 3r/l + (r/l)^2$. Since $\psi(0,l) = 1$ and $\psi(l,l) = -1$, it follows that

$$|P_{l-1,l}(z)(1+z/l)|^2 = \left[\sum_{r=0}^{l-1} \frac{(2l-1-r)!l!(-1)^r\psi(r,l)}{(2l-1)!r!(l-r)!}\rho^r\cos r\phi\right]^2$$

$$+ \left[\sum_{r=1}^{l-1} \frac{(2l-1-r)!l!(-1)^r\psi(r,l)}{(2l-1)!r!(l-r)!}\rho^r\sin r\phi\right]^2$$

$$= \sum_{r=0}^{l-1} \left[\frac{(2l-1-r)!l!(-1)^r\psi(r,l)}{(2l-1)!r!(l-r)!}\right]^2\rho^{2r}$$

$$+2\sum_{r=0}^{l}\sum_{i=r+1}^{l}(-1)^{i+r}\frac{(2l-1-r)!l!\psi(r,l)}{(2l-1)!r!(l-r)!}\frac{(2l-1-i)!l!\psi(i,l)}{(2l-1)!i!(l-i)!}\rho^{i+r}\cos(i-r)\phi. \tag{0.26}$$

Using the inequality $-1 \leq \psi(r, l) \leq 1$, which holds for all $1 \leq r \leq l$, we get

$$|P_{l-1,l}(z)(1 + z/l)|^2 \leq \sum_{r=0}^{l-1} \left[\frac{(2l - 1 - r)!l!}{(2l - 1)!r!(l - r)!} \right]^2 \rho^{2r}$$

$$+ 2 \sum_{r=0}^{l} \sum_{i=r+1}^{l} \left[\frac{l!}{(2l-1)!} \right]^2 \frac{(2l - 1 - r)!(2l - 1 - i)!}{r!(l - r)!i!(l - i)!} \rho^{i+r} \cos(i - r)\phi = |Q_{l-1,l}(z)|^2.$$

Lemma (0.5) is proved.

Lemma 0.6. *The inequality*

$$|R_{l-2,l}(z)(1 + z/(2l - 1))| \leq 1 \tag{0.27}$$

holds for all z, $z = \rho e^{\pm i\phi}$, $0 \leq \rho < \infty$, $0 \leq \phi \leq \pi/2l$.

Proof. We shall follow the scheme of the proof of Lemma 0.5. By (0.4),

$$P_{l-2,l}(z)(1 + z/(2l - 1))$$

$$= \sum_{r=0}^{l-2} \frac{(2l - 2 - r)!(l - 2)!(-1)^r}{(2l - 2)!r!(l - 2 - r)!} z^r + \sum_{r=0}^{l-2} \frac{(2l - 2 - r)!(l - 2)!(-1)^r}{(2l - 2)!r!(l - 2 - r)!(2l - 1)} z^{r+1}$$

$$= 1 + \sum_{r=1}^{l-2} \frac{(2l - 2 - r)!l!(-1)^r z^r \psi(r, l)}{(2l - 2)!r!(l - r)!} + (-1)^{l-2} \frac{l!}{(2l - 2)!(2l - 1)} z^{l-1},$$

where

$$\psi(r, l) = \left(1 - \frac{r}{l}\right) \left[1 - \frac{r}{l - 1} - \frac{r}{l - 1}\left(1 - \frac{r}{2l - 1}\right)\right].$$

Since $\psi(0, l) = 1$ and $\psi(l, l) = 0$, we have that

$$|P_{l-2,l}(z)(1 + z/(2l - 1))|^2 = \left[\sum_{r=0}^{l} \frac{(2l - 2 - r)!l!!(-1)^r \psi(r, l)}{(2l - 2)!r!(l - r)!} \rho^r \cos r\phi \right]^2$$

$$+ \left[\sum_{r=0}^{l} \frac{(2l - 2 - r)!l!!(-1)^r \psi(r, l)}{(2l - 2)!r!(l - r)!} \rho^r \sin r\phi \right]^2$$

$$= \sum_{r=0}^{l} \left[\frac{(2l - 2 - r)!l!!\psi(r, l)}{(2l - 2)!r!(l - r)!} \right]^2 \rho^{2r}$$

$$+2\sum_{r=0}^{l}\sum_{i=r+1}^{l}\rho^{i+r}\left[\frac{l!}{(2l-2)!}\right]^{2}\frac{(2l-2-r)!(2l-2-i)!\psi(r,l)\psi(i,l)}{r!(l-r)!i!(l-i)!}\cos(i-r)\phi.$$

$$(0.28)$$

Using the inequality $-1\leq\psi(r,l)\leq 1$, which holds for all $0\leq r\leq l$, we get

$$|P_{l-2,l}(z)(1+z/(2l-1))|^{2}\leq\sum_{r=0}^{l}\left[\frac{(2l-2-r)!l!}{(2l-2)!r!(l-r)!}\right]^{2}\rho^{2r}$$

$$+2\sum_{r=0}^{l}\sum_{i=r+1}^{l}\rho^{i+r}\left[\frac{l!}{(2l-2)!}\right]^{2}\frac{(2l-2-r)!(2l-2-i)!}{r!(l-r)!i!(l-i)!}\cos(i-r)\phi=|Q_{l-2,l}(z)|^{2}.$$

Lemma 0.6 is proved.

Lemma 0.7. *The inequality*

$$\left|z^{\alpha}\left(\frac{z+z_{1}}{z-\bar{z}_{1}}\right)^{k}\frac{1}{z-\bar{z}_{1}}\right|\leq Mk^{-\alpha},\quad k\geq 1,\ 0\leq\alpha\leq\frac{1}{2}$$

$$(0.29)$$

holds for all z, $z=\rho e^{\pm i\phi}$, $0\leq\rho<\infty$, $0\leq\phi<\pi/2$ and all z_{1} with $\mathrm{Re}\,z_{1}<0$.

Proof. Put $|z_{1}|=\rho_{1}$ and let $[zvz_{1}]$ denote the angle between the vectors z and z_{1}. Since $0\leq\phi<\pi/2$ and $\mathrm{Re}\,z_{1}<0$,

$$\cos\phi_{2}\equiv\cos[zvz_{1}]<\cos[zv(-\bar{z}_{1})]\equiv\cos\phi_{3}.$$

Further, we have that

$$\left|z^{\alpha}\left(\frac{z+z_{1}}{z-\bar{z}_{1}}\right)^{k}\frac{1}{z-\bar{z}_{1}}\right|=(\rho/\rho_{1})^{\alpha}\left(\frac{(\rho/\rho_{1})^{2}+2(\rho/\rho_{1})\cos\phi_{2}+1}{(\rho/\rho_{1})^{2}+2(\rho/\rho_{1})\cos\phi_{3}+1}\right)^{-k/2}\times$$

$$\times\frac{\rho_{1}^{-1+\alpha}}{\left((\rho/\rho_{1})^{2}+2(\rho/\rho_{1})\cos\phi_{3}+1\right)^{1/2}}.$$

Let us consider the function

$$\Psi_{\alpha}(r)=r^{\alpha}\left(\frac{r^{2}+2r\cos\phi_{2}+1}{r^{2}+2r\cos\phi_{3}+1}\right)^{-k/2}\frac{1}{\left(r^{2}+2r\cos\phi_{3}+1\right)^{1/2}}.$$

Clearly, $\Psi_{\alpha}(r)\geq 0$, $\Psi_{\alpha}(0)=\Psi_{\alpha}(\infty)=0$ and

$$\sup_{0\leq r\leq\infty}\Psi_{\alpha}(r)=\max\left\{\max_{0\leq r\leq 1}\Psi_{\alpha}(r),\ \sup_{1\leq r<\infty}\Psi_{\alpha}(r)\right\}.$$

First let us estimate the expression $\max_{0 \leq r \leq 1} \Psi_\alpha(r)$. Since

$$\Psi_\alpha(r) \leq \left(1 - \frac{\cos\phi_3 - \cos\phi_2}{\cos\phi_3 + 1} r\right)^{k/2} r^\alpha \leq M_1 \max_{0 \leq s \leq 1} (1 - s)^{k/2} s^\alpha,$$

we have that

$$\max_{0 \leq r \leq 1} \Psi_\alpha(r) \leq M_1 \max_{0 \leq s \leq 1} (1 - s)^{k/2} s^\alpha. \tag{0.30}$$

It is readily verified that

$$\max_{0 \leq s \leq 1} (1 - s)^{k/2} s^\alpha \leq M_2 k^{-\alpha}. \tag{0.31}$$

Consequently,

$$\max_{0 \leq r \leq 1} \Psi_\alpha(r) \leq M_3 k^{-\alpha}.$$

Now let us estimate the expression $\sup_{1 \leq r < \infty} \Psi_\alpha(r)$. Making the substitution $r = 1/t$, we have that

$$\sup_{1 \leq r < \infty} \Psi_\alpha(r) = \max_{0 \leq t \leq 1} \left(\frac{t^2 + 2t\cos\phi_2 + 1}{t^2 + 2t\cos\phi_3 + 1}\right)^{-k/2} \frac{t^{1-\alpha}}{\left(t^2 + 2t\cos\phi_3 + 1\right)^{1/2}}$$

$$\leq M_1 \max_{0 \leq t \leq 1} (1 - t)^{k/2} t^{1-\alpha}.$$

By (0.31), $\sup_{1 \leq r < \infty} \Psi_\alpha(r) \leq M_4 k^{-1+\alpha}$. Therefore,

$$\sup_{0 \leq r < \infty} \Psi_\alpha(r) \leq M_5 \max\left\{k^{-\alpha}, k^{-1+\alpha}\right\} = M_5 k^{-\alpha}. \tag{0.32}$$

Lemma 0.7 is proved.

Lemma 0.8. *Let $j = l\text{-}2,\ l\text{-}1,\ l$. Then the inequality*

$$\left|z^{-\delta}\left[R_{j,l}^{k+r}(z) - R_{j,l}^k(z)\right](1 + z)^{-\theta}\right| \leq M \min\left\{(k + r)^\delta, rk^{-1+\delta}\right\},$$

$$0 \leq \delta \leq 1,\ 1 \leq k \leq k + r \leq N \tag{0.33}$$

holds for any $z = \rho e^{\pm i\phi}$, $0 \leq \rho < \infty$, $0 \leq \phi < \pi/2$. Here $\theta = 1$ if $j = l$ and $\theta = 0$ if $j \neq l$.

Proof. By (0.7), (0.10), (0.18), and (0.29), we have

$$\left|z^\alpha R_{l,l}^k(z) Q_{l,l}^{-1}(z)(1 + z)^{-1}\right| \leq M k^{-\alpha}, \quad 0 \leq \alpha \leq \frac{l + 1}{2} \tag{0.34}$$

and

$$\left|z^{\alpha} R_{j,l}^{k}(z) Q_{l,j}^{-1}(z)\right| \leq M k^{-\alpha}, \quad 0 \leq \alpha \leq \frac{l+1}{2}, \quad j = l-2, \, l-1. \tag{0.35}$$

Using the identity

$$\left|z^{-\delta}\left[R_{j,l}^{k+r}(z) - R_{j,l}^{k}(z)\right](1+z)^{-\theta}\right|$$

$$= \left|z^{1-\delta} \sum_{\tau=0}^{r-1} R_{j,l}^{k+\tau}(z) \frac{Q_{j,l}(z) - P_{j,l}(z)}{z Q_{j,l}(z)}(1+z)^{-\theta}\right| \tag{0.36}$$

and the estimates (0.34) and (0.35), we obtain

$$\left|z^{-\delta}\left[R_{j,l}^{k+r}(z) - R_{j,l}^{k}(z)\right](1+z)^{-\theta}\right| \leq M r k^{-1+\delta}. \tag{0.37}$$

Now let us establish the inequality

$$\left|z^{-\delta}\left[R_{j,l}^{k+r}(z) - R_{j,l}^{k}(z)\right](1+z)^{-\theta}\right| \leq M(k+r)^{\delta}. \tag{0.38}$$

We have to examine two cases: $\rho \leq 1/(k+r)$ and $\rho \geq 1/(k+r)$.

In the first case the estimates (0.34), (0.35) and the identity (0.36) yield

$$\left|z^{-\delta}\left[R_{j,l}^{k+r}(z) - R_{j,l}^{k}(z)\right](1+z)^{-\theta}\right| \leq M r(k+r)^{\delta-1} \leq M(k+r)^{\delta}.$$

In the second case, the triangle inequality and the estimates (0.34), (0.35) yield

$$\left|z^{-\delta}\left[R_{j,l}^{k+r}(z) - R_{j,l}^{k}(z)\right](1+z)^{-\theta}\right| \leq M(k+r)^{\delta}.$$

Inequality (0.38) is thus established. Inequality (0.33) follows from (0.37) and (0.38). Lemma 0.8 is proved.

2. Difference schemes of Padé class.

Here we shall construct difference schemes of high order of accuracy for the approximate solution of the Cauchy problem (1.1) of Chapter 1. On the segment $[0,1]$ we consider a uniform grid $[0,1]_{\tau} = \{t_k = k\tau, \ k = 0, 1, \cdots, N, \ N\tau = 1\}$ with step $\tau > 0$. Using formula (1.19) of Chapter 1, we obtain the following relation between $v(t_k)$ and $v(t_{k-1})$:

$$v(t_k) = \exp\{-\tau A\} v(t_{k-1}) + \int_{t_{k-1}}^{t_k} \exp\{-(t_k - s)A\} f(s) ds. \tag{0.39}$$

This yields the equality

$$\tau^{-1}(v(t_k) - v(t_{k-1})) + \tau^{-1}(I - \exp\{-\tau A\})v(t_{k-1}) = \varphi_k,$$

$$\varphi_k = \tau^{-1} \int_{t_{k-1}}^{t_k} \exp\{-(t_k - s)A\}f(s)ds, \quad 1 \le k \le N. \tag{0.40}$$

Relation (0.39) and equality (0.40) are equivalent. The latter will be referred to as *the exact two-step scheme for the Cauchy problem* (1.1) *of Chapter 1*. From (0.40) it is clear that for the approximate solution of the Cauchy problem (1.1) of Chapter 1 it is neccesary to approximate the bounded linear operator $\exp\{-\tau A\}$ in the Banach space E and the elements φ_k of E.

Replacing the operator $\exp\{-\tau A\}$ by its Padé approximant $R_{j,l}(\tau A)$, $l-4 \le j \le l$, and the elements φ_k by close (simpler) elements $\varphi_k^{j,l}$ such that

$$\|\varphi_k - \varphi_k^{j,l}\|_E \le M\tau^{j+l}, \tag{0.41}$$

we obtain the Padé difference scheme

$$\tau^{-1}(u_k - u_{k-1}) + \tau^{-1}[I - R_{j,l}(\tau A)]u_{k-1} = \varphi_k^{j,l}, \quad u_0 = v_0, \quad 1 \le k \le N. \tag{0.42}$$

From the strong positivity of the operator A and the way the roots of the polynomials $Q_{j,l}(z)$, $l - 4 \le j \le l$, are positioned it follows that the operator $[Q_{j,l}(\tau A)]^{-1}$ exists and is uniformly bounded with respect to τ in E. Furthermore, for $k = 0, 1 \cdots, l$ the operators $(\tau A)^k[Q_{j,l}(\tau A)]^{-1}$ are uniformly bounded in τ. Hence, the operator $R_{j,l}(\tau A)$, $l - 4 \le j \le l$, is uniformly bounded in τ. In the case where the spectral angle $\phi(A)$ of the operator A is sufficiently small ($\phi(A) < \pi/l$) these facts will be proved for any $j \le l$. The operator $R_{j,l}(\tau A)$, which defines the difference scheme, is usually called the *step operator*.

Let us remark that in constructing such difference schemes it is important to know how to construct a right-hand side $\varphi_k^{j,l}$ that satisfies (0.41) and is sufficiently simple. The choice of $\varphi_k^{j,l}$ is not unique. Below we shall see that $\varphi_k^{j,l}$ can be defined by the formula

$$\varphi_k^{j,l} = \sum_{r=1}^{j+l-1} J_r f^{(r)}(t_{k-1}), \tag{0.43}$$

where

$$J_r = (-1)^r A^{-r} J_0 + \sum_{i=0}^{r}(-1)^{r-i} A^{-(r-i+1)}\frac{\tau^{i-1}}{i!},$$

$$J_0 = (\tau A)^{-1}[I - R_{j,l}(\tau A)], \quad 1 \le r \le j + l - 1,$$

provided the function $f(t)$ has a $(j+l)$-th continuous derivative.

Definition 0.1. We say that *the difference scheme* (0.42) *is of approximation order m on the solutions* $v(t)$ *of the Cauchy problem* (1.1) *of Chapter 1 if the* expressions

$$R_k = \tau^{-1}[\exp\{\tau A\} - R_{j,l}(\tau A)]v(t_{k-1}) - \varphi_k^{j,l} + \varphi_k \qquad (0.44)$$

obey the estimate

$$\|R_k\|_E \le M\tau^m \quad M = \text{const}, \quad k = 1, \cdots, N. \qquad (0.45)$$

R_k is called the *approximation error* of the difference scheme (0.42).

Theorem 0.1. *For any* $1 \le n \le j + l + 1$ *and any* $x \in D(A^n)$ *we have the* inequality

$$\|[\exp\{-\tau A\} - R_{j,l}(\tau A)]x\|_E \le M\tau^n \|A^n x\|_E, \qquad (0.46)$$

where M does not depend on τ *and* x.

Proof. We will follow the scheme of the proof of Theorem 1.1 in Chapter 2. Using the identity

$$R_{j,l}(\tau A)x - \exp\{-\tau A\}x = \int_0^1 [R'_{j,l}(z)|_{z=s\tau A} + R(s\tau A)]\tau A \exp\{-(1-s)\tau A\}x \, ds$$

and Lemma 0.1, we obtain the identity

$$R_{j,l}(\tau A)x - \exp\{-\tau A\}x$$

$$= \int_0^1 \frac{(-1)^j l! j!}{[(j+l)!]^2} s^{j+l}\tau^{j+l+1} A^{j+l+1}[Q_{j,l}(s\tau A)]^{-2}\exp\{-(1-s)\tau A\}x \, ds.$$

Estimate (0.46) follows this identity, the uniform boundedness with respect to τ of the norm of the operators $(\tau A)^n[Q_{j,l}(\tau A)]^{-2}$ for $0 \le n \le 2l$, and the estimate (1.10) in Chapter 1. Theorem 0.1 is proved.

Now let us investigate the approximation error of the difference scheme (0.42) for the approximate solution of problem (1.1) of Chapter 1. From the formula (1.19) for its solution $v(t)$ it follows that $v(t) \in D(A^{j+l+1})$ and the function $A^{j+l+1}v(t)$ is continuous if $v_0 \in D(A^{j+l+1})$ and the function $A^{j+l}f(t)$ satisfies some Hölder condition in E. This allows us to estimate the first part of the approximation error. Indeed, from Theorem 0.1 (with $n = j+l+1$) it follows that

$$\|\tau^{-1}[\exp\{-\tau A\} - R_{j,l}(\tau A)]v(t_{k-1})\|_E \leq M\tau^{j+l}\|A^{j+l+1}v(t_{k-1})\|_E \leq M_1\tau^{j+l}.$$
$$(0.47)$$

Let us assume in addition that $f^{(k)}(t) \in C([0,1], D(A^{j+l-k}))$ for $k = 0, 1, \cdots, j+l$. Using Taylor's formula and integration by parts, we obtain the representation

$$\tau^{-1}\int_{t_{k-1}}^{t_k} \exp\{-(t_k-s)A\}f(s)ds = \sum_{r=0}^{j+l-1} T_r f^{(r)}(t_{k-1})$$

$$+\tau^{-1}\int_{t_{k-1}}^{t_k} \exp\{-(t_k-s)A\} \int_{t_{k-1}}^{s} \frac{(s-z)^{j+l-1}}{(j+l-1)!} f^{(j+l)}(z)dz\,ds,$$

in which

$$T_r = (-1)^r A^{-r}T_0 + \sum_{i=1}^{r}(-1)^{r-i}A^{-(r-i+1)}\frac{\tau^{i-1}}{i!}, \quad 1 \leq r \leq j+l-1,$$

$$T_0 = (\tau A)^{-1}[I - \exp\{-\tau A\}].$$

From this relation and (0.43) it follows that

$$\varphi_k - \varphi_k^{j,l} = \tau^{-1}\int_{t_{k-1}}^{t_k} \exp\{-(t_k-s)A\} \int_{t_{k-1}}^{s} \frac{(s-z)^{j+l-1}}{(j+l-1)!} f^{(j+l)}(z)dz\,ds$$

$$+ \sum_{r=0}^{j+l-1}(T_r - J_r)f^{(r)}(t_{k-1}).$$

Since

$$T_r - J_r = (-1)^r A^{-(r+1)}\tau^{-1}[R_{j,l}(\tau A) - \exp\{-\tau A\}], \quad 0 \leq r \leq j+l-1,$$

we finally obtain the relation

$$\varphi_k - \varphi_k^{j,l} = \tau^{-1}\int_{t_{k-1}}^{t_k} \exp\{-(t_k-s)A\} \int_{t_{k-1}}^{s} \frac{(s-z)^{j+l-1}}{(j+l-1)!} f^{(j+l)}(z)dz\,ds$$

$$+ \sum_{r=0}^{j+l-1}(-1)^r A^{-(r+1)}\tau^{-1}[R_{j,l}(\tau A) - \exp\{-\tau A\}]f^{(r)}(t_{k-1}).$$

Using the estimate (0.46) for $n = j + l + 1$, we derive from this relation the inequality

$$\|\varphi_k - \varphi_k^{j,l}\|_E \leq M_0 \tau^{j+l} \max_{0 \leq z \leq 1} \|f^{(j+l)}(z)\|_E + \sum_{r=0}^{j+l-1} M_r \tau^{j+l} \max_{0 \leq z \leq 1} \|A^{j+l-r} f^{(r)}(z)\|_E.$$

This means that the second part of the approximation error is estimated by

$$\|\varphi_k - \varphi_k^{j,l}\|_E \leq M \tau^{j+l}.$$

Thus, under the assumptions made above on the smoothness of the data involved in the problem (1.1) of Chapter 1 the approximation error R_k of the difference scheme (0.42) obeys the estimate

$$\|R_k\|_E \leq M \tau^{j+l}, \tag{0.48}$$

where M does not depend on τ.

Note that the (l, l)- and $(l-1, l)$-difference schemes of Padé class include difference schemes of arbitrary order of approximation. Moreover, the corresponding functions $R_{j,l}(z)$, $j = l-1, l$ are bounded at infinity. Such difference schemes are the simplest, in the sense that the degrees of the denominators of the corresponding Padé approximants of the function $\exp\{-z\}$ are minimal for a fixed order of approximation of the difference schemes.

Consider again the difference schemes (0.42). Let us reduce such schemes to an operator problem in the space $E(\tau)$. In addition to the operator $\mathcal{D} = \mathcal{D}_\tau^1$, acting from the space $E \times E(\tau)$ of vectors $u = (u_0, u_1, \cdots, u_N)$ into the spaces $E(\tau)$ of vectors $v = (v_1, \cdots, v_N)$ by the rule

$$v = \mathcal{D}u, \quad v_k = \tau^{-1}(u_k - u_{k-1}), \quad k = 1, \cdots, N,$$

define the operator $A_{j,l}$ from the space $E(\tau)$ of vectors $(u_0, ..., u_{N-1})$ to the space $E(\tau)$ of vectors $v = (v_1, \cdots, v_N)$ by the rule

$$v = A_{j,l}u, \quad v_k = \tau^{-1}[I - R_{j,l}(\tau A)]u_{k-1}, \quad k = 1, \cdots, N.$$

Then the difference schemes (0.42) can obviously be rewritten as the equivalent operator equation

$$\mathcal{D}\Pi(u_0)u^\tau + A_{j,l}u^\tau = \varphi_{j,l}^\tau, \tag{0.49}$$

where $\varphi_{j,l}^{\tau} = (\varphi_1^{j,l}, \cdots, \varphi_N^{j,l})$.

As we noted above, the strong positivity of the operator A implies the existence of a bounded operator $R_{j,l}(\tau A)$, defined on the entire space E. Consequently, for any $\varphi_{j,l}^{\tau}$ and u_0 the solution of (0.49) exists, and we have the formula

$$u_k = R_{j,l}^k(\tau A)u_0 + \sum_{r=1}^{k} R_{j,l}^{k-r}(\tau A)\varphi_r^{j,l}, \quad k = 1, \cdots, N. \tag{0.50}$$

The operator problem (0.49) will be considered in the space $E(\tau)$. From its unique solvability for arbitrary $u_0 \in E$ and $\varphi_{j,l}^{\tau} \in E(\tau)$ it follows that its solution u^{τ} defines an additive and homogeneous operator $u^{\tau}(\varphi_{j,l}^{\tau}, u_0)$ from $E(\tau) \times E$ into $E(\tau)$. As in Chapter 2, the vector spaces $E(\tau)$ and $E(\tau) \times E$ can be equipped with norms that make them into Banach spaces. Then obviously the boundedness of the operator step will imply the continuity of the operator $u^{\tau}(\varphi_{j,l}^{\tau}, u_0)$. In this case it is natural to speak about the difference problem (0.49) being *stable*. Since the operator $u^{\tau}(\varphi_{j,l}^{\tau}, u_0)$ is additive and homogeneous, problem (0.49) is stable if and only if the following inequality holds:

$$\|u^{\tau}(\varphi_{j,l}^{\tau}, u_0)\|_{E(\tau)} \leq M \left[\|u_0\|_E + \|\varphi_{j,l}^{\tau}\|_{E(\tau)} \right],$$

where M does not depend on u_0 and $\varphi_{j,l}^{\tau}$, but, generally speaking, depends on τ.

Finally, let us consider problem (0.49) as an operator problem in the space $\mathcal{E}(E)$. To this end let us introduce the operators $\overline{\mathcal{D}}$, $\overline{\Pi}$, and $\overline{A}_{j,l}$, which act in $\mathcal{E}(E)$ componentwise as $\mathcal{D}$, Π, and $A_{j,l}$, respectively. This yields the operator equation

$$\overline{\mathcal{D}}\,\overline{\Pi}(u_0)u + \overline{A}_{j,l}u = \varphi_{j,l} \tag{0.51}$$

in the vector space $\mathcal{E}(E)$. This equation is clearly uniquely solvable for any $u_0 \in E$ and $\varphi_{j,l} \in \mathcal{E}(E)$. This solvability is equivalent to the unique solvability of problem (0.49) for all $0 < \tau \leq \tau_0$. Solving problem (0.51) by means of formula (0.50), we obtain an additive and homogeneous operator $u(\varphi_{j,l}, u_0)$ acting from the space $\mathcal{E}(E)$ into itself. Proceeding by analogy with Chapter 2, we equip the spaces $\mathcal{E}(E)$ and $\mathcal{E}(E) \times E$ with norms, transforming them into Banach spaces.

Definition 0.2. We say that the problem (0.51) is *stable in the Banach space* $\mathcal{E}(E)$ if $u(\varphi_{j,l}, u_0)$, regarded as an operator from $\mathcal{E}(E) \times E$ into $\mathcal{E}(E)$, is continuous. Since the operator $u(\varphi_{j,l}, u_0)$ is linear, problem (0.51) is stable in $\mathcal{E}(E)$ if and only if the inequality

$$\|u^{\tau}\|_{E(\tau)} \leq M \left[\|u_0\|_E + \|\varphi_{j,l}^{\tau}\|_{E(\tau)} \right] \tag{0.52}$$

holds, where M not only does not depend on u_0, $\varphi_{j,l}^\tau$, but also does not depend on τ.

Next, let us consider problem (0.51) in the space $\mathcal{C}(E)$.

Theorem 0.2. *For the difference problem* (0.51) *to be stable in* $\mathcal{C}(E)$ *it is necessary and sufficient that the following estimate hold for any* $k = 1, \cdots, N$ *and* $\tau > 0$:

$$\|R_{j,l}^k(\tau A)\|_{E \to E} \leq M, \tag{0.53}$$

where M does not depend on τ and k.

Now let us consider problem (0.51) in the space $\mathcal{C}_0^\alpha(E)$ $(0 < \alpha < 1)$.

Theorem 0.3. *For the difference problem* (0.51) *to be stable in* $\mathcal{C}_0^\alpha(E)$ *it is necessary and sufficient that the estimates* (0.53) *and*

$$\|R_{j,l}^k(\tau A) - R_{j,l}^{k+r}(\tau A)\|_{E \to E} \leq M_1 r^\alpha k^{-\alpha}, \quad 1 \leq k < k+r \leq N, \tag{0.54}$$

hold, where M_1 does not depend on τ, k and r.

Finally, let us formulate the following result.

Theorem 0.4. *For the difference problem* (0.51) *to be stable in* $\mathcal{L}_p(E)$, $1 < p < \infty$, *it is necessary and sufficient that the following estimate hold:*

$$\left(\sum_{k=1}^N \|R_{j,l}^k u_0\|_E^p \tau \right)^{1/p} \leq M\|u_0\|_E, \tag{0.55}$$

where M does not depend on τ and $u_0 \in E$.

The proofs of these theorems are identical to those of the analogous assertions in Chapter 2.

1.   WELL-POSEDNESS OF THE DIFFERENCE PROBLEM IN $\mathcal{C}(E)$

1.   The homogeneous problem.

Here we shall consider equation (0.51) in the Banach space $\mathcal{C}(E)$. An element $u \in \mathcal{C}(E)$ will now be called a *solution of* (0.51) if, in addition, the elements

$\overline{\mathcal{D}}\,\overline{\Pi}(u_0)u$ and $\overline{A}_{j,l}u$ belong to $\mathcal{C}(E)$. If problem (0.51) is solvable in $\mathcal{C}(E)$, then obviously $\varphi_{j,l} \in \mathcal{C}(E)$. As in the case of the simpler difference problem (0.6) of Chapter 2, here we are not able to find a necessary condition on u_0.

We shall assume that $u_0 \in D(A)$.

Definition 1.1. We say that problem (0.51) is *well posed in* $\mathcal{C}(E)$ if the following conditions are satisfied:

1) For any $u_0 \in D(A)$ and $\varphi_{j,l} \in \mathcal{C}(E)$ there exists a unique solution of (0.51) in $\mathcal{C}(E)$.

2) Problem (0.51) is stable in $\mathcal{C}(E)$.

As we remarked above, this kind of well-posedness is equivalent to the well-posedness of problem (0.49) in $C(\tau, E)$ uniformly in τ for $0 < \tau \leq \tau_0$.

Suppose problem (0.51) is well posed in $\mathcal{C}(E)$. This means, in particular, that the corresponding homogeneous problem ($\varphi_{j,l} \equiv 0$) is well posed. From condition 2) it follows that the homogeneous problem is stable in $\mathcal{C}(E)$. As we established in Section 0, this is the case if and only if estimate (0.53) holds. But estimate (0.53) with $j = 0$ and $l = 1$ is the criterion for the operator $-A$ to generate a strongly continuous semigroup $\exp\{-tA\}$, $t \geq 0$ (see Section 0, Chapter 1). Hence, in the present case a necessary and sufficient condition for the well-posedness in $\mathcal{C}(E)$ of the homogeneous problem is the strong continuity of the semigroup $\exp\{-tA\}$, $t \geq 0$. It turns out that, in the general case, too, a necessary condition for the well-posedness in $\mathcal{C}(E)$ of the homogeneous difference problem (0.51) (with $\varphi_{j,l} \equiv 0$) is the strong continuity of the semigroup $\exp\{-tA\}$, $t \geq 0$.

Theorem 1.1. *For the operator $-A$ with dense domain $D(A)$ to be the generator of a strongly continuous semigroup it is necessary that the estimate (0.53) hold.*

Proof. Let us introduce the operators

$$A_\tau = \frac{1}{\tau}[I - R_{j,l}(\tau A)], \quad 0 < \tau \leq \tau_0, \tag{1.1}$$

which are obviously bounded. Let us show that on the elements $x \in D(A)$ the operators A_τ converge to A:

$$\lim_{\tau \to 0} \|A_\tau x - Ax\|_E = 0. \tag{1.2}$$

To this end we first observe that

$$\lim_{\tau \to 0} \|A^{-1}A_\tau x - x\|_E = 0 \tag{1.3}$$

for any $x \in E$. This equality is obvious whenever $x \in D(A)$, because in this case

$$A^{-1}A_\tau x - x = (\tau A)^{-1}[I - R_{j,l}(\tau A)]x - x$$

$$= Q_{j,l}^{-1}(\tau A)\left[\sum_{r=2}^{l}(\tau A)^{r-2} \frac{(j+l-r)!l!}{(j+l)!r!(l-r)!}\left(1 - \frac{r(j+l-r+1)}{l-r+1}\right)\right.$$

$$\left. - \sum_{r=1}^{j}(-\tau A)^{r-1}\frac{(j+l-r)!j!}{(j+l)!r!(j-r)!}\right]\tau A x \tag{1.4}$$

and, by (0.53),

$$\|A^{-1}A_\tau x - x\|_E \le M\tau\|Ax\|_E.$$

The validity of equality (1.3) for any $x \in E$ follows from the fact that $D(A)$ is dense in E and the uniform boundedness of the norms of the operators $A^{-1}A_\tau$. The latter in turn follows from representation (1.4) and the uniform boundedness of the operators $(\tau A)^{r-1}Q_{j,l}^{-1}(\tau A)$ for $r \le l$. Now (1.2) is established with no difficulty, since for $x \in D(A)$ we have that $A_\tau x - Ax = A^{-1}A_\tau Ax - Ax$.

Now let us define the operators $\exp\{-tA_\tau\}$ by the formula

$$\exp\{-tA_\tau\} = \sum_{i=0}^{\infty}\frac{t^i}{i!}(-A_\tau)^i.$$

Since $A_\tau = \tau^{-1}I - \tau^{-1}R_{j,l}(\tau A)$, it have that

$$\exp\{-tA_\tau\} = \exp\{-t[\tau^{-1}I - \tau^{-1}R_{j,l}(\tau A)]\} = e^{-\frac{t}{\tau}}\sum_{i=0}^{\infty}\frac{t^i}{i!\tau^i}R_{j,l}^i.$$

From the estimate (0.53) it follows that

$$\|\exp\{-tA_\tau\}\|_{E\to E} \le e^{-\frac{t}{\tau}}\sum_{i=0}^{\infty}\frac{t^i}{i!\tau^i}\|R_{j,l}^i\|_{E\to E} \le Me^{-\frac{t}{\tau}}\sum_{i=0}^{\infty}\frac{t^i}{i!\tau^i} = M. \tag{1.5}$$

Thus, we have constructed a family of operators $\{A_\tau\}$, $0 < \tau \le \tau_0$, with the properties (1.2) and (1.5). Furthermore, the operators A_τ with different values of τ commute, and they also commute with A on $D(A)$. Since the spectrum of A

lies in the left half-plane by assumption, this implies that the operator $-A$ is the generator of a semigroup $\exp\{-tA\}$, which can be defined as the limit

$$\lim_{\tau \to 0+} \exp\{-tA_\tau\}x = \exp\{-tA\}x \qquad (1.6)$$

for every $x \in D(A)$. The proof follows the scheme of the proof of the Hille-Phillips-Yosida-Miyadera theorem. Theorem 1.1 is proved.

This theorem provides us with a necessary condition for the stability of the difference problem (0.51) for the cases $l - 4 \le j \le l$. Earlier the following result was established.

Theorem 1.2. *For the operator $-A$ with dense domain $D(A)$ to be the generator of a strongly continuous semigroup $\exp\{-tA\}$ it is sufficient that the estimate (0.53) hold for $j = l - 1$.*

The same assertion holds for $j = 0$, $l = 2$. The authors do not know whether it holds in the general case $j = l - 2$.

Let us show that from the stability of the homogeneous difference problem (0.51) we can infer its well-posedness. In fact, the solution $u(0, u_0)$ of the homogeneous problem satisfies

$$\overline{\mathcal{D}}\,\overline{\Pi}(u_0)u(0, u_0) = -\overline{A}_{j,l}u(0, u_0) = -u(0, \overline{A}_{j,l}u_0).$$

Consequently,

$$\|\overline{\mathcal{D}}\,\overline{\Pi}(u_0)u\|_{C(E)} + \|\overline{A}_{j,l}u\|_{C(E)} \le M_0\|\overline{A}_{j,l}u_0\|_E. \qquad (1.7)$$

By (0.46),

$$\|A_{j,l}(\tau A)^{-1}\|_{E \to E} \le M. \qquad (1.8)$$

Hence, (1.7) yields the coercivity inequality

$$\|\overline{\mathcal{D}}\,\overline{\Pi}(u_0)u\|_{C(E)} + \|\overline{A}_{j,l}u\|_{C(E)} \le MM_0\|Au_0\|_E. \qquad (1.9)$$

Further, the stability of the difference problem (0.51) implies the stability of the general difference problem. Indeed, formula (0.50) and the estimate (0.53) imply the inequality (see Section 0)

$$\|u\|_{C(E)} \le M \left[\|u_0\|_E + \|\varphi_{j,l}\|_{C(E)}\right]. \qquad (1.10)$$

Let us use this inequality to investigate the rate of convergence of the solutions of the difference problem (0.51) to the solution of the differential problem (1.1) of Chapter 1.

Theorem 1.3. *Let $v_0 \in D(A^{j+l+1})$. Suppose that the function $f(t)$ has $l + j$ derivatives such that $f^{(n)}(t) \in D(A^{j+l+1-n})$, $n = 0, 1, \cdots, j + l$ and the function $A^{j+l+1-n} f^{(n)}(t)$ satisfies a Hölder condition. Let u_k be the components of the solution of problem (0.51) with right-hand sides $\varphi_k^{j,l}$ defined by formulas (0.43). Let $z_k = v(t_k) - u_k$ be the components of the error of the solution $v(t)$ to problem (1.1) of Chapter 1. Then the following estimate of the rate of convergence holds:*

$$\|z\|_{\mathcal{C}(E)} \leq M \tau^{j+l}, \tag{1.11}$$

where M does not depend on τ.

Proof. The components z_k satisfy the relations

$$\tau^{-1}(z_k - z_{k-1}) + \tau^{-1}[I - R_{j,l}(\tau A)]z_{k-1} = R_k, \quad k = 1, \cdots, N; \quad z_0 = 0. \tag{1.12}$$

Here the components R_k of the approximation vector are defined by formula (0.44).

Using the stability inequality (1.10) and the estimates (0.48) of the components R_k, we obtain the estimate (1.11). Theorem 1.3 is proved.

An examination of the proof shows that the quantity M in the estimate (1.11) depends linearly on the quantitites

$$\|A^{j+l+1}v_0\|_E, \quad \max_{0 \leq t \leq 1} \|A^{j+l+1-n}f^{(n)}(t)\|_E, \quad n = 0, 1, \cdots, j + l.$$

In particular, we have the estimate

$$\|[R_{j,l}^k(\tau A) - \exp\{-k\tau A\}]v_0\|_E \leq M_0 \|A^{j+l+1}v_0\|_E \, \tau^{j+l}, \tag{1.13}$$

which characterizes the rate of convergence of the solutions of the difference problem to the solution of the corresponding differential equation.

Since the stability condition (1.10) is satisfied and $D(A^{j+l+1})$ is dense in E, inequality (1.13) implies that

$$\lim_{\tau \to 0, \ k\tau \to t} \|[R_{j,l}^k(\tau A) - \exp\{-tA\}]v_0\|_E = 0 \tag{1.14}$$

for all $v_0 \in E$, $t \in [0,1]$, uniformly in t.

2.   The nonhomogeneous problem.

Let us consider now the nonhomogeneous problem (0.51) with null initial data ($u_0 = 0$). Its solution is given by the operator $u(\varphi_{j,l}, 0)$. From condition 2) of well-posedness it follows that this operator is continuous in $\mathcal{C}(E)$. Condition 1) means that $\overline{A}_{j,l} u(\varphi_{j,l}, 0)$ is an additive, homogeneous, everywhere defined operator in $\mathcal{C}(E)$. Since the operator A is closed in E and the operators $A_{j,l} A^{-1}$ are uniformly bounded with respect to τ in the norm of E (estimate (1.8)), $\overline{A}_{j,l}$ is closed in $\mathcal{C}(E)$. Hence, Banach's theorem on the boundedness of a linear operator defined on the entire space applies. It follows that the solution $u(\varphi_{j,l}, 0)$ obeys the coercivity inequality

$$\|\overline{\mathcal{D}}\,\overline{\Pi}(0)u\|_{\mathcal{C}(E)} + \|\overline{A}_{j,l}u\|_{\mathcal{C}(E)} \leq M_1 \|\varphi_{j,l}\|_{\mathcal{C}(E)}. \tag{1.15}$$

Inequalities (1.9) and (1.15) yield the coercivity inequality

$$\|\overline{\mathcal{D}}\,\overline{\Pi}(0)u\|_{\mathcal{C}(E)} + \|\overline{A}_{j,l}u\|_{\mathcal{C}(E)} \leq M \left[\|Au_0\|_E + \|\varphi_{j,l}\|_{\mathcal{C}(E)} \right] \tag{1.16}$$

for the solutions of the general problem (0.51).

On the other hand, from (1.16) it follows that a stable problem (0.51) is well posed in $\mathcal{C}(E)$. Therefore, the coercivity inequality is a sufficient condition for the well-posedness in $\mathcal{C}(E)$ of a stable problem (0.51).

Note that in the case of problem (0.6) of Chapter 2 the coercivity inequality in $\mathcal{C}(E)$ implies the stability of the difference problem. The authors do not know whether this is true in the case of the general problem (0.51).

At the end of the preceding subsection we have shown that a necessary condition for the stability of the difference problem (0.51) in $\mathcal{C}(E)$ is the strong continuity of the semigroup $\exp\{-tA\}$, $t \geq 0$. Let us show that a necessary condition for the stronger requirement of well-posedness of problem (0.51) in $\mathcal{C}(E)$ is the analyticity of this semigroup.

In the case where $l - 4 \leq j \leq l$, the denominators of the Padé fractions $R_{j,l}(z) = P_{j,l}(z)/Q_{j,l}(z)$ used in the construction of difference schemes have their roots in the left complex half-plane $\mathbf{C}^-$. For this reason, when one definines the operators $R_{j,l}(\tau A)$ one assumes that the spectrum of A lies in the right complex half-plane $\mathbf{C}^+$. (We esentially already made use of this fact in proving the bound-edness (for $t \to +\infty$) of the semigroup $\exp\{-tA\}$.) It follows that the equation

$$\lambda w + Aw = \psi \tag{1.17}$$

has a unique solution $w = w(\lambda, \psi)$ for any λ with $\operatorname{Re}\lambda \geq 0$ and $\psi \in E$. To prove that the semigroup $\exp\{-tA\}$ is analytic, it suffices to establish the estimate

$$(1 + |\lambda|)\|w\|_E \leq M_0 \|\psi\|_E \tag{1.18}$$

with some M_0 that does not depend on ψ.

Consider the function $v(t) = e^{\lambda t} w$. Clearly, we have the identity ($1 \leq k \leq N$, $N\tau = 1$):

$$\tau^{-1}(v(t_k) - v(t_{k-1})) + \tau^{-1}[I - R_{j,l}(\tau A)]v(t_{k-1}) = \varphi_k, \tag{1.19}$$

with

$$\varphi_k = \tau^{-1}(v(t_k) - v(t_{k-1})) + \tau^{-1}[I - R_{j,l}(\tau A)]v(t_{k-1}).$$

Relation (1.19) is precisely the difference problem (0.49). Hence, from the well-posedness of this problem one infers the coercivity inequality

$$\max_{1 \leq k \leq N} \|\tau^{-1}(v(t_k) - v(t_{k-1}))\|_E + \max_{1 \leq k \leq N} \|\tau^{-1}[I - R_{j,l}(\tau A)]v(t_{k-1})\|_E$$

$$\leq M \left[\|Av(t_0)\|_E + \max_{1 \leq k \leq N} \|\varphi_k\|_E \right]. \tag{1.20}$$

The functions $v'(t) = \lambda e^{\lambda t} w$ and $Av(t) = e^{\lambda t} Aw$ are continuous. Let $\tau \to 0$ and $k\tau \to t \in [0,1]$. Then $\tau^{-1}(v(t_k) - v(t_{k-1})) \to v'(t) = \lambda e^{\lambda t} w$, and, by (1.3), $\tau^{-1}[I - R_{j,l}(\tau A)]v(t_{k-1}) \to Av(t) = e^{\lambda t} Aw$, the convergence being uniform in $t \in [0,1]$. Hence, $\varphi_k \to v'(t) + Av(t) = e^{\lambda t}(\lambda w + Aw) = e^{\lambda t}\psi$. Passing to the limit in (1.20), we obtain the inequality

$$\max_{0 \leq t \leq 1} \|\lambda e^{\lambda t} w\|_E + \max_{0 \leq t \leq 1} \|e^{\lambda t} Aw\|_E \leq M \left[\|Aw\|_E + \max_{0 \leq t \leq 1} \|e^{\lambda t}\psi\|_E \right].$$

Thus,

$$|\lambda|\,\|w\|_E + \|Aw\|_E \leq M \left[\|\psi\|_E + e^{-\operatorname{Re}\lambda}\|Aw\|_E \right].$$

Now let $\operatorname{Re}\lambda \geq G_0 > 0$ and $e^{G_0} > M$. Then the last inequality yields

$$|\lambda|\,\|w\|_E + (1 - Me^{-G_0})\|Aw\|_E \leq M\|\psi\|_E.$$

Finally, since the operator A^{-1} is bounded by assumption, there is an M_1 such that the inequality

$$(|\lambda| + 1)\|w\|_E \leq M_1\|\psi\|_E \tag{1.21}$$

holds for all λ with $\operatorname{Re}\lambda \geq G_0$. Now let $\lambda = G_0 + \rho e^{\pm i(\frac{\pi}{2}+\phi)}$ for $0 \leq \phi \leq \phi_0$. Then in equation (1.17) we can make the substitution $z = (G_0 + i\tau + A)^{-1}w$ and arrive at the relation

$$z - \tau \tan\phi\,(G_0 + i\tau + A)^{-1}z = \psi.$$

Since, by (1.21),

$$\|\tau\tan\phi\,(G_0+i\tau+A)^{-1}\|_{E\to E} \leq |\tau|\tan\phi\,M_1\big(G_0^2+\tau^2\big)^{1/2} \leq M_1\tan\phi,$$

for sufficiently small $\phi_0 > 0$ we have that $\|z\|_E \leq [1 - M_1\tan\phi]^{-1}\|\psi\|_E$. Hence, (1.21) gives

$$\|w\|_E \leq \frac{M_1[1 - M_1\tan\phi]^{-1}}{\big(G_0^2 + \tau^2\big)^{1/2} + 1}\|\psi\|_E.$$

This obviously yields the inequality

$$(|\lambda| + 1)\|w\|_E \leq M_2\|\psi\|_E,$$

with $M_2 = M_1[\cos\phi_0\,(1 - M_1\tan\phi_0)]^{-1}$ for all λ in the domain $\mathcal{G}_0 = \mathcal{G}(G_0, \phi) \subset \mathbf{C}^+$ bounded to the left by the contour $G_0 + \rho e^{\pm i(\frac{\pi}{2}+\phi_0)}$. Since $\mathbf{C}^+ \backslash \mathcal{G}_0$ is a bounded set and since, by assumption, the operator $\lambda I + A$ has a bounded inverse for all $\lambda \in \mathbf{C}^+$, the estimate (1.18) holds for some $M_0 > 0$. As we know, this estimate means that the semigroup $\exp\{-tA\}$ is analytic and its norm decays exponentially.

The "complex-field" proof given above did not use the stability in $\mathcal{C}(E)$ of the difference problem (0.51); rather, it relied only on the coercivity inequality for the solutions of this problem. There exists another, "real-field" proof of the analyticity of the semigroup $\exp\{-tA\}$, which relies on the well-posedness of the difference problem (0.51). That proof is based on the following assertion.

Theorem 1.4. *If the difference problem (0.51) is well-posed in $\mathcal{C}(E)$, then the differential problem (1.1) of Chapter 1 is well-posed in $C(E)$.*

Proof. As we have shown earlier, from the stability of problem (0.51) it follows that the semigroup $\exp\{-tA\}$ is strongly continuous. This allows us to construct a unique solution in $C(E)$ to problem (1.1) of Chapter 1. Specifically, if $v_0 \in D(A)$

and the function $Af(t)$ is continuous, then, as we have shown in Chapter 1, this unique solution $v(t)$ exists and is given by the formula (1.19) of Chapter 1:

$$v(t) = \exp\{-tA\}v_0 + \int_0^t \exp\{-(t-s)A\}f(s)ds.$$

Next, let us use the method applied in estimating the solution of equation (1.19). Given the function $v(t)$, consider the identity

$$\tau^{-1}(v(t_k) - v(t_{k-1})) + \tau^{-1}[I - R_{j,l}(\tau A)]v(t_{k-1}) = r_k,$$

$$r_k = \tau^{-1}(v(t_k) - v(t_{k-1})) + \tau^{-1}[I - R_{j,l}(\tau A)]v(t_{k-1}).$$

This yields the estimate

$$\max_{1 \le k \le N} \|\tau^{-1}(v(t_k) - v(t_{k-1}))\|_E + \max_{1 \le k \le N} \|\tau^{-1}[I - R_{j,l}(\tau A)]v(t_{k-1})\|_E$$

$$\le M \left[\|Av_0\|_E + \max_{1 \le k \le N} \|r_k\|_E \right].$$

Since the functions $v'(t)$ and $Av(t)$ are continuous and relation (1.3) holds, passing to the limit in the last inequality gives the coercivity inequality in the norm of $\mathcal{C}(E)$ for the solutions $v(t)$ of the differential problem (1.1) of Chapter 1:

$$\max_{0 \le t \le 1} \|v'(t)\|_E + \max_{0 \le t \le 1} \|Av(t)\|_E \le M \left[\|Av_0\|_E + \max_{0 \le t \le 1} \|f(t)\|_E \right].$$

This inequality was established for the set of functions $f(t)$ with the property that the function $Af(t)$ is continuous; since the operator A^{-1} is bounded, in this case $f(t)$, too, is continuous. The set of such functions is dense in $\mathcal{C}(E)$, because $D(A)$ is dense in E. As we remarked in Chapter 1, if the coercivity inequality holds on a dense subset of $\mathcal{C}(E)$, then problem (1.1) Chapter 1 is well-posed in $\mathcal{C}(E)$. Theorem 1.4 is proved.

In Chapter 1 we have shown that a necessary condition for the well-posedness of problem (1.1) is the analyticity of the semigroup $\exp\{-tA\}$. Hence, the analyticity of this semigroup is a necessary condition for the well-posedness of the difference problem (0.51). This is a new, "real-field" proof of the analyticity of the semigroup $\exp\{-tA\}$. However, Theorem 1.4 gives considerably more information.

In Chapter 1 we gave an example of an operator A for which problem (1.1) is not well-posed in $\mathcal{C}(E)$, although the semigroup $\exp\{-tA\}$ is analytic in E. This means that the analyticity of such a semigroup cannot be a sufficient condition for the well-posedness of the difference problem (0.51) in $\mathcal{C}(E)$ for arbitrary E and A.

3.   Sufficient conditions for almost-well-posedness. A real-field criterion for analyticity.

In Chapter 2 it was shown that estimates of the step operator of the simplest difference scheme (0.6) allow one to establish stability and to prove an almost coercive inequality for the solutions of this difference problem, as well as a real-field criterion for the analyticity of the semigroup $\exp\{-tA\}$. It turns out that similar estimates of the operator step hold for a wide class of Padé difference schemes, and this allows one to extend to this class the aforementioned results of Chapter 2.

In what follows it will be assumed that $-A$ is the generator of an analytic semigroup $\exp\{-tA\}$.

First let us consider difference schemes (0.51) generated by the Padé fractions $R_{j,l}(z)$ with $j = l-2$, $l-1$. We have the estimates $(1 \leq k \leq N$, $\tau = 1/N)$

$$\|R_{j,l}^k(\tau A)\|_{E \to E} \leq M, \tag{1.22}$$

$$\|A R_{j,l}^k(\tau A)\|_{E \to E} \leq M(k\tau)^{-1}, \tag{1.23}$$

where M does not depend on τ. These estimates will be proved below.

They allow us to obtain the following results.

Theorem 1.5. *The difference problem* (0.51) *is stable in the space* $\mathcal{C}(E)$.

Theorem 1.6. *The solutions of the difference problem* (0.51) *satisfy the almost-coercive inequality*

$$\|\mathcal{D}\Pi(u_0)u^\tau\|_{C(\tau,E)} + \|A_{j,l}u^\tau\|_{C(\tau,E)}$$

$$\leq M\left[\|Au_0\|_E + \min\left\{\ln(1/\tau), 1 + |\ln\|A\|_{E \to E}|\right\}\|\varphi_{j,l}^\tau\|_{C(\tau,E)}\right], \tag{1.24}$$

where M does not depend on τ (and A).

(Here $\|A\|_{E \to E} = \infty$ in the case of the method of lines, see Section 1, Chapter 1.)

Theorem 1.7. (Real-field criterion for analyticity). *Estimate* (1.23) *is not only necessary but also sufficient for the analyticity of the semigroup* $\exp\{-tA\}$.

The proofs of Theorems 1.5 to 1.7 are carried out according to the same scheme as the proofs of the corresponding results in Chapter 2.

Now let us turn to the difference scheme (0.51) generated by the Padé fractions $R_{l,l}(z)$. The step operator $R_{l,l}(\tau A)$ has "worse" properties than the step operators $R_{l-2,l}(\tau A)$ and $R_{l-1,l}(\tau A)$.

The following estimates hold for $1 \leq k \leq N$:

$$\|R_{l,l}^k(\tau A)(I + \tau A)^{-1}\|_{E \to E} \leq M, \tag{1.25}$$

$$\|AR_{l,l}^k(\tau A)(I + \tau A)^{-2}\|_{E \to E} \leq M(k\tau)^{-1}, \tag{1.26}$$

where M does not depend on τ. These estimates will be proved below.

They allow us to establish the following results.

Theorem 1.8. *The solutions of the difference scheme (0.51) satisfy the stability inequality*

$$\|u^\tau\|_{C(\tau,E)} \leq M\Big[\|(I + \tau A)u_0\|_E + \|(I + \tau A)\varphi_{l,l}^\tau\|_{C(\tau,E)}\Big]. \tag{1.27}$$

Theorem 1.9. *The solutions of the difference scheme (0.51) satisfy the almost-coercivity inequality*

$$\|\mathcal{D}\Pi(u_0)u^\tau\|_{C(\tau,E)} + \|A_{J,l}u^\tau\|_{C(\tau,E)}$$

$$\leq M\Big[\|Au_0\|_E + \min\big\{\ln(1/\tau), 1 + \ln\|A\|_{E \to E}\big\}\|(I + \tau A)\varphi_{l,l}^\tau\|_{C(\tau,E)}\Big]. \tag{1.28}$$

The proofs of Theorems 1.8 and 1.9 follow the same scheme as the proofs of Theorems 1.5 and 1.6, with the estimates (1.25) and (1.26) replacing (1.22) and (1.23), respectively.

In applications to the estimation of the rate of convergence (when $u_0 = 0$ and $\varphi_{l,l}^\tau$ is replaced by the approximation error of the difference scheme R^τ) the indicated theorems yield estimates of the error z^τ of the solutions of difference schemes. In comparison to Theorems 1.5 and 1.6, in Theorems 1.8 and 1.9 one requires an estimate of the approximant R^τ not only in the norm of E, but also in the norm of $D(A)$, but with an order of decay for $\tau \to 0$ smaller by one.

It turns out that Theorems 1.8 and 1.9 can be sharpened, in the sense that one can eliminate the operators $I + \tau A$ from the right-hand sides of the estimates (1.27) and (1.28).

Theorem 1.10. *The solutions of the difference scheme* (0.51) *satisfy the stability inequality*

$$\|u^\tau\|_{C(\tau,E)} \le M \, \min \left\{ \ln(1/\tau), 1 + |\ln\|A\|_{E\to E}| \right\} \left[\|u_0\|_E + \|\varphi_{l,l}^\tau\|_{C(\tau,E)} \right]. \quad (1.29)$$

Proof. By formula (0.5) and the boundedness of the operator $R_{l,l}(\tau A)$, to prove (1.29) it suffices to establish the estimate

$$\|R_{l,l}^{2k}(\tau A)\|_{E\to E} \le M \, \min \left\{ \ln(1/\tau), 1 + |\ln\|A\|_{E\to E}| \right\}, \quad 1 \le k \le N/2. \quad (1.30)$$

To this end we shall use the identity

$$R_{l,l}^{2k}(\tau A) = R_{l,l}^2(\tau A) + \sum_{i=1}^{k-1} [R_{l,l}^2(\tau A) - I] R_{l,l}^{2i}(\tau A), \quad k \ge 2. \quad (1.31)$$

Since, by the definition of the Padé fractions (0.4),

$$I - R_{l,l}^2(\tau A) = \left[Q_{l,l}^2(z) - Q_{l,l}^2(-z) \right]/Q_{l,l}^2(z), \quad (1.32)$$

it follows that for l odd and even we have the estimate

$$\|(I + \tau A)^2 [I - R_{l,l}^2(\tau A)] A^{-1}\|_{E\to E} \le M\tau, \quad (1.33)$$

where M does not depend on τ. From (1.33) and (1.26) we obtain the estimate

$$\|[I - R_{l,l}^2(\tau A)] R_{l,l}^{2i}(\tau A)\|_{E\to E} \le M(i\tau)^{-1}\tau. \quad (1.34)$$

On the other hand, from (1.32) we obtain

$$\|(I + \tau A)^2 [I - R_{l,l}^2(\tau A)]\|_{E\to E} \le M\|A\|_{E\to E} \tau. \quad (1.35)$$

Further, estimates (1.35) and (1.25) yield

$$\|[I - R_{l,l}^2(\tau A)] R_{l,l}^{2i}(\tau A)\|_{E\to E} \le M\tau\|A\|_{E\to E}. \quad (1.36)$$

Hence, the following estimate holds:

$$\|[I - R_{l,l}^2(\tau A)] R_{l,l}^{2i}(\tau A)\|_{E\to E} \le M\tau \min \left\{ 1/(i\tau), \|A\|_{E\to E} \right\}. \quad (1.37)$$

From (1.31) it follows that

$$\|R_{l,l}^{2k}(\tau A)\|_{E\to E} \le \|R_{l,l}^2\|_{E\to E} + \sum_{i=1}^{k-1} \|[I - R_{l,l}^2(\tau A)]R_{l,l}^{2i}(\tau A)\|_{E\to E}$$

$$\le \|R_{l,l}^2\|_{E\to E} + \sum_{i=1}^{N} \|[I - R_{l,l}^2(\tau A)]R_{l,l}^{2i}(\tau A)\|_{E\to E}.$$

We have to examine three possibilities for the norm $\|A\|_{E\to E}$: $\|A\|_{E\to E} < 1$, $\|A\|_{E\to E} > N$, and $1 \le \|A\|_{E\to E} \le N$. If $\|A\|_{E\to E} < 1$, then, by (1.37),

$$\sum_{i=1}^{N} \|[I - R_{l,l}^2(\tau A)]R_{l,l}^{2i}(\tau A)\|_{E\to E} \le M \sum_{i=1}^{N} \|A\|_{E\to E}\,\tau \le M. \tag{1.38}$$

If $\|A\|_{E\to E} > N$, then, by (1.37),

$$\sum_{i=1}^{N} \|[I - R_{l,l}^2(\tau A)]R_{l,l}^{2i}(\tau A)\|_{E\to E} \le M \sum_{i=1}^{N} \frac{\tau}{i\tau} \le M \ln\frac{1}{\tau}. \tag{1.39}$$

Now let $1 \le \|A\|_{E\to E} \le N$. Then, by (1.37),

$$\sum_{i=1}^{N} \|[I - R_{l,l}^2(\tau A)]R_{l,l}^{2i}(\tau A)\|_{E\to E} \le M \sum_{i=1}^{N_0-1} \|A\|_{E\to E}\,\tau + M \sum_{i=N_0}^{N} \frac{\tau}{i\tau}$$

$$\le M\|A\|_{E\to E}\frac{N_0}{N} + M\ln(N/N_0) \le M + M\ln(2\|A\|_{E\to E}).$$

Here $N_0 = [N/\|A\|_{E\to E}]$, with $[x]$ denoting the integer part of x. Hence,

$$\sum_{i=1}^{N} \|[I - R_{l,l}^2(\tau A)]R_{l,l}^{2i}(\tau A)\|_{E\to E} \le M_1 + M\ln\|A\|_{E\to E}. \tag{1.40}$$

The estimates (1.38)–(1.40) and the obvious estimate $\|R_{l,l}^2(\tau A)\|_{E\to E} \le M$ yield (1.30). Theorem 1.10 is proved.

Theorem 1.11. *If l is even, then the solutions of the difference problem (0.51) obey the coercivity inequality*

$$\|\mathcal{D}\Pi(u_0)u^\tau\|_{C(\tau,E)} + \|A_{l,l}u^\tau\|_{C(\tau,E)}$$

$$\le M\Big[\|Au_0\|_E + \min\Big\{\ln(1/\tau) + |\ln\|A\|_{E\to E}|\Big\}\|\varphi_{l,l}^\tau\|_{C(\tau,E)}\Big]. \tag{1.41}$$

188 Padé difference schemes Chap. 3

Proof. If l is even, then the identity

$$I - R_{l,l}(z) = [Q_{l,l}(z) - Q_{l,l}(-z)]/Q_{l,l}(z) \qquad (1.42)$$

obviously yields the estimate

$$\|(I + \tau A)^2[I - R_{l,l}(\tau A)]A^{-1}\|_{E \to E} \leq M\tau, \qquad (1.43)$$

where M does not depend on τ. This allows us to eliminate the operator $I + \tau A$ in the right-hand sides of an inequality of the form (1.27). Here the method of proof is the same as in the case of problem (0.6) in Chapter 2.

We conclude this subsection by the following remarks.

Remark 1. As we mentioned above (Subsection 1), in the case $j = l - 1$ the estimate (1.22) was established earlier for the wider class of operators A that generate a strongly continuous semigroup $\exp\{-tA\}$. Below we shall give another proof, which makes essential use of the analyticity of this semigroup.

Remark 2. The stability theorems 1.8 and 1.10 are based on estimate (1.25), which holds for any operator A that generates an analytic semigroup $\exp\{-tA\}$. Recently the following sharper estimate was obtained (for a smaller class of operators):

$$\|R_{l,l}^k(\tau A)\|_{E \to E} \leq M; \qquad (1.44)$$

this allows us to investigate the stability of difference schemes not only with respect to the right-hand side, but also with respect to the initial conditions.

4.   Estimates of powers of the operator step.

In this subsection we give the proofs of the estimates (1.22), (1.23), (1.25), and (1.26). These proofs are based on an auxiliary estimate for powers of the operator step. Judging by the preceding subsection, it may seem that the case $j = l$ is more difficult than the cases $j = l - 2$ and $j = l - 1$. Here the investigation will start with the case $j = l$ and it will be shown that the cases $j = l - 2$ and $j = l - 1$ reduce to it by means of estimates given in Section 0. The proof of (1.22), (1.23), (1.25), and (1.26) is carried out by comparing the powers of the operator step with the semigroup.

Lemma 1.1. *The following estimate holds for any $1 \leq k \leq N$:*

$$\|(\tau A)^{1/2} R_{l,l}^k(\tau A)(I + \tau A)^{-1}\|_{E \to E} \leq Mk^{-1/2}, \qquad (1.45)$$

where M does not depend on τ and k.

Proof. By the Cauchy-Riesz formula (4.2) of Chapter 2,

$$(\tau A)^{1/2} R_{l,l}^k(\tau A)(I + \tau A)^{-1} = \frac{1}{2\pi i} \int_{S_1 \cup S_2} z^{1/2} R_{l,l}^k(z) \frac{1}{1+z} (z - \tau A)^{-1} dz.$$

Using the estimate (4.1) of Chapter 2, the representation (0.8), and the estimate (0.9), we obtain

$$\|(\tau A)^{1/2} R_{l,l}^k(\tau A)(I + \tau A)^{-1}\|_{E \to E} \leq M \int_0^\infty |z|^{1/2} \left| \frac{z + z_1}{z - \bar{z}_1} \right|^k \frac{|dz|}{|z - \bar{z}_1||z|}$$

$$\leq M\rho_1^{-1/2} \int_0^\infty \Psi(\rho) d\rho.$$

Here $z_1 \in \mathbf{C}$ with $\operatorname{Re} z_1 < 0$ and $|z_1| = \rho_1$,

$$\Psi(\rho) = \rho^{1/2} \left(\frac{\rho^2 + 2\rho \cos \phi_2 + 1}{\rho^2 + 2\rho \cos \phi_3 + 1} \right)^{-k/2} \frac{1}{\rho(\rho^2 + 2\rho \cos \phi_3 + 1)^{1/2}},$$

where ϕ_2 [resp., ϕ_3] is the angle between the vectors z and z_1 [resp., z and $-\bar{z}_1$]. To prove (1.45) it suffices to establish the estimate

$$\int_0^\infty \Psi(\rho) d\rho \leq Mk^{-1/2}. \qquad (1.46)$$

We have

$$\int_0^\infty \Psi(\rho) d\rho = \int_0^1 \Psi(\rho) d\rho + \int_1^\infty \Psi(\rho) d\rho = I_1 + I_2.$$

First let us estimate I_1. By (0.30),

$$I_1 \leq M_1 \int_0^1 (1-r)^{k/2} r^{-1/2} dr$$

$$= M_1 \left(\int_0^{1/k} (1-r)^{k/2} r^{-1/2} dr + \int_{1/k}^1 (1-r)^{k/2} r^{-1/2} dr \right)$$

$$\leq M_1 \left(\int_0^{1/k} r^{-1/2} dr + k^{1/2} \int_0^1 (1-r)^{k/2} dr \right) \leq M_2 k^{-1/2}. \qquad (1.47)$$

To estimate I_2, let us make the substitution $\rho = 1/y$. We obtain

$$\int_1^\infty \Psi(\rho)d\rho = \int_0^1 \Psi(y)dy,$$

which in conjunction with (1.47) implies (1.45). Lemma 1.1 is proved.

Lemma 1.2. *The following estimate holds for any $1 \le k \le N$ and for $j = l - 2, l - 1$:*

$$\|(\tau A)^{1/2}R_{j,l}^k(\tau A)\|_{E\to E} \le Mk^{-1/2}, \tag{1.48}$$

where M does not depend on τ and k.

Proof. For $k = 1$ the estimate (1.48) is obvious. Since

$$\|R_{j,l}(\tau A)(I + \tau A)\|_{E\to E} \le M, \tag{1.49}$$

to prove (1.48) is suffices to establish the estimate

$$\|(\tau A)^{1/2}R_{j,l}^{k-1}(\tau A)(I + \tau A)^{-1}\|_{E\to E} \le Mk^{-1/2}. \tag{1.50}$$

By the Cauchy-Riesz formula (4.2) of Chapter 2,

$$(\tau A)^{1/2}R_{j,l}^{k-1}(\tau A)(I + \tau A)^{-1} = \frac{1}{2\pi i}\int_{S_1\cup S_2} z^{1/2}R_{j,l}^{k-1}(z)\frac{1}{1+z}(z - \tau A)^{-1}dz.$$

Using the estimate (4.1) of Chapter 2, and the estimates (0.10) and (0.18), we obtain

$$\|(\tau A)^{1/2}R_{j,l}^{k-1}(\tau A)(I + \tau A)^{-1}\|_{E\to E}$$

$$\le M\int_0^\infty |z|^{1/2}|Q_{l,j}(-z)/Q_{l,j}(z)|^{k-1}\frac{|dz|}{|1 + z||z|}.$$

Next, using the representation (0.8) and the estimate (0.9), we obtain

$$\|(\tau A)^{1/2}R_{j,l}^{k-1}(\tau A)(I + \tau A)^{-1}\|_{E\to E} \le M_1\int_0^\infty |z|^{-1/2}\left|\frac{z + z_1}{z - \overline{z}_1}\right|^{k-1}\frac{|dz|}{|z - \overline{z}_1|}.$$

The remaining part of the proof of (1.50) is identical to the proof of (1.45). Lemma 1.2 is proved.

Now let us establish (1.22), (1.23), (1.25) and (1.26).

Lemma 1.3. *The estimates* (1.22) *and* (1.23) *hold.*

Proof. We use the identity

$$R_{j,l}^k(\tau A) - \exp\{-k\tau A\} = \int_0^1 [\psi_{j,l}'(s\tau A)]_s ds, \qquad (1.51)$$

where

$$\psi_{j,l}(s\tau A) = R_{j,l}^k(s\tau A)\exp\{-k(1-s)\tau A\}.$$

The derivative $\psi_{j,l}'(s\tau A)|_s$ is given by

$$\psi_{j,l}'(s\tau A)|_s = \{R_{j,l}(s\tau A) + R_{j,l}'(s\tau A)|_s\} k\tau A R_{j,l}^{k-1}(s\tau A)\exp\{-k(1-s)\tau A\}. \quad (1.52)$$

Using equality (0.5), identity (1.51), and formula (1.52), we obtain

$$R_{j,l}^k(\tau A) - \exp\{-k\tau A\} = \int_0^1 k(\tau A)^{l+j+1} s^{l+j} \frac{(-1)^j l! j!}{((j+l)!)^2} R_{j,l}^{k-1}(s\tau A) \times$$

$$\times Q_{j,l}^{-2}(s\tau A)\exp\{-k(1-s)\tau A\}ds. \qquad (1.53)$$

Since

$$\|(s\tau A)^{l+j} Q_{j,l}^{-2}\|_{E\to E} \le M, \qquad (1.54)$$

it follows that

$$\|R_{j,l}^k(\tau A) - \exp\{-k\tau A\}\|_{E\to E}$$

$$\le M \int_0^1 k\|(\tau A)^{1/2} R_{j,l}^{k-1}(\tau A)\|_{E\to E}\|(\tau A)^{1/2}\exp\{-k(1-s)\tau A\}\|_{E\to E}ds.$$

Using the estimates (1.10) and (1.14) of Chapter 1, we obtain

$$\|(\tau A)^{1/2}\exp\{-k(1-s)\tau A\}\|_{E\to E} \le Mk^{-1/2}(1-s)^{-1/2}, \qquad (1.55)$$

which in conjunction with (1.45) gives

$$\|R_{j,l}^k(\tau A) - \exp\{-k\tau A\}\|_{E\to E} \le M_1 \int_0^1 \frac{k}{((k-1)k)^{1/2}} \frac{ds}{(s(1-s))^{1/2}} \le M_2$$

for $k > 1$. By the triangle inequality, the estimate (1.10) of Chapter 1, and the estimate (4.1) of Chapter 2, (1.22) holds for $k > 1$. For $k = 1$ (1.22) is obvious.

Turning now to the estimate (1.23), we observe that for $k = 1$ it is obvious, while for $k > 1$ it follows from the identity

$$\tau A R_{j,l}^k(\tau A) = (\tau A)^{1/2} R_{j,l}^{[k/2]}(\tau A)(\tau A)^{1/2} R_{j,l}^{k-[k/2]}(\tau A)$$

and the estimate (1.48). Lemma 1.3 is proved.

Lemma 1.4. *The estimates* (1.25) *and* (1.26) *hold.*

Proof. Using identity (1.53), we obtain,

$$[R_{l,l}^k(\tau A) - \exp\{-k\tau A\}](I + \tau A)^{-1}$$

$$= \int_0^1 k(\tau A)^{2l+1} s^{2l} \frac{(-1)^l (l!)^2}{((2l)!)^2} R_{l,l}^{k-1}(s\tau A)(I+\tau A)^{-1} Q_{l,l}^{-2}(s\tau A)\exp\{-k(1-s)\tau A\}ds.$$

By (1.54),

$$\|[R_{l,l}^k(\tau A) - \exp\{-k\tau A\}](I + \tau A)^{-1}\|_{E \to E}$$

$$\leq M_1 \int_0^1 k\|(\tau A)^{1/2} R_{l,l}^{k-1}(s\tau A)(I+\tau A)^{-1}\|_{E\to E}\|(\tau A)^{1/2}\exp\{-k(1-s)\tau A\}\|_{E\to E}ds.$$

Using the estimates (1.45) and (1.46), we obtain

$$\|[R_{l,l}^k(\tau A) - \exp\{-k\tau A\}](I + \tau A)^{-1}\|_{E\to E} \leq$$

$$M_1 \int_0^1 \frac{k}{((k-1)k)^{1/2}} \frac{ds}{(s(1-s))^{1/2}} \leq M_2$$

for $k > 1$. By the triangle inequality, the estimate (1.10) of Chapter 1, and the estimate (4.1) of Chapter 2, (1.25) holds for $k > 1$. For $k = 1$ (1.25) is obvious.

Turning now to the estimate (1.26), we observe that for $k = 1$ it is obvious, while for $k > 1$ it follows from the identity

$$\tau A R_{l,l}^k(\tau A)(I + \tau A)^{-2}$$

$$= (\tau A)^{1/2} R_{l,l}^{[k/2]}(\tau A)(I + \tau A)^{-1}(\tau A)^{1/2} R_{j,l}^{k-[k/2]}(\tau A)(I + \tau A)^{-1}$$

and the estimate (1.45). Lemma 1.4 is proved.

2.   WELL-POSEDNESS OF THE DIFFERENCE PROBLEM IN $C_0^\alpha(E)$

1.   The case of a general space $C_0^\alpha(E)$.

In Chapter 2 we have shown that the analyticity of the semigroup $\exp\{-tA\}$ is a necessary and sufficient condition for the well-posedness of the simplest difference problem (0.6) if the space $C(E)$ is extended or restricted in certain ways. As it turns out, this holds true for a broad class of Padé difference schemes.

A solution u of problem (0.51) is called *solution in* $C_0^\alpha(E)$ if $\overline{\mathcal{D}}\,\overline{\Pi}(u_0)u \in C_0^\alpha(E)$ and $\overline{A}_{j,l}u \in C_0^\alpha(E)$. As in the case of the space $C(E)$, one shows that for the problem (0.51) to be solvable in $C_0^\alpha(E)$ it is necessary that $\varphi_{j,l} \in C_0^\alpha(E)$. Here again we were not able to find a necessary condition for u_0, and we shall assume that $u_0 \in D(A)$.

Definition 2.1. We say that problem (0.51) *is well posed in the space* $C_0^\alpha(E)$ if the following conditions are satisfied:

1) For any $\varphi_{j,l} \in C_0^\alpha(E)$ and $u_0 \in D(A)$ there exists a unique solution $u = u(\varphi_{j,l}, u_0)$ in $C_0^\alpha(E)$ of problem (0.51).

2) Problem (0.51) is stable in $C_0^\alpha(E)$.

From the well-posedness in $C_0^\alpha(E)$ of the nonhomogeneous problem (0.51) with $u_0 = 0$ it follows that the operator $\overline{A}_{j,l}u(\varphi_{j,l}, 0)$ is bounded in $C_0^\alpha(E)$. This leads to the coercivity inequality

$$\|\overline{\mathcal{D}}\,\overline{\Pi}(u_0)u\|_{C_0^\alpha(E)} + \|\overline{A}_{j,l}u\|_{C_0^\alpha(E)} \le M(\alpha)\|\varphi_{j,l}\|_{C_0^\alpha(E)} \tag{2.1}$$

for the solutions of the difference problem (0.51) with $u_0 = 0$. Further, the solvability in $C_0^\alpha(E)$ of problem (0.51) with $u_0 = 0$ implies, via the substitution $u = w + u_0$, the unique solvability in $C_0^\alpha(E)$ of the general problem (0.51) when $u_0 \in D(A)$, and the coercivity inequality

$$\|\overline{\mathcal{D}}\,\overline{\Pi}(u_0)u\|_{C_0^\alpha(E)} + \|\overline{A}_{j,l}u\|_{C_0^\alpha(E)} \le M(\alpha)\left[\|Au_0\|_E + \|\varphi_{j,l}\|_{C_0^\alpha(E)}\right]. \tag{2.2}$$

Inequality (2.2) means that the following inequality holds:

$$\|\mathcal{D}\Pi(u_0)u^\tau\|_{C_0^\alpha(\tau,E)} + \|A_{j,l}u^\tau\|_{C_0^\alpha(\tau,E)} \le M(\alpha)\left[\|Au_0\|_E + \|\varphi_{j,l}^\tau\|_{C_0^\alpha(\tau,E)}\right],$$

where $M(\alpha)$ does not depend on τ. On the other hand, from this last inequality for the solutions of a stable problem (0.49) we derive the well-posedness in $C_0^\alpha(E)$ of problem (0.51). Hence, the coercivity inequality is a necessary and sufficient condition for the well-posedness in $C_0^\alpha(E)$ of a stable problem (0.51).

Note that in the case of problem (0.6) of Chapter 2, the coercivity inequality in $C_0^\alpha(E)$ implies the stability of the difference problem. The authors do not know whether this is true for the general problem (0.51). As in the case of the space $\mathcal{C}(E)$, from the well-posedness in $\mathcal{C}_0^\alpha(E)$ of the difference problem (0.51) one infers the well-posedness in $C_0^\alpha(E)$ of the differential problem (1.1) of Chapter 1. Therefore, the analyticity of the semigroup $\exp\{-tA\}$ is a necessary condition for the well-posedness of the difference problem (0.51) in $\mathcal{C}_0^\alpha(E)$. It turn out that this condition is also sufficient for the well-posedness in $\mathcal{C}_0^\alpha(E)$ of a broad class of Padé difference schemes. For this reason, in what follows we shall assume that $-A$ is the generator of an analytic semigroup $\exp\{-tA\}$.

First let us consider the difference problem (0.51) generated by the Padé fractions $R_{j,l}(z)$ with $j = l - 2, l - 1$. We begin by deriving some smoothness estimates for powers of the operator step.

Lemma 2.1. *For any $1 \le k < k + r \le N$ and $0 \le \alpha \le 1$ one has the estimates*

$$\|R_{j,l}^k(\tau A) - R_{j,l}^{k+r}(\tau A)\|_{E \to E} \le M \frac{r^\alpha}{k^\alpha}, \tag{2.3}$$

$$\|\tau A[R_{j,l}^k(\tau A) - R_{j,l}^{k+r}(\tau A)]\|_{E \to E} \le M \frac{r^\alpha}{k^{1+\alpha}}, \tag{2.4}$$

where M does not depend on τ, α, k and r.

Proof. We shall use the formula

$$R_{j,l}^k(\tau A) - R_{j,l}^{k+r}(\tau A) = \sum_{i=0}^{r-1}[I - R_{j,l}(\tau A)]R_{j,l}^{k+i}(\tau A). \tag{2.5}$$

From (1.8) we obtain the inequality

$$\|R_{j,l}^k(\tau A) - R_{j,l}^{k+r}(\tau A)\|_{E \to E} \le M \sum_{i=0}^{r-1} \|AR_{j,l}^{k+i}(\tau A)\|_{E \to E}\,\tau.$$

By the estimate (1.23),

$$\|R_{j,l}^k(\tau A) - R_{j,l}^{k+r}(\tau A)\|_{E \to E} \le M_1 \sum_{i=0}^{r-1} \frac{1}{k+i} \le M_1 \frac{r}{k}. \tag{2.6}$$

The estimate

$$\|R_{j,l}^k(\tau A) - R_{j,l}^{k+r}(\tau A)\|_{E \to E} \le M_1 \tag{2.7}$$

is an obvious consequence of (1.22) and the triangle inequality. Estimate (2.3) follows from (2.6) and (2.7). Using the semigroup property and the estimates (1.8) and (1.23), we obtain

$$\|\tau A[R_{j,l}^k(\tau A) - R_{j,l}^{k+r}(\tau A)]\|_{E \to E}$$

$$\leq \|[I - R_{j,l}(\tau A)](\tau A)^{-1}\|_{E \to E} \|\tau A R^{[k/2]}(\tau A)\|_{E \to E} \|\tau A R^{k-[k/2]}(\tau A)\|_{E \to E}$$

$$\leq M \frac{1}{k^2} \tag{2.8}$$

for $k > 1$. For $k = 1$ the estimate (2.8) is obvious. Using (2.5) and (2.8), we obtain

$$\|\tau A[R_{j,l}^k(\tau A) - R_{j,l}^{k+r}(\tau A)]\|_{E \to E} \leq \sum_{i=0}^{r-1} \|\tau A[I - R_{j,l}(\tau A)]R_{j,l}^{k+i}(\tau A)\|_{E \to E}$$

$$\leq M \sum_{i=0}^{r-1} \frac{1}{(k+i)^2} \leq M \frac{r}{k^2}. \tag{2.9}$$

The estimate

$$\|\tau A[R_{j,l}^k(\tau A) - R_{j,l}^{k+r}(\tau A)]\|_{E \to E} \leq M \frac{1}{k} \tag{2.10}$$

is an obvious consequence of (1.23) and the triangle inequality. Lemma 2.1 is proved.

The estimates (2.3) and (2.4) allow us to establish the following results.

Theorem 2.1. *The difference problem (0.51) is stable in $C_0^\alpha(E)$.*

Theorem 2.2. *The difference problem (0.51) is well posed in $C_0^\alpha(E)$.*

The proofs of these two theorems follow the same scheme as the proofs of the corresponding results in Chapter 2.

Now let us turn to the difference problem (0.51) generated by the Padé fractions $R_{l,l}(z)$. We already know that the operator step $R_{l,l}(\tau A)$ has "worse" properties than the operator steps $R_{l-2,l}(\tau A)$ and $R_{l-1,l}(\tau A)$. Nevertheless, the following result holds.

Lemma 2.2. *For any $1 \leq k < k+r \leq N$ and $0 \leq \alpha \leq 1$ one has the estimates*

$$\|[R_{l,l}^k(\tau A) - R_{l,l}^{k+r}(\tau A)](I + \tau A)^{-1}\|_{E \to E} \leq M \frac{r^\alpha}{k^\alpha}, \tag{2.11}$$

$$\|\tau A[R_{l,l}^k(\tau A) - R_{l,l}^{k+r}(\tau A)](I + \tau A)^{-3}\|_{E \to E} \le M \frac{r^\alpha}{k^{1+\alpha}}, \qquad (2.12)$$

where M does not depend on τ, α, k and r.

The proof of the estimates (2.11) and (2.12) is carried in the same way as that of the estimates (2.3) and (2.4) and relies on the bound

$$\|(I + \tau A)[I - R_{l,l}(\tau A)](\tau A)^{-1}\|_{E \to E} \le M \qquad (2.13)$$

and the estimates (1.25) and (1.26).

The estimates (2.11) and (2.12) allow us to establish the following results.

Theorem 2.3. Let $\varphi_{l,l}^\tau \in D(A)$. Then the solutions of the difference problem (0.51) obey the stability inequality

$$\|u^\tau\|_{C_0^\alpha(\tau,E)} \le M \left[\|(I + \tau A)u_0\|_E + \|(I + \tau A)\varphi_{l,l}^\tau\|_{C_0^\alpha(\tau,E)} \right], \qquad (2.14)$$

where M does not depend on τ, α, u_0, and $\varphi_{l,l}^\tau$.

Theorem 2.4. Let $\varphi_{l,l}^\tau \in D(A^2)$. Then the solutions of the difference problem (0.51) obey the coercivity inequality

$$\|\mathcal{D}\Pi(u_0)u^\tau\|_{C_0^\alpha(\tau,E)} + \|A_{l,l}u^\tau\|_{C_0^\alpha(\tau,E)}$$

$$\le M\left[\|Au_0\|_E + \frac{1}{\alpha(1-\alpha)} \|(I + \tau A)^2 \varphi_{l,l}^\tau\|_{C_0^\alpha(\tau,E)} \right], \qquad (2.15)$$

where M does not depend on τ, α, u_0, and $\varphi_{l,l}^\tau$.

The proofs of Theorems 2.3 and 2.4, like those of Theorems 2.1 and 2.2, follow the scheme of the proofs of the corresponding results in Chapter 2.

Since for even l the estimate (1.43), which is stronger than (2.13), holds, we have the following

Lemma 2.3. For any $1 \le k < k + r \le N$ and $0 \le \alpha \le 1$ and even l we have the estimate

$$\|\tau A[R_{l,l}^k(\tau A) - R_{l,l}^{k+r}(\tau A)](I + \tau A)^{-2}\|_{E \to E} \le M \frac{r^\alpha}{k^{1+\alpha}}, \qquad (2.16)$$

where M does not depend on τ, α, k, and r.

This yields the following result.

Theorem 2.5. *Let $\varphi_{l,l}^\tau \in D(A)$. Then for even l the solutions of the difference problem* (0.51) *obey the coercivity inequality*

$$\|\mathcal{D}\Pi(u_0)u^\tau\|_{C_0^\alpha(\tau,E)} + \|A_{l,l}u^\tau\|_{C_0^\alpha(\tau,E)}$$

$$\leq M\left[\|Au_0\|_E + \frac{1}{\alpha(1-\alpha)}\|(I+\tau A)\varphi_{l,l}^\tau\|_{C_0^\alpha(\tau,E)}\right], \tag{2.17}$$

where M does not depend on τ, α, u_0, and $\varphi_{l,l}^\tau$.

Proof. Clearly, it suffices to estimate $\|A_{l,l}u^\tau\|_{C_0^\alpha(\tau,E)}$. By formula (0.50), we have that

$$A_{l,l}u_k = A_{l,l}w_k + A_{l,l}g_k$$

$$= A_{l,l}R_{l,l}^k(\tau A)u_0 + \sum_{i=1}^k A_{l,l}R_{l,l}^{k-i}(\tau A)\varphi_i^{l,l}\tau, \quad 1 \leq k \leq N. \tag{2.18}$$

Hence, it suffices to estimate each term. Since

$$A_{l,l}w_k = R_{l,l}^k(\tau A)(I+\tau A)^{-1}(I+\tau A)A_{l,l}A^{-1}Au_0, \quad 1 \leq k \leq N,$$

the estimate

$$\|A_{l,l}w^\tau\|_{C_0^\alpha(\tau,E)} \leq M\|Au_0\|_E \tag{2.19}$$

is a consequence of (1.33) and (2.11).

Now let us estimate $A_{l,l}g^\tau$. Using the relation

$$A_{l,l}g_k = [I - R_{l,l}^k(\tau A)]\varphi_k^{l,l} + \sum_{i=1}^k A_{l,l}R_{l,l}^{k-i}(\tau A)(\varphi_i^{l,l} - \varphi_k^{l,l})\tau, \quad k > 1, \tag{2.20}$$

and the estimates (1.33), (1.25), (1.26), we obtain

$$\|A_{l,l}g_k\|_E \leq \|[I - R_{l,l}^k(\tau A)](I+\tau A)^{-1}\|_{E\to E}\|(I+\tau A)\varphi_k^{l,l}\|_E$$

$$+ \sum_{i=1}^{k-1}\|(I+\tau A)^2 A_{l,l}A^{-1}\|_{E\to E}\|AR_{l,l}^{k-i}(\tau A)(I+\tau A)^{-2}\|_{E\to E}\|\varphi_i^{l,l} - \varphi_k^{l,l}\|_E\,\tau$$

$$\leq M \left[\|(I + \tau A)\varphi_k^{l,l}\|_E + \sum_{i=1}^{k-1-i} \frac{1}{(k-i)^{1-\alpha}i^\alpha} \|\varphi_{l,l}^\tau\|_{C_0^\alpha(\tau, E)} \right]$$

$$\leq M \int_0^{k\tau} \frac{ds}{(k\tau - s)^{1-\alpha}s^\alpha} \|(I + \tau A)\varphi_{l,l}^\tau\|_{C_0^\alpha(\tau, E)}$$

$$\leq M_1 \frac{1}{\alpha(1-\alpha)} \|(I + \tau A)\varphi_{l,l}^\tau\|_{C_0^\alpha(\tau, E)} \tag{2.21}$$

for $k > 1$. The estimate (2.21) with $k = 1$ is obvious.

Now let us estimate the difference $A_{l,l}g_{k+r} - A_{l,l}g_k$ for $1 \leq k < k + r \leq N$. We shall examine separately the cases $k \leq 2r$ and $k > 2r$. If $k \leq 2r$, then (2.21) yields

$$\|A_{l,l}g_{k+r} - A_{l,l}g_k\|_E \leq \|A_{l,l}g_{k+r}\|_E + \|A_{l,l}g_k\|_E$$

$$\leq \frac{2M}{\alpha(1-\alpha)} \|(I + \tau A)\varphi_{l,l}^\tau\|_{C_0^\alpha(\tau, E)} r^\alpha r^{-\alpha}$$

$$\leq \frac{2^{1+\alpha}M}{\alpha(1-\alpha)} \|(I + \tau A)\varphi_{l,l}^\tau\|_{C_0^\alpha(\tau, E)} r^\alpha k^{-\alpha}.$$

Now let $k > 2r$. Since $1 \leq k < k + r \leq N$, we have $k + r > 1$, and then (2.20) implies that

$$A_{l,l}g_{k+r} - A_{l,l}g_k = \left\{ \varphi_{k+r}^{l,l} - \varphi_k^{l,l} + R_{l,l}^k(\tau A)\varphi_k^{l,l} - R_{l,l}^{k+r}(\tau A)\varphi_{k+r}^{l,l} \right\}$$

$$+ \sum_{i=k-r+1}^{k+r-1} A_{l,l}R_{l,l}^{k+r-i}(\tau A)(\varphi_i^{l,l} - \varphi_{k+r}^{l,l})\tau + \sum_{i=k-r+1}^{k-1} A_{l,l}R_{l,l}^{k-i}(\tau A)(\varphi_k^{l,l} - \varphi_i^{l,l})\tau$$

$$+ \sum_{i=1}^{k-r} A_{l,l}R_{l,l}^{k+r-i}(\tau A)(\varphi_k^{l,l} - \varphi_{k+r}^{l,l})\tau + \sum_{i=1}^{k-r} A_{l,l}[R_{l,l}^{k-i}(\tau A) - R_{l,l}^{k+r-i}(\tau A)](\varphi_i^{l,l} - \varphi_k^{l,l})\tau$$

$$= I_1 + I_2 + I_3 + I_4 + I_5.$$

Since

$$I_1 = [I - R_{l,l}^{k+r}(\tau A)](I + \tau A)^{-1}(I + \tau A)(\varphi_{k+r}^{l,l} - \varphi_k^{l,l})$$

$$+ [R_{l,l}^k(\tau A) - R_{l,l}^{k+r}(\tau A)](I + \tau A)^{-1}(I + \tau A)\varphi_k^{l,l},$$

using the estimates (1.25) and (2.11) we see that

$$\|I_1\|_E \leq M \|(I + \tau A)\varphi_{l,l}^\tau\|_{C_0^\alpha(\tau, E)} r^\alpha k^{-\alpha}.$$

Next, by (1.26) and (1.33),

$$\|I_2\|_E \leq \sum_{i=k-r+1}^{k+r-1} \|(I+\tau A)^2 A_{l,l} A^{-1}\|_{E\to E} \|AR_{l,l}^{k+r-i}(\tau A)(I+\tau A)^{-2}\|_{E\to E}\times$$

$$\times\|\varphi_i^{l,l} - \varphi_{k+r}^{l,l}\|_E \, \tau \leq M \sum_{i=k-r+1}^{k+r-1} \frac{\tau}{((k+r-i)\tau)^{1-\alpha}(i\tau)^\alpha}\|\varphi_{l,l}^\tau\|_{C_0^\alpha(\tau,E)}$$

$$\leq M \int_{(k-r+1)\tau}^{(k+r)\tau} \frac{ds}{((k+r)\tau - s)^{1-\alpha}s^\alpha}\|\varphi_{l,l}^\tau\|_{C_0^\alpha(\tau,E)}$$

$$\leq \frac{M}{((k-r)\tau)^\alpha}\int_{(k-r)\tau}^{(k+r)\tau} \frac{ds}{((k+r)\tau - s)^{1-\alpha}}\|\varphi_{l,l}^\tau\|_{C_0^\alpha(\tau,E)}$$

$$\leq \frac{M_1}{\alpha}\|\varphi_{l,l}^\tau\|_{C_0^\alpha(\tau,E)}r^\alpha(k-r)^{-\alpha}.$$

Since $k - r \geq k/2$,

$$\|I_2\|_E \leq \frac{2^\alpha M_1}{\alpha}\|\varphi_{l,l}^\tau\|_{C_0^\alpha(\tau,E)}r^\alpha k^{-\alpha}.$$

In a similar way one proves that

$$\|I_3\|_E \leq \frac{2^\alpha M_1}{\alpha}\|\varphi_{l,l}^\tau\|_{C_0^\alpha(\tau,E)}r^\alpha k^{-\alpha}.$$

Next, using the identity

$$I_4 = [R_{l,l}^k(\tau A) - R_{l,l}^{2r}(\tau A)](\varphi_{k+r}^{l,l} - \varphi_k^{l,l})$$

and the estimates (1.33), (2.11), we obtain

$$\|I_4\|_E \leq \|[R_{l,l}^k(\tau A) - R_{l,l}^{2r}(\tau A)](I+\tau A)^{-1}\|_{E\to E}\|(I+\tau A)(\varphi_{k+r}^{l,l} - \varphi_k^{l,l})\|_E$$

$$\leq M\|(I+\tau A)\varphi_{l,l}^\tau\|_{C_0^\alpha(\tau,E)}r^\alpha k^{-\alpha}.$$

Finally, the estimates (1.33) and (2.16) with $\alpha = 1$ yield

$$\|I_5\|_E \leq \sum_{i=1}^{k-r} \|(I+\tau A)^2 A_{l,l} A^{-1}\|_{E\to E}\times$$

$$\times\|A[R_{l,l}^{k-i}(\tau A) - R_{l,l}^{k+r-i}(\tau A)](I+\tau A)^{-2}\|_{E\to E}\|\varphi_i^{l,l} - \varphi_k^{l,l}\|_E\tau$$

$$\leq M\sum_{i=1}^{k-r} \frac{r\tau\tau}{((k-i)\tau)^{2-\alpha}(i\tau)^\alpha}\|\varphi_{l,l}^\tau\|_{C_0^\alpha(\tau,E)}$$

$$\leq M \int_0^{(k-r)\tau} r\tau(k\tau - s)^{\alpha-2} s^{-\alpha} ds \, \|\varphi_{l,l}^\tau\|_{C_0^\alpha(\tau,E)}.$$

Since for $t > 2\tau > 0$ we have the estimate

$$\int_0^{t-\tau} \tau(t-s)^{\alpha-2} s^{-\alpha} ds \leq \frac{M}{1-\alpha} \tau^\alpha t^{-\alpha},$$

we conclude that

$$\|I_5\|_E \leq \frac{M}{1-\alpha} \|\varphi_{l,l}^\tau\|_{C_0^\alpha(\tau,E)} r^\alpha k^{-\alpha}.$$

Thus, we have shown that the inequality

$$\|A_{l,l}g_{k+r} - A_{l,l}g_k\|_E \leq \frac{M}{\alpha(1-\alpha)} \|(I+\tau A)\varphi_{l,l}^\tau\|_{C_0^\alpha(\tau,E)} r^\alpha k^{-\alpha} \qquad (2.22)$$

holds for all $1 \leq k < k + r \leq N$. From (2.21) and (2.22) it follows that

$$\|A_{l,l}g^\tau\|_{C_0^\alpha(\tau,E)} \leq \frac{M}{\alpha(1-\alpha)} \|(I+\tau A)\varphi_{l,l}^\tau\|_{C_0^\alpha(\tau,E)},$$

which in conjunction with (2.19) yields the needed inequality

$$\|A_{l,l}u^\tau\|_{C_0^\alpha(\tau,E)} \leq M\left[\|Au_0\|_E + \frac{1}{\alpha(1-\alpha)} \|(I+\tau A)\varphi_{l,l}^\tau\|_{C_0^\alpha(\tau,E)}\right]. \qquad (2.23)$$

Theorem 2.5 is proved.

From the proof given above one can see that

$$\|A_{l,l}u^\tau\|_{C_0^\alpha(\tau,E)} \leq \frac{M}{\alpha(1-\alpha)} \|\varphi_{l,l}^\tau\|_{C_0^\alpha(\tau,E)} \qquad (2.24)$$

provided that

$$\|R_{l,l}^k(\tau A)\|_{E\to E} \leq M, \quad 1 \leq k \leq N. \qquad (2.25)$$

Below we shall give examples of spaces for which this last condition is satisfied. This yields the following result.

Theorem 2.6. *Let l be even. If the difference problem* (0.51) *is stable in $C(E)$, then it is well posed in $C_0^\alpha(E)$.*

The authors do not know whether analogous results are valid for the difference problem (0.51) in $C_0^\alpha(E)$ when l is odd. However, close results can be established for the difference problem (0.51) when l is odd, but in a space smaller than $\mathcal{C}_0^\alpha(E)$. This is the objective of the next subsection.

2.   The case of the special space $\tilde{\mathcal{C}}_0^\alpha(E)$.

Let us consider the difference scheme (0.51) generated by the Padé fractions $R_{l,l}(z)$ with l odd. We define the space $\tilde{\mathcal{C}}_0^\alpha(\tau, E)$, $0 < \alpha < 1$, as the vector space $E(\tau)$ of grid functions φ^τ, equipped with the norm

$$\|\varphi^\tau\|_{\tilde{\mathcal{C}}_0^\alpha(\tau,E)} = \|\varphi^\tau\|_{C(\tau,E)} + \max_{1 \le k < k+2r \le N} \|\varphi_{k+2r} - \varphi_k\|_E (2r)^{-\alpha} k^\alpha. \qquad (2.26)$$

Then, according to the definition of the space $\mathcal{E}(E)$ (see Chapter 2), $\tilde{\mathcal{C}}_0^\alpha(E) = \tilde{\mathcal{C}}_0^\alpha(\mathcal{E}(E))$. Unfortunately, when l is odd we only have the estimate

$$\|(I + \tau A)[I - R_{l,l}(\tau A)](\tau A)^{-1}\|_{E \to E} \le M, \qquad (2.27)$$

which is "worse" then (1.43), since we lose a resolvent $(I + \tau A)^{-1}$. However, the lost resolvent can be recovered from the estimate

$$\|(I + \tau A)[I + R_{l,l}(\tau A)]\|_{E \to E} \le M. \qquad (2.28)$$

This allows us to obtain some "improved" smoothness estimates for the operator step $R_{l,l}(\tau A)$ in the case where l is odd as well.

Lemma 2.4. *For any $1 \le k \le N$ we have the estimate*

$$\| \sum_{r=0}^{k-1} R_{l,l}^r(\tau A)\|_{E \to E} \le Mk, \qquad (2.29)$$

where M does not depend on τ and k.

Proof. The needed inequality (2.29) is a consequence of the identities

$$\sum_{r=0}^{2m-1} R_{l,l}^r(\tau A) = \sum_{r=0}^{m-1} [I + R_{l,l}(\tau A)]R_{l,l}^{2r}(\tau A), \quad m = 1, 2, \cdots,$$

$$\sum_{r=0}^{2m} R_{l,l}^r(\tau A) = I + \sum_{r=0}^{m-1} [I + R_{l,l}(\tau A)]R_{l,l}^{2r+1}(\tau A), \quad m = 1, 2, \cdots,$$

and the estimates (2.28) and (1.25). Lemma 2.4 is proved.

Lemma 2.5. *For any $1 \leq k < k + 2r \leq N$ and $0 \leq \alpha \leq 1$ we have the estimate*

$$\|\tau A[R_{l,l}^{k+2r}(\tau A) - R_{l,l}^{k}(\tau A)](I + \tau A)^{-2}\|_{E \to E} \leq M \frac{r^\alpha}{k^{1+\alpha}}, \qquad (2.30)$$

where M does not depend on τ, α, k, and r.

Proof. Let us use the identity

$$R_{l,l}^{k+2r}(\tau A) - R_{l,l}^{k}(\tau A) = \sum_{i=0}^{r-1} R_{l,l}^{k+2i}(\tau A)[R_{l,l}^{2}(\tau A) - I]. \qquad (2.31)$$

By (1.54) and the semigroup property of $R_{l,l}^{k}(\tau A)$, we have

$$\|\tau A R_{l,l}^{m}(\tau A)[R_{l,l}^{2}(\tau A) - I](I + \tau A)^{-2}\|_{E \to E}$$

$$\leq \|(I + \tau A)^{2}[R_{l,l}^{2}(\tau A) - I](\tau A)^{-1}\|_{E \to E} \|\tau A R_{l,l}^{[m/2]}(\tau A)(I + \tau A)^{-2}\|_{E \to E} \times$$

$$\times \|\tau A R_{l,l}^{m-[m/2]}(\tau A)(I + \tau A)^{-2}\|_{E \to E} \leq M/m^{2} \qquad (2.32)$$

for $m > 1$. For $m = 1$ the estimate (2.32) is obvious. Using identity (2.31) and inequality (2.32), we obtain

$$\|\tau A[R_{l,l}^{k+2r}(\tau A) - R_{l,l}^{k}(\tau A)](I + \tau A)^{-2}\|_{E \to E} \leq M \sum_{i=0}^{r-1} \frac{1}{(k+2i)^{2}} \leq M \int_{0}^{r} \frac{ds}{(k+2s)^{2}}$$

$$\leq M \frac{r}{k^{2}}. \qquad (2.33)$$

By (1.26) and the triangle inequality,

$$\|\tau A[R_{l,l}^{k+2r}(\tau A) - R_{l,l}^{k}(\tau A)](I + \tau A)^{-2}\|_{E \to E} \leq M \frac{r}{k},$$

which in conjunction with (2.33) yields the estimate (2.30). Lemma 2.5 is proved.

Lemma 2.6. *For any $1 \leq k < k + 2r \leq N$ we have the estimate*

$$\|R_{l,l}^{k+2r}(\tau A) - R_{l,l}^{k}(\tau A)\|_{E \to E} \leq M \frac{r}{k}, \qquad (2.34)$$

where M does not depend on τ, k, and r.

Proof. Using the identity (2.31) and the estimate (1.54) we obtain

$$\|R_{l,l}^{k+2r}(\tau A) - R_{l,l}^k(\tau A)\|_{E\to E} \leq \sum_{i=0}^{r-1} \|(I+\tau A)^2[I - R_{l,l}^2(\tau A)](\tau A)^{-1}\|_{E\to E}\times$$

$$\times\|\tau A R_{l,l}^{k+2i}(\tau A)(I+\tau A)^{-2}\|_{E\to E} \leq M\sum_{i=0}^{r-1}\frac{1}{k+2i} \leq M\frac{r}{k}.$$

Lemma 2.6 is proved.

The smoothness estimates obtained above for the operator step $R_{l,l}(\tau A)$ allow us to establish the following result.

Theorem 2.7. *Let condition (2.25) be satisfied. Suppose the operator $I+R_{l,l}(\tau A)$ has an inverse $[I+R_{l,l}(\tau A)]^{-1}$ and $\varphi_{l,l}^\tau \in D([I+R_{l,l}(\tau A)]^{-1})$. Then the solutions of the difference problem (0.51) obey the coercivity inequality*

$$\|\mathcal{D}\Pi(u_0)u^\tau\|_{\tilde{C}_0^\alpha(\tau,E)} + \|A_{l,l}u^\tau\|_{\tilde{C}_0^\alpha(\tau,E)}$$

$$\leq M\left[\|Au_0\|_E + \frac{1}{\alpha(1-\alpha)}\|[(I+R_{l,l}(\tau A))^{-1}\varphi_{l,l}^\tau\|_{\tilde{C}_0^\alpha(\tau,E)}\right], \qquad (2.35)$$

where M does not depend on τ, α, u_0, and $\varphi_{l,l}^\tau$.

Proof. By the definition of the space $\tilde{C}_0^\alpha(\tau,E)$ and inequality (2.15),

$$\|A_{l,l}w^\tau\|_{\tilde{C}_0^\alpha(\tau,E)} \leq \|A_{l,l}w^\tau\|_{C_0^\alpha(\tau,E)} \leq M\|Au_0\|_E.$$

Hence, in order to prove the theorem is suffices to establish the estimate

$$\|A_{l,l}g^\tau\|_{\tilde{C}_0^\alpha(\tau,E)} \leq \frac{M}{\alpha(1-\alpha)}\|[I+R_{l,l}(\tau A)]^{-1}\varphi_{l,l}^\tau\|_{\tilde{C}_0^\alpha(\tau,E)}. \qquad (2.36)$$

To this end let us prove the estimates

$$\|A_{l,l}g_k\|_E \leq \frac{M}{\alpha(1-\alpha)}\|[I+R_{l,l}(\tau A)]^{-1}\varphi_{l,l}^\tau\|_{\tilde{C}_0^\alpha(\tau,E)}, \quad 1\leq k\leq N \qquad (2.37)$$

and

$$\|A_{l,l}(g_{k+2r} - g_k)\|_E \leq \frac{M}{\alpha(1-\alpha)}\|[I+R_{l,l}(\tau A)]^{-1}\varphi_{l,l}^\tau\|_{\tilde{C}_0^\alpha(\tau,E)}k^{-\alpha}(2r)^\alpha,$$

$$1 \le k < k + 2r \le N \tag{2.38}$$

From indentity (2.20) we derive the formulas

$$A_{l,l}g_{2k} = [I + R_{l,l}(\tau A)]^{-1}\big\{[R_{l,l}(\tau A) - R_{l,l}^{2k+1}(\tau A)]\varphi_{2k}^{l,l} + [I - R_{l,l}^{2k}(\tau A)]\varphi_{2k-1}^{l,l}\big\}$$

$$+ \sum_{i=1}^{k-1} A_{l,l}R_{l,l}^{2k-2i+1}(\tau A)(\varphi_{2i-1}^{l,l} - \varphi_{2k-1}^{l,l})\tau + \sum_{i=1}^{k-1} A_{l,l}R_{l,l}^{2k-2i}(\tau A)(\varphi_{2i}^{l,l} - \varphi_{2k}^{l,l})\tau,$$

$$2 \le k \le N,$$

$$A_{l,l}g_2 = A_{l,l}R_{l,l}(\tau A)\varphi_1^{l,l}\tau + A_{l,l}\varphi_2^{l,l}\tau, \tag{2.39}$$

and

$$A_{l,l}g_{2k-1} = [I + R_{l,l}(\tau A)]^{-1}\big\{[I - R_{l,l}^{2k}(\tau A)]\varphi_{2k-1}^{l,l} + [R_{l,l}(\tau A) - R_{l,l}^{2k-1}(\tau A)]\varphi_{2k-2}^{l,l}\big\}$$

$$+ \sum_{i=1}^{k} A_{l,l}R_{l,l}^{2k-2i}(\tau A)(\varphi_{2i-1}^{l,l} - \varphi_{2k-1}^{l,l})\tau + \sum_{i=1}^{k-1} A_{l,l}R_{l,l}^{2k-2i-1}(\tau A)(\varphi_{2i}^{l,l} - \varphi_{2k}^{l,l})\tau,$$

$$3 \le k \le N,$$

$$A_{l,l}g_1 = A_{l,l}\varphi_1^{l,l}, \quad A_{l,l}g_3 = A_{l,l}R_{l,l}^2(\tau A)\varphi_1^{l,l}\tau + A_{l,l}R_{l,l}(\tau A)\varphi_2^{l,l}\tau + A_{l,l}\varphi_3^{l,l}\tau. \tag{2.40}$$

For $k = 1, 2, 3$ the estimate (2.37) is obvious. Let $k \ge 4$. By (1.26), (1.25), and (2.28), we have

$$\|A_{l,l}g_{2k}\|_E \le \|R_{l,l}(\tau A) - R_{l,l}^{2k+1}(\tau A)\|_{E \to E} \|[I + R_{l,l}(\tau A)]^{-1}\varphi_{2k}^{l,l}\|_E$$

$$+ \|I - R_{l,l}^{2k}(\tau A)\|_{E \to E} \|[I + R_{l,l}(\tau A)]^{-1}\varphi_{2k-1}^{l,l}\|_E$$

$$+ \sum_{i=1}^{k-1} \|(I + \tau A)A_{l,l}A^{-1}\|_{E \to E} \|AR_{l,l}^{2k-2i+1}(\tau A)(I + \tau A)^{-2}\|_{E \to E} \times$$

$$\times \|(I + \tau A)[I + R_{l,l}(\tau A)]\|_{E \to E} \|[I + R_{l,l}(\tau A)]^{-1}(\varphi_{2i-1}^{l,l} - \varphi_{2k-1}^{l,l})\|_E \tau$$

$$+ \sum_{i=1}^{k-1} \|(I + \tau A)A_{l,l}A^{-1}\|_{E \to E} \|AR_{l,l}^{2k-2i}(\tau A)(I + \tau A)^{-2}\|_{E \to E} \times$$

$$\times \|(I + \tau A)[I + R_{l,l}(\tau A)]\|_{E \to E} \|[I + R_{l,l}(\tau A)]^{-1}(\varphi_{2i}^{l,l} - \varphi_{2k}^{l,l})\|_E \tau$$

$$\le M\left[1 + \sum_{i=1}^{k-1} \frac{1}{(2k - 2i + 1)^{1-\alpha}(2i - 1)^\alpha} + \sum_{i=1}^{k-1} \frac{1}{(2k - 2i)^{1-\alpha}(2i)^\alpha}\right] \times$$

$$\times \|[I + R_{l,l}(\tau A)]^{-1}\varphi_{l,l}^\tau\|_{\tilde{C}_0^\alpha(\tau,E)}$$

$$\leq M \int_0^{2k\tau} \frac{ds}{(2k\tau - s)^{1-\alpha}s^\alpha} \|[I + R_{l,l}(\tau A)]^{-1}\varphi_{l,l}^\tau\|_{\tilde{C}_0^\alpha(\tau,E)}$$

$$\leq \frac{M}{\alpha(1-\alpha)}\|[I + R_{l,l}(\tau A)]^{-1}\varphi_{l,l}^\tau\|_{\tilde{C}_0^\alpha(\tau,E)}. \tag{2.41}$$

Estimate (2.37) for k even is thus proved.

To prove estimate (2.37) for k odd one proceeds as in the proof of (2.41). Now let us establish estimate (2.38). For $k \leq 2r$ it follows from (2.37). Indeed, by the triangle inequality, we have

$$\|A_{l,l}(g_{k+2r} - g_k)\|_E \leq \|A_{l,l}g_{k+2r}\|_E + \|A_{l,l}g_k\|_E$$

$$\leq \frac{2M}{\alpha(1-\alpha)}\|[I + R_{l,l}(\tau A)]^{-1}\varphi_{l,l}^\tau\|_{\tilde{C}_0^\alpha(\tau,E)}(2r)^\alpha(2r)^{-\alpha}$$

$$\leq \frac{2^{1+\alpha}M}{\alpha(1-\alpha)}\|[I + R_{l,l}(\tau A)]^{-1}\varphi_{l,l}^\tau\|_{\tilde{C}_0^\alpha(\tau,E)}(2r)^\alpha k^{-\alpha}.$$

Now let $k \geq 2r$. First we will prove (2.38) for even k. From formula (2.39) it follows that

$$A_{l,l}g_{2k+2r} - A_{l,l}g_{2k} = [I + R_{l,l}(\tau A)]^{-1}\{[R_{l,l}^{2k+1}(\tau A) - R_{l,l}^{2k+2r+1}(\tau A)]\varphi_{2k+2r}^{l,l}$$

$$+ [R_{l,l}^{2k}(\tau A) - R_{l,l}^{2k+2r}(\tau A)]\varphi_{2k+2r-1}^{l,l} + [R_{l,l}(\tau A) - R_{l,l}^{2k+1}(\tau A)$$

$$+ R_{l,l}^{2k+2r}(\tau A) - R_{l,l}^{4r}(\tau A)](\varphi_{2k+2r}^{l,l} - \varphi_{2k}^{l,l}) + [R_{l,l}^{2k+2r+1}(\tau A) - R_{l,l}(\tau A)$$

$$+ I - R_{l,l}^{2k}(\tau A)](\varphi_{2k+2r-1}^{l,l} - \varphi_{2k-1}^{l,l})\}$$

$$\left\{ \sum_{i=k-r+1}^{k+r-1} A_{l,l}R_{l,l}^{2k+2r-2i+1}(\tau A)(\varphi_{2i-1}^{l,l} - \varphi_{2k+2r-1}^{l,l})\tau \right.$$

$$\left. + \sum_{i=k-r+1}^{k+r-1} A_{l,l}R_{l,l}^{2k+2r-2i}(\tau A)(\varphi_{2i}^{l,l} - \varphi_{2k+2r}^{l,l})\tau \right\}$$

$$+ \left\{ \sum_{i=k-r+1}^{k-1} A_{l,l}R_{l,l}^{2k-2i+1}(\tau A)(\varphi_{2k-1}^{l,l} - \varphi_{2i-1}^{l,l})\tau \right.$$

$$\left. + \sum_{i=k-r+1}^{k-1} A_{l,l}R_{l,l}^{2k-2i}(\tau A)(\varphi_{2k}^{l,l} - \varphi_{2i}^{l,l})\tau \right\}$$

$$+\left\{\sum_{i=1}^{k-r} A_{l,l}[R_{l,l}^{2k+2r-2i+1}(\tau A) - R_{l,l}^{2k-2i+1}(\tau A)](\varphi_{2i-1}^{l,l} - \varphi_{2k-1}^{l,l})\tau\right.$$

$$\left.+\sum_{i=1}^{k-r} A_{l,l}[R_{l,l}^{2k+2r-2i}(\tau A) - R_{l,l}^{2k-2i}(\tau A)](\varphi_{2i}^{l,l} - \varphi^{l,l})2k\tau\right\} = I_1 + I_2 + I_3 + I_4.$$

By (2.25), (2.11), (1.25), (2.34), and (2.28), we have

$$\|I_1\|_E \le M\|[I + R_{l,l}(\tau A)]^{-1}\varphi_{l,l}^{\tau}\|_{\tilde{C}_0^{\alpha}(\tau,E)}(2r)^{\alpha}(2k)^{-\alpha}.$$

Further, by (1.26) and (2.28) we have

$$\|I_2\|_E \le \sum_{i=k-r+1}^{k+r-1} \|(I+\tau A)A_{l,l}A^{-1}\|_{E\to E}\|AR_{l,l}^{2k+2r-2i+1}(\tau A)(I+\tau A)^{-2}\|_{E\to E}\times$$

$$\times\|(I+\tau A)[I+R_{l,l}(\tau A)]\|_{E\to E}\|[I+R_{l,l}(\tau A)]^{-1}(\varphi_{2i-1}^{l,l} - \varphi_{2k+2r-1}^{l,l})\|_E\tau$$

$$+\sum_{i=k-r+1}^{k+r-1} \|(I+\tau A)A_{l,l}A^{-1}\|_{E\to E}\|AR_{l,l}^{2k+2r-2i}(\tau A)(I+\tau A)^{-2}\|_{E\to E}\times$$

$$\times\|(I+\tau A)[I+R_{l,l}(\tau A)]\|_{E\to E}\|[I+R_{l,l}(\tau A)]^{-1}(\varphi_{2i}^{l,l} - \varphi_{2k+2r}^{l,l})\|_E\tau$$

$$\le M\left[\sum_{i=k-r+1}^{k+r-1}\frac{1}{(2k+2r-2i+1)^{1-\alpha}(2i-1)^{\alpha}}+\sum_{i=k-r+1}^{k+r-1}\frac{1}{(2k+2r-2i)^{1-\alpha}(2i)^{\alpha}}\right]\times$$

$$\times\|[I+R_{l,l}(\tau A)]^{-1}\varphi_{l,l}^{\tau}\|_{\tilde{C}_0^{\alpha}(\tau,E)}$$

$$\le M\int_{(k-r)\tau}^{(k+r)\tau}\frac{ds}{(2k\tau - s)^{1-\alpha}s^{\alpha}}\|[I+R_{l,l}(\tau A)]^{-1}\varphi_{l,l}^{\tau}\|_{\tilde{C}_0^{\alpha}(\tau,E)}$$

$$\le \frac{M}{\alpha}\|[I+R_{l,l}(\tau A)]^{-1}\varphi_{l,l}^{\tau}\|_{\tilde{C}_0^{\alpha}(\tau,E)}(2r)^{\alpha}(2k)^{-\alpha}.$$

The norm of I_3 is estimated in a similar manner. Finally, using (2.30) and (2.28), we obtain

$$\|I_4\|_E \le \sum_{i=1}^{k-r}\|(I+\tau A)A_{l,l}A^{-1}\|_{E\to E}\times$$

$$\times\|A[R_{l,l}^{2k+2r-2i+1}(\tau A) - R_{l,l}^{2k-2i+1}(\tau A)](I+\tau A)^{-2}\|_{E\to E}\times$$

$$\times\|(I+\tau A)[I+R_{l,l}(\tau A)]\|_{E\to E}\|[I+R_{l,l}(\tau A)]^{-1}(\varphi_{2i-1}^{l,l} - \varphi_{2k-1}^{l,l})\|_E\tau$$

$$+\sum_{i=1}^{k-r}\|(I+\tau A)A_{l,l}A^{-1}\|_{E\to E}\|A[R_{l,l}^{2k+2r-2i}(\tau A) - R_{l,l}^{2k-2i}(\tau A)](I+\tau A)^{-2}\|_{E\to E}\times$$

$$\times \|(I + \tau A)[I + R_{l,l}(\tau A)]\|_{E \to E} \|[I + R_{l,l}(\tau A)]^{-1}(\varphi_{2i-1}^{l,l} - \varphi_{2k-1}^{l,l})\|_E \tau$$

$$\leq M 2 r \tau \left[\sum_{i=1}^{k-r} \frac{\tau}{((2k-2i+1)\tau)^{2-\alpha}((2i-1)\tau)^\alpha} + \sum_{i=1}^{k-r} \frac{\tau}{((2k-2i)\tau)^{2-\alpha}(2i\tau)^\alpha} \right] \times$$

$$\times \|[I + R_{l,l}(\tau A)]^{-1} \varphi_{l,l}^\tau\|_{\tilde{C}_0^\alpha(\tau,E)}$$

$$\leq M 2 r \tau \sum_{i=1}^{2k-2r} \frac{\tau}{((2k-i)\tau)^{2-\alpha}(i\tau)^\alpha} \|[I + R_{l,l}(\tau A)]^{-1} \varphi_{l,l}^\tau\|_{\tilde{C}_0^\alpha(\tau,E)}$$

$$\leq M 2 r \tau \int_0^{(2k-2r)\tau} \frac{ds}{(2k\tau - s)^{2-\alpha} s^\alpha} \|[I + R_{l,l}(\tau A)]^{-1} \varphi_{l,l}^\tau\|_{\tilde{C}_0^\alpha(\tau,E)}.$$

Since

$$2 r \tau \int_0^{(2k-2r)\tau} \frac{ds}{(2k\tau - s)^{2-\alpha} s^\alpha} \leq \frac{M}{\alpha(1-\alpha)} (2r)^\alpha (2k)^{-\alpha},$$

we conclude that

$$\|I_4\|_E \leq \frac{M}{\alpha(1-\alpha)} \|[I + R_{l,l}(\tau A)]^{-1} \varphi_{l,l}^\tau\|_{\tilde{C}_0^\alpha(\tau,E)} (2r)^\alpha (2k)^{-\alpha}.$$

Combining the estimates for I_1, I_2, I_3, and I_4 we obtain (2.38) for even k. Using (2.40) one can similarly establish (2.38) for odd k. Theorem 2.7 is proved.

The condition in Theorem 2.7 that the inverse of the operator $I + R_{l,l}(\tau A)$ exists is obviously satisfied for $l = 1$, when $[I + R_{l,l}(\tau A)]^{-1} = I + \frac{\tau}{2} A$. In the next section we shall give a condition on the spectrum of A under which the inverse in question exists for l odd and is given by a power of the operator $I + \tau A$.

To conclude this subsection, let us give a theorem in which the stringent requirements of stability and existence of the operator $[I + R_{l,l}(\tau A)]^{-1}$ are dropped at the expense of allowing a logarithmic factor in the right-hand side of the coercivity inequality.

Theorem 2.8. *Let $\varphi_{l,l}^\tau \in D(A)$. Then the solutions of problem (0.51) obey the inequality*

$$\|\mathcal{D}\Pi(u_0)u^\tau\|_{\tilde{C}_0^\alpha(\tau,E)} + \|A_{l,l}u^\tau\|_{\tilde{C}_0^\alpha(\tau,E)}$$

$$\leq M \left[\|Au_0\|_E + \left(\frac{1}{\alpha(1-\alpha)} + \ln \frac{1}{\tau} \right) \|(I + \tau A)\varphi_{l,l}^\tau\|_{\tilde{C}_0^\alpha(\tau,E)} \right],$$

where M does not depend on τ, α, u_0 and $\varphi_{l,l}^\tau$.

The proof follows the scheme of the proof of Theorem 2.7, with a difference in the estimates of the first terms in the representation (2.37). Namely,

$$[I + R_{l,l}(\tau A)]^{-1}\{[R_{l,l}(\tau A) - R_{l,l}^{2k+1}(\tau A)]\varphi_{2k}^{l,l} + [I - R_{l,l}^{2k}(\tau A)]\varphi_{2k-1}^{l,l}\}$$

$$= (I + \tau A)[I - R_{l,l}(\tau A)]A^{-1}\sum_{i=0}^{k-1} AR_{l,l}^{2i}(I + \tau A)^{-2}(I + \tau A)[R_{l,l}(\tau A)\varphi_{2k}^{l,l} + \varphi_{2k-1}^{l,l}].$$

As in Theorem 1.9, the estimation of the right-hand side in this last equality in the norm of $\tilde{C}_0^\alpha(\tau, E)$ leads to a supplementary logarithmic term. The remainig terms are estimated as in Theorem 2.7.

3.   WELL-POSEDNESS OF THE DIFFERENCE PROBLEM IN $\mathcal{L}_p(E)$

1.   Definition of the well-posedness of the difference problem in $\mathcal{L}_p(E)$. Stability of the difference problem.

In Chapter 2 we have established the well-posedness of the simplest difference problem (0.6) in the Lebesgue (Bochner) difference space $\mathcal{L}_p(E)$, which is larger than $\mathcal{C}(E)$. As it turns out, well-posedness in $\mathcal{L}_p(E)$ holds for a broad class of Padé difference schemes, generated by the fractions $R_{j,l}(z)$ with $j = l - 2,\ l - 1$. A solution of the problem (0.51) is called *a solution of this problem in* $\mathcal{L}_p(E)$ if $\overline{\mathcal{D}}\,\overline{\Pi}(u_0)(u)$ and $\overline{A}_{j,l}u$ belong to $\mathcal{L}_p(E)$. For the solvability of problem (0.51) in $\mathcal{L}_p(E)$ it is obviously necessary that $\varphi_{j,l} \in \mathcal{L}_p(E)$. As in the case of the simplest difference problem (0.6), the necessary condition that u_0 must satisfy is more complicated. For this reason we will begin our analysis with the nonhomogeneous problem with null initial condition ($u_0 = 0$).

Definition 3.1. We will say that *the problem* (0.51) *with* $u_0 = 0$ *is well-posed in* $\mathcal{L}_p(E)$ if the following conditions are satisfied.

1) For any $\varphi_{j,l} \in \mathcal{L}_p(E)$ there exists a unique solution $u = u(\varphi_{j,l}, 0)$ in $\mathcal{L}_p(E)$ of this problem.

2) Problem (0.51) is stable in $\mathcal{L}_p(E)$.

As in the case of the space $\mathcal{C}(E)$, one can show that the well-posedness in $\mathcal{L}_p(E)$ of the nonhomogeneous problem (0.51) with $u_0 = 0$ implies that the operator $\overline{A}_{j,l}u(\varphi_{j,l}, 0)$ is bounded in $\mathcal{L}_p(E)$. Furthermore, from the solvability in $\mathcal{L}_p(E)$ of problem (0.51) with $u_0 = 0$ it follows, via the substitution $u = w + u_0$, that the problem (0.51) is uniquely solvable for any $u_0 \in D(A)$ and that the coercivity inequality holds:

$$\|\overline{\mathcal{D}}\,\overline{\Pi}(u_0)(u)\|_{\mathcal{L}_p(E)} + \|\overline{A}_{j,l}u\|_{\mathcal{L}_p(E)} \leq M(p)\Big[\|Au_0\|_E + \|\varphi_{j,l}\|_{\mathcal{L}_p(E)}\Big]. \qquad (3.1)$$

This in turn implies the analyticity of the semigroup $\exp\{-tA\}$. For this reason in what follows we shall assume that $-A$ is the generator of an analytic semigroup.

The estimate (1.23) enables us to establish the following fact.

Theorem 3.1. *The difference problem (0.51) with $j = l - 2$, $l - 1$ is stable in the space $\mathcal{L}_p(E)$.*

The analyticity of the semigroup $\exp\{-tA\}$ allows us to establish the stability of diagonal Padé schemes. Specifically, from estimate (1.25) one derives the following result.

Theorem 3.2. *Let $u_0 \in D(A)$ and $\varphi_{j,l}^\tau \in D(A)$. Then for $j = l$ the solution of the difference problem (0.51) obeys the stability inequality*

$$\|u^\tau\|_{L_p(\tau,E)} \leq M\Big[\|(I + \tau A)u_0\|_E + \|(I + \tau A)\varphi_{l,l}^\tau\|_{L_p(\tau,E)}\Big],$$

where M does not depend on τ, p, u_0 and $\varphi_{l,l}^\tau$.

The method of the proof of Theorem 1.9 allows us to eliminate the operator $I + \tau A$ from the right-hand side of the last inequality.

Theorem 3.3. *The solutions of the difference problem (0.51) with $j = l$ obey the stability inequality*

$$\|u^\tau\|_{L_p(\tau,E)} \leq M\Big[\min\Big\{\ln\frac{1}{\tau}, 1 + |\ln\|A\|_{E\to E}|\Big\}\Big]\Big[\|u_0\|_E + \|\varphi_{l,l}^\tau\|_{L_p(\tau,E)}\Big],$$

where M does not depend on τ, p, u_0 and $\varphi_{l,l}^\tau$.

2.   Spaces of initial data. Well-posedness of the difference problem.

It follows from formula (0.50) that the solution of the homogeneous problem (0.49) with $\varphi_{j,l}^\tau = 0$ has the form

$$w_k = R_{j,l}^k(\tau A)u_0, \quad 1 \le k \le N.$$

Hence, in order that w be a solution of problem (0.51) in $\mathcal{L}_p(E)$ it is necessary and sufficient that

$$\sup_{0<\tau\le\tau_0} \sum_{k=1}^N \|A_{j,l}R_{j,l}^k(\tau A)u_0\|_E^p \tau < \infty.$$

By (1.23), the set of all elements $u_0 \in E$ satisfying this condition is a linear set that contains $D(A)$. It becomes a Banach space $E''_{1-\frac{1}{p}}$ when it is equipped with the norm

$$< u_0 >_{1-\frac{1}{p}} = \sup_{0<\tau\le\tau_0} \left(\sum_{k=1}^N \|A_{j,l}R_{j,l}^k(\tau A)u_0\|_E^p \tau \right)^{1/p} + \|u_0\|_E. \qquad (3.2)$$

Let u^τ be a solution of the general problem (0.49). Since for $u_0 = 0$ that problem has a solution g^τ for any $\varphi_{j,l}^\tau \in L_p(\tau, E)$, the grid function $w^\tau = u^\tau - g^\tau$ will be a solution in $L_p(\tau, E)$ of problem (0.49) with $\varphi_{j,l}^\tau = 0$. Consequently, $u_0 \in E''_{1-\frac{1}{p}}$. Thus, for the solvability in $\mathcal{L}_p(E)$ of the difference problem (0.51) (under the assumption that problem (0.51) with $u_0 = 0$ is well posed) it is necessary and sufficient that $u_0 \in E''_{1-\frac{1}{p}}$.

In Chapter 2 we have shown that in the case of the simplest difference problem (0.6) the spaces E''_α coincide with the spaces E_α introduced in Chapter 1, Section 3. It turns out that, under some constraints, such an assertion is true for the general difference problem (0.51) when $j = l-2, l-1$. As we already observed in Section 3 of Chapter 2, we can assume with no loss of generality that the norm of the semigroup $\exp\{-tA\}$ decays exponentially. Then in the case of the simplest problem (0.6) the norm of the powers of the operator step is decreasing. Such an assertion also holds for the general problem (0.51) for $j = 0$ and $j = l-2$. The authors do not know whether this is true for the general problem (0.51) with an arbitrary positive operator A. As it turn out, such a fact can be established for the powers of the operator step $R_{j,l}(\tau A)$ when $j = l-2, l-1$ under the condition that A is a strongly positive operator with spectral angle $\phi(A, E) < \frac{\pi}{2l}$. Accordingly, from now on we shall assume that A has this last property. Then we have the estimates

$$\|R_{j,l}^k(\tau A)\|_{E \to E} \le M(1 + \delta\tau)^{-k}, \quad k = 1, 2, \cdots, \quad \delta > 0, \quad M > 0, \qquad (3.3)$$

and

$$\|AR_{j,l}^k(\tau A)\|_{E \to E} \le \frac{M}{k\tau}(1 + \delta\tau)^{-k}, \quad k = 1, 2, \cdots, \quad \delta > 0, \quad M > 0. \qquad (3.4)$$

These estimates will be proved below. They allow us to establish the following results.

Theorem 3.4. *The spaces E_α'' and E_α are equal for all $0 < \alpha < 1$.*

Theorem 3.5. *The solutions of the difference problem* (0.51) *obey the estimate*

$$\sup_{0 < \tau \le \tau_0} \max_{1 \le k \le N} |u_k|_{1 - \frac{1}{p}} \le \frac{Mp^2}{p - 1}\Big[|u_0|_{1 - \frac{1}{p}} + \|\varphi_{j,l}\|_{\mathcal{L}_p(E)}\Big]. \qquad (3.5)$$

The estimate (3.5) allows us to sharpen inequality (3.1) and establish the well-posedness in $\mathcal{L}_p(E)$ of problem (0.51) and the coercivity inequality

$$\|\overline{\mathcal{D}}\,\overline{\Pi}(u_0)u\|_{\mathcal{L}_p(E)} + \|\overline{A}_{j,l}u\|_{\mathcal{L}_p(E)} + \|u\|_{C(E_{1 - \frac{1}{p}})} \le M(p)\Big[|u_0|_{1 - \frac{1}{p}} + \|\varphi_{j,l}\|_{\mathcal{L}_p(E)}\Big], \qquad (3.6)$$

where $M(p)$ does not depend on u_0 and $\varphi_{j,l}$.

Theorem 3.6. *If the difference problem* (0.51) *is well posed in $\mathcal{L}_{p_0}(E)$ for some $1 < p_0 < \infty$, then it is well posed in $\mathcal{L}_p(E)$ for any $1 < p < \infty$.*

The proofs of Theorems 3.4–3.6 follow the scheme of the proofs of the corresponding theorems in Chapter 2. As in the case of the simplest difference problem (0.6) of Chapter 2, the estimate (3.5) is established for arbitrary strongly positive operators A with a spectral angle $\phi(A, E) < \frac{\pi}{2l}$. In contrast, the complete inequality (3.6) in the space $\mathcal{L}_p(E)$ is established under the assumption that it holds in $\mathcal{L}_{p_0}(E)$ for some p_0, $1 < p_0 < \infty$. One can ask when is this last condition satisfied? We have an answer in the case where $E = H$ is a Hilbert space.

Theorem 3.7. *The difference problem* (0.51) *with a positive operator A with spectral angle $\phi(A, E) < \frac{\pi}{2l}$ is well posed in $\mathcal{L}_2(H)$.*

Proof. Let us define a convolution operator $B = B_\tau$ in the space $L_2^\infty(H)$ by the rule (see Section 3 of Chapter 2)

$$Bf(t) = \int_{-\infty}^{\infty} \tilde{B}\left(\left[\frac{t}{\tau}\right] - \left[\frac{s}{\tau}\right]\right) W_\tau f(s)ds, \tag{3.7}$$

where

$$\tilde{B}\left(\left[\frac{t}{\tau}\right]\right) = \begin{cases} A_{j,l} R_{j,l}^{\left[\frac{t}{\tau}\right]}(\tau A), & t \geq \tau, \\ 0, & t < \tau, \end{cases} \tag{3.8}$$

and

$$W_\tau f(t) = \left((W_\tau f)_k = \tau^{-1} \int_{t_{k-1}}^{t_k} f(s)ds, \quad t_{k-1} < t \leq t_k, \quad k = \left[\frac{t}{\tau}\right], \quad t_k = k\tau\right).$$

To prove the theorem it suffices to establish the estimate

$$\max_{0 \leq t \leq 1} \|\hat{B}(t)\|_{H \to H} \leq M. \tag{3.9}$$

Here

$$\hat{B}(t) = \sum_{s=-\infty}^{\infty} \tau \tilde{B}(s)e^{2\pi i t s}.$$

Using the definition (3.8) of the kernel of B, we deduce that

$$\hat{B}(t) = \sum_{s=1}^{\infty} \tau A_{j,l} R_{j,l}^s(\tau A)e^{2\pi i t s} = [I - R_{j,l}(\tau A)] \sum_{s=1}^{\infty} (R_{j,l}(\tau A)e^{2\pi i t})^s$$

$$= [I - R_{j,l}(\tau A)][I - R_{j,l}(\tau A)e^{2\pi i t}]^{-1} R_{j,l}(\tau A)e^{2\pi i t}.$$

By the estimates (1.8) and (1.23), we have

$$\|\hat{B}(t)\|_{H \to H} \leq \|(I + \tau A)[I - R_{j,l}(\tau A)](\tau A)^{-1}\|_{H \to H}\|R_{j,l}(\tau A)\|_{H \to H} \times$$

$$\times \|(I + \tau A)^{-1}\tau A[I - R_{j,l}(\tau A)e^{2\pi i t}]^{-1}\|_{H \to H}$$

$$\leq M\|(I + \tau A)^{-1}\tau A[I - R_{j,l}(\tau A)e^{2\pi i t}]^{-1}\|_{H \to H}.$$

Hence, to prove (3.9) it suffices to verify that

$$\|(I + \tau A)^{-1}\tau A[I - R_{j,l}(\tau A)e^{2\pi i t}]^{-1}\|_{H \to H} \leq M$$

for all $t \in [0,1]$ and $0 < \tau \le \tau_0$. Consider the polynomial $S_t(z) = Q_{j,l}(z) - P_{j,l}(z)e^{2\pi it}$. First let us show that $S_t(z)$ has no roots in the sector $\Gamma = \{z \in \mathbb{C}^+ : z = \rho e^{i\phi},\ 0 < \rho < \infty,\ 0 \le \phi \le \frac{\pi}{2(l-1)}\}$. By the inverse triangle inequality,

$$|S_t(z)| \ge |\,|Q_{j,l}(z)| - |P_{j,l}(z)|\,| = \frac{|\,|Q_{j,l}(z)|^2 - |P_{j,l}(z)|^2\,|}{|Q_{j,l}(z)| + |P_{j,l}(z)|}.$$

By formula (0.3), we have

$$|Q_{j,l}(z)|^2 = \sum_{i=0}^{l}\left(\frac{(l+j-i)!l!}{(l+j)!i!(l-i)!}\right)^2 \rho^{2i}$$

$$+2\sum_{i=0}^{l-1}\sum_{r=i+1}^{l}\frac{(l+j-i)!(l!)^2(l+j-r)!}{((l+j)!)^2 i!(l-i)!r!(l-r)!}\rho^{i+r}\cos(r-i)\phi$$

and

$$|P_{j,l}(z)|^2 = \sum_{i=0}^{j}\left(\frac{(l+j-i)!j!}{(l+j)!i!(j-i)!}\right)^2 \rho^{2i}$$

$$+2\sum_{i=0}^{j-1}\sum_{r=i+1}^{j}\frac{(l+j-i)!(j!)^2(l+j-r)!(-1)^{i+r}}{((l+j)!)^2 i!(j-i)!r!(j-r)!}\cos(r-i)\phi.$$

Since

$$\frac{l!}{(l-i)!} \ge \frac{j!}{(j-i)!},\ 0 \le i \le j,\quad \cos(r-i)\phi \ge 0,\ 0 < r-i \le l-1,$$

we have

$$\frac{(l+j-i)!l!}{(l+j)!i!(l-i)!} \ge \frac{(l+j-i)!j!}{(l+j)!i!(j-i)!},\ 0 \le i \le j,$$

and

$$|Q_{j,l}(z)|^2 \ge |P_{j,l}(z)|^2.$$

Thus, we obtain the estimate

$$|S_t(z)| \ge \frac{1}{2|Q_{j,l}(z)|}\left\{\sum_{i=1}^{j}\left(\frac{(l+j-i)!}{(l+j)!i!}\right)^2\left[\left(\frac{l!}{(l-i)!}\right)^2 - \left(\frac{j!}{(j-i)!}\right)^2\right]\rho^{2i}\right.$$

$$+\sum_{i=0}^{j-1}\sum_{r=i+1}^{j}\frac{(l+j-i)!(l+j-r)!}{((l+j)!)^2 i!r!}\left[\frac{(l!)^2}{(l-i)!(l-r)!}\right.$$

$$
-(-1)^{i+r}\frac{(j)^2!}{(j-i)!(j-r)!}\Bigg]\rho^{i+r}\cos(r-i)\phi+\sum_{i=j+1}^{l}\left(\frac{(l+j-i)!l!}{(l+j)!i!(l-i)!}\right)^2\rho^{2i}
$$

$$
+\left(\sum_{i=0}^{j-1}\sum_{r=j+1}^{l}+\sum_{i=j}^{l-1}\sum_{r=i+1}^{l}\right)\rho^{i+r}\frac{(l+j-i)!(l!)^2(l+j-r)!\cos(r-i)\phi}{((l+j)!)^2 i!(l-i)!r!(l-r)!}\Bigg\}
$$

$$
\geq \rho\psi(\rho). \tag{3.10}
$$

Here $\psi(\rho)$ is a continuous function such that $0 < m \leq \psi(\rho) \leq M_1\rho^{l-1} + M_2$, with $m, M_1, M_2 > 0$. Consequently, the roots $z_1, z_2, \cdots, z_l$ of the polynomial $S_t(z)$ lie outside the sector Γ. It follows that the operator $[S_{t_0}(\tau A)]^{-1}$ exists and is bounded for fixed $t_0 \in [0, 1]$, and is uniformly bounded for $0 < \tau \leq \tau_0$. Since the coefficients of the polynomial $S_t(z)$ are bounded uniformly in $t \in [0, 1]$, its roots obey the inequality

$$
|z_k(t)| \leq M, \quad 0 \leq k \leq l. \tag{3.11}
$$

By Viète's formula,

$$
z_1(t)z_2(t)\cdots z_l(t) = (-1)^l\frac{(1-e^{2\pi it})}{j!}(j+l)!. \tag{3.12}
$$

Therefore, for any $\varepsilon \in (0, 1)$ and any $t \in [\varepsilon, 1-\varepsilon]$ we have that $|z_k(t)| \geq \delta(\varepsilon) > 0$. On the other hand, from (3.12) it follows that for any $\eta > 0$ one can find an $\varepsilon \in (0, 1)$ such that for any $t \in [0, \varepsilon] \cup [1-\varepsilon, 1]$ there is a number $k = k(t)$ such that $|z_k(t)| \leq \eta$. With no loss of generality we may assume that $|z_1(t)| \leq \eta$. Let us show that for sufficiently small $\eta > 0$ the remaining roots $z(t)$ are separated from zero by a constant that does not depend on η and t. To this end we use the representation

$$
S_t(z) = \frac{j!}{(j+l)!}(z - z_1(t))(z - z_2(t))\cdots(z - z_l(t)).
$$

Since $|z_1(t)| \leq \eta$, from this representation and (3.11) it follows that

$$
|S_t(z)| \leq \frac{j!}{(j+l)!}2\eta|\eta - z_k(t)|(\eta + M)^{l-2}
$$

for $k = 2, \cdots, l$. Using (3.10), this yields

$$
|\eta - z_k(t)| \geq m\frac{(j+l)!}{2j!}(\eta + M)^{2-l} \geq m\frac{(j+l)!}{2j!}M^{2-l} = M_1.
$$

Hence, if $\eta \leq M_1/2$, we have the estimate $|z_k(t)| \geq M_1/2$.

Let $c_1, c_2, \cdots, c_l$ be the roots of the polynomial $Q_{j,l}(z)$. Then we can write

$$\mathcal{J} = \tau A(I + \tau A)^{-1}[I - R_{j,l}(\tau A)e^{2\pi it}]^{-1} = \tau A(I + \tau A)^{-1}Q_{j,l}(\tau A)S_{j,l}^{-1}(\tau A)$$

$$= \left\{\tau A[z_1(t) - \tau A]^{-1}\right\}\left[(c_1 - \tau A)(I + \tau A)^{-1}\right]\left\{(c_2 - \tau A)[z_2(t) - \tau A]^{-1}\right\} \times \cdots \times$$

$$\times \left\{(c_l - \tau A)[z_l(t) - \tau A]^{-1}\right\}M_2.$$

Further, since

$$\tau A[z_1(t) - \tau A]^{-1} = -I + z_1(t)[z_1(t) - \tau A]^{-1},$$

$$(c_1 - \tau A)(I + \tau A)^{-1} = (c_1 - 1)(I + \tau A)^{-1} - I,$$

and

$$(c_k - \tau A)[z_k(t) - \tau A]^{-1} = [c_k - z_k(t)][z_k(t) - \tau A]^{-1} + I, \quad k \geq 2,$$

we obtain the inequality

$$\|\mathcal{J}\|_{H \to H} \leq M_2\left[1 + \frac{M|z_1(t)|}{|z_1(t)| + \tau}\right]\left[1 + \frac{M|c_1 - 1|}{1 + \tau}\right]\left[1 + \frac{M|c_2 - z_2(t)|}{|z_2(t)| + \tau}\right] \times \cdots \times$$

$$\times \left[1 + \frac{M|c_l - z_l(t)|}{|z_l(t)| + \tau}\right].$$

Since we can consider that the roots $z_1(t), z_2(t), \cdots, z_l(t)$ are separated from zero uniformly in $t \in [0, 1]$, the last inequality yields the estimate

$$\|\tau A(I + \tau A)^{-1}[I - R_{j,l}(\tau A)e^{2\pi it}]^{-1}\|_{H \to H} \leq M_3.$$

Theorem 3.7 is proved.

In Section 4 we will prove the well-posedness of problem (0.51) in $\mathcal{L}_p(E)$ for a wide class of Banach spaces E.

3.   Estimates of powers of the operator step.

In this subsection we prove the estimates (3.3) and (3.4). The proofs rely on an auxiliary estimate for the powers of the operator step.

Lemma 3.1. *The following estimate holds:*

$$\|(\tau A)^{1/2} R_{l-1,l}^k(\tau A)(I + \frac{\tau}{l}A)^{k-1}\|_{E\to E} \le Mk^{-1/2}, \quad k = 1, 2, \cdots, \qquad (3.13)$$

where M does not depend on τ and k.

Proof. For $k = 1$ the estimate (3.13) is obvious. Suppose $k > 1$. By the Cauchy-Riesz formula (4.2) of Chapter 2, we have

$$(\tau A)^{1/2} R_{l-1,l}^k(\tau A)\left(I + \frac{\tau}{l}A\right)^{k-1}$$

$$= \frac{1}{2\pi i}\int_{S_1\cup S_2} z^{1/2} R_{l-1,l}^k(z)\left(1 + \frac{z}{l}\right)^{k-1}(z - \tau A)^{-1}dz.$$

Using the estimate (4.1) of Chapter 2 and the estimate (0.26), we obtain

$$\|(\tau A)^{1/2} R_{l-1,l}^k(\tau A)\left(I + \frac{\tau}{l}A\right)^{k-1}\|_{E\to E}$$

$$\le M\int_0^\infty |z|^{1/2}\left|\frac{P_{l-1,l}(z)\left(1 + \frac{z}{l}\right)}{Q_{l-1,l}(z)}\right|^k \frac{d|z|}{|1 + \frac{z}{l}|(|z| + \tau)} \le M\int_0^\infty \theta(\rho)d\rho. \qquad (3.14)$$

Here

$$\theta(\rho) = \frac{1}{\rho^{1/2}(1 + 2(\rho/l)\cos\phi + (\rho/l)^2)^{1/2}}\left(\frac{\theta_1(\rho)}{\theta_2(\rho)}\right)^{1/2},$$

where

$$\theta_1(\rho) = \sum_{r=0}^{l}\left(\frac{(2l - 1 - r)!l!\psi(r,l)}{(2l - 1)!r!(l - r)!}\right)^2 \rho^{2r}$$

$$+2\sum_{r=0}^{l-1}\sum_{i=r+1}^{l}\frac{(2l - 1 - r)!(l!)^2(2l - 1 - i)!\psi(r,l)\psi(i,l)\rho^{i+r}\cos(i - r)\phi}{((2l - 1)!)^2r!(l - r)!i!(l - i)!}$$

and

$$\theta_2(\rho) = \sum_{r=0}^{l}\left(\frac{(2l - 1 - r)!l!}{(2l - 1)!r!(l - r)!}\right)^2 \rho^{2r}$$

$$+2\sum_{r=0}^{l-1}\sum_{i=r+1}^{l}\frac{(2l - 1 - r)!(l!)^2(2l - 1 - i)!\rho^{i+r}\cos(i - r)\phi}{((2l - 1)!)^2r!(l - r)!i!(l - i)!}.$$

To prove (3.13) it suffices to establish the bound

$$\int_0^\infty \theta(\rho)d\rho \le Mk^{-1/2}. \qquad (3.15)$$

We have

$$\int_0^\infty \theta(\rho)d\rho = \int_0^1 \theta(\rho)d\rho + \int_1^\infty \theta(\rho)d\rho = I_1 + I_2.$$

First let us estimate I_1. It is easily seen that if $0 \leq \rho \leq 1$ then

$$\theta(\rho) \leq \rho^{-1/2}(1 - \mu\rho)^{-k/2}, \tag{3.16}$$

where

$$\mu = \frac{\frac{2l}{2l-1}(1 + \psi(1, l)) \cos \phi}{\theta_2(1)}$$

and

$$\psi(r, l) = 1 - 3\frac{r}{l} + \frac{r^2}{l^2}$$

(see Section 0). From this it is seen that $\mu \in (0, 1)$. Consequently,

$$I_1 \leq Mk^{-1/2}. \tag{3.17}$$

Now let us estimate I_2. Making the substitution $\rho = 1/y$, we obtain

$$I_2 = \int_0^1 \theta(1/y)y^{-2}dy.$$

It is readily verified that

$$\theta(1/y)y^{-2} \leq y^{-1/2}(1 - \mu_1 y)^{-k/2}, \tag{3.18}$$

where

$$\mu_1 = \frac{2\left(\frac{l!}{(2l)!}\right)^2 [1 - (-1)^{l-1}\psi(l-1, l)] \cos \phi}{\theta_2(1)}.$$

Since $\mu_1 \in (0, 1)$, (3.18) yields

$$I_2 \leq Mk^{-1/2},$$

which in conjunction with (3.17) yields (3.14). Using (3.14) and the estimate (3.3) of Chapter 2, we obtain (3.13). Lemma 3.1 is proved.

Lemma 3.2. *The following estimate holds:*

$$\left\|(\tau A)^{1/2} R_{l-2,l}^k(\tau A)\left(I + \frac{\tau}{l}A\right)^{k-1}\right\|_{E \to E} \leq Mk^{-1/2}, \quad k = 1, 2, \cdots, \tag{3.19}$$

where M does not depend on τ and k.

The **proof** follows the scheme of the proof of Lemma 3.1 and relies of the estimate (3.3) of Chapter 2 and Lemma 0.6.

Now we can prove the following result.

Lemma 3.3. *The estimates* (3.3) *and* (3.4) *hold.*

Proof. First let $j = l - 1$. We shall use the identity

$$R_{l-1,l}^k(\tau A)\left(I + \frac{\tau}{l}A\right)^{k-1} - \exp\left\{-\left(k\tau - (k-1)\frac{\tau}{l}\right)A\right\} = \int_0^1 \psi_s'(s\tau A)ds.$$

Here

$$\psi(s\tau A) = R_{l-1,l}^k(s\tau A)\left(I + \frac{s\tau}{l}A\right)^{k-1}\exp\left\{-\left(k\tau - (k-1)\frac{\tau}{l}\right)(1-s)A\right\},$$

and hence the derivative $\psi'(s\tau A)_s$ is given by the expression

$$\psi_s'(s\tau A) = \left[k\tau A R_{l-1,l}^{k-1}(s\tau A)\left(I + \frac{s\tau}{l}A\right)^{k-1}[R_{l-1,l}'(s\tau A) + R_{l-1,l}(s\tau A)]\right.$$

$$\left.- \frac{(k-1)\tau}{l}AR_{l-1,l}^k(s\tau A)\left(I + \frac{s\tau}{l}A\right)^{k-2}\frac{s\tau}{l}A\right] \times$$

$$\times \exp\left\{-\left(k\tau - (k-1)\frac{\tau}{l}\right)(1-s)A\right\}.$$

Using the equality (0.5), the identity (1.51), and the expression (I.52), we obtain

$$R_{l-1,l}^k(\tau A)\left(I + \frac{\tau}{l}A\right)^{k-1} - \exp\left\{-\left(k\tau - (k-1)\frac{\tau}{l}\right)A\right\} =$$

$$\int_0^1 \frac{ks^{2l-1}(-1)^{l-1}l!(l-1)!}{((2l-1)!)^2}(\tau A)^{2l}R_{l-1,l}^{k-1}(s\tau A)Q_{l-1,l}^{-2}(s\tau A) \times$$

$$\times \left(I + \frac{s\tau}{l}A\right)^{k-1}\exp\left\{-\left(k\tau - (k-1)\frac{\tau}{l}\right)(1-s)A\right\} ds$$

$$- \int_0^1 \frac{(k-1)s}{2l}(\tau A)^2 R_{l-1,l}^k(s\tau A)\left(I + \frac{s\tau}{l}A\right)^{k-2} \times$$

$$\times \exp\left\{-\left(k\tau - (k-1)\frac{\tau}{l}\right)(1-s)A\right\} ds.$$

Using the estimates (1.54), (1.55), and (1.45), we obtain

$$\left\| R_{l-1,l}^k(\tau A)\left(I + \frac{\tau}{l}A\right)^{k-1} - \exp\left\{-\left(k\tau - (k-1)\frac{\tau}{l}\right)A\right\}\right\|_{E\to E}$$

$$\leq M \int_0^1 k \left\|(\tau A)^{1/2} R_{l-1,l}^{k-1}(s\tau A)\left(I + \frac{s\tau}{l}A\right)^{k-2}\right\|_{E\to E} \times$$

$$\times \left\|(\tau A)^{1/2} \exp\left\{-\left(k\tau - (k-1)\frac{\tau}{l}\right)(1-s)A\right\}\right\|_{E\to E} ds$$

$$+M \int_0^1 (k-1)\left\|(\tau A)^{1/2} R_{l-1,l}^k(s\tau A)\left(I + \frac{s\tau}{l}A\right)^{k-1}\right\|_{E\to E} \times$$

$$\times \left\|(\tau A)^{1/2} \exp\left\{-\left(k\tau - (k-1)\frac{\tau}{l}\right)(1-s)A\right\}\right\|_{E\to E} ds$$

$$\leq M_1 \int_0^1 \frac{ds}{((s(1-s))^{1/2}} \cdot \frac{k}{((k-2)k)^{1/2}} \leq M_2$$

if $k > 2$. By the triangle inequality and the estimate (1.10) of Chapter 1, we have that

$$\left\| R_{l-1,l}^k(\tau A)\left(I + \frac{\tau}{l}A\right)^{k-1}\right\|_{E\to E} \leq M \qquad (3.20)$$

for $k > 2$. For $k = 1$, 2 the estimate (3.20) is obvious. Thus, (3.20) holds for any $k \geq 1$. From (3.20) and the estimate (3.3) of Chapter 1 we obtain the estimate

$$\|R_{l-1,l}^k(\tau A)\|_{E\to E} \leq M\left(1 + \frac{\delta}{l}\tau\right)^{-(k-1)} \leq M\left(1 + \frac{\delta}{l}\tau_0\right)\left(1 + \frac{\delta}{l}\tau\right)^{-k}.$$

The estimate (3.3) is thus established for $j = l - 1$; the proof for $j = l - 2$ is similar.

Now let us prove the estimate (3.4). For $k = 1$ and $k = 2$ it is obvious. If $k > 1$, then using the identity

$$\tau A R_{j,l}^k(\tau A)\left(I + \frac{\tau}{l}A\right)^{k-2} = (\tau A)^{1/2} R_{j,l}^{[k/2]}(\tau A)\left(I + \frac{\tau}{l}A\right)^{[k/2]-1} \times$$

$$\times (\tau A)^{1/2} R_{j,l}^{k-[k/2]}(\tau A)\left(I + \frac{\tau}{l}A\right)^{k-[k/2]-1}$$

and the estimate (3.13), we get

$$\left\|\tau A R_{j,l}^k(\tau A)\left(I + \frac{\tau}{l}A\right)^{k-2}\right\|_{E\to E} \leq Mk^{-1}.$$

From this and the estimate (3.4) of Chapter 2 we obtain the estimate

$$\|\tau A R_{j,l}^k(\tau A)\|_{E \to E} \leq \frac{M}{k}\left(1 + \frac{\delta}{l}\tau\right)^{-(k-2)} \leq \frac{M}{k}\left(1 + \frac{\delta}{l}\tau_0\right)^2 \left(1 + \frac{\delta}{l}\tau\right)^{-k}.$$

Lemma 3.3 is proved.

Finally, let us prove the assertion that was used in Section 2 in the investigation of the well-posedness of the difference problem (0.51) generated by the Padé fractions $R_{l,l}(z)$ with l odd.

Lemma 3.4. *If A is a strongly positive operator with spectral angle $\phi(A, E) < \frac{\pi}{2l}$, then the operator $I + R_{l,l}(\tau A)$ is invertible for l odd and one has the estimate*

$$\|[I + R_{l,l}(\tau A)]^{-1}(I + \tau A)^{-1}\|_{E \to E} \leq M.$$

Proof. Since $[1 + R_{l,l}(z)](1 + z)^{-1}$ is a rational function, it suffices to show that the function

$$\psi(z) = 2 \sum_{r=0}^{(l-1)/2} \frac{(2l - 2r)!l!}{(2l)!(2r)!(l - 2r)!} z^{2r}$$

has no zeros in the sector $\Gamma = \{z \in \mathbf{C}^+ : z = \rho e^{\pm i\phi}, 0 \leq \rho < \infty, 0 \leq \phi \leq \frac{\pi}{2l}\}$. Since

$$\left|\sum_{r=0}^{(l-1)/2} \frac{(2l - 2r)!l!}{(2l)!(2r)!(l - 2r)!} z^{2r}\right|^2 = \sum_{r=0}^{(l-1)/2} \left[\frac{(2l - 2r)!l!}{(2l)!(2r)!(l - 2r)!}\rho^{2r}\right]^2$$

$$+ 2 \sum_{r=0}^{((l-1)/2)-1} \sum_{s=r+1}^{(l-1)/2} \frac{(2l - 2r)!(l!)^2(2l - 2s)!}{((2l)!)^2(l - 2r)!(2s)!(l - 2s)!}\rho^{2r+2s} \cos 2(s - r)\phi \geq 1,$$

we obviously have $|\psi(z)| \geq 2$. Lemma 3.4 is proved.

4. WELL-POSEDNESS OF THE DIFFERENCE PROBLEM IN $\mathcal{L}_p(E'_{\alpha,q})$

1. Stability of the difference problem.

In Section 3 we have shown that well-posed solvability in $\mathcal{L}_p(E)$ implies the analyticity of the semigroup $\exp\{-tA\}$ in E. The authors do not know whether this analyticity, as in the case of the differential problem, is a sufficient condition of well-posed solvability of problem (0.51) in $\mathcal{L}_p(E)$ for arbitary E and A.

In Chapter 2 we proved that the simplest difference problem (0.6) is well-posed solvable in $\mathcal{L}_p(E'_{\alpha,q})$. It turns out that this remains true for the general problem (0.51) in $\mathcal{L}_p(E'_{\alpha,q})$, $0 < \alpha < 1$, $1 \leq p, q \leq \infty$. We have

Theorem 4.1. *The difference problem* (0.51) *with* $l - 2 \leq j \leq l$ *is stable in* $\mathcal{L}_p(E'_{\alpha,q})$.

Proof. By formula (0.50) and Theorem 0.5, to prove this assertion it suffices to establish the estimate

$$\|R^k_{j,l}(\tau A)\|_{E'_{\alpha,q} \to E'_{\alpha,q}} \leq M(\alpha), \quad 1 \leq k \leq N. \tag{4.1}$$

From the fact that the operators $R^k_{j,l}(\tau A)$ and A commute it follows that

$$\|R^k_{j,l}(\tau A)\|_{E'_{\alpha,q} \to E'_{\alpha,q}} \leq \|R^k_{j,l}(\tau A)\|_{E \to E}, \tag{4.2}$$

which in conjunction with (1.22) gives the estimate (4.1) for $j = l - 2$, $l - 1$, with $M(\alpha) = M$. Such a simple approach is not applicable in the case $j = l$, since a bound $\|R^k_{l,l}(\tau A)\|_{E \to E} \leq M$ that holds uniformly in τ is not available for an arbitrary space E and an arbitrary strongly positive operator A. We shall therefore use another approach. Since the function $(\lambda + (z/\tau))^{-1}R^k_{l,l}(z)$ decreases at infinity and is analytic in the right half-plane $\mathbf{C}^+ = \{z : \operatorname{Re} z \geq 0\}$, then in view of the strong positivity of A the Cauchy-Riesz formula (4.2) of Chapter 2 yields the representation

$$(\lambda + A)^{-1}R^k_{l,l}(\tau A) = \frac{1}{2\pi i} \int_{S_1 \cup S_2} R^k_{l,l}(z)\left(\lambda + \frac{z}{\tau}\right)^{-1}(z - \tau A)^{-1}dz. \tag{4.3}$$

By the estimate (4.1) of Chapter 2,

$$\|A(\lambda + A)^{-1}R^k_{l,l}(\tau A)x\|_E \leq M \int_0^\infty |R^k_{l,l}(z)|\left(\lambda + \frac{|z|}{\tau}\right)^{-1}\|A(|z| + \tau A)^{-1}x\|_E d|z|,$$

which in conjunction with (0.8) and (0.9) yields

$$\|A(\lambda + A)^{-1}R^k_{l,l}(\tau A)x\|_E \leq M \int_0^\infty \frac{1}{\lambda + \rho}\|A(\rho + A)^{-1}x\|_E d\rho$$

$$= M \int_0^\infty \frac{1}{1+r} \|(A(r\lambda + A)^{-1}x\|_E dr. \tag{4.4}$$

Now if $x \in E'_{\alpha,q}$, then, by the Minkowski inequality, we have

$$\left(\int_0^\infty \|\lambda^\alpha A(\lambda + A)^{-1} R_{l,l}^k(\tau A)x\|_E^q \frac{d\lambda}{\lambda} \right)^{1/q}$$

$$\leq M \int_0^\infty \frac{1}{1+r} \left(\int_0^\infty \|\lambda^\alpha A(r\lambda + A)^{-1}x\|_E^q \frac{d\lambda}{\lambda} \right)^{1/q} dr$$

$$= M \int_0^\infty \frac{1}{(1+r)r^\alpha} \left(\int_0^\infty \|(\lambda r)^\alpha A(r\lambda + A)^{-1}x\|_E^q \frac{d(\lambda r)}{\lambda r} \right)^{1/q} dr$$

$$= M \int_0^\infty \frac{dr}{(1+r)r^\alpha} \|x\|'_{\alpha,q}$$

for any q, $q \neq \infty$. Since

$$\int_0^\infty \frac{dr}{(1+r)r^\alpha} = \frac{\pi}{\sin \pi\alpha},$$

we conclude that for any $x \in E'_{\alpha,q}$, $q \neq \infty$ and $k = 1, \cdots, N$,

$$\|R_{l,l}^k(\tau A)x\|'_{\alpha,q} \leq \frac{M}{\alpha(1-\alpha)} \|x\|'_{\alpha,q}. \tag{4.5}$$

The estimate (4.5) with $q = \infty$ can be established by passing to the limit in (4.5) (with $q \neq \infty$). This yields (4.1) for $j = l$. Theorem (4.1) is proved.

2.   Well-posedness of the difference problem.

In this subsection we establish the coercivity inequality for the solutions of the difference problem (0.51) in $\mathcal{L}_p(E'_{\alpha,q})$. First we shall consider the nonhomogeneous difference problem (0.51) with $u_0 = 0$.

Theorem 4.2. *The nonhomogeneous difference problem* (0.51) *with* $u_0 = 0$, *generated by the Padé fractions* $R_{j,l}(z)$ *for* $j = l - 2$, $l - 1$ *is well-posed solvable in* $\mathcal{L}_p(E'_{\alpha,q})$.

Proof. First let us consider the case $q \neq \infty$. It suffices to establish the coercivity inequality

$$\|A_{j,l}u^\tau\|_{L_q(\tau, E'_{\alpha,q})} \leq \frac{M}{\alpha(1-\alpha)} \|\varphi_{j,l}^\tau\|_{L_q(\tau, E'_{\alpha,q})}, \tag{4.6}$$

where M does not depend on τ, α, q, and $\varphi^\tau_{j,l}$. By formula (0.50), we have

$$A_{j,l}u_{k-1} = [I - R_{j,l}(\tau A)] \sum_{r=1}^{k-2} R_{j,l}^{k-1-r}(\tau A)\varphi_r^{j,l} + [I - R_{j,l}(\tau A)]\varphi_{k-1}^{j,l} = I_1 + I_2.$$

Since

$$\|I - R_{j,l}(\tau A)\|_{E \to E} \leq M,$$

we have

$$\|I_2\|_{L_q(\tau, E'_{\alpha,q})} \leq \|I - R_{j,l}(\tau A)\|_{E \to E}\|\varphi^\tau_{j,l}\|_{L_q(\tau, E'_{\alpha,q})} \leq M\|\varphi^\tau_{j,l}\|_{L_q(\tau, E'_{\alpha,q})}. \quad (4.7)$$

Now let us estimate I_1 in the norm of $L_q(\tau, E'_{\alpha,q})$. To this end we use the Cauchy-Riesz formula (4.2) of Chapter 2 to write

$$A(\lambda + A)^{-1}I_1$$

$$= \frac{1}{2\pi i}\int_{S_1 \cup S_2} \sum_{r=1}^{k-2}(1 - R_{j,l}(z))R_{j,l}^{k-1-r}(z)\left(\lambda + \frac{z}{\tau}\right)^{-1} A(z - \tau A)^{-1}\varphi_r^{j,l}dz.$$

By the estimate (4.1) of Chapter 2,

$$\|A(\lambda + A)^{-1}I_1\|_E$$

$$\leq M\int_0^\infty \sum_{r=1}^{k-2}|(1 - R_{j,l}(z))R_{j,l}^{k-1-r}(z)|\left(\lambda + \frac{|z|}{\tau}\right)^{-1}\|A(|z| + \tau A)^{-1}\varphi_r^{j,l}\|_E d|z|.$$

From this inequality, (0.8), and the estimates (0.9), (0.10), and (0.18) it follows that

$$\|A(\lambda + A)^{-1}I_1\|_E$$

$$\leq M\int_0^\infty \sum_{r=1}^{k-2}\frac{1}{\rho + \lambda\tau}\left(\frac{1 - 2a\rho\cos\phi + \rho^2}{1 + 2a\rho\cos\phi + \rho^2}\right)^{\frac{k-2-r}{2}}\frac{1}{1 + 2a\rho\cos\phi + \rho^2}\times$$

$$\times\|A(\rho\tau^{-1} + A)^{-1}\varphi_r^{j,l}\|_E d\rho$$

$$\leq M\int_0^\infty \sum_{i=1}^{N}\frac{1}{\rho + \lambda\tau}\left(\frac{1 - 2a\rho\cos\phi + \rho^2}{1 + 2a\rho\cos\phi + \rho^2}\right)^{i/2}\frac{1}{1 + 2a\rho\cos\phi + \rho^2}\times$$

$$\times\|A(\rho\tau^{-1} + A)^{-1}\varphi_{k-2-i}^*\|_E d\rho.$$

Here $\varphi_i^* = \varphi_i^{j,l}$ if $i = 1, \cdots, N$ and $\varphi_i^* = 0$ otherwise. From the sum Minkowski inequality (with respect to k) we obtain

$$\left(\sum_{k=1}^{N} \|A(\lambda + A)^{-1} I_1\|_E^q \, \tau \right)^{1/q}$$

$$\leq M \int_0^\infty \sum_{i=1}^{N} \frac{1}{\rho + \lambda\tau} \left(\frac{1 - 2a\rho\cos\phi + \rho^2}{1 + 2a\rho\cos\phi + \rho^2} \right)^{i/2} \frac{1}{1 + 2a\rho\cos\phi + \rho^2} \times$$

$$\times \left(\sum_{k=1}^{N} \|A(\rho\tau^{-1} + A)^{-1} \varphi_{k-2-i}^*\|_E^q \tau \right)^{1/q} d\rho. \tag{4.8}$$

By the definition of the grid function φ_i^*,

$$\sum_{k=1}^{N} \|A(\rho\tau^{-1} + A)^{-1} \varphi_{k-2-i}^*\|_E^q \tau \leq \sum_{s=1}^{N} \|A(\rho\tau^{-1} + \tau A)^{-1} \varphi_s^{j,l}\|_E^q \tau,$$

which in conjunction with (4.8) yields

$$\left(\sum_{k=1}^{N} \|A(\lambda + A)^{-1} I_1\|_E^q \, \tau \right)^{1/q}$$

$$\leq M \int_0^\infty \sum_{i=1}^{N} \frac{1}{\rho + \lambda\tau} \left(\frac{1 - 2a\rho\cos\phi + \rho^2}{1 + 2a\rho\cos\phi + \rho^2} \right)^{i/2} \frac{1}{1 + 2a\rho\cos\phi + \rho^2} \times$$

$$\times \left(\sum_{s=1}^{N} \|A(\rho\tau^{-1} + A)^{-1} \varphi_s^{j,l}\|_E^q \tau \right)^{1/q} d\rho.$$

Summing the geometric progression (with respect to i), we see that

$$\left(\sum_{k=1}^{N} \|A(\lambda + A)^{-1} I_1\|_E^q \, \tau \right)^{1/q}$$

$$\leq \frac{M}{\cos\phi} \int_1^\infty \frac{1}{\rho + \lambda\tau} \left(\sum_{s=1}^{N} \|A(\rho\tau^{-1} + \tau A)^{-1} \varphi_s^{j,l}\|_E^q \tau \right)^{1/q} d\rho.$$

The substitution $\rho = \tau r \lambda$ yields

$$\left(\sum_{k=1}^{N} \|A(\lambda + A)^{-1} I_1\|_E^q \, \tau \right)^{1/q}$$

$$\leq \frac{M}{\cos\phi} \int_0^\infty \frac{1}{1+r} \left(\sum_{s=1}^N \|A(r\lambda+A)^{-1}\varphi_s^{j,l}\|_E^q \tau \right)^{1/q} dr.$$

Next, using the Minkowski inequality we obtain

$$\left(\int_0^\infty \sum_{k=1}^N \tau \|\lambda^\alpha A(\lambda+A)^{-1}I_1\|_E^q \frac{d\lambda}{\lambda} \right)^{1/q}$$

$$\leq \frac{M}{\cos\phi} \int_0^\infty \frac{dr}{(1+r)r^\alpha} \left(\sum_{s=1}^N \int_0^\infty \tau \|(r\lambda)^\alpha A(r\lambda+A)^{-1}\varphi_s^{j,l}\|_E^q \frac{dr\lambda}{r\lambda} \right)^{1/q}$$

$$\leq \frac{M\pi}{\cos\phi \sin\pi\alpha} \|\varphi_{j,l}^\tau\|_{L_q(\tau,E'_{\alpha,q})},$$

whence

$$\|I_1\|_{L_q(\tau,E'_{\alpha,q})} \leq \frac{M_1}{\alpha(1-\alpha)} \|\varphi_{j,l}^\tau\|_{L_q(\tau,E'_{\alpha,q})}. \tag{4.9}$$

From (4.7) and (4.9) we obtain the inequality (4.6) when $q \neq \infty$. Inequality (4.6) with $q = \infty$ is obtained by passing to the limit $q \to \infty$. Theorem 4.2 is proved.

Theorem 4.3. *Let $\varphi_{l,l}^\tau \in D(A)$. Then the solutions of the nonhomogeneous difference problem* (0.49) *with $u_0 = 0$ generated by the Padé fractions $R_{l,l}(z)$ obey the coercivity inequality*

$$\|\mathcal{D}\Pi(u_0)u^\tau\|_{L_q(\tau,E'_{\alpha,q})} + \|A_{l,l}u^\tau\|_{L_q(\tau,E'_{\alpha,q})} \leq \frac{M}{\alpha(1-\alpha)} \|(I+\tau A)\varphi_{l,l}^\tau\|_{L_q(\tau,E'_{\alpha,q})},$$

where M does not depend on τ, α, q, and $\varphi_{l,l}^\tau$.

The proof of this assertion folllows the scheme of the proof of Theorem 4.2.

We already know that the operator step of a diagonal difference difference scheme with l even has "better" properties than the operator step for l odd. This fact allows us to obtain the following result.

Theorem 4.4. *The nonhomogeneous difference problem* (0.51) *with $u_0 = 0$ generated by the Padé fractions $R_{l,l}(z)$ for even l is well-posed solvable in $\mathcal{L}_p(E_{\alpha,q})$.*

The proof of this assertion also follows the scheme of the proof of Theorem 4.2; here one uses the estimate (1.33), which holds for even l.

From Theorems 3.6 and 4.2 we derive the following result.

Theorem 4.5. *Let A be a strongly positive operator with spectral angle $\phi(A, E) < \frac{\pi}{2l}$. Then the solutions of the nonhomogeneous difference problem (0.51) with $u_0 = 0$ obey the coercivity inequality*

$$\|\overline{\mathcal{D}}\,\overline{\Pi}(u_0)u^\tau\|_{L_p(E'_{\alpha,q})} + \|\overline{A}_{j,l}u^\tau\|_{\mathcal{L}_p(E'_{\alpha,q})}$$

$$\leq \frac{Mp^2}{(p-1)\alpha(1-\alpha)}\|\varphi_{j,l}\|_{\mathcal{L}_p(E'_{\alpha,q})}, \quad 0 < \alpha < 1, \ 1 < p, q < \infty, \tag{4.10}$$

where M does not depend on α, p, q, and $\varphi_{j,l}$.

Notice that the meaning of Theorems 4.2 and 4.5 is that the nonhomogenous difference problem (0.51) with $u_0 = 0$ generated by the Padé fractions $R_{j,l}(z)$ for $j = l - 2$, $l - 1$ is well-posed solvable in the space $\mathcal{L}_p(E'_{\alpha,q})$ whenever $p, q \in (1, \infty)$, $0 < \alpha < 1$ or $p = q = 1$ and $p = q = \infty$, $0 < \alpha < 1$.

Now let us turn to the general problem (0.51) generated by the Padé fractions with $j = l - 2$, $l - 1$. By (0.50), $u = w + g$, where g is the solution of the nonhomogeneous problem (0.51) with $u_0 = 0$. In order that w be a solution in $\mathcal{L}_p(E'_{\alpha,q})$ of problem (0.51) with $\varphi_{j,l} = 0$ it is necessary and sufficient that

$$\sup_{0 < \tau \leq \tau_0} \sum_{k=1}^{N} \left(\|A_{j,l}R_{j,l}^k(\tau A)u_0\|'_{\alpha,q}\right)^p \tau < \infty.$$

The quantity (see Section 3)

$$\langle u_0 \rangle''_{1+\alpha-\frac{1}{p},q} = \sup_{0 < \tau \leq \tau_0} \left(\sum_{k=1}^{N} \left[\|A_{j,l}R_{j,l}^k(\tau A)u_0\|'_{\alpha,q}\right]^p \tau\right)^{1/p}$$

is a norm in the space of initial data $E''_{1+\alpha-\frac{1}{p},q}$ consisting of all $u_0 \in E'_{\alpha,q}$ for which this quantity is finite.

Theorem 4.6. *Let $1 < p, q < \infty$ or $p = q = \infty$. Then the following coercivity inequality holds:*

$$\|\overline{\mathcal{D}}\,\overline{\Pi}(u_0)u^\tau\|_{\mathcal{L}_p(E'_{\alpha,q})} + \|\overline{A}_{j,l}u^\tau\|_{\mathcal{L}_p(E'_{\alpha,q})} + \max_{1 \leq k \leq N}\langle u_k\rangle''_{1+\alpha-\frac{1}{q},p}$$

$$\leq M\left[\langle u_0\rangle''_{1+\alpha-\frac{1}{q},p} + \frac{M_1(p,q)}{\alpha(1-\alpha)}\|\varphi_{j,l}\|_{\mathcal{L}_p(E'_{\alpha,q})}\right], \tag{4.11}$$

where $M_1(p,q) = q^2/(q-1)$ if $p \neq q$ and $M_1(p,p) = 1$.

The proof is carried out according to the scheme of the proof of Theorem 4.4 of Chapter 2.

Theorem 4.6 does not cover all cases $p, q \in [1, \infty]$. Theorems 4.2 and 4.6 are supplemented by

Theorem 4.7. 1) *Let $p = q = 1$ or $p = q = \infty$. Then the solutions of the difference problem* (0.51) *obey the coercivity inequality*

$$\|\overline{\mathcal{D}}\,\overline{\Pi}(u_0)u^\tau\|_{\mathcal{L}_p(E'_{\alpha,q})} + \|\overline{A}_{j,l}u^\tau\|_{\mathcal{L}_p(E'_{\alpha,q})}$$

$$\leq M\left[\langle u_0\rangle''_{1+\alpha-\frac{1}{q},p} + \frac{1}{\alpha(1-\alpha)}\|\varphi_{j,l}\|_{\mathcal{L}_p(E'_{\alpha,q})}\right]. \tag{4.12}$$

2) *Let $1 < p < \infty$. Then the solutions of the difference problem* (0.51) *obey the inequality*

$$\max_{1\leq k\leq N}\langle u_k\rangle''_{1+\alpha-\frac{1}{q},\infty} \leq M\left[\langle u_0\rangle''_{1+\alpha-\frac{1}{q},\infty} + \frac{p^2}{p-1}\|\varphi_{j,l}\|_{\mathcal{L}_p(E_{\alpha,\infty})}\right]. \tag{4.13}$$

The proof of this assertion follows the scheme of the proof of Theorem 4.5 of Chapter 2.

5. WELL-POSEDNESS OF THE DIFFERENCE PROBLEM IN DIFFERENCE ANALOGUES OF SPACES OF SMOOTH FUNCTIONS

1. Well-posedness of the difference problem in $C_0^{\beta,\gamma}(E)$.

In Chapter 2 we have shown that the simplest difference problem (0 6) is well posed in the space $C_0^{\beta,\gamma}(E)$. It turns out that this is true for the general difference problem (0.51) generated by a broad class of Padé difference schemes.

First let us consider the special nonhomogeneous difference problem (0.51) with $u_0 = 0$ and $\varphi_1^{j,l} = 0$. A solution u of this problem is called a *solution in* $C_0^{\beta,\gamma}(E)$ if $\overline{\mathcal{D}}\,\overline{\Pi}(u_0)u$ and $\overline{A}_{j,l}u$ belong to $C_0^{\beta,\gamma}(E)$. If u is a solution of problem

(0.51) with $u_0 = 0$ and $\varphi_1^{j,l} = 0$, the obviously $\varphi_{j,l} \in C_0^{\beta,\gamma}(E)$. As above, we introduce the following

Definition 5.1. The difference problem (0.51) with $u_0 = 0$ and $\varphi_1^{j,l} = 0$ is said to be *well posed in* $C_0^{\beta,\gamma}(E)$ if two conditions are satisfied:

1) For any $\varphi_{j,l} \in C_0^{\beta,\gamma}(E)$ there exists a unique solution $u(\varphi_{j,l}, 0)$ of this problem.

2) The operator $u(\varphi_{j,l}, 0)$ is continuous in $C_0^{\beta,\gamma}(E)$.

As in the case of the space $C(E)$ one establishes that for the well-posedness of this difference problem in $C_0^{\beta,\gamma}(E)$ it is necessary and sufficient that the coercivity inequality hold:

$$\|\mathcal{D}\Pi(0)u^\tau\|_{C_0^{\beta,\gamma}(\tau,E)} + \|A_{j,l}u^\tau\|_{C_0^{\beta,\gamma}(\tau,E)} \leq M(\beta,\gamma)\|\varphi_{j,l}^\tau\|_{C_0^{\beta,\gamma}(\tau,E)} \qquad (5.1)$$

where $M(\beta,\gamma)$ does not depend on τ and $\varphi_{j,l}^\tau$.

By passing to the limit (when $\tau \to 0$) we derive from (5.1) the corresponding inequality for the differential problem. It follows that $-A$ must be the generator of an analytic semigroup in E. As it turns out, this condition is not only necessary, but also sufficient for the well-posedness of problem (0.51) with $u_0 = 0$ and $\varphi_1^{j,l} = 0$. Accordingly, from now on we shall assume that $-A$ generates an analytic semigroup in E.

Theorem 5.1. *The difference problem (0.51) with $u_0 = 0$ and $\varphi_1^{j,l} = 0$ generated by the Padé fractions $R_{j,l}(z)$ with $j = l-2$, $l-1$ is well-posed in $C_0^{\beta,\gamma}(E)$. The solutions of this problem obey the coercivity inequality (5.1) in which $M(\beta,\gamma) = \frac{M}{\beta(1-\beta)}$, where M does not depend on β, γ, and $\varphi_{j,l}$.*

The proof of this assertion follows the scheme of the proof of Theorem 5.1 of Chapter 2 and is based on Theorem 2.2 and the estimates (1.22), (1.23) and (2.3), (2.4).

Theorem 5.2. *Let $j = l$ and $\varphi_{l,l}^\tau \in D(A^2)$. Then the solutions of the difference problem (0.51) with $u_0 = 0$ and $\varphi_1^{j,l} = 0$ obey the coercivity inequality*

$$\|\mathcal{D}\Pi(0)u^\tau\|_{C_0^{\beta,\gamma}(\tau,E)} + \|A_{l,l}u^\tau\|_{C_0^{\beta,\gamma}(\tau,E)} \leq \frac{M}{\beta(1-\beta)}\|(I+\tau A)^2\varphi_{l,l}^\tau\|_{C_0^{\beta,\gamma}(\tau,E)}, \quad (5.2)$$

where M does not depend on β, γ, τ, and $\varphi_{l,l}^\tau$.

The proof of this assertion follows the scheme of the proof of Theorem 5.1 of Chapter 2 and of Theorem 4.3, and relies on the estimates (1.25), (1.26) and (2.11), (2.12).

If $j = l$ is an even number, estimate (5.2) can be improved. More precisely, we have the following result.

Theorem 5.3. Let $\varphi^\tau_{l,l} \in D(A)$. Then the solutions of the difference problem (0.51) with $u_0 = 0$ and $\varphi^{j,l}_1 = 0$ obey the coercivity inequality

$$\|\mathcal{D}\Pi(0)u^\tau\|_{C^{\beta,\gamma}_0(\tau,E)} + \|A_{l,l}u^\tau\|_{C^{\beta,\gamma}_0(\tau,E)} \leq \frac{M}{\beta(1-\beta)}\|(I+\tau A)\varphi^\tau_{l,l}\|_{C^{\beta,\gamma}_0(\tau,E)}, \quad (5.3)$$

where M does not depend on β, γ, τ, and $\varphi^\tau_{l,l}$.

Theorem 5.4. If the difference problem (0.51) with $u_0 = 0$ and $\varphi^{j,l}_1 = 0$ is stable in $C(E)$, then it is well posed in $C^{\beta,\gamma}_0(E)$.

The proofs of Theorems 5.3 and 5.4 follow the scheme of the proofs of Theorem 5.1 of Chapter 2 and of Theorems 4.4, 4.5, and rely on the estimates (1.25), (1.26), (2.11), and (2.16).

Now let us consider the general difference problem (0.51). Formula (0.50) shows that the solution of the homogeneous problem (0.49) with $\varphi^\tau_{j,l} = 0$ and $w_0 = u_0 - \tau[I - R_{j,l}(\tau A)]^{-1}\varphi^{j,l}_1$ has the form

$$w_{k-1} = R^{k-1}_{j,l}(\tau A)w_0 = R^{k-1}_{j,l}\left(u_0 - \tau[I - R_{j,l}(\tau A)]^{-1}\varphi^{j,l}_1\right), \quad 1 \leq k \leq N.$$

Hence, for w to be a solution in $C^{\beta,\gamma}_0(E)$ of problem (0.51) with $\varphi_{j,l} = 0$ it is necessary and sufficient that

$$\sup_{0<\tau\leq\tau_0} \|\{A_{j,l}R^{i-1}_{j,l}(\tau A)w_0\}^N_1\|_{C^{\beta,\gamma}_0(\tau,E)} < \infty.$$

Clearly, by the estimates (1.22) and (1.25), the set of all $w_0 \in E$ satisfying this condition is a linear set containing $D(A)$. It becomes a Banach space when equipped with the norm

$$\langle w_0 \rangle^{\beta,\gamma}_1 = \sup_{0<\tau\leq\tau_0}\left[\max_{1\leq i\leq N}\|A_{j,l}R^{i-1}_{j,l}(\tau A)w_0\|_E\right.$$

$$+ \sup_{1 \leq \imath < \imath + r \leq N} \frac{(\imath + r)\tau)^{\gamma} \|A_{\jmath,l}[R_{\jmath,l}^{\imath+r-1}(\tau A) - R_{\jmath,l}^{\imath-1}(\tau A)]w_0\|_E}{(r\tau)^{\beta}} \Bigg]. \qquad (5.4)$$

Let $u^{\tau} = (u_1, \cdots, u_N)$ be a solution of the general problem (0.49). We can write

$$u_k = g_k + w_k + \tau[I - R_{\jmath,l}(\tau A)]^{-1}\varphi_1^{\jmath,l}, \quad 1 \leq k \leq N,$$

where g^{τ} is a solution of problem (0.49) with $g_0 = 0$ and right-hand side $\varphi_k^{\jmath,l} - \varphi_1^{\jmath,l}$ and w^{τ} is a solution of the homogeneous problem (0.49) with $w_0 = u_0 - \tau[I - R_{\jmath,l}(\tau A)]^{-1}\varphi_1^{\jmath,l}$. From this representation and the solvability in $C_0^{\beta,\gamma}(E)$ one concludes that the general problem (0.51) is uniquely solvable whenever $u_0 - \tau[I - R_{\jmath,l}(\tau A)]^{-1}\varphi_1^{\jmath,l} \in E_1^{\beta,\gamma}$ and that its solutions obey the coercivity inequality

$$\|\overline{\mathcal{D}}\,\overline{\Pi}(u_0)u\|_{C_0^{\beta\,\gamma}(E)} + \|\overline{A}_{\jmath,l}u\|_{C_0^{\beta\,\gamma}(E)}$$

$$\leq M(\beta,\gamma)\|\varphi_{\jmath,l}\|_{C_0^{\beta\,\gamma}(E)} + M\langle u_0 - \tau[I - R_{\jmath,l}(\tau A)]^{-1}\varphi_1^{\jmath,l}\rangle_1^{\beta,\gamma}. \qquad (5.5)$$

We say that the general difference problem (0.51) is *well-posed in* $C_0^{\beta,\gamma}(E)$ if whenever $u_0 \in D(A)$, $u_0 - \tau[I - R_{\jmath,l}(\tau A)]^{-1}\varphi_1^{\jmath,l} \in E_1^{\beta,\gamma}$ and $\varphi_{\jmath,l} \in C_0^{\beta,\gamma}(E)$ it has a unique solution and the coercivity inequality (5.5) holds.

The foregoing arguments show that the analyticity of the semigroup $\exp\{-tA\}$ is not only necessary, but also sufficient for the well-posedness in $C_0^{\beta,\gamma}(E)$ of the difference problem (0 51) The conditions formulated above on the data of the problem (0.51) are such that the element $u_0 - \tau[I - R_{\jmath,l}(\tau A)]^{-1}$ is required to belong to a smaller space than the elements $u_{k-1} - \tau[I - R_{\jmath,l}(\tau A)]^{-1}\varphi_k^{\jmath,l}$, $k = 2, \cdots, N$. Is this really the case? The answer is provided by the following assertions.

Theorem 5.5. *Let* $\jmath = l - 2$, $l - 1$ *and* $u_0 - \tau[I - R_{\jmath,l}(\tau A)]^{-1}\varphi_1^{\jmath,l} \in E_1^{\beta,\gamma}$. *Then the solutions of the difference problem* (0.49) *satisfy the inequality*

$$\max_{1 \leq k \leq N}\langle u_{k-1} - \tau[I - R_{\jmath,l}(\tau A)]^{-1}\varphi_k^{\jmath,l}\rangle_1^{\beta,\gamma}$$

$$\leq M\left[\langle u_0 - \tau[I - R_{\jmath,l}(\tau A)]^{-1}\varphi_1^{\jmath,l}\rangle_1^{\beta,\gamma} + \frac{1}{\beta(1-\beta)}\|\varphi_{\jmath,l}\|_{C_0^{\beta\,\gamma}(E)}\right],$$

where M *does not depend on* β, γ, u_0, *and* $\varphi_{\jmath,l}$.

Theorem 5.6. *Let* $\jmath = l$ *and* $u_0 - \tau[I - R_{l,l}(\tau A)]^{-1}\varphi_1^{l,l} \in E_1^{\beta,\gamma}$, $\varphi_{l,l}^{\tau} \in D(A^2)$. *Then the solutions of the difference problem* (0.49) *satisfy the inequality*

$$\max_{1 \leq k \leq N}\langle u_{k-1} - \tau[I - R_{l,l}(\tau A)]^{-1}\varphi_k^{l,l}\rangle_1^{\beta,\gamma}$$

$$\leq M\left[\langle u_0 - \tau[I - R_{l,l}(\tau A)]^{-1}\varphi_1^{l,l}\rangle_1^{\beta,\gamma} + \frac{1}{\beta(1-\beta)}\|(I+\tau A)^2\varphi_{l,l}^{\tau}\|_{C_0^{\beta,\gamma}(\tau,E)}\right],$$

where M does not depend on τ, β, γ, u_0, and $\varphi_{l,l}^{\tau}$.

Theorem 5.7. *Let* $j = l$ *be an even number, and let* $u_0 - \tau[I - R_{l,l}(\tau A)]^{-1}\varphi_1^{l,l} \in E_1^{\beta,\gamma}$. *Then the solutions of the difference problem* (0.49) *satisfy the following two inequalities:* 1) *If* $\varphi_{l,l}^{\tau} \in D(A)$,

$$\max_{1 \leq k \leq N} \langle u_{k-1} - \tau[I - R_{l,l}(\tau A)]^{-1}\varphi_k^{l,l}\rangle_1^{\beta,\gamma}$$

$$\leq M\left[\langle u_0 - \tau[I - R_{l,l}(\tau A)]^{-1}\varphi_1^{l,l}\rangle_1^{\beta,\gamma} + \frac{1}{\beta(1-\beta)}\|(I+\tau A)\varphi_{l,l}^{\tau}\|_{C_0^{\beta,\gamma}(\tau,E)}\right],$$

where M does not depend on τ, β, γ, u_0, and $\varphi_{l,l}^{\tau}$.

2) *If the difference problem* (0.51) *is stable,*

$$\max_{1 \leq k \leq N} \langle u_{k-1} - \tau[I - R_{l,l}(\tau A)]^{-1}\varphi_k^{l,l}\rangle_1^{\beta,\gamma}$$

$$\leq M\left[\langle u_0 - \tau[I - R_{l,l}(\tau A)]^{-1}\varphi_1^{l,l}\rangle_1^{\beta,\gamma} + \frac{1}{\beta(1-\beta)}\|\varphi_{l,l}^{\tau}\|_{C_0^{\beta,\gamma}(\tau,E)}\right],$$

where M does not depend on τ, β, γ, u_0, and $\varphi_{l,l}^{\tau}$.

The proofs of Theorems 5.5–5.7 follow the scheme of the proof of Theorem 5.2 of Chapter 2.

The authors do not know whether analogues of Theorems 5.3, 5.4, and 5.7 are valid for the difference problem (0.51) in $C_0^{\beta,\gamma}(E)$ if l is odd. However, as in the case of the space $C_0^{\alpha}(E)$, close results can be established in the last case in a space $\tilde{C}_0^{\beta,\gamma}(E)$ that is smaller than $C_0^{\beta,\gamma}(E)$.

Let us consider the difference problem (0.51) generated by the Padé fractions $R_{l,l}(z)$ with l odd. Construct a space $\tilde{C}_0^{\beta,\gamma}(\tau, E)$, where $0 \leq \gamma \leq \beta$, by endowing the space $E(\tau)$ of grid functions with the norm

$$\|\varphi^{\tau}\|_{\tilde{C}_0^{\beta,\gamma}(\tau,E)} = \|\varphi^{\tau}\|_{C(\tau,E)} +$$

$$\sup_{1 \leq k < k+2r \leq N} \tau^{\gamma-\beta}(k+2r)^{\gamma}(2r)^{-\beta}\|\varphi_{k+2r} - \varphi_k\|_E. \tag{5.6}$$

Then, by the definition of the space $\mathcal{C}(E)$ (see Chapter 2), $\tilde{\mathcal{C}}_0^{\beta,\gamma}(E) = \tilde{C}_0^{\beta,\gamma}(\mathcal{C}(E))$. To these spaces there correspond the spaces of traces $\tilde{E}_1^{\beta,\gamma}$, which consist of the elements $w_0 \in E$ for which the norm

$$\langle \tilde{w}_0 \rangle_1^{\beta,\gamma} = \sup_{0 < \tau \leq \tau_0} \left[\max_{1 \leq i \leq N} \|A_{j,l} R_{j,l}^{i-1} w_0\|_E \right.$$

$$+ \sup_{1 \leq i < i+2r \leq N} ((i+2r)\tau)^\gamma (2r\tau)^{-\beta} \|A_{j,l}[R_{j,l}^{i+2r-1}(\tau A) - R_{j,l}^{i-1}(\tau A)] w_0\|_E \right]$$

is finite.

Theorem 5.8. *Suppose condition (2.25) is satisfied. Suppose further that the operator $I + R_{l,l}(\tau A)$ has an inverse and that $\varphi_{l,l}^\tau \in D([I + R_{l,l}(\tau A)]^{-1})$. Then the solutions of the difference problem (0.51) obey the coercivity inequality*

$$\|\mathcal{D}\Pi(u_0)u^\tau\|_{\tilde{C}_0^{\beta,\gamma}(\tau,E)} + \|A_{l,l}u^\tau\|_{\tilde{C}_0^{\beta,\gamma}(\tau,E)} + \max_{1 \leq k \leq N} \langle u_{k-1} - \tau[I - \widetilde{R_{l,l}}(\tau A)]^{-1}\varphi_k^{l,l} \rangle_1^{\beta,\gamma}$$

$$\leq M \left[\langle u_0 - \tau[I - \widetilde{R_{l,l}}(\tau A)]^{-1}\varphi_1^{l,l} \rangle_1^{\beta,\gamma} + \frac{1}{\beta(1-\beta)}\|[I + R_{l,l}(\tau A)]^{-1}\varphi_{l,l}^\tau\|_{\tilde{C}_0^{\beta,\gamma}(\tau,E)} \right],$$

where M does not depend on τ, β, γ, u_0, and $\varphi_{l,l}^\tau$.

Theorem 5.9. *Let $\varphi_{l,l}^\tau \in D(A)$. Then the solutions of problem (0.51) obey the inequality*

$$\|\mathcal{D}\Pi(u_0)u^\tau\|_{\tilde{C}_0^{\beta,\gamma}(\tau,E)} + \|A_{l,l}u^\tau\|_{\tilde{C}_0^{\beta,\gamma}(\tau,E)} + \max_{1 \leq k \leq N} \langle u_{k-1} - \tau[I - \widetilde{R_{l,l}}(\tau A)]^{-1}\varphi_k^{l,l} \rangle_1^{\beta,\gamma}$$

$$\leq M \left[\langle u_0 - \tau[I - \widetilde{R_{l,l}}(\tau A)]^{-1}\varphi_1^{l,l} \rangle_1^{\beta,\gamma} \right.$$

$$\left. + \left(\frac{1}{\beta(1-\beta)} + \ln\frac{1}{\tau} \right) \|(I + \tau A)^{-1}\varphi_{l,l}^\tau\|_{\tilde{C}_0^{\beta,\gamma}(\tau,E)} \right],$$

where M does not depend on τ, β, γ, u_0, and $\varphi_{l,l}^\tau$.

The proofs of Theorems 5.8 and 5.9 follow the scheme of the proofs of Theorems 2.6, 2.7 above and Theorems 5.1, 5.2 of Chapter 2.

The estimates given above show that the exponent γ of the weight factor can be any number in the segment $[0, \beta]$. In particular, this means that problem (0.51) is well-posed in various analogues of Hölder spaces (with respect to time). The Hölder exponent β must lie in $(0,1)$, which does not allow one to establish the well-posedness of problem (0.51) in difference analogues of spaces of continuous functions. These conclusions concern the case of an arbitrary Banach space E. In Chapter 2 we established the well-posedness of the simplest difference scheme (0.6) for $\beta = \gamma = 0$ in various restrictions of the arbitrary space E. It turns out that such a result is valid for the difference problem (0.51) generated by a broad class of Padé fractions. This result relies on Hölder estimates for powers of the operator step.

2.   Estimates of powers of the operator step. The coercivity inequality for the general problem.

Let us establish some smoothness estimates for powers of the operator step in the fractional norm spaces E'_α that will be needed below.

Lemma 5.1. *Let* $j = l - 2,\ l - 1$. *Then the following estimates hold for any* $1 \le k < k + r \le N$:

$$\|R_{j,l}^k(\tau A) - R_{j,l}^{k+r}(\tau A)\|_{E'_{\alpha-\gamma} \to E'_{\alpha-\beta}} \le M(\alpha,\gamma)((k+r)\tau)^{\beta-\gamma}, \qquad (5.7)$$

in which

$$M(\alpha,\gamma) = \begin{cases} \frac{M}{(\alpha-\gamma)(1-\alpha)}, & 0 \le \gamma \le \beta \le 1,\ 0 < \alpha < 1,\ \alpha \ne \gamma, \\ M, & \alpha = \gamma = \beta, \end{cases}$$

where M *does not depend on* $\tau,\ \alpha,\ \beta,\ \gamma,\ k,$ *and* r.

Proof. By the Cauchy-Riesz formula (4.2) of Chapter 2, we have

$$\lambda^{\alpha-\beta} A(\lambda + A)^{-1}[R_{j,l}^k(\tau A) - R_{j,l}^{k+r}(\tau A)]x$$

$$= \frac{\lambda^{\alpha-\beta}}{2\pi i} \int_{S_1 \cup S_2} \frac{1}{\lambda + z}[R_{j,l}^k(\tau z) - R^{k+r}(\tau z)]A(z - A)^{-1}x\,dz. \qquad (5.8)$$

Using this identity and the estimate (4.1) of Chapter 2, we obtain

$$\lambda^{\alpha-\beta}\|A(\lambda + A)^{-1}[R_{j,l}^k(\tau A) - R_{j,l}^{k+r}(\tau A)]x\|_E$$

$$\le M \int_0^\infty \frac{\lambda^{\alpha-\beta}}{(\lambda+\rho)\rho^{\alpha-\beta}}\rho^{-(\beta-\gamma)}|R_{j,l}^k(\tau\rho e^{i\phi}) - R_{j,l}^{k+r}(\tau\rho e^{i\phi})|\rho^{\alpha-\gamma}\|A(\rho+A)^{-1}x\|_E d\rho$$

$$\leq \|x\|'_{\alpha-\gamma} \int_0^\infty \frac{\lambda^{\alpha-\beta}}{(\lambda+\rho)\rho^{\alpha-\beta}} \rho^{-(\beta-\gamma)} |R_{j,l}^k(\tau\rho e^{i\phi}) - R_{j,l}^{k+r}(\tau\rho e^{i\phi})| d\rho. \qquad (5.9)$$

From this and the estimate (0.32) with $\delta = \beta - \gamma$ it follows that

$$\lambda^{\alpha-\beta} \|A(\lambda+A)^{-1}[R_{j,l}^k(\tau A) - R_{j,l}^{k+r}(\tau A)]x\|_E$$

$$\leq M \int_0^\infty \frac{\lambda^{\alpha-\beta} d\rho}{(\lambda+\rho)\rho^{\alpha-\beta}} ((k+r)\tau)^{\beta-\gamma} \|x\|'_{\alpha-\gamma} \leq \frac{M}{(\alpha-\beta)(1-\alpha+\beta)} \|x\|'_{\alpha-\gamma}$$

for all $\lambda > 0$. Consequently,

$$\|[R_{j,l}^k(\tau A) - R_{j,l}^{k+r}(\tau A)]x\|'_{\alpha-\beta} \leq \frac{M}{(\alpha-\beta)(1-\alpha+\beta)} \|x\|'_{\alpha-\gamma}. \qquad (5.10)$$

On the other hand, using (0.10), (0.18), (0.29), (0.32), and the inequality

$$\frac{\lambda^{\alpha-\beta}\rho^{1-\alpha+\beta}}{\lambda+\rho} \leq 1,$$

we obtain the estimate

$$\lambda^{\alpha-\beta} \|A(\lambda+A)^{-1}[R_{j,l}^k(\tau A) - R_{j,l}^{k+r}(\tau A)]x\|_E$$

$$\leq M \int_0^\infty \frac{1}{\rho^{1+\beta-\gamma}} |R_{j,l}^k(\tau\rho e^{i\phi}) - R_{j,l}^{k+r}(\tau\rho e^{i\phi})| d\rho \, \|x\|'_{\alpha-\gamma}$$

$$= M \left(\int_0^{\frac{1}{(k+r)\tau}} + \int_{\frac{1}{(k+r)\tau}}^{\frac{1}{k\tau}} + \int_{\frac{1}{k\tau}}^\infty \right) \frac{1}{\rho^{1+\beta-\gamma}} |R_{j,l}^k(\tau\rho e^{i\phi}) - R_{j,l}^{k+r}(\tau\rho e^{i\phi})| d\rho \, \|x\|'_{\alpha-\gamma} \leq$$

$$M_1 \left[\int_0^{\frac{1}{(k+r)\tau}} d\rho \, ((k+r)\tau)^{1+\beta-\gamma} + \int_{\frac{1}{(k+r)\tau}}^{\frac{1}{k\tau}} \frac{d\rho}{\rho^{1+\beta-\gamma}} + \int_{\frac{1}{k\tau}}^\infty \frac{d\rho}{\rho^2} \frac{1}{(k\tau)^{1-\beta+\gamma}} \right] \|x\|'_{\alpha-\gamma}$$

$$\leq \frac{M_2}{\beta-\gamma} ((k+r)\tau)^{1+\beta-\gamma} \|x\|'_{\alpha-\gamma}.$$

Thus, we have shown that

$$\lambda^{\alpha-\beta} \|A(\lambda+A)^{-1}[R_{j,l}^k(\tau A) - R_{j,l}^{k+r}(\tau A)]x\|_E \leq \frac{M_2}{\beta-\gamma} ((k+r)\tau)^{\beta-\gamma} \|x\|'_{\alpha-\gamma}$$

for any $\lambda > 0$. Therefore,

$$\|[R_{j,l}^k(\tau A) - R_{j,l}^{k+r}(\tau A)]x\|'_{\alpha-\beta} \leq \frac{M_2}{\beta-\gamma} ((k+r)\tau)^{\beta-\gamma} \|x\|'_{\alpha-\gamma}.$$

From this and (5.10) we obtain the estimates (5.7) for $0 \leq \gamma \leq \beta \leq \alpha$, $\alpha \neq \gamma$. For $\alpha = \gamma$ the estimate (5.6) follows from the triangle inequality. Lemma 5.1 is proved.

Lemma 5.2. *Let $j = l - 2$, $l - 1$. Then the following estimates hold for any $1 \leq k < k + r \leq N$:*

$$\|R_{j,l}^k(\tau A) - R_{j,l}^{k+r}(\tau A)\|_{E'_{\alpha-\gamma} \to E'_{\alpha-\beta}} \leq \frac{Mr\tau}{(k\tau)^{1+\gamma-\beta}}, \quad 0 \leq \gamma \leq \beta \leq \alpha, \ 0 < \alpha < 1,$$

(5.11)

where M does not depend on τ, α, β, γ, k, and r.

Proof. Using the inequalities (0.10), (0.18), (0.29) and the estimates (0.32), we obtain

$$\lambda^{\alpha-\beta}\|A(\lambda + A)^{-1}[R_{j,l}^k(\tau A) - R_{j,l}^{k+r}(\tau A)]x\|_E$$

$$\leq M\left[\int_0^{\frac{1}{k\tau}} \frac{1}{\rho^{1+\beta-\gamma}}|R_{j,l}^k(\tau \rho e^{i\phi}) - R_{j,l}^{k+r}(\tau \rho e^{i\phi})|d\rho\right.$$

$$\left. + \int_{\frac{1}{k\tau}}^\infty \frac{1}{\rho^2}\rho^{1+\beta-\gamma}|R_{j,l}^k(\tau \rho e^{i\phi}) - R_{j,l}^{k+r}(\tau \rho e^{i\phi})|d\rho\right]\|x\|'_{\alpha-\gamma}$$

$$\leq M_1\left[\int_0^{\frac{1}{k\tau}} d\rho \frac{r\tau}{(k\tau)^{\gamma-\beta}} + \int_{\frac{1}{k\tau}}^\infty \frac{d\rho}{\rho^2}\frac{r\tau}{(k\tau)^{2+\gamma-\beta}}\right]\|x\|'_{\alpha-\gamma} \leq M_1 \frac{2r\tau}{(k\tau)^{1-\beta+\gamma}}\|x\|'_{\alpha-\gamma}.$$

Thus, we have proved that

$$\lambda^{\alpha-\beta}\|A(\lambda + A)^{-1}[R_{j,l}^k(\tau A) - R_{j,l}^{k+r}(\tau A)]x\|_E \leq M\frac{r\tau}{(k\tau)^{1-\beta+\gamma}}\|x\|'_{\alpha-\gamma}$$

for any $\lambda > 0$. This yields the estimates (5.11). Lemma 5.2 is proved.

Lemma 5.3. *Let $j = l$. Then the following estimates hold for all $1 \leq k < k + r \leq N$:*

$$\|[R_{l,l}^k(\tau A) - R_{l,l}^{k+r}(\tau A)](I + \tau A)^{-2}\|_{E'_{\alpha-\gamma} \to E'_{\alpha-\beta}} \leq M(\alpha, \gamma)((k+r)\tau)^{\beta-\gamma}, \quad (5.12)$$

in which

$$M(\alpha, \gamma) = \begin{cases} \frac{M}{(\alpha-\gamma)(1-\alpha)}, & 0 \leq \gamma \leq \beta \leq \alpha, \ 0 < \alpha < 1, \ \alpha \neq \gamma, \\ M, & \alpha = \beta = \gamma, \end{cases}$$

where M does not depend on τ, α, β, γ, k, and r.

Lemma 5.4. *Let* $j = l$. *Then the following estimates hold for all* $1 \leq k < k+r \leq N$:

$$\|[R_{l,l}^{k}(\tau A) - R_{l,l}^{k+r}(\tau A)](I + \tau A)^{-(2+\theta)}\|_{E'_{\alpha-\gamma} \to E'_{\alpha-\beta}} \leq M \frac{r\tau}{(k\tau)^{1-\beta+\gamma}},$$

$$0 \leq \gamma \leq \beta \leq \alpha, \ 0 < \alpha < 1, \tag{5.13}$$

where M *does not depend on* τ, α, β, γ, k, *and* r. *Here* $\theta = 0$ *if* l *is even and* $\theta = 1$ *if* l *is odd.*

The proofs of Lemmas 5.3 and 5.4 follow the scheme of the proofs of Lemmas 5.1 and 5.2, and rely on the estimates (1.25), (1.26), and (0.32).

Now let us study the problem (0.51) in the spaces $C_0^{\beta,\gamma}(E_{\alpha-\beta})$, $0 \leq \gamma \leq \beta \leq \alpha$, $0 < \alpha < 1$. To these spaces there correspond the spaces of traces $E_{1+\alpha-\beta}^{\beta,\gamma}$, which consist of all elements $w_0 \in E$ for which the norm

$$\langle w_0 \rangle_{1+\alpha-\gamma}^{\beta,\gamma} = \sup_{0 < \tau \leq \tau_0} \left[\max_{1 \leq i \leq N} \|A_{j,l} R_{j,l}^{i-1}(\tau A) w_0\|'_{\alpha-\beta} \right.$$

$$\left. + \sup_{1 \leq i < i+r \leq N} ((i+r)\tau)^{\gamma}(r\tau)^{-\beta}\|A_{j,l}[R_{j,l}^{i+r-1}(\tau A) - R_{j,l}^{i-1}(\tau A)]w_0\|'_{\alpha-\beta} \right]. \tag{5.14}$$

is finite.

Theorem 5.10. *Let* $j = l-1$, $l-2$ *and* $u_0 - \tau[I - R_{j,l}(\tau A)]^{-1}\varphi_1^{j,l} \in E_{1+\alpha-\beta}^{\beta,\gamma}$. *Then the solutions of the difference problem* (0.51) *obey the coercivity inequality*

$$\|\mathcal{D}\Pi(u_0)u^{\tau}\|_{C_0^{\beta,\gamma}(\tau, E'_{\alpha-\beta})} + \|A_{j,l}u^{\tau}\|_{C_0^{\beta,\gamma}(\tau, E'_{\alpha-\beta})}$$

$$+ \max_{1 \leq k \leq N} \langle u_{k-1} - \tau[I - R_{j,l}(\tau A)]^{-1}\varphi_k^{j,l} \rangle_{1+\alpha-\beta}^{\beta,\gamma}$$

$$\leq M \left[\langle u_0 - \tau[I - R_{j,l}(\tau A)]^{-1}\varphi_1^{j,l} \rangle_{1+\alpha-\beta}^{\beta,\gamma} + \frac{1}{\alpha(1-\alpha)}\|\varphi_{j,l}^{\tau}\|_{C_0^{\beta,\gamma}(\tau, E'_{\alpha-\beta})} \right], \tag{5.15}$$

where M *does not depend on* τ, α, β, γ, u_0 *and* $\varphi_{j,l}^{\tau}$.

Theorem 5.11. *Let* $j = l$ *and* $u_0 - \tau[I - R_{l,l}(\tau A)]^{-1}\varphi_1^{l,l} \in E_{1+\alpha-\beta}^{\beta,\gamma}$, $\varphi_{l,l}^{\tau} \in D(A^2)$. *Then the solutions of the difference problem* (0.49) *obey the coercivity inequality*

$$\|\mathcal{D}\Pi(u_0)u^{\tau}\|_{C_0^{\beta,\gamma}(\tau, E'_{\alpha-\beta})} + \|A_{l,l}u^{\tau}\|_{C_0^{\beta,\gamma}(\tau, E'_{\alpha-\beta})}$$

$$+ \max_{1 \leq k \leq N} \langle u_{k-1} - \tau[I - R_{l,l}(\tau A)]^{-1} \varphi_k^{l,l} \rangle_{1+\alpha-\beta}^{\beta,\gamma}$$

$$\leq M \left[\langle u_0 - \tau[I - R_{l,l}(\tau A)]^{-1} \varphi_1^{l,l} \rangle_{1+\alpha-\beta}^{\beta,\gamma} \right.$$

$$\left. + \frac{1}{\alpha(1-\alpha)} \|(I + \tau A)^2 \varphi_{l,l}^\tau\|_{C_0^{\beta,\gamma}(\tau, E'_{\alpha-\beta})} \right],$$

where M does not depend on τ, α, β, γ, u_0 and $\varphi_{l,l}^\tau$.

Theorem 5.12. *Let $j = l$ be an even number and $u_0 - \tau[I - R_{l,l}(\tau A)]^{-1} \varphi_1^{l,l} \in E_{1+\alpha-\beta}^{\beta,\gamma}$. Then the solutions of the difference problem (0.49) obey the following inequalities:*

1) *If $\varphi_{l,l}^\tau \in D(A)$, then*

$$\|\mathcal{D}\Pi(u_0)u^\tau\|_{C_0^{\beta,\gamma}(\tau, E'_{\alpha-\beta})} + \|A_{l,l}u^\tau\|_{C_0^{\beta,\gamma}(\tau, E'_{\alpha-\beta})}$$

$$+ \max_{1 \leq k \leq N} \langle u_{k-1} - \tau[I - R_{l,l}(\tau A)]^{-1} \varphi_k^{l,l} \rangle_{1+\alpha-\beta}^{\beta,\gamma}$$

$$\leq M \left[\langle u_0 - \tau[I - R_{l,l}(\tau A)]^{-1} \varphi_1^{l,l} \rangle_{1+\alpha-\beta}^{\beta,\gamma} + \frac{1}{\alpha(1-\alpha)} \|(I + \tau A)\varphi_{l,l}^\tau\|_{C_0^{\beta,\gamma}(\tau, E'_{\alpha-\beta})} \right].$$

2) *If the difference problem (0.51) is stable, then*

$$\|\mathcal{D}\Pi(u_0)u^\tau\|_{C_0^{\beta,\gamma}(\tau, E'_{\alpha-\beta})} + \|A_{l,l}u^\tau\|_{C_0^{\beta,\gamma}(\tau, E'_{\alpha-\beta})}$$

$$+ \max_{1 \leq k \leq N} \langle u_{k-1} - \tau[I - R_{l,l}(\tau A)]^{-1} \varphi_k^{l,l} \rangle_{1+\alpha-\beta}^{\beta,\gamma}$$

$$\leq M \left[\langle u_0 - \tau[I - R_{l,l}(\tau A)]^{-1} \varphi_1^{l,l} \rangle_{1+\alpha-\beta}^{\beta,\gamma} + \frac{1}{\alpha(1-\alpha)} \|\varphi_{l,l}^\tau\|_{C_0^{\beta,\gamma}(\tau, E'_{\alpha-\beta})} \right].$$

Here M does not depend on τ, α, β, γ, u_0 and $\varphi_{l,l}^\tau$.

The proofs of Theorems 5.10–5.12 follow the scheme of the proofs of the corresponding problems in Section 2 of this chapter and Section 5 of Chapter 2.

Now let us consider the difference problem (0.51) generated by the Padé fractions $R_{l,l}(z)$ with l odd. We shall study it in the spaces $\tilde{C}_0^{\beta,\gamma}(E_{\alpha-\beta})$, $0 \leq \gamma \leq \beta \leq \alpha$, $0 < \alpha < 1$. To them there correspond the spaces of traces $\tilde{E}_{1+\alpha-\beta}^{\beta,\gamma}$, which consist of all elements $w_0 \in E$ for which the norm

$$\langle \widetilde{w_0} \rangle_{1+\alpha-\beta}^{\beta,\gamma} = \sup_{0 < \tau \leq \tau_0} \left[\max_{1 \leq i \leq N} \|A_{l,l} R_{l,l}^{i-1}(\tau A) w_0\|'_{\alpha-\beta} \right.$$

$$+ \sup_{1 \le i < i+2r \le N} ((i+2r)\tau)^\gamma (2r\tau)^{-\beta} \|A_{l,l}[R_{l,l}^{i+2r-1}(\tau A) - R_{l,l}^{i-1}(\tau A)]w_0\|_{\alpha-\beta}'\Big].$$

is finite.

Theorem 5.13. *Suppose that the stability condition (condition (2.25)) is satisfied. Suppose further that the operator $I + R_{l,l}(\tau A)$ has an inverse and $\varphi_{l,l}^\tau \in D([I + R_{l,l}(\tau A)]^{-1})$. Then the solutions of the difference problem (0.49) obey the coercivity inequality*

$$\|\mathcal{D}\Pi(u_0)u^\tau\|_{\tilde{C}_0^{\beta,\gamma}(\tau, E_{\alpha-\beta}')} + \|A_{l,l}u^\tau\|_{\tilde{C}_0^{\beta,\gamma}(\tau, E_{\alpha-\beta}')}$$

$$+ \max_{1 \le k \le N} \langle u_{k-1} - \tau[I - \widetilde{R_{l,l}(\tau A)}]^{-1}\varphi_k^{l,l}\rangle_{1+\alpha-\beta}^{\beta,\gamma}$$

$$\le M \Bigg[\langle u_0 - \tau[I - \widetilde{R_{l,l}(\tau A)}]^{-1}\varphi_1^{l,l}\rangle_{1+\alpha-\beta}^{\beta,\gamma}$$

$$+ \frac{1}{\alpha(1-\alpha)}\|[I + R_{l,l}(\tau A)]^{-1}\varphi_{l,l}^\tau\|_{\tilde{C}_0^{\beta,\gamma}(\tau, E_{\alpha-\beta}')} \Bigg],$$

where M does not depend on τ, α, β, γ, u_0 and $\varphi_{l,l}^\tau$.

Theorem 5.14. *Let $\varphi_{l,l}^\tau \in D(A)$. Then the solutions of the difference problem (0.49) obey the coercivity inequality*

$$\|\mathcal{D}\Pi(u_0)u^\tau\|_{\tilde{C}_0^{\beta,\gamma}(\tau, E_{\alpha-\beta}')} + \|A_{l,l}u^\tau\|_{\tilde{C}_0^{\beta,\gamma}(\tau, E_{\alpha-\beta}')}$$

$$+ \max_{1 \le k \le N} \langle u_{k-1} - \tau[I - \widetilde{R_{l,l}(\tau A)}]^{-1}\varphi_k^{l,l}\rangle_{1+\alpha-\beta}^{\beta,\gamma}$$

$$\le M \Bigg[\langle u_0 - \tau[I - \widetilde{R_{l,l}(\tau A)}]^{-1}\varphi_1^{l,l}\rangle_{1+\alpha-\beta}^{\beta,\gamma}$$

$$+ \left(\frac{1}{\alpha(1-\alpha)} + \ln\frac{1}{\tau}\right)\|(I + \tau A)\varphi_{l,l}^\tau\|_{\tilde{C}_0^{\beta,\gamma}(\tau, E_{\alpha-\beta}')} \Bigg],$$

where M does not depend on τ, α, β, γ, u_0 and $\varphi_{l,l}^\tau$.

The definition of the spaces $\tilde{E}_{1+\alpha-\beta}^{\beta,\gamma}$ is quite complicated. If $j = l-1$, $l-2$ or if $j = l$ and l is even, then from (5.7) and (5.11)–(5.13) it follows that $Aw_0 \in E_{\alpha-\gamma}'$ implies $w_0 \in \tilde{E}_{1+\alpha-\beta}^{\beta,\gamma}$ and we have the inequality

$$\langle \widetilde{w_0}\rangle_{1+\alpha-\beta}^{\beta,\gamma} \le M(\alpha,\gamma)\|Aw_0\|_{\alpha-\gamma}',$$

where

$$M(\alpha, \gamma) = \begin{cases} \frac{M}{(\alpha - \gamma)(1 - \alpha)}, & 0 \leq \gamma \leq \beta \leq \alpha, \ 0 < \alpha < 1, \ \alpha \neq \gamma, \\ \frac{M}{\alpha(1 - \alpha)}, & \alpha = \beta = \gamma. \end{cases} \tag{5.16}$$

We not able to establish the equivalence of these two norms. Nevertheless, we have the following result.

Theorem 5.15. *Let* $j = l - 2$, $l - 1$ *or* $j = l$ *and* l *be an even number. Let* $\tau^{-1}[I - R_{j,l}(\tau A)]u_0 - \varphi_1^{j,l} \in E'_{\alpha - \gamma}$. *Then the solutions of the difference problem* (0.49) *obey the coercivity inequality*

$$\|\mathcal{D}\Pi(u_0)u^\tau\|_{C_0^{\beta,\gamma}(\tau, E'_{\alpha-\beta})} + \|A_{j,l}u^\tau\|_{C_0^{\beta,\gamma}(\tau, E'_{\alpha-\beta})} + \max_{1 \leq k \leq N} \|A_{j,l}u_{k-1} - \varphi_k^{j,l}\|'_{\alpha-\gamma}$$

$$\leq M(\alpha, \gamma) \left[\|A_{j,l}u_0 - \varphi_1^{j,l}\|'_{\alpha-\gamma} + \|\varphi_{j,l}^\tau\|_{C_0^{\beta,\gamma}(\tau, E'_{\alpha-\beta})} \right],$$

in which $M(\alpha, \gamma)$ *is given by* (5.16), *where* M *does not depend on* τ, α, β, γ, u_0 *and* $\varphi_{j,l}^\tau$.

The proof of Theorem 5.15 follows the scheme of the proof of Theorem 5.4 of Chapter 2.

If $j = l$ and l is an odd number, then by (5.12) and (5.13) we have

$$\langle \widetilde{w_0} \rangle_{1+\alpha-\gamma}^{\beta,\gamma} \leq M(\alpha, \gamma) \|(I + \tau A)Aw_0\|'_{\alpha-\gamma}$$

whenever $A^2 w_0 \in E'_{\alpha-\gamma}$, where $M(\alpha, \gamma)$ is given by (5.16). We have the following results.

Theorem 5.16. *Suppose* $j = l$, l *is an odd number, and* $(I + \tau A)^{-1}\{\tau^{-1}[I - R_{l,l}(\tau A)]u_0 - \varphi_1^{l,l}\} \in E'_{\alpha-\gamma}$. *Suppose further that the operator* $I + R_{l,l}(\tau A)$ *has an inverse and* $\varphi_{l,l}^\tau \in D([I + R_{l,l}(\tau A)]^{-1})$. *Then the solutions of the difference problem* (0.49) *obey the coercivity inequality*

$$\|\mathcal{D}\Pi(u_0)u^\tau\|_{\tilde{C}_0^{\beta,\gamma}(\tau, E'_{\alpha-\beta})} + \|A_{l,l}u^\tau\|_{\tilde{C}_0^{\beta,\gamma}(\tau, E'_{\alpha-\beta})} + \max_{1 \leq k \leq N} \|A_{l,l}u_{k-1} - \varphi^{l,l}\|'_{\alpha-\gamma}$$

$$\leq M(\alpha, \gamma) \left[\|A_{l,l}u_0 - \varphi_1^{l,l}\|'_{\alpha-\gamma} + \|[I + R_{l,l}(\tau A)]^{-1}\varphi_{l,l}^\tau\|_{\tilde{C}_0^{\beta,\gamma}(\tau, E'_{\alpha-\beta})} \right],$$

in which $M(\alpha, \gamma)$ *is given by* (5.16) *where* M *does not depend on* τ, α, β, γ, u_0 *and* $\varphi_{l,l}^\tau$.

Theorem 5.17. *Suppose $j = l$, l is an odd number, and $\varphi_{l,l}^\tau \in D(A)$. Then the solutions of the difference problem* (0.49) *obey the coercivity inequality*

$$\|\mathcal{D}\Pi(u_0)u^\tau\|_{\tilde{C}_0^{\beta,\gamma}(\tau,E'_{\alpha-\beta})} + \|A_{l,l}u^\tau\|_{\tilde{C}_0^{\beta,\gamma}(\tau,E'_{\alpha-\beta})} + \max_{1\le k\le N}\|A_{l,l}u_{k-1} - \varphi_k^{l,l}\|'_{\alpha-\gamma}$$

$$\le M(\alpha,\gamma)\left[\|A_{l,l}u_0 - \varphi_1^{l,l}\|'_{\alpha-\gamma} + \left(\frac{1}{\alpha(1-\alpha)} + \ln\frac{1}{\tau}\right)\|(I+\tau A)\varphi_{l,l}^\tau\|_{\tilde{C}_0^{\beta,\gamma}(\tau,E'_{\alpha-\beta})}\right],$$

in which $M(\alpha,\gamma)$ is given by (5.16), *where M does not depend on τ, α, β, γ, u_0 and $\varphi_{l,l}^\tau$.*

CHAPTER 4

DIFFERENCE SCHEMES FOR PARABOLIC EQUATIONS

1. ELLIPTIC DIFFERENCE OPERATORS WITH CONSTANT COEFFICIENTS

1. The definition of an elliptic difference operator and properties of its symbol.

Let us consider a differential operator with constant coefficients of the form

$$A = \sum_{r=|m|} a_r \frac{\partial^{r_1 + \cdots + r_n}}{\partial x_1^{r_1} \cdots \partial x_n^{r_n}}, \tag{1.1}$$

acting on functions defined on the entire space $\mathbf{R}^n$. Here $r \in \mathbf{R}^n$ is a vector with nonnegative integer components, $|r| = r_1 + \cdots + r_n$. If $\varphi(y)$ $(y = (y_1, \cdots, y_n) \in \mathbf{R}^n)$ is an infinitely differentiable function that decays at infinity together with all its derivatives, then by means of the Fourier transformation one establishes the equality

$$\mathcal{F}(A\varphi)(\xi) = A(\xi)\mathcal{F}(\varphi)(\xi).$$

Here the Fourier transform operator is defined by the rule

$$\mathcal{F}(\varphi)(\xi) = (2\pi)^{-n/2} \int_{\mathbf{R}^n} \exp\{-i(y, \xi)\}\varphi(y)dy,$$

$$(y, \xi) = y_1\xi_1 + \cdots + y_n\xi_n.$$

The function $A(\xi)$ is called the *symbol* of the operator A and is given by

$$A(\xi) = \sum_{|r|=m} a_r (i\xi)^{r_1} \cdots (i\xi)^{r_n}. \tag{1.2}$$

We shall assume that

$$0 < M_1 |\xi|^m \le |A(\xi)| \le M_2 |\xi|^m < \infty \tag{1.3}$$

for $\xi \ne 0$. As is known, if $n \ge 2$, then m is necessarily even.

Let us define the grid space $\mathbf{R}_h^n$ $(0 < h \le h_0)$ as the set of all points of the Euclidean space $\mathbf{R}^n$ whose coordinates are given by

$$x_k = s_k h, \quad s_k = 0, \pm 1, \pm 2, \cdots, \quad k = 1, \cdots, n. \tag{1.4}$$

The number h is called the *step* of the grid space. A function defined on $\mathbf{R}_h^n$ will be called a *grid function*. To the operator A we assign the difference operator with constant coefficients

$$A_h = h^{-m} \sum_{m \le |s| \le S} b_s \Delta_{1-}^{s_1} \Delta_{1+}^{s_2} \cdots \Delta_{n-}^{s_{2n-1}} \Delta_{n+}^{s_{2n}}, \tag{1.5}$$

which acts on functions defined on the entire space $\mathbf{R}_h^n$. Here $s \in \mathbf{R}^{2n}$ is a vector with nonnegative integer coordinates,

$$\Delta_{k\pm} f^h(x) = \pm (f^h(x \pm e_k h) - f^h(x)), \tag{1.6}$$

and e_k is the unit vector of the axis x_k.

An infinitely differentiable function of the continuous argument $y \in \mathbf{R}^n$ that is continuous and bounded together with all its derivatives is said to be *smooth*. Let $\varphi(y)$ be a smooth function on $\mathbf{R}^n$. Using the Taylor expansion of $\varphi(y)$, one can show that

$$\sup_{x \in \mathbf{R}_h^n} \left| h^{-1} \Delta_{k\pm} \varphi(x) - \frac{\partial}{\partial y_k} \varphi(x) \right| \le M(\varphi) h. \tag{1.7}$$

Here the grid functions $\varphi(x)$ and $\frac{\partial}{\partial y_k} \varphi(x)$ are the traces of the functions $\varphi(y)$ and $\frac{\partial}{\partial y_k} \varphi(y)$, respectively. The last inequality means that the difference operator $h^{-1} \Delta_{k\pm}$ is a first-order approximation for the differential operator $\frac{\partial}{\partial y_k}$.

We say that the difference operator A_h is *a t-th order $(t > 0)$ approximation of the differential operator A* if the inequality

$$\sup_{x \in \mathbf{R}_h^n} |A_h \varphi(x) - A\varphi(x)| \le M(\varphi) h^t \tag{1.8}$$

holds for any smooth function $\varphi(y)$. We shall assume that the operator A_h approximates the differential operator A with any prescribed order.

A function of a continuous [resp., discrete] argument that decays at infinity faster than any negative power of $|y|$ [resp., $|x|$] is said to be *rapidly decreasing*. Let us define the Fourier transform of a grid function $f^h(x)$ by the formula

$$\tilde{f}(\xi) = (2\pi)^{-n} \sum_{x \in \mathbf{R}_h^n} \exp\{-i(x,\xi)\} f^h(x) h^n, \quad \xi \in \mathbf{R}^n. \tag{1.9}$$

This formula defines a $2\pi h^{-1}$-periodic smooth function of the continuous argument ξ whenever $f^h(x)$ is a rapidly decreasing grid function. In this last case (1.9) is just the Fourier series expansion of the function $\tilde{f}(\xi)$, and the numbers $f^h(x)$ are the Fourier coefficients, given by the formula

$$f^h(x) = \int_{|\xi_1| \le \pi h^{-1}} \cdots \int_{|\xi_n| \le \pi h^{-1}} \exp\{i(x,\xi)\} \tilde{f}(\xi) d\xi_1 \cdots d\xi_n. \tag{1.10}$$

The inverse Fourier transform of a $2\pi h^{-1}$-periodic function $\varphi(\xi)$ is defined to be the grid function $\hat{\varphi}^h(x)$ given by the formula

$$\hat{\varphi}^h(x) = \int_{|\xi_1| \le \pi h^{-1}} \cdots \int_{|\xi_n| \le \pi h^{-1}} \exp\{i(x,\xi)\} \varphi(\xi) d\xi_1 \cdots d\xi_n. \tag{1.11}$$

Formulas (1.10) and (1.11) establish a one-to-one correspondence between rapidly decreasing grid functions and smooth periodic functions of a continuous argument. In particular, if $f^h(x)$ is a rapidly decreasing grid function, then

$$\widehat{[\tilde{f}]}^h(x) = f^h(x).$$

If $f^h(x)$ is a rapidly decreasing grid function, then the grid function $A_h f^h(x)$ is also rapidly decreasing. Hence, the Fourier transform of $A_h f^h(x)$ exists and is given by (1.9), and we have the equality

$$\tilde{A}_h f(\xi) = A(\xi h, h) \tilde{f}(\xi). \tag{1.12}$$

The function $A(\xi h, h)$ is obtained by replacing the operator $\Delta_{k\pm}$ in the right-hand side of equality (1.5) with the expression $\pm(\exp\{\pm i\xi_k h\} - 1)$, respectively, and is called the *symbol of the difference operator*. Since $\exp\{\pm i\xi_k h\}$ are bounded analytic $2\pi h^{-1}$-periodic functions, the symbol $A(\xi h, h)$ is a bounded analytic $2\pi h^{-1}$-periodic function. It follows that for large $|\xi|$ one has the estimate

$$|A(\xi h, h)| \le M(h) |\xi|^m, \quad |\xi|^2 = |\xi_1|^2 + \cdots + |\xi_n|^2.$$

Let us show that such an estimate holds for all ξ such that $|\xi_k h| \leq \pi$.

Lemma 1.1. *The following estimate holds for $\xi \in \mathbf{R}^n$ such that $|\xi_k h| \leq \pi$:*

$$|A(\xi h, h)| \leq M|\xi|^m, \tag{1.13}$$

where M does not depend on h.

Proof. Consider the function

$$B(w) = \sum_{m \leq |s| \leq S} b_s (1 - \exp\{-i w_1\})^{s_1} (\exp\{i w_1\} - 1)^{s_2} \times$$

$$\times \cdots (1 - \exp\{-i w_n\})^{s_{2n-1}} (\exp\{i w_n\} - 1)^{s_{2n}}, \quad |w_k| \leq \pi. \tag{1.14}$$

Then $A(\xi h, h) = h^{-m} B(\xi h)$. Note that $B(w)$ is a polynomial in the functions $\pm(\exp\{\pm i w_k\} - 1)$. We have

$$|\exp\{\pm i w_k\} - 1| \leq |w_k|, \tag{1.15}$$

$$|w_1|^{s_1 + s_2} \cdots |w_n|^{s_{2n-1} + s_{2n}} \leq |w_1^2 + \cdots + w_n^2|^{|s|/2} = |w|^{|s|}. \tag{1.16}$$

Consequently, for small $\varepsilon_0 > 0$ and $|w_k| \leq \varepsilon_0$ the following estimate holds:

$$|B(w)| \leq M|w|^m. \tag{1.17}$$

For $\varepsilon_0 \leq |w_k| \leq \pi$ (1.17) holds thanks to the boundedness of the function $B(w)$; moreover, M does not depend on w. Since $|\xi_k h| \leq \pi$, we have

$$|A(\xi h, h)| \leq h^{-m} |B(\xi h)| \leq M h^{-m} |\xi h|^m = M|\xi|^m,$$

where M does not depend on h. Lemma 1.1 is proved.

In the sequel we shall be interested in the behavior of the function $A(\xi h, h)$ not only for real ξ, but also for complex z. Let us consider the domain in the complex n-dimensional space $\mathbf{C}^n$ given by

$$\Omega = \{z : z_k = \xi_k + i \eta_k, \ |\xi_k h| \leq \pi, \ |\eta_k h| \leq b, \ k = 1, \cdots, n\},$$

where $b > 0$, and the function of complex variables $B(w + i\theta)$ for $|w_k| \leq \pi$ and $|\theta_k| \leq b$. Using the complex analogues of the inequalities (1.15), (1.16) and the boundedness of $B(w + i\theta)$, we obtain the estimate

$$|B(w + i\theta)| \leq M(b)|w + i\theta|^m.$$

Hence, for $z \in \Omega$ we have

$$|A(zh, h)| \leq M(b)|z|^m, \quad |z|^2 = |z_1|^2 + \cdots + |z_n|^2. \tag{1.18}$$

This means that in complex space the symbol of the difference operator obeys an estimate of the same form as that for the symbol of the operator A.

However, while the function $A(\xi)$ is homogeneous, the function $A(\xi h, h)$ does not enjoy this property. Homogeneity is convenient since then $\arg A(t\xi) = \arg A(\xi)$ for all $t > 0$. Hence, if $A(\xi)$ is separated from zero uniformly in ξ, then in the complex space the function $\arg A(tz)$ will be jointly uniformly continuous. The separation of $A(\xi)$ from zero for $|\xi| \geq \xi_0$ guarantees a lower bound for the symbol of the operator A. It turns out that if one requires an analogous lower bound for the modulus of the function $A(\xi h, h)$, then under some constraints on the domain of variation of the arguments the lower bound is preserved when we pass to the complex space. In this case the function $\arg A(tzh, h)$ will be jointly uniformly continuous in the variables t and z.

Lemma 1.2. *Suppose $|A(\xi h, h)| \geq M|\xi|^m$ if $|\xi_k h| \leq \pi$, where $M > 0$ does not depend on h. Let*

$$\Omega = \{z : \; z_k = \xi_k + i\eta_k, \; 0 < \xi_0 \leq |\xi_k| \leq \pi h^{-1}, \; |\eta_k| \leq a, \; k = 1, \cdots, n\},$$

where $a > 0$ is sufficiently small. Then $|A(zh, h)| \geq M_1|z|^m$ for all $z \in \Omega$, where $M_1 > 0$ does not depend on h.

Proof. For $z \in \Omega$ and all $0 < a \leq b$ we have $\delta|z_k| \leq |\xi_k| \leq |z_k|$, where $\delta = \delta(b) > 0$ does not depend on h. Hence,

$$\bar{\delta}|z| \leq |\xi| \leq |z|, \quad \bar{\delta} = \bar{\delta}(b, n) > 0. \tag{1.19}$$

Consider the function $\frac{\partial}{\partial w_k} B(w)$ for $|w_k| \leq \pi$. Using the inequality

$$\left| \frac{\partial}{\partial w_k} (\exp\{-iw_k\} - 1)^r \right| \leq M(r)|\exp\{-iw_k\} - 1|^{r-1} \leq M(r)|w_k|^{r-1},$$

the inequality (1.16), and the boundedness of the function $\frac{\partial}{\partial w_k} B(w)$, and proceeding as in the proof of Lemma 1.1, we obtain the estimate

$$\left| \frac{\partial}{\partial w_k} B(w + i\theta) \right| \leq \overline{M}|w + i\theta|^{m-1}, \quad \theta_k = \eta_k h, \quad \overline{M} = \overline{M}(b) > 0. \tag{1.20}$$

Consider the function $\varphi(t) = B(w + it\theta)$. Since $B(w + i\theta)$ is an entire function,

$$\varphi(1) - \varphi(0) = \int_0^1 \varphi'(t)dt = \int_0^1 \sum_{k=1}^n \frac{\partial}{\partial w_k} B(w + it\theta)i\theta_k dt.$$

Using the assumptions of the lemma and the estimate (1.18), we obtain

$$|B(w + i\theta)| \geq |B(w)| - \sum_{k=1}^n \int_0^1 \left| \frac{\partial}{\partial w_k} B(w + it\theta) \right| |\theta_k| dt$$

$$\geq M|w|^m - \sum_{k=1}^n \overline{M} \int_0^1 |w + it\theta|^{m-1} |\theta_k| dt \geq M\overline{\delta}|w + i\theta|^m - \overline{M}|w + i\theta|^{m-1}|\theta|.$$

This yields the estimate

$$|A(zh, h)| = h^{-m}|B(zh)| \geq M\overline{\delta}|\xi h + i\eta h|^m h^{-m} - \overline{M}|\xi h + i\eta h|^{m-1}|\eta h| h^{-m}$$

$$\geq M\overline{\delta}|z|^m - \overline{M}|z|^{m-1}a.$$

If we take a small enough, then

$$M\overline{\delta}|z|^m - \overline{M}|z|^{m-1}a \geq M_1|z|^m > 0, \quad M_1 = M_1(a) > 0.$$

Lemma 1.2 is proved.

Now let us show that the function $\arg A(zh, h)$ is continuous in z.

Lemma 1.3. *Suppose $|A(\xi h, h)| \geq M|\xi|^m$ for $|\xi h| \leq \pi$. Let $0 < rh < L$, $L > 0$ and $|r\xi h| \leq \pi$. Then for any $\varepsilon > 0$ the following inequality holds for all $z \in \Omega$, provided $|\eta_k| \leq a(\varepsilon, \xi_0)$:*

$$|\arg A(rzh, h) - \arg A(r\xi h, h)| \leq \varepsilon.$$

Proof. We can assume that $r = 1$, since

$$\arg A(rzh, h) = \arg \frac{B(rzh)}{h^m} = \arg \frac{B(rzh)}{r^m h^m} = \arg \frac{B(sz)}{s^m}$$

and $0 < s < L$. By analogy with the proof of Lemma 1.2, we have

$$A(zh, h) = A(\xi h, h) + h^{-m} \sum_{k=1}^n \int_0^1 \frac{\partial}{\partial w_k} B(\xi h + ith\eta)ih\eta_k \, dt$$

$$= A(\xi h, h) + \sum_{k=1}^{n} i\eta_k \tilde{B}_k h^{1-m}.$$

Here the coefficients $\tilde{B}_k$ obey the inequalities $|\tilde{B}_k| \le M_k h^{m-1}|z|^{m-1}$. Since $|A(\xi h, h)| \ge M\xi_0^m > 0$ for $\xi \in \Omega$, it follows that

$$\arg A(zh, h) = \arg A(\xi h, h) + \arg \left[1 + \sum_{k=1}^{n} i\eta_k \tilde{B}_k h^{1-m}(A(\xi h, h))^{-1} \right].$$

Further,

$$\left| \sum_{k=1}^{n} i\eta_k \tilde{B}_k h^{1-m}(A(\xi h, h))^{-1} \right| \le \sum_{k=1}^{n} a M_k h^{m-1}|z|^{m-1} h^{1-m}(M\xi_0)^{-1} \le a\tilde{M}\xi_0^{-1}.$$

Let us choose a such that $\arctan(a\tilde{M}\xi_0^{-1}) < \varepsilon$. Then

$$\arg \left[1 + \sum_{k=1}^{n} i\eta_k \tilde{B}_k h^{1-m}(A(\xi h, h))^{-1} \right] < \varepsilon$$

because

$$\operatorname{Im} \left[1 + \sum_{k=1}^{n} i\eta_k \tilde{B}_k h^{1-m}(A(\xi h, h))^{-1} \right] \le a\tilde{M}\xi_0^{-1}.$$

Lemma 1.3 is proved.

2.   A formula for the solution of the resolvent equation.

Here we are interested in finding the resolvent of the operator $-A_h$, i.e., in solving the equation

$$A_h u^h(x) + \lambda u^h(x) = f^h(x), \quad x \in \mathbf{R}_h^n. \tag{1.21}$$

Let $C_h = C(\mathbf{R}_h^n)$ denote the Banach space of bounded grid functions $u^h(x)$ defined on $\mathbf{R}_h^n$, equipped with the norm

$$\|u^h\|_{C_h} = \sup_{x \in \mathbf{R}_h^n} |u^h(x)|.$$

Since in what follows we will be concerned with the positivity of the operator A_h in C_h, it is natural to establish a formula for the solution of equation (1.21) for $f^h \in C_h$. The simplest way to solve this equation is by means of the Fourier transformation. However, this transformation does not exist for every function $f^h \in C_h$. For this reason we shall first establish a formula for the solution of (1.21) in the class of rapidly decreasing functions.

Lemma 1.4. *Let $A(\xi h, h) + \lambda \neq 0$ for $|\xi_k h| \leq \pi$ and let $f^h(x)$ be a rapidly decreasing function. Then in the class of rapidly decreasing functions equation (1.21) is uniquely solvable, and the following formula holds:*

$$u^h(x) = (2\pi)^{-n} \sum_{y \in \mathbf{R}_h^n} \mathcal{G}^h(x - y, \lambda) f^h(y) h^n, \tag{1.22}$$

where

$$\mathcal{G}^h(x, \lambda) = \int_{|\xi_1| \leq \frac{\pi}{h}} \cdots \int_{|\xi_n| \leq \frac{\pi}{h}} (A(\xi h, h) + \lambda)^{-1} \exp\{-i(\xi, x)\} d\xi_1 \cdots d\xi_n. \tag{1.23}$$

Proof. First let us show that if equation (1.21) has a solution in the class of rapidly decreasing functions, then that solution is unique. Consider the equation

$$A_h u^h(x) + \lambda u^h(x) = 0, \quad x \in \mathbf{R}_h^n. \tag{1.24}$$

Let $u^h(x)$ be a rapidly decreasing solution of (1.24). Then $A_h u^h(x) + \lambda u^h(x)$ is a rapidly decreasing grid function and $(A_h u^h(x) + \lambda u^h(x))\widetilde{} = (A(\xi h, h) + \lambda)\tilde{u}(\xi) = 0$. Since $A(\xi h, h) + \lambda \neq 0$, it follows that $\tilde{u}(\xi) = 0$, whence $u^h(x) \equiv \hat{\tilde{u}}^h(\xi) \equiv 0$, which proves the claimed uniqueness.

Let us establish the formula for the solution. Consider the function

$$v(\xi) = (A(\xi h, h) + \lambda)^{-1} \tilde{f}(\xi).$$

Since $A(\xi h, h) + \lambda \neq 0$ and $A(\xi h, h)$ and $\tilde{f}(\xi)$ are smooth $2\pi h^{-1}$-periodic functions, the inverse Fourier transform $\hat{v}^h(x)$ exists and is a rapidly decreasing grid function. Let us show that $\hat{v}^h(x)$ is a solution of equation (1.21). Consider the difference

$$A_h \hat{v}^h(x) + \lambda \hat{v}^h(x) - f^h(x) \equiv z^h(x).$$

Since $\hat{v}^h(x)$ and $f^h(x)$ are rapidly decreasing grid functions,

$$\tilde{z}(\xi) = (A_h \hat{v}^h(x) + \lambda \hat{v}^h(x) - f^h(x))\widetilde{} = (A(\xi h, h) + \lambda)v(\xi) - \tilde{f}(\xi).$$

Using the definition of the function $v(\xi)$, we see that $\tilde{z}(\xi) = 0$. Since $z^h(x)$ is a rapidly decreasing function, it follows that $z^h(x) \equiv 0$. Hence, $\hat{v}^h(x)$ is the solution of equation (1.21). Using the transformations (1.9) and (1.11), we obtain

$$u^h(x) = \hat{v}^h(x) = \int_{|\xi_1 h| \leq \pi} \cdots \int_{|\xi_n h| \leq \pi} (A(\xi h, h) + \lambda)^{-1} \exp\{i(x, \xi)\} \tilde{f}(\xi)\, d\xi_1 \cdots d\xi_n$$

$$= (2\pi)^{-n} \sum_{y \in \mathbf{R}_h^n} \int_{|\xi_1 h| \leq \pi} \cdots \int_{|\xi_n h| \leq \pi} (A(\xi h, h) + \lambda)^{-1} \exp\{i(x - y, \xi)\} \times$$

$$\times d\xi_1 \cdots d\xi_n\, f^h(y) h^n = (2\pi)^{-n} \sum_{y \in \mathbf{R}_h^n} \mathcal{G}^h(x - y, \lambda) f^h(y) h^n.$$

Lemma 1.4 is proved.

The function $(2\pi)^{-n} \mathcal{G}^h(x, \lambda)$ is called the *fundamental solution* (or the *Green function*) of the resolvent equation (1.21). By definition, $\mathcal{G}^h(x, \lambda)$ is the inverse Fourier transform of the smooth $2\pi h^{-1}$-periodic function $(A(\xi h, h) + \lambda)^{-1}$. Therefore, if h and λ are such that $A(\xi h, h) + \lambda \neq 0$, then formula (1.23) gives a rapidly decreasing grid function $\mathcal{G}^h(x, \lambda)$. Consequently, the series

$$(2\pi)^{-n} \sum_{y \in \mathbf{R}_h^n} \mathcal{G}^h(x - y, \lambda) f^h(y) h^n \tag{1.25}$$

is absolutely convergent for any function $f^h \in C_h$ and gives a grid function that belongs to the space C_h. Let us show that this function is a solution of equation (1.21).

Lemma 1.5. *Let $A(\xi h, h) + \lambda \neq 0$ for $|\xi_k h| \leq \pi$ and let $f^h \in C_h$. Then equation (1.21) is uniquely solvable in C_h and its solution $u^h(x)$ is given by formula (1.22).*

Proof. Let us consider a sequence of functions $\{f_k^h(x)\}$ with compact support such that $f_k^h(x) = f^h(x)$ for $|x| \leq M(k)$ and $f_k^h(x) = 0$ for $|x| > M(k)$, where $M(k) \to \infty$. The difference $A_h u^h(x) + \lambda u^h(x) - f^h(x) \equiv z^h(x)$ can be written in the form

$$z^h(x) = A_h u^h(x) - A_h u_k^h(x) + \lambda u^h(x) - \lambda u_k^h(x) + f_k^h(x) - f^h(x),$$

where $u_k^h(x)$ is the solution of equation (1.21) with right-hand side $f_k^h(x)$, which by Lemma 1.4 exists and is unique. Using the definition of the operator A_h, we obtain

$$|z^h(x)| \leq M h^{-m} |u^h(x) - u_k^h(x)| + \lambda |u^h(x) - u_k^h(x)| + |f^h(x) - f_k^h(x)|.$$

Let us estimate the difference $u^h(x) - u_k^h(x)$. We have

$$|u^h(x) - u_k^h(x)|$$

$$\leq (2\pi)^{-n} \left(\sum_{y\in\mathbf{R}_h^n:\ |y|\leq M(k)} + \sum_{y\in\mathbf{R}_h^n:\ |y|>M(k)} \right) |\mathcal{G}^h(x-y,\lambda)|\,|f^h(y) - f_k^h(y)|h^n.$$

By definition, $f^h(y) - f_k^h(y) = 0$ for $|y| \leq M(k)$. Hence, the first sum in the last inequality vanishes. The second sum tends to zero when $k \to \infty$ since $\mathcal{G}^h(x-y,\lambda)$ is a rapidly decreasing grid function and $f^h(y) - f_k^h(y) \equiv f^h(y) \in C_h$. Therefore, $|u^h(x)-u_k^h(x)| \to 0$ as $k \to \infty$, for all $x \in \mathbf{R}_h^n$. Again, recall that $f^h(x)-f_k^h(x) = 0$ for all $x \in \mathbf{R}_h^n$ provided k is large enough. But $z^h(x)$ does not depend on k. We conclude that $z^h(x) \equiv 0$, and hence $u^h(x)$ is a solution of equation (1.21).

Let us show that the solution of equation (1.21) is unique. Consider equation (1.24) and assume that it has a bounded nonzero solution $u_0^h(x)$. Let $\{u_k^h(x)\}$ be a sequence of functions with compact support such that $u_k^h(x) = u_0^h(x)$ for $|x| \leq M(k)$ and $u_k^h(x) = 0$ for $|x| > M(k)$, where $M(k) \to \infty$. Define the function with compact support $f_k^h(x)$ by $f_k^h(x) = A_h u_k^h(x) + \lambda u_k^h(x)$. By Lemma 1.4, equation (1.21) with right-hand side $f_k^h(x)$ is uniquely solvable, and its solution is given by formula (1.22). Therefore,

$$u_k^h(x) = (2\pi)^{-n} \sum_{y\in\mathbf{R}_h^n} \mathcal{G}^h(x-y,\lambda) f_k^h(y)h^n.$$

Since $u_0^h(x)$ is a solution of (1.24),

$$f_k^h(x) = A_h(u_k^h(x) - u_0^h(x)) + \lambda(u_k^h(x) - u_0^h(x)).$$

From the definition of the functions $u_k^h(x)$ it follows that the difference $u_k^h(x) - u_0^h(x)$ converges uniformly to zero on any bounded subset of $\mathbf{R}_h^n$ as $k \to \infty$. Consequently, $u_0^h(x) \equiv 0$. Lemma 1.5 is proved.

What Lemma 1.5 asserts is that under the condition $A(\xi h, h)+\lambda \neq 0$ the resolvent of the operator $-A_h$ in the space C_h exists, and one has the representation

$$(\lambda + A_h)^{-1}f^h(x) = (2\pi)^{-n} \sum_{y\in\mathbf{R}_h^n} \mathcal{G}^h(x-y,\lambda)f^h(y)h^n.$$

From this equality it follows that

$$\|(\lambda + A_h)^{-1}\|_{C_h \to C_h} \le (2\pi)^{-n} \sum_{y \in \mathbf{R}_h^n} |\mathcal{G}^h(y, \lambda)| h^n. \tag{1.26}$$

Let us show that the norm of the operator $(\lambda + A_h)^{-1}$ is in fact equal to the right-hand side of inequality (1.26).

Lemma 1.6. *Let λ and h be such that $A(\xi h, h) + \lambda \ne 0$. Then the resolvent of $-A_h$ in C_h exists and we have the equality*

$$\|(\lambda + A_h)^{-1}\|_{C_h \to C_h} = (2\pi)^{-n} \sum_{y \in \mathbf{R}_h^n} |\mathcal{G}^h(y, \lambda)| h^n.$$

Proof. It suffices to establish the converse of inequality (1.26). Let $x_0 \in \mathbf{R}_h^n$. Then for any function $f^h(x) \in C_h$,

$$|(\lambda + A_h)^{-1} f^h(x_0)| = \left| (2\pi)^{-n} \sum_{y \in \mathbf{R}_h^n} \mathcal{G}^h(x_0 - y, \lambda) f^h(y) h^n \right|.$$

Using the definition of the norm of an operator, we have

$$\left| (2\pi)^{-n} \sum_{y \in \mathbf{R}_h^n} \mathcal{G}^h(x_0 - y, \lambda) f^h(y) h^n \right| \le \sup_{x \in \mathbf{R}_h^n} |(\lambda + A_h)^{-1} f^h(x)|$$

$$= \|(\lambda + A_h)^{-1} f^h\|_{C_h} \le \|(\lambda + A_h)^{-1}\|_{C_h \to C_h} \|f^h\|_{C_h}.$$

Since the last relation holds for any function $f^h(x) \in C_h$, we can take for $f^h(x)$ the following function:

$$f_{x_0}^h(y) = \begin{cases} \exp\{-i \arg \mathcal{G}^h(x_0 - y, \lambda)\}, & \text{if } \mathcal{G}^h(x_0 - y, \lambda) \ne 0, \\ 0, & \text{if } \mathcal{G}^h(x_0 - y, \lambda) = 0. \end{cases}$$

Then $f_{x_0}^h(y) \in C_h$ and $\|f_{x_0}^h\|_{C_h} = 1$. Consequently,

$$\left| (2\pi)^{-n} \sum_{y \in \mathbf{R}_h^n} \mathcal{G}^h(x_0 - y, \lambda) f_{x_0}^h(y) h^n \right| = (2\pi)^{-n} \sum_{y \in \mathbf{R}_h^n} |\mathcal{G}^h(x_0 - y, \lambda)| h^n$$

$$= (2\pi)^{-n} \sum_{y \in \mathbf{R}_h^n} |\mathcal{G}^h(y, \lambda)| h^n,$$

which in conjunction with the last inequality proves the converse of (1.26). Lemma 1.6 is proved.

Our next objective is to obtain estimates for the norm of the operator $(\lambda + A_h)^{-1}$ in C_h that are uniform in h. As follows from Lemma 1.6, to obtain such estimates we have to estimate the series

$$(2\pi)^{-n} \sum_{y \in \mathbf{R}_h^n} |\mathcal{G}^h(y, \lambda)| h^n$$

uniformly in h. To this end we shall establish estimates for the modulus of the function $\mathcal{G}^h(y, \lambda)$.

3. Point estimates for the fundamental solution of the resolvent equation.

In this subsection we establish point estimates for the function $\mathcal{G}^h(y, \lambda)$ defined by (1.23). The fundamental solution of the resolvent equation (1.21) is $(2\pi)^{-n}\mathcal{G}^h(y, \lambda)$. Hence, the point estimates of the function $\mathcal{G}^h(y, \lambda)$ and of the fundamental solution will differ only by the factor $(2\pi)^{-n}$, and for this reason the estimates of $\mathcal{G}^h(y, \lambda)$ will be referred to as estimates of the fundamental solution. The methods for deriving these estimates depend on the ratio between the numbers h and $|\lambda|$, as well as on the ratio between the dimension n of the space and the order m of the operator, and are based on passing from integration over an n-dimensional cube in the real space to integration over a similar domain in the n-dimensional complex space. Such a passage is possible whenever the integrand in equality (1.23) is analytic in the domain in question. For analyticity it is necessary that $A(\xi h, h) + \lambda$ be separated from zero uniformly in h. The latter can be achieved either thanks to the difference between $|A(\xi h, h)|$ and $|\lambda|$, or thanks to the relation between $\arg A(zh, h)$ and $\arg \lambda$, which does not allow the vectors $A(zh, h)$ and λ to have opposite directions.

The first case is realized for large values of the parameter λ.

Lemma 1.7. Let $|\lambda| \geq Lh^{-m}$, where $L > 0$ is large enough. Then

$$|\mathcal{G}^h(y, \lambda)| \leq Mh^{-n} \exp\{-bh^{-1}(|y_1| + \cdots + |y_n|)\}|\lambda|^{-1}. \qquad (1.27)$$

Proof. Recall that we put $(b > 0)$

$$\Omega = \{z : z_k = \xi_k + i\eta_k, \; |\xi_k h| \leq \pi, \; |\eta_k h| \leq b, \; k = 1, \cdots, n\}.$$

For $z \in \Omega$, $|z| \leq M_\Omega h^{-1}$. Hence, by the estimate (1.18),

$$|(Azh, h)| \leq M(b)|z|^m \leq M(b)M_\Omega^m h^{-m} = 2^{-1}Lh^{-m}.$$

It follows that for $z \in \Omega$

$$|A(zh, h)| \geq |\lambda| - |A(zh, h)| \geq |\lambda| - 2^{-1}Lh^{-m} \geq 2^{-1}|\lambda|. \qquad (1.28)$$

Let us make the substitution $\overline{\xi}_k = \xi_k + i(\operatorname{sign} y_k)bh^{-1}$ in the integral appearing in (1.23). Using the analyticity and the periodicity of the functions $(A(zh, h) + \lambda)^{-1}$ and $\exp\{i(y, z)\}$, we obtain

$$\mathcal{G}^h(y, \lambda) = \int_{|\xi_1 h| \leq \pi} \cdots \int_{|\xi_n h| \leq \pi} (A(\overline{\xi}h, h) + \lambda)^{-1} \exp\{i(y, \overline{\xi})\}d\xi_1 \cdots d\xi_n.$$

By the estimate (1.28),

$$|\mathcal{G}^h(y, \lambda)| \leq 2|\lambda|^{-1} \exp\{-bh^{-1}(|y_1| + \cdots + |y_n|)\} \int_{|\xi_1 h| \leq \pi} \cdots \int_{|\xi_n h| \leq \pi} d\xi_1 \cdots d\xi_n$$

$$\leq Mh^{-n} \exp\{-bh^{-1}(|y_1| + \cdots + |y_n|)\}|\lambda|^{-1}.$$

Lemma 1.7 is proved.

Lemma 1.8. *Let* $|\lambda| \geq Lh^{-m}$, *where* $L > 0$ *is large enough. Then*

$$|\mathcal{G}^h(y, \lambda)| \leq Mh^{-n-m+1} \exp\{-bh^{-1}(|y_1| + \cdots + |y_n|)\}|\lambda|^{-2}|y|^{-1}, \quad y \neq 0. \quad (1.29)$$

Proof. By (1.14),

$$A(\xi h, h) = \sum_{|r|=m} a_r(i\xi)^r + h^{-m} \sum_{m+1 \leq |r| \leq S} d_r(i\xi h)^r. \qquad (1.30)$$

Setting $\overline{\xi}_k = \xi_k + (i \operatorname{sign} y_k)bh^{-1}$, we have

$$\mathcal{G}^h(y, \lambda) = \exp\{-bh^{-1}(|y_1| + \cdots + |y_n|)\}J(y),$$

where

$$J(y) = \int_{|\xi_1 h| \leq \pi} \cdots \int_{|\xi_n h| \leq \pi} (A(\overline{\xi}h, h) + \lambda)^{-1} \exp\{i(y, \overline{\xi})\}d\xi_1 \cdots d\xi_n.$$

It is now clear that in order to prove inequality (1.29) it suffices to establish the estimate

$$|J(y)| \leq M h^{-n-m+1} |\lambda|^{-2} |y|^{-1}, \quad y \neq 0. \tag{1.31}$$

Let $y \neq 0$, and choose an index r such that $y_r = \max_{1 \leq k \leq n} |y_k| \neq 0$. With no loss of generality we may assume that $r = n$. Let us integrate in $J(y)$ once with respect to the n-th variable. By the periodicity of the functions $(A(\xi h, h) + \lambda)^{-1}$ and $\exp\{i(y, \xi)\}$,

$$J(y) = (iy_n)^{-1} \int_{|\xi_1 h| \leq \pi} \cdots \int_{|\xi_n h| \leq \pi} \frac{\partial}{\partial \bar{\xi}_n} \left(A(\bar{\xi} h, h) + \lambda \right)^{-1} \times$$

$$\times \exp\{i(y, \bar{\xi})\} d\xi_1 \cdots d\xi_n. \tag{1.32}$$

Note that for $|\lambda| \geq L h^{-m}$ we have the inequality

$$\left| \frac{\partial}{\partial \bar{\xi}_n} (A(\bar{\xi} h, h) + \lambda)^{-1} \right| \leq M_1 |\lambda|^{-2} h^{-m+1}. \tag{1.33}$$

Indeed, this follows from the fact that

$$\frac{\partial}{\partial \bar{\xi}_n} (A(\bar{\xi} h, h) + \lambda)^{-1} = (A(\bar{\xi} h, h) + \lambda)^{-2} h^{-m} \frac{\partial}{\partial \bar{\xi}_n} B(\bar{\xi} h),$$

where

$$B(y) = \sum_{|r|=m} a_r (iy)^r + \sum_{m+1 \leq |r| \leq S} d_r (iy)^r,$$

and for $|\xi_k h| \leq \pi$,

$$\left| (A(\bar{\xi} h, h) + \lambda)^{-2} \right| \leq M_1 |\lambda|^{-2}, \quad \left| \frac{\partial}{\partial \bar{\xi}_n} B(\bar{\xi} h) \right| \leq M_1 h.$$

Identity (1.32) and inequality (1.33) yield the inequality (1.31). Lemma 1.8 is proved.

From (1.27) and (1.29) we obtain the estimate

$$|\mathcal{G}^h(y, \lambda)| \leq M h^{-n} \exp\{-b h^{-1} (|y_1| + \cdots + |y_n|)\} |\lambda|^{-(1+m\alpha)} h^{-m\alpha(m-1)} |y|^{-m\alpha},$$

$$y \neq 0, \quad 0 \leq \alpha \leq m^{-1}. \tag{1.34}$$

In the proof of the estimate of $|\mathcal{G}^h(y, \lambda)|$ for large values of the parameter λ we imposed no restrictions on the symbol of the operator A_h. Nevertheless, the function under the integral sign in equality (1.23) remained analytic in a sufficiently large domain of the complex space. The reason for this is that when we pass from the real variable ξ to the complex one $\overline{\xi}$, the effects of the real and imaginary parts of the latter on $|A(\xi h, h)|$ are equivalent. In contrast, for small values of λ we have to use the relation between the arguments of the vectors $A(\overline{\xi} h, h)$ and λ, and hence impose restrictions on $\arg A(\xi h, h)$. Furthermore, to ensure that $\arg A(\xi h, h)$ will be close to $\arg A(\overline{\xi} h, h)$ it is necessary that the function $\arg A(\xi h, h)$ be continuous, which in turn is ensured by the lower estimate for the modulus of the function $A(\xi h, h)$.

First let us prove the estimate of $|\mathcal{G}^h(y, \lambda)|$ for small values of $|\lambda|$ in the case where the dimension of the space is smaller than the order of the operator.

Lemma 1.9. *Suppose that for $0 \leq |\xi_k h| \leq \pi$,*
   a) $|A(\xi h, h)| \geq M_0 |\xi|^m$, where $M_0 > 0$;
   b) $|\arg A(\xi h, h)| \leq \phi < \phi_0 < \pi$.
   Let $0 < |\lambda| \leq L h^{-m}$ and $|\arg \lambda| \leq \pi - \phi_0$. Then for $m > n$,

$$|\mathcal{G}^h(y, \lambda)| \leq M \exp\{-a_0 |\lambda|^{\frac{1}{m}}(|y_1| + \cdots + |y_n|)\}|\lambda|^{\frac{n}{m}-1}, \quad a_0 > 0. \qquad (1.35)$$

Proof. Let us make the substitution $\xi_k = |\lambda|^{\frac{1}{m}} \overline{\xi}_k$ in the integral appearing in formula (1.23), and set $w = \lambda |\lambda|^{-1}$. Then

$$\mathcal{G}^h(y, \lambda) = \int_{|\overline{\xi}_1| \leq \pi(|\lambda|^{\frac{1}{m}} h)^{-1}} \cdots \int_{|\overline{\xi}_n| \leq \pi(|\lambda|^{\frac{1}{m}} h)^{-1}} (A(|\lambda|^{\frac{1}{m}} \overline{\xi} h, h) + \lambda)^{-1} \times$$

$$\times \exp\{i|\lambda|^{\frac{1}{m}} (\overline{\xi}, y)\}|\lambda|^{\frac{n}{m}} d\overline{\xi}_1 \cdots d\overline{\xi}_n$$

$$= |\lambda|^{\frac{n}{m}-1} \int_{|\overline{\xi}_1| \leq \pi(|\lambda|^{\frac{1}{m}} h)^{-1}} \cdots \int_{|\overline{\xi}_n| \leq \pi(|\lambda|^{\frac{1}{m}} h)^{-1}} (A(|\lambda|^{\frac{1}{m}} \overline{\xi} h, h)|\lambda|^{-1} + w)^{-1} \times$$

$$\times \exp\{i|\lambda|^{\frac{1}{m}} (\overline{\xi}, y)\}|\lambda|^{\frac{n}{m}} d\overline{\xi}_1 \cdots d\overline{\xi}_n. \qquad (1.36)$$

We now use the estimate (1.18) and choose ξ_0 and a_1 such that for $|\overline{\xi}_k| < \xi_0$ and $|\tilde{\eta}_k| < a_1$ we have

$$\left| A(|\lambda|^{\frac{1}{m}} (\overline{\xi} + i\tilde{\eta})h, h)|\lambda|^{-1} \right| \leq M|\overline{\xi} + i\tilde{\eta}|^m < 2^{-1}.$$

Then

$$\left| A(|\lambda|^{\frac{1}{m}} (\overline{\xi} + i\tilde{\eta})h, h)|\lambda|^{-1} + w \right| \geq 1 - 2^{-1} = 2^{-1}. \qquad (1.37)$$

By Lemma 1.3, we can choose $a_2 > 0$ such that if we set

$$\tilde{\Omega} = \{\tilde{z}_k : \tilde{z}_k = \overline{\xi}_k + i\tilde{\eta}_k, \ \xi_0 \leq |\overline{\xi}_k| \leq \pi(|\lambda|^{\frac{1}{m}}h)^{-1}, \ |\tilde{\eta}_k| \leq a_2\},$$

then for $\tilde{z} \in \tilde{\Omega}$ one has the bound $|\arg A(|\lambda|^{\frac{1}{m}}\tilde{z}h, h)| < \phi_0$. Since $|\arg w| = |\arg \lambda| \leq \pi - \phi_0$, this implies that for $\tilde{z} \in \tilde{\Omega}$,

$$\left| A(|\lambda|^{\frac{1}{m}}\tilde{z}h, h)|\lambda|^{-1} + w \right| \geq M(\phi_0) \left| A(|\lambda|^{\frac{1}{m}}\tilde{z}h, h)|\lambda|^{-1} + w \right|^{-1}.$$

Using Lemma 1.2, we see that for $\tilde{z} \in \tilde{\Omega}$,

$$\left| A(|\lambda|^{\frac{1}{m}}\tilde{z}h, h)|\lambda|^{-1} + w \right| \geq \tilde{M}(\phi_0)(|\tilde{z}|^m + 1) \geq \overline{M}(\phi_0)(|\overline{\xi}|^m + 1). \tag{1.38}$$

Let $a_0 = \min(a_1, a_2)$. Then, combining the estimates (1.37) and (1.38) we conclude that (1.38) holds for all $\tilde{z}_k$ such that $0 \leq |\overline{\xi}_k| \leq \pi(|\lambda|^{\frac{1}{m}}h)^{-1}$, $|\tilde{\eta}_k| \leq a_0$, $k = 1, \cdots, n$. Consequently, the integrand in (1.36) is analytic in the domain $\tilde{\Omega}$ if $a_2 = a_0$. Let us make the substitution $\tilde{\xi}_k = \overline{\xi}_k + i(\text{sign } y_k)a_0$ in the integrand. By the analyticity and periodicity of this integrand, we have

$$\mathcal{G}^h(y, \lambda) = |\lambda|^{\frac{n}{m}-1} \int_{|\overline{\xi}_1| \leq \pi(|\lambda|^{\frac{1}{m}}h)^{-1}} \cdots \int_{|\overline{\xi}_n| \leq \pi(|\lambda|^{\frac{1}{m}}h)^{-1}} \times$$

$$\times \left(A(|\lambda|^{\frac{1}{m}}\tilde{\xi}h, h)|\lambda|^{-1} + w \right)^{-1} \exp\{i|\lambda|^{\frac{1}{m}}(\overline{\xi}, y)\}d\overline{\xi}_1 \cdots d\overline{\xi}_n. \tag{1.39}$$

Using estimate (1.38), this yields

$$|\mathcal{G}^h(y, \lambda)| \leq M|\lambda|^{\frac{n}{m}-1} \exp\{-a_0|\lambda|^{\frac{1}{m}}(|y_1| + \cdots + |y_n|)\} \times$$

$$\times \int_{|\overline{\xi}_1| \leq \pi(|\lambda|^{\frac{1}{m}}h)^{-1}} \cdots \int_{|\overline{\xi}_n| \leq \pi(|\lambda|^{\frac{1}{m}}h)^{-1}} (1 + |\overline{\xi}|^m)^{-1}d\overline{\xi}_1 \cdots d\overline{\xi}_n$$

$$\leq M|\lambda|^{\frac{n}{m}-1} \exp\{-a_0|\lambda|^{\frac{1}{m}}(|y_1| + \cdots + |y_n|)\} \int_{\mathbf{R}^n} (1 + |\overline{\xi}|^m)^{-1}d\overline{\xi}.$$

Since $m > n$, the last integral is bounded. Lemma 1.9 is proved.

In the case $m \leq n$ one can use (1.39) and (1.38) to establish the inequality

$$|\mathcal{G}^h(y, \lambda)| \leq M|\lambda|^{\frac{n}{m}-1} \exp\{-a_0|\lambda|^{\frac{1}{m}}(|y_1| + \cdots + |y_n|)\} \times$$

$$\times \int_{|\xi_1| \leq \pi|\lambda|^{\frac{1}{m}}h|^{-1}} \cdots \int_{|\xi_n| \leq \pi|\lambda|^{\frac{1}{m}}h|^{-1}} (1 + |\xi|^m)^{-1}d\xi_1 \cdots d\xi_n.$$

The last integral can be calculated, but its value depends on λ and h and grows unboundedly when the product $|\lambda|^{\frac{1}{m}} h$ decreases. The estimate of the fundamental solution obtained in this manner is rather crude, since the function $\exp\{i|\lambda|^{\frac{1}{m}}(\xi, y)\}$ figuring under the integral sign in (1.39) is estimated in modulus regardless of its oscillation properties. To find an estimate that takes these properties into account one needs to regularize the integral by means of integation by parts.

Lemma 1.10. *Suppose that the assumptions of Lemma 1.9 hold. Then for $m \leq n$ one has the estimate*

$$|\mathcal{G}^h(y, \lambda)| \leq M|\lambda|^{-\frac{1}{m}}(|y| + h)^{m-n-1}\exp\{-a_0|\lambda|^{\frac{1}{m}}(|y_1| + \cdots + |y_n|)\}.$$

Proof. Proceeding by analogy with Lemma 1.9, we obtain the representation (1.39) and write it in the form

$$\mathcal{G}^h(y, \lambda) = |\lambda|^{\frac{n}{m}-1}\exp\{-a_0|\lambda|^{\frac{1}{m}}(|y_1| + \cdots + |y_n|)\}\times$$

$$\times \int_{|\xi_1|\leq\pi|\lambda^{\frac{1}{m}}h|^{-1}} \cdots \int_{|\xi_n|\leq\pi|\lambda^{\frac{1}{m}}h|^{-1}} \exp\{i|\lambda|^{\frac{1}{m}}(\xi, y)\}f(\xi)d\xi_1\cdots d\xi_n, \qquad (1.40)$$

where

$$f(\xi) = \left(A(|\lambda|^{\frac{1}{m}}(\xi + i(\operatorname{sign} y)a_0)h, h)|\lambda|^{-1} + w\right)^{-1}.$$

The estimates (1.37) and (1.38) mean that the inequality

$$|f(\xi)| \leq M_1(1 + |\xi|^m)^{-1} \qquad (1.41)$$

holds for $0 \leq |\xi_k| \leq \pi|\lambda^{\frac{1}{m}}h|^{-1}$. From the definition of the symbol $A(\xi h, h)$ it follows that under the assumptions of the lemma the function $f(\xi)$ is infinitely differentiable. Recall that $A(\xi h, h) = h^{-m}B(\xi h)$, where $B(w)$ is defined by (1.14). Similarly to the way the estimate of the modulus of the derivative $\frac{\partial}{\partial w_k}B(w + i\theta)$ was obtained in Lemma 1.2, one can establish the estimate

$$\left|\left(\frac{\partial}{\partial w_k}\right)^r B(w + i\theta)\right| \leq M_r|w + i\theta|^{m-r}$$

for any $r > 0$. It follows that

$$\left|\left(\frac{\partial}{\partial w_k}\right)^r B(\xi h + i\eta h)\right| \leq M_r|\xi h + i\eta h|^{m-r}h^r = M|\xi + i\eta|^{m-r}h^m.$$

Hence, letting $z = \xi + i(\text{sign } y)a_0$, we obtain

$$\left|\left(\frac{\partial}{\partial \xi_k}\right)^r (A(|\lambda|^{\frac{1}{m}} zh, h)|\lambda|^{\frac{1}{m}} + w)\right| = |\lambda|^{-1} \left|\left(\frac{\partial}{\partial \xi_k}\right)^r A(|\lambda|^{\frac{1}{m}} zh, h)\right|$$

$$= |\lambda h^m|^{-1} \left|\left(\frac{\partial}{\partial \xi_k}\right)^r B(|\lambda|^{\frac{1}{m}} zh)\right| \le M_r |\lambda h^m|^{-1} |\lambda^{\frac{1}{m}} zh|^{m-r} |\lambda^{\frac{1}{m}} h|^r$$

$$= M_r |z|^{m-r} \le \tilde{M}_r (1 + |\xi|)^{m-r}. \tag{1.42}$$

The estimate (1.18) yields

$$\left|A(|\lambda|^{\frac{1}{m}} zh, h) + w\right| \le M(1 + |z|^m) \le M_1(1 + |\xi|^m). \tag{1.43}$$

Successively differentiating the function $f(\xi)$ and estimating the numerator [resp., denominator] of the resulting fraction by means of (1.42) and (1.43) [resp., (1.38)], we conclude that

$$\left|\left(\frac{\partial}{\partial \xi_k}\right)^r f(\xi)\right| \le M_r (1 + |\xi|^{m+r})^{-1} \tag{1.44}$$

for $|\xi_k| \le \pi |\lambda^{\frac{1}{m}} h|^{-1}$. Let $y \ne 0$. Choose an index j such that $|y_j| = \max_{1 \le k \le n} |y_k|$. With no loss of generality we may take $j = n$. In the integral appearing in the right-hand side of equality (1.40) let us integrate by parts $n - m + 1$ times with respect to the n-th variable. Since the integrand is periodic, we obtain

$$\mathcal{G}^h(y, \lambda) = |\lambda|^{\frac{n}{m} - 1} \exp\{-a_0 |\lambda|^{\frac{1}{m}} (|y_1| + \cdots + |y_n|)\} (i |\lambda|^{\frac{1}{m}} y_n)^{m-n-1} \times$$

$$\times \int_{|\xi_1| \le \pi |\lambda^{\frac{1}{m}} h|^{-1}} \cdots \int_{|\xi_n| \le \pi |\lambda^{\frac{1}{m}} h|^{-1}} \exp\{i |\lambda|^{\frac{1}{m}} (\xi, y)\} \left(\frac{\partial}{\partial \xi_n}\right)^{n-m+1} f(\xi)\, d\xi_1 \cdots d\xi_n, \tag{1.45}$$

whence

$$|\mathcal{G}^h(y, \lambda)| \le |\lambda|^{-\frac{1}{m}} |y_n|^{m-n-1} \exp\{-a_0 |\lambda|^{\frac{1}{m}} (|y_1| + \cdots + |y_n|)\} \times$$

$$\times \int_{\mathbf{R}^n} (1 + |\xi|^{n+1})^{-1} d\xi \le M |\lambda|^{-1/m} |y|^{m-n-1} \exp\{-a_0 |\lambda|^{\frac{1}{m}} (|y_1| + \cdots + |y_n|)\}. \tag{1.46}$$

If $y = 0$, then, using formula (1.23) we obtain

$$\mathcal{G}^h(0, \lambda) = \int_{|\xi_1| \le \pi h^{-1}} \cdots \int_{|\xi_n| \le \pi h^{-1}} (A(\xi h, h) + \lambda)^{-1} d\xi_1 \cdots d\xi_n. \tag{1.47}$$

Making the substitution $\xi_k = |\lambda|^{\frac{1}{m}}\tilde{\xi}_k$ and using the estimate (1.38), we obtain

$$|\mathcal{G}^h(0,\lambda)| \leq M|\lambda|^{\frac{n}{m}-1} \int_{|\tilde{\xi}_1|\leq\pi|\lambda^{\frac{1}{m}}h|^{-1}} \cdots \int_{|\tilde{\xi}_n|\leq\pi|\lambda^{\frac{1}{m}}h|^{-1}} (1+|\tilde{\xi}|^m)^{-1}d\tilde{\xi}_1\cdots d\tilde{\xi}_n$$

$$\leq M_1|\lambda|^{\frac{n}{m}-1} \int_{|\tilde{\xi}_1|\leq\pi|\lambda^{\frac{1}{m}}h|^{-1}} \cdots \int_{|\tilde{\xi}_n|\leq\pi|\lambda^{\frac{1}{m}}h|^{-1}} (1+|\tilde{\xi}|^{-\frac{m}{n}})^n d\tilde{\xi}_1\cdots d\tilde{\xi}_n$$

$$\leq M_2|\lambda|^{\frac{n}{m}-1} \left(\int_{|\xi|\leq\pi|\lambda^{\frac{1}{m}}h|^{-1}} (1+|\xi|)^{-\frac{m}{n}}d\xi \right)^n.$$

Let $m < n$. Then, since $|\lambda^{\frac{1}{m}}h|^{-1} \geq L^{-\frac{1}{m}} > 0$, we have

$$|\mathcal{G}^h(0,\lambda)| \leq M|\lambda|^{\frac{n}{m}-1}((|\lambda|^{\frac{1}{m}}h)^{-1})^{n-m} = Mh^{m-n} \leq ML^{\frac{1}{m}}|\lambda|^{-\frac{1}{m}}h^{m-n-1}. \quad (1.48)$$

If $m = n$, then

$$|\mathcal{G}^h(0,\lambda)| \leq M|\ln^n|\lambda^{\frac{1}{m}}h|| \leq M|\lambda^{\frac{1}{m}}h|^{-1}|\lambda^{\frac{1}{m}}h|\ln^n|\lambda^{\frac{1}{m}}h|| \leq M_1|\lambda|^{-\frac{1}{m}}h^{-1}. \quad (1.49)$$

Combining the estimates (1.48), (1.49) with the estimate (1.46), we complete the proof of Lemma 1.10.

4. Sharpening of the point estimates of the fundamental solution of the resolvent equation.

In this subsection we shall obtain sharper estimates for the fundamental solution as $|\lambda^{\frac{1}{m}}h| \to 0$ in the case $m \leq n$. First let us study one property of the Fourier coefficients of a periodic function. Let $f(y)$ be a $2T$-periodic function of the one-dimensional real argument y. The Fourier coefficients of $f(y)$ are defined as

$$f_n = \int_{-T}^{T} \exp\{-i\pi T^{-1}ny\}f(y)dy. \quad (1.50)$$

If $f(y)$ is integrable, then $f_n \to 0$ as $n \to \infty$. If $f(y)$ is absolutely continuous, then in (1.50) one can integrate by parts. Using the fact that $f(y)$ is periodic, we obtain

$$f_n = iT(\pi n)^{-1} \int_{-T}^{T} \exp\{i\pi T^{-1}ny\}f'(y)dy \quad (n \neq 0).$$

From this it follows that $nf_n \to 0$ as $n \to \infty$. As it turns out, in the right-hand side of (1.50) one can carry out a fractional integration by parts, and not only an integer one.

Lemma 1.11.   *Let $f(y)$ be a continuously differentiable $2T$-periodic function. Then for any α, $0 < \alpha < 1$, and $n \neq 0$ we have the equality*

$$f_n = \frac{\alpha(iT)^\alpha}{(\pi n)^\alpha \Gamma(1-\alpha)} \int_{-T}^{T} \exp\{i\pi T^{-1} ny\} \int_0^\infty \frac{f(y) - f(y-t)}{t^{1+\alpha}} dt\, dy. \qquad (1.51)$$

Proof. In the space $C(2T)$ of continuous $2T$-periodic functions the rule $U(t)\varphi(y) = \varphi(y+t)$ defines a strongly continuous group of translations. Its generator is defined by the formula $-B\varphi(y) = \varphi'(y)$ on the continuously differentiable functions. Since the group $U(t)$ is unitary, the resolvent $(\lambda - B)^{-1}$ of the operator B obeys the estimate (4.1) of Chapter 2 for λ in the left half-plane ($|\arg \lambda| = \phi > \pi/2$). Hence, if $\varepsilon > 0$, the operator $B_\varepsilon = B + \varepsilon$ is positive, and its negative fractional powers can be defined by the rule (see [31])

$$B_\varepsilon^{-\alpha} = \frac{1}{\Gamma(\alpha)} \int_0^\infty t^{\alpha-1} \exp\{-tB_\varepsilon\} dt. \qquad (1.52)$$

From this formula it follows that for any element f in the domain $D(B_\varepsilon)$ of the operator B_ε and any α, $0 < \alpha < 1$ we have the equality

$$B_\varepsilon^\alpha f = \frac{1}{\Gamma(1-\alpha)} \int_0^\infty t^{-\alpha} B_\varepsilon \exp\{-tB_\varepsilon\} f\, dt. \qquad (1.53)$$

For $0 < \alpha < 1$ we have the embedding $D(B_\varepsilon^\alpha) \supset D(B_\varepsilon)$. Consequently, $f(y) = B_\varepsilon^{-\alpha} B_\varepsilon^\alpha f(y) = B_\varepsilon^{-\alpha}\varphi(y)$. Using formulas (1.50) and (1.52), we obtain

$$f_n = \int_{-T}^{T} \exp\{i\pi T^{-1} ny\} B_\varepsilon^{-\alpha}\varphi(y)\, dy$$

$$= \frac{1}{\Gamma(\alpha)} \int_0^\infty t^{\alpha-1} \int_{-T}^{T} \exp\{i\pi T^{-1} ny\} \exp\{-tB_\varepsilon\}\varphi(y)\, dy\, dt.$$

Next, using the definition of the operator B_ε and the fact that $\exp\{-tB_\varepsilon\}$ is a semigroup of translations, we obtain

$$f_n = \frac{1}{\Gamma(\alpha)} \int_0^\infty t^{\alpha-1} \exp\{-t\varepsilon\} \int_{-T}^{T} \exp\{i\pi T^{-1} ny\} \exp\{-tB\}\varphi(y)\, dy\, dt$$

$$= \frac{1}{\Gamma(\alpha)} \int_0^\infty t^{\alpha-1} \exp\{-t\varepsilon\} \int_{-T}^{T} \exp\{i\pi T^{-1} ny\}\varphi(y-t)\, dy\, dt.$$

Let us make the substitution $y - t = \tau$ in the inner integral. Using the periodicity of the functions $\exp\{i\pi T^{-1}ny\}$ and $\varphi(y)$, we deduce that

$$f_n = \frac{1}{\Gamma(\alpha)} \int_0^\infty t^{\alpha-1} \exp\{-t(\varepsilon - i\pi T^{-1}n)\}dt \int_{-T}^T \exp\{i\pi T^{-1}n\tau\}\varphi(\tau)d\tau$$

$$= (\varepsilon - i\pi T^{-1}n)^{-\alpha} \int_{-T}^T \exp\{i\pi T^{-1}ny\}\varphi(y)dy.$$

Consider the function $\varphi(y)$. Using formula (1.53), we can write

$$\varphi(y) = B_\varepsilon^\alpha f(y) = \frac{1}{\Gamma(1-\alpha)} \int_0^\infty t^{-\alpha} B_\varepsilon \exp\{-tB_\varepsilon\}f(y)dt.$$

Since for $\varepsilon > 0$ the semigroup $\exp\{-tB_\varepsilon\}$ decays exponentially, integrating by parts we obtain

$$\varphi(y) = \frac{t^{-\alpha}}{\Gamma(1-\alpha)} \int_0^t B_\varepsilon \exp\{-sB_\varepsilon\}f(y)ds \Big|_0^\infty$$

$$+ \frac{\alpha}{\Gamma(1-\alpha)} \int_0^\infty t^{-\alpha-1} \int_0^t B_\varepsilon \exp\{-sB_\varepsilon\}f(y)ds\, dt$$

$$= \frac{\alpha}{\Gamma(1-\alpha)} \int_0^\infty t^{-\alpha-1}(1 - \exp\{-tB_\varepsilon\})f(y)dt$$

$$= \frac{\alpha}{\Gamma(1-\alpha)} \int_0^\infty t^{-\alpha-1}(f(y) - \exp\{-t\varepsilon\}f(y-t))dt.$$

Thus,

$$f_n = (\varepsilon - i\pi T^{-1}n)^{-\alpha} \frac{\alpha}{\Gamma(1-\alpha)} \int_{-T}^T \exp\{i\pi T^{-1}ny\} \times$$

$$\times \int_0^\infty \frac{f(y) - \exp\{-t\varepsilon\}f(y-t)}{t^{1+\alpha}} dt\, dy.$$

Since $n \neq 0$ and $f(y) \in C^1(2T)$, in the right-hand side of the last equality one can pass to the limit $\varepsilon \to 0$. This yields (1.51). Lemma 1.11 is proved.

From equality (1.51) it follows that if $f \in W_1^\alpha(2T)$ then $n^\alpha f_n \to 0$ as $n \to \infty$. Here $W_1^\alpha(2T)$ denotes the completion of the space of smooth periodic functions in the norm

$$\|f\|_{W_1^\alpha} = \int_{-T}^T |f(y)|dy + \int_{-T}^T \int_0^\infty |f(y) - f(y-t)|t^{-\alpha-1}dt\, dy.$$

In the proof of Lemma 1.10 the integral was regularized by means of integration by parts $n - m + 1$ times, although in order to guarantee that the integral will converge it would have been enough to integrate by parts $n - m + \alpha$ times. This observation will be implemented here by using formula (1.51).

Lemma 1.12. *Let the assumptions of Lemma 1.10 hold. Then for $0 < \alpha \le 1$ we have the estimate*

$$|\mathcal{G}^h(y, \lambda)| \le M(\alpha)|\lambda|^{-\frac{\alpha}{m}}(h + |y|)^{m-n-\alpha}\exp\{-a_0|\lambda|^{\frac{1}{m}}(|y_1| + \cdots |y_n|)\}, \quad (1.54)$$

where $M(\alpha) = M/\alpha$.

Proof. As in the proof of Lemma 1.10, we may assume that $|y_n| = \max_{1 \le k \le n}|y_k|$ for $y \ne 0$. Let us use the representation (1.40) and integrate by parts $n - m$ times with respect to the n-th variable in the right-hand side integral. By the periodicity of the functions $\exp\{i|\lambda|^{\frac{1}{m}}(\xi, y)\}$ and $f(\xi)$, we obtain

$$\mathcal{G}^h(y, \lambda) = |\lambda|^{\frac{n}{m}-1}\exp\{-a_0|\lambda|^{\frac{1}{m}}(|y_1| + \cdots |y_n|)\}(i|\lambda|^{\frac{1}{m}}y_n)^{m-n} \times$$

$$\times \int_{|\xi_1| \le \pi|\lambda|^{\frac{1}{m}}h|^{-1}} \cdots \int_{|\xi_n| \le \pi|\lambda|^{\frac{1}{m}}h|^{-1}} \exp\{i|\lambda|^{\frac{1}{m}}(\xi, y)\}\varphi(\xi)d\xi_1 \cdots d\xi_n. \quad (1.55)$$

Here $\varphi(\xi) = (\frac{\partial}{\partial \xi_n})^{n-m}f(\xi)$. Using the estimate (1.41), we see that

$$|\varphi(\xi)| \le M(1 + |\xi|)^{-n} \quad \text{for } |\xi_k| \le \pi|\lambda^{\frac{1}{m}}h|^{-1}. \quad (1.56)$$

Let us denote the integral in the right-hand side of (1.55) by J. Using formula (1.51) of the integral with respect to ξ_n, we obtain

$$J = \frac{\alpha}{\Gamma(1-\alpha)}(|\lambda|^{\frac{1}{m}}y_n)^{-\alpha}\int_{-T}^{T} \cdots \int_{-T}^{T} \exp\{i|\lambda|^{\frac{1}{m}}(\xi, y)\} \times$$

$$\times \int_0^\infty (\varphi(\xi_1, \cdots, \xi_n) - \varphi(\xi_1, \cdots, \xi_{n-1}, \xi_n - t))t^{-\alpha-1}dt \, d\xi_1 \cdots d\xi_n.$$

Here $T = \pi|\lambda^{\frac{1}{m}}h|^{-1}$, $y_n = jh$, $j \ne 0$. Note that T cannot approach zero because $|\lambda^{\frac{1}{m}}h| < L$. Further,

$$|J| \le \frac{M\alpha}{\Gamma(1-\alpha)}|\lambda^{\frac{1}{m}}y_n|^{-\alpha}\int_0^\infty t^{-\alpha-1} \times$$

$$\times \int_{-T}^{T} \cdots \int_{-T}^{T} |\varphi(\xi_1, \cdots, \xi_n) - \varphi(\xi_1, \cdots, \xi_{n-1}, \xi_n - t)| d\xi_1 \cdots d\xi_n \, dt. \qquad (1.57)$$

To prove the lemma we need to show that the integral in the last inequality is bounded by $M(\alpha) = M/\alpha^2$, where M does not depend on α and T. We shall estimate this integral for small as well as for large values of t. To this end we write it as a sum of two integrals:

$$\int_0^T t^{-\alpha-1} \int_{-T}^{T} \cdots \int_{-T}^{T} |\varphi(\xi_1, \cdots, \xi_n) - \varphi(\xi_1, \cdots, \xi_{n-1}, \xi_n - t)| d\xi_1 \cdots d\xi_n \, dt$$

$$+ \int_T^{\infty} t^{-\alpha-1} \int_{-T}^{T} \cdots \int_{-T}^{T} |\varphi(\xi_1, \cdots, \xi_n) - \varphi(\xi_1, \cdots, \xi_{n-1}, \xi_n - t)| d\xi_1 \cdots d\xi_n \, dt$$

$$= J_1 + J_2.$$

First let us consider J_1. If the integral with respect to the variables $\xi_1, \cdots, \xi_n$ were a bounded quantity that does not depend on t, then J_1 would diverge because the singularity with respect to the variable t at zero is not integrable. For this reason the estimation of the integrand will be carried out not only with the help of inequality (1.56), but also with that of the inequality

$$\left| \frac{\partial}{\partial \xi_n} \varphi(\xi) \right| \leq M(1 + |\xi|)^{-n-1}, \quad |\xi_k| \leq T. \qquad (1.58)$$

The restriction on $|\xi_k|$ in these inequalites forces us to divide the integral with respect to the variable ξ_n into a sum of two integrals as follows:

$$J_1 = \int_0^T t^{-\alpha-1} \int_{-T}^{T} \cdots \int_{-T}^{T} \left(\int_{-T}^{-T+t} + \int_{-T+t}^{T} \right) \times$$

$$\times |\varphi(\xi_1, \cdots, \xi_n) - \varphi(\xi_1, \cdots, \xi_{n-1}, \xi_n - t)| d\xi_n \cdots d\xi_1 \, dt$$

$$= J_{11} + J_{12}.$$

To estimate J_{12} we can use the inequalities (1.56) and (1.58), since $|\xi_n| \leq T$ and $|\xi_n - t| \leq T$. To estimate J_{11} we shall exploit the periodicity of the integrand and the fact that the length of the interval of integration with respect to ξ_n is equal to t. We obtain

$$|J_{11}| \leq \int_0^T t^{-\alpha-1} \int_{-T}^{T} \cdots \int_{-T}^{T} \int_{-T}^{-T+t} \times$$

$$\times (|\varphi(\xi_1, \cdots, \xi_n)| + |\varphi(\xi_1, \cdots, \xi_{n-1}, \xi_n - t)|) d\xi_n \cdots d\xi_1 \, dt = J_{111} + J_{112}.$$

Let us estimate the integral J_{112} (J_{111} is simpler to deal with). Using the periodicity of the function $\varphi(\xi)$ and inequality (1.56), we obtain

$$J_{112} = \int_0^T t^{-\alpha-1} \int_{-T}^T \cdots \int_{-T}^T \int_{-T}^{-T+t} |\varphi(\xi_1, \cdots, \xi_{n-1}, 2T + \xi_n - t)| d\xi_n \cdots d\xi_1 dt$$

$$\leq M \int_0^T t^{-\alpha-1} \int_{-T}^T \cdots \int_{-T}^T \int_{-T}^{-T+t} (1 + |(\xi_1, \cdots, \xi_{n-1}, 2T + \xi_n - t)|^n)^{-1} d\xi_n \cdots d\xi_1 dt$$

since $|2T + \xi_n - t| \leq T$. Further,

$$J_{112} \leq M_1 \int_0^T t^{-\alpha-1} \int_{-T}^T \cdots \int_{-T}^T \int_{-T}^{-T+t} \times$$

$$\times ((1 + |2T + \xi_n - t|)^n + |(\xi_1, \cdots, \xi_{n-1})|^n)^{-1} d\xi_n \cdots d\xi_1 dt$$

$$\leq M_2 \int_0^T t^{-\alpha-1} \int_{-T}^{-T+t} \int_0^\infty ((1 + |2T + \xi_n - t|)^n + r^n)^{-1} r^{n-2} dr \, d\xi_n \, dt$$

$$\leq M_3 \int_0^T t^{-\alpha-1} \int_{-T}^{-T+t} (1 + |2T + \xi_n - t|)^{-1} d\xi_n \, dt$$

$$= M_3 \int_0^T t^{-\alpha-1} \ln\{(1 + T)(1 + T - t)^{-1}\} dt \leq M_3 \int_0^\infty t^{-\alpha-1} \ln(1 + t) dt$$

$$= M_3 \left[(-\alpha)^{-1} t^{-\alpha} \ln(1 + t) \Big|_0^\infty + \alpha^{-1} \int_0^\infty t^{-\alpha} (1 + t)^{-1} dt \right]$$

$$= M_3 (\alpha \sin \pi\alpha)^{-1} \leq M_4 \alpha^{-2}.$$

Here M_4 does not depend on α or T.

The integral J_{12} is also divided into two terms such that in the first $|\xi_n - t| \geq |\xi_n|$, while in the second $|\xi_n - t| \leq |\xi_n|$:

$$J_{12} = \int_0^T t^{-\alpha-1} \int_{-T}^T \cdots \int_{-T}^T \left(\int_{-T+t}^{t/2} + \int_{t/2}^T \right) \times$$

$$\times |\varphi(\xi_1, \cdots, \xi_n) - \varphi(\xi_1, \cdots, \xi_{n-1}, \xi_n - t)| d\xi_1 \cdots d\xi_n \, dt$$

$$= J_{121} + J_{122}.$$

Let us estimate the integral J_{121}. By the inequalities (1.56), (1.58), and $|\xi_n - t| \geq |\xi_n|$, the integrand obeys the estimate

$$|\theta(\xi, t)| \equiv t^{-\alpha-1} |\varphi(\xi_1, \cdots, \xi_n) - \varphi(\xi_1, \cdots, \xi_{n-1}, \xi_n - t)|$$

$$\leq M \min\{t^{-\alpha}(1+|\xi|)^{-n-1},\ t^{-\alpha-1}(1+|\xi|)^{-n}\}.$$

From this it follows that for $0 < \rho < 1$

$$|\theta(\xi,t)| \leq M t^{-\alpha\rho}(1+|\xi|)^{-n\rho-\rho} t^{-(1+\alpha)(1-\rho)}(1+|\xi|)^{-n(1-\rho)}$$

$$= M t^{-\alpha-(1-\rho)}(1+|\xi|)^{-n-\rho}.$$

Let $0 < \rho_1 < \alpha < \rho_2 < 1$. Then

$$|\theta(\xi,t)| \leq M(1+|\xi|)^{-n-\rho_1} \min\{t^{-\alpha-1+\rho_1}, t^{-\alpha-1+\rho_2}\}.$$

Taking $\rho_1 = \frac{\alpha}{2}$ and $\rho_2 = \frac{1}{2} + \frac{\alpha}{2}$, we obtain

$$J_{121} \leq M \int_0^T \min\{t^{-\frac{\alpha}{2}-1}, t^{-\frac{\alpha}{2}-\frac{1}{2}}\} \int_{-T}^T \cdots \int_{-T}^T \int_{-T+t}^{t/2} (1+|\xi|)^{-n-\frac{\alpha}{2}} d\xi_1 \cdots d\xi_n\, dt$$

$$\leq M \int_0^T \min\{t^{-\frac{\alpha}{2}-1}, t^{-\frac{\alpha}{2}-\frac{1}{2}}\} dt \int_{\mathbf{R}^n} (1+|\xi|)^{-n-\frac{\alpha}{2}} d\xi \leq M_2(\alpha).$$

Here $M_2(\alpha) = M/\alpha^2$ and M does not depend on α and T. The integral J_{122} is estimated in a similar manner, with the point ξ replaced by $(\xi_1, \cdots, \xi_{n-1}, \xi_n - t)$ in inequality (1.59).

Consider now the integral J_2. Here $t > T$, and hence if the integral over ξ is bounded, then the integral over t converges, and its value tends to zero as $T \to \infty$. On the other hand, here one cannot use inequality (1.58), since $\xi_n - t$ does not satisfy the restrictions imposed on $|\xi_n|$. Using inequality (1.56), we estimate the integral over ξ by a quantity that depends on T. The value of this integral will also grow unboundedly as $T \to \infty$, but at a smaller rate than the rate of decay of the integral with respect to t, which guarantees that the whole integral is bounded. Let us estimate J_2. Using the periodicity of the function $\varphi(\xi)$ and inequality (1.56), we obtain

$$J_2 \leq \int_T^\infty t^{-\alpha-1} \int_{-T}^T \cdots \int_{-T}^T (|\varphi(\xi_1, \cdots, \xi_n)| + |\varphi(\xi_1, \cdots, \xi_{n-1}, \xi_n - t)|) d\xi_1 \cdots d\xi_n\, dt$$

$$\leq 2 \int_T^\infty t^{-\alpha-1} \int_{-T}^T \cdots \int_{-T}^T |\varphi(\xi_1, \cdots, \xi_n)| d\xi_1 \cdots d\xi_n\, dt$$

$$\leq 2M \frac{1}{\alpha T^\alpha} \int_{-T}^T \cdots \int_{-T}^T (1+|\xi|)^{-n} d\xi \leq M_1 \frac{1}{\alpha T^\alpha} \ln(1+T) \leq M_2/\alpha,$$

where M_2 does not depend on α and T.

Combining the estimates of the integrals J_1 and J_2 we conclude that the integral in the inequality (1.57) is bounded by $M(\alpha) = M/\alpha^2$. Hence, for $y \neq 0$ we have

$$|\mathcal{G}^h(y,\lambda)| \leq \alpha M(\alpha)|\lambda|^{-\frac{\alpha}{m}}|y|^{m-n-\alpha}\exp\{-a_0|\lambda|^{\frac{1}{m}}(|y_1| + \cdots + |y_n|)\}. \qquad (1.60)$$

If $y = 0$, then similarly to the way we proceeded in Lemma 1.10, we obtain the estimates

$$|\mathcal{G}^h(0,\lambda)| \leq Mh^{m-n} \leq ML^{\frac{\alpha}{m}}|\lambda|^{-\frac{\alpha}{m}}h^{m-n-\alpha}, \quad \text{if } m < n,$$

$$|\mathcal{G}^h(0,\lambda)| \leq M\left|\ln|\lambda^{\frac{1}{m}}h|\right|$$

$$\leq M|\lambda^{\frac{1}{m}}h|^{-\alpha}(|\lambda|^{\frac{1}{m}}h)^{\alpha}\ln|\lambda^{\frac{1}{m}}h| \leq M_1|\lambda|^{-\frac{\alpha}{m}}h^{-\alpha}, \quad \text{if } m = n.$$

Combining these estimates with the estimate (1.60), we obtain the assertion of Lemma 1.12.

To conclude this subsection, let us formulate a result on point estimates of the fundamental solution of the resolvent equation. These estimate will enable us in the next subsection to establish the positivity of elliptic difference operators with constant coefficients.

Theorem 1.1. *Suppose that for $0 \leq |\xi_k h| \leq \pi$ the following conditions are satisfied: a) $|A(\xi h, h)| \geq M_0|\xi|^m$ with $M_0 > 0$, and b) $|\arg A(\xi h, h)| \leq \phi < \phi_0 < \pi$. Then for some $M > 0$ and $a > 0$ the following estimates hold:*
1) *If $|\lambda| \geq Lh^{-m}$, then*

$$|\mathcal{G}^h(y,\lambda)| \leq M\exp\{-ah^{-1}(|y_1| + \cdots + |y_n|)\}h^{-n}\frac{1}{|\lambda|^{1+m\alpha}h^{m\alpha(m-1)}|y|^{m\alpha}},$$

$$0 \leq \alpha \leq m^{-1}, \ y \neq 0.$$

2) *If $0 < |\lambda| \leq Lh^{-m}$ and $|\arg \lambda| \leq \pi - \phi_0$, then for $m > n$,*

$$|\mathcal{G}^h(y,\lambda)| \leq M\exp\{-a|\lambda|^{\frac{1}{m}}(|y_1| + \cdots + |y_n|)\}|\lambda|^{\frac{n}{m}-1}.$$

3) *If $0 < |\lambda| \leq Lh^{-m}$ and $|\arg \lambda| \leq \pi - \phi_0$, then for $m \leq n$,*

$$|\mathcal{G}^h(y,\lambda)| \leq M\alpha^{-1}\exp\{-a|\lambda|^{\frac{1}{m}}(|y_1| + \cdots + |y_n|)\}\times$$

$$\times |\lambda|^{-\frac{\alpha}{m}}(h+|y|)^{m-n-\alpha}, \quad 0 < \alpha < 1.$$

5. Positivity of homogeneous elliptic difference operators with constant coefficients.

By Lemma 1.6, to establish the positivity of the difference operator A_h in the space C_h we need to prove that the inequality

$$\sum_{y \in \mathbf{R}_h^n} |\mathcal{G}^h(y, \lambda)| h^n \equiv M_h(\lambda) \le M|\lambda|^{-1},$$

where M does not depend on h.

In accordance with the three estimates of the fundamental solution given in Theorem 1.1, we shall consider three cases of estimation of $M_h(\lambda)$. In the case where $|\lambda| \ge Lh^{-m}$ we obtain

$$M_h(\lambda) \le Mh^{-n}|\lambda|^{-1} \sum_{y \in \mathbf{R}_h^n} \exp\{-ah^{-1}(|y_1| + \cdots + |y_n|)\}h^n$$

$$\le M|\lambda|^{-1} \sum_{k \in \mathbf{R}_1^n} \exp\{-a(|k_1| + \cdots + |k_n|)\} = M|\lambda|^{-1}\left(\sum_{k=0}^{\infty} e^{-ak}\right)^n = M_1|\lambda|^{-1}.$$

In the case where $0 < |\lambda| \le Lh^{-m}$ and $m > n$ we obtain

$$M_h(\lambda) \le M|\lambda|^{-1} \sum_{y \in \mathbf{R}_h^n} \exp\{-a|\lambda|^{\frac{1}{m}}(|y_1| + \cdots + |y_n|)\}|\lambda|^{\frac{n}{m}}h^n$$

$$\le M_1|\lambda|^{-1}\left(\sum_{k=0}^{\infty} \exp\{-a|\lambda|^{\frac{1}{m}}kh\}|\lambda|^{\frac{1}{m}}h\right)^n$$

$$\le M_2|\lambda|^{-1}\left(\int_0^{\infty} \exp\{-a|\lambda|^{\frac{1}{m}}t\}|\lambda|^{\frac{1}{m}}dt\right)^n \le M_3|\lambda|^{-1}.$$

Finally, in the case where $0 < |\lambda| \le Lh^{-m}$, $m \le n$ and $0 < \alpha < 1$ we obtain

$$M_h(\lambda) \le M|\lambda|^{-\frac{\alpha}{m}} \sum_{y \in \mathbf{R}_h^n} \exp\{-a|\lambda|^{\frac{1}{m}}(|y_1| + \cdots + |y_n|)\}(h + |y|)^{m-n-\alpha}h^n$$

$$\le M|\lambda|^{-1} \sum_{y \in \mathbf{R}_h^n} \exp\{-a|\lambda|^{\frac{1}{m}}(|y_1| + \cdots + |y_n|)\}(|\lambda|^{\frac{1}{m}}(h + |y|))^{m-n-\alpha}|\lambda|^{\frac{n}{m}}h^n$$

$$\leq M|\lambda|^{-1}\left(\sum_{k=0}^{\infty}\exp\{-a|\lambda|^{\frac{1}{m}}kh\}(|\lambda|^{\frac{1}{m}}kh)^{\frac{m-n-\alpha}{n}}|\lambda|^{\frac{1}{m}}h\right)^{n}$$

$$\leq M|\lambda|^{-1}\left(\int_{0}^{\infty}\exp\{-a|\lambda|^{\frac{1}{m}}t\}(|\lambda|^{\frac{1}{m}}t)^{\frac{m-n-\alpha}{n}}|\lambda|^{\frac{1}{m}}dt\right)^{n}$$

$$\leq M|\lambda|^{-1}\left(\int_{0}^{\infty}\exp\{-a\tau\}\tau^{\frac{m-n-\alpha}{n}}d\tau\right)^{n}\leq M_{1}|\lambda|^{-1},$$

because $(m-n-\alpha)n^{-1} > -1$. Hence, in all three cases $M_h(\lambda) \leq M|\lambda|^{-1}$. Thus, we have the following

Theorem 1.2. *Let the assumptions of Theorem 1.1 be satisfied. Then the operator $A_h + \varepsilon$ is strongly positive for $\varepsilon > 0$, and its spectral angle $\phi(A_h + \varepsilon, C_h) \leq \phi_0$.*

We conclude this subsection by indicating a way of constructing positive difference operators. Let us consider the difference operator

$$\Delta_{ks} = h^{-1}\sum_{j=1}^{s}\alpha_{j}[(\Delta_{k+} + I)^{j} - (I - \Delta_{k-})^{j}].$$

The real coefficients α_j can be chosen so that Δ_{ks} will approximate the operator $\partial/\partial x_k$ to order $2s$. The symbol of the operator Δ_{ks} is

$$h^{-1}\sum_{j=1}^{s}\alpha_{j}(\exp\{ij\xi_{k}h\} - \exp\{-j\xi_{k}h\}) = 2h^{-1}i\sum_{j=1}^{s}\alpha_{j}\sin(j\xi_{k}h) = id(\xi_{k}h, h).$$

Since the α_j are real numbers, $d(\xi_k h, h)$ is real. Furthermore, since $2\sum_{j=1}^{s}\alpha_j j = 1$, $\lim_{h\to 0} d(\xi_k h, h) = \xi_k$. It follows that $d(\xi_k h, h) \neq 0$ for small h, provided that $\xi_k \neq 0$. Let us associate to the operator (1.1) the difference operator

$$A_{hs} = \sum_{|r|=m} a_{r}\Delta_{1s}^{r_{1}}\cdots\Delta_{ns}^{r_{n}},$$

which approximates the operator A to order no less than $2s$. Suppose that the symbol of A satisfies condition (1.3) for $\xi \neq 0$. Then the symbol $A(\xi h, h)$ of the difference operator A_{hs} will also satisfy the conditions of Theorem 1.1, since $A(\xi h, h)$ is obtained from $A(\xi)$ by substituting $d(\xi_k h, h)$ in place of ξ_k. Therefore, the conditions of Theorem 1.2 are satisfied, and consequently the operator $A_{hs} + \varepsilon$ is positive.

6. Point estimates of the fundamental solution of the resolvent equation in the case $m \leq n$.

The estimates obtained in Theorem 1.1 allowed us to establish the positivity of the difference operator A_h in the space C_h. However, the estimate 3) of that theorem can be sharpened further. In the case $m = n$, to this end it suffices to choose the parameter α in the aforementioned estimate in an optimal manner.

Lemma 1.13. *Let the assumptions of Theorem 1.1 be satisfied and $m = n$. Then for $0 < |\lambda| \leq Lh^{-m}$ and $|\arg \lambda| \leq \pi - \phi_0$ the following estimate holds:*

$$|\mathcal{G}^h(y, \lambda)| \leq M \exp\{-a|\lambda|^{\frac{1}{m}}(|y_1| + \cdots |y_n|)\}(\ln(1 + \gamma^{-1}) + 1).$$

Here $\gamma = |\lambda|^{\frac{1}{m}}(h + |y|)$.

Proof. In the right-hand side of the estimate 3) of Theorem 1.1, let us isolate the factor that depends on α. It has the form $\alpha^{-1}|\lambda|^{-\frac{\alpha}{m}}(h + |y|)^{-\alpha} = (\alpha\gamma^\alpha)^{-1}$. If $\gamma \geq e^{-1}$, then $\min(\alpha\gamma^\alpha)^{-1} \leq e$. If $0 < \gamma < e^{-1}$, then taking $\alpha = |\ln \gamma|^{-1}$ we obtain $(\alpha\gamma^\alpha)^{-1} \leq e^{-1}\ln\gamma^{-1}$. Combining these estimates, we see that $\min(\alpha\gamma^\alpha)^{-1} \leq M(\ln(1 + \gamma^{-1}) + 1)$, which concludes the proof of the lemma.

Let us sharpen the estimate 3) of Theorem 1.1 in the case where $m < n$. For any positive operator A we have

$$(\lambda + A)^{-1} = \int_0^\infty rt^{r-1}(\lambda + t + A)^{-(r+1)}dt. \tag{1.61}$$

Here $r \geq 1$ is an integer. This formula allows us to express the fundamental solution as an integral of a convolution of fundamental solutions. Using the inequalities of three kinds given in Theorem 1.1 to estimate the convolution, we succeed in eliminating the parameter α.

Lemma 1.14. *Let the assumptions of Lemma 1.10 be satisfied. Then for $m < n$ the following inequality holds:*

$$|\mathcal{G}^h(y, \lambda)| \leq M \exp\{-a|\lambda|^{\frac{1}{m}}|y|\}(h + |y|)^{m-n}. \tag{1.62}$$

Proof. Using formulas (1.61) and (1.22) we obtain

$$(2\pi)^{-n} \sum_{y \in \mathbf{R}_h^n} \mathcal{G}^h(x - y, \lambda) f^h(y) h^n$$

$$= \int_0^\infty (2\pi)^{-n} \sum_{z \in \mathbf{R}_h^n} \mathcal{G}^h(x - z, \lambda + t) h^n \cdot (2\pi)^{-n} \sum_{y \in \mathbf{R}_h^n} \mathcal{G}^h(z - y, \lambda + t) f^h(y) h^n dt$$

for any $f^h(y) \in C_h$. Hence, the following identity holds:

$$\mathcal{G}^h(y, \lambda) = (2\pi)^{-n} \int_0^\infty \sum_{z \in \mathbf{R}_h^n} \mathcal{G}^h(y - z, \lambda + t) \cdot \mathcal{G}^h(z, \lambda + t) h^n dt. \tag{1.63}$$

For $L > 0$ we have

$$|\mathcal{G}^h(y, \lambda)| \leq (2\pi)^{-n} \left(\int_{|\lambda + t| \leq Lh^{-m}} + \int_{|\lambda + t| \geq Lh^{-m}} \right) \times$$

$$\times \sum_{z \in \mathbf{R}_h^n} |\mathcal{G}^h(y - z, \lambda + t)| \cdot |\mathcal{G}^h(z, \lambda + t)| h^n dt = J_1 + J_2$$

Let us estimate J_1. Using (1.54) ($a_0 = a > 0$) and the assumption $m < n$, we have

$$J_1 \leq \int_{|\lambda + t| \leq Lh^{-m}} \sum_{z \in \mathbf{R}_h^n} M^2(\alpha) \exp\{-a|\lambda + t|^{\frac{1}{m}}(|y - z| + |z|)\} \times$$

$$\times |\lambda + t|^{-\frac{2\alpha}{m}} (h + |y - z|)^{m-n-\alpha} (h + |z|)^{m-n-\alpha} h^n dt.$$

For $0 < \varepsilon < 1 - \alpha$ we obtain

$$J_1 \leq M^2(\alpha) \int_{|\lambda + t| \leq Lh^{-m}} \exp\left\{-\frac{a}{2}|\lambda + t|^{\frac{1}{m}}|y|\right\} |\lambda + t|^{-\frac{2\alpha}{m}} \times$$

$$\times \left(\sum_{|y - z| < |z|} \exp\{-a|\lambda + t|^{\frac{1}{m}}|y - z|\}(h + |y - z|)^{m-n-\alpha-\varepsilon}(h + |z|)^{m-n-\alpha+\varepsilon} h^n \right.$$

$$\left. + \sum_{|y - z| \geq |z|} \exp\{-a|\lambda + t|^{\frac{1}{m}}|z|\}(h + |y - z|)^{m-n-\alpha+\varepsilon}(h + |z|)^{m-n-\alpha-\varepsilon} h^n \right) dt.$$

Since $|y/2| \leq |z|$ when $|y - z| \leq |z|$ and $|y/2| \leq |y - z|$ when $|y - z| \geq |z|$, we have

$$J_1 \leq M^2(\alpha) \int_{|\lambda + t| \leq Lh^{-m}} \exp\left\{-\frac{a}{2}|\lambda + t|^{\frac{1}{m}}|y|\right\} |\lambda + t|^{-\frac{2\alpha}{m}} \times$$

$$\times \left(\sum_{|y - z| < |z|} \exp\{-a|\lambda + t|^{\frac{1}{m}}|y - z|\}(h + |y - z|)^{m-n-\alpha-\varepsilon}(h + |y/2|)^{m-n-\alpha+\varepsilon} h^n \right.$$

$$+ \sum_{|y-z|\geq|z|} \exp\{-a|\lambda + t|^{\frac{1}{m}}|z|\}(h+|y/2|)^{m-n-\alpha+\varepsilon}(h+|z|)^{m-n-\alpha-\varepsilon}h^n \Bigg) dt$$

$$\leq M^2(\alpha)\int_{|\lambda+t|\leq Lh^{-m}} \exp\left\{-\frac{a}{2}|\lambda + t|^{\frac{1}{m}}|y|\right\}|\lambda+t|^{-\frac{2\alpha}{m}}(h+|y|)^{m-n-\alpha+\varepsilon}\times$$

$$\times \sum_{z\in\mathbf{R}_h^n} \exp\{-a|\lambda+t|^{\frac{1}{m}}|z|\}(h+|z|)^{m-n-\alpha-\varepsilon}h^n dt.$$

Let us estimate the sum under the integral sign. Since $|z|\geq|z_j|$, we have

$$\sum_{z\in\mathbf{R}_h^n} \exp\{-a|\lambda+t|^{\frac{1}{m}}|z|\}(h+|z|)^{m-n-\alpha-\varepsilon}h^{-n}$$

$$\leq \sum_{z\in\mathbf{R}_h^n}\prod_{j=1}^n (h+|z_j|)^{\frac{(m-n-\alpha-\varepsilon)}{n}} \exp\left\{-\frac{a}{n}|\lambda+t|^{\frac{1}{m}}|z_j|\right\}h.$$

Since the function $x^{-\gamma}\exp\{-ax\}$ is monotonically decreasing for $\gamma>0$, $x>0$,

$$\sum_{k=0}^\infty \exp\left\{-\frac{a}{n}|\lambda+t|^{\frac{1}{m}}k\right\}(1+k)^{(m-n-\alpha-\varepsilon)n^{-1}}$$

$$\leq \int_0^\infty \exp\left\{-\frac{a}{n}|\lambda+t|^{\frac{1}{m}}\tau\right\}\tau^{\frac{m-n-\alpha-\varepsilon}{n}}d\tau$$

$$= \int_0^\infty \exp\left\{-\frac{a}{n}s\right\}s^{\frac{m-n-\alpha-\varepsilon}{n}}ds\,|\lambda+t|^{-\frac{1}{m}}|\lambda+t|^{-\frac{m-n-\alpha-\varepsilon}{nm}}$$

$$= M|\lambda+t|^{-\frac{1}{m}}|\lambda+t|^{-\frac{m-n-\alpha-\varepsilon}{nm}},$$

whence

$$\sum_{z\in\mathbf{R}_h^n}\exp\{-a|\lambda+t|^{\frac{1}{m}}|z|\}(h+|z|)^{m-n-\alpha-\varepsilon}h^n$$

$$\leq M^n|\lambda+t|^{-\frac{n}{m}}|\lambda+t|^{-\frac{m-n-\alpha-\varepsilon}{m}} = M^n|\lambda+t|^{\frac{-m+\alpha+\varepsilon}{m}}.$$

Since $|\lambda+t|^{\frac{1}{m}}h<L^{\frac{1}{m}}$, we have

$$J_1 \leq M^n M^2(\alpha)\int_{|\lambda+t|\leq Lh^{-m}} \exp\left\{-\frac{a}{2}|\lambda+t|^{\frac{1}{m}}(h+|y|)\right\}\times$$

$$\times|\lambda+t|^{-\frac{2\alpha}{m}}(h+|y|)^{m-n-\alpha+\varepsilon}|\lambda+t|^{\frac{-m+\alpha+\varepsilon}{m}}dt.$$

Further, since $|\arg\lambda|\leq\pi-\phi_0$, $|\lambda+t|^{\frac{1}{m}}\geq M(\phi_0,m)(|\lambda|^{\frac{1}{m}}+t^{\frac{1}{m}})$. Consequently,

$$J_1 \leq M_1\exp\{-a_1|\lambda|^{\frac{1}{m}}(h+|y|)\}(h+|y|)^{m-n-\alpha+\varepsilon}\times$$

$$\times \int_0^\infty t^{\frac{-m-\alpha+\varepsilon}{m}} \exp\{-a_1 t^{\frac{1}{m}}(h+|y|)\}dt$$

$$\le M_2 \exp\{-a_2|\lambda|^{\frac{1}{m}}(h+|y|)\}(h+|y|)^{m-n-\alpha+\varepsilon}\times$$

$$\times \int_0^\infty \exp\{-a_2\tau^{\frac{1}{m}}\}\tau^{\frac{-m-\alpha+\varepsilon}{m}}d\tau\,(h+|y|)^{-m}(h+|y|)^{m+\alpha-\varepsilon}$$

$$= M_2 \exp\{-a_2|\lambda|^{\frac{1}{m}}(h+|y|\}(h+|y|)^{m-n}\int_0^\infty m\exp\{-a_2 s\}s^{-1+\varepsilon-\alpha}ds.$$

Taking $\alpha \in (0, 1/2)$ and $\varepsilon \in (\alpha, 1)$, we obtain

$$J_1 \le M_3 \exp\{-a_3|\lambda|^{\frac{1}{m}}|y|\}(h+|y|)^{m-n}$$

for some $M_3 > 0$ and $a_3 > 0$. Now let us estimate J_2 using Lemma 1.7. We obtain

$$J_2 \le \int_{|\lambda+t|\ge Lh^{-m}} \sum_{z\in\mathbf{R}_h^n} M^2 \times$$

$$\times \exp\{-ah^{-1}(|z_1| + \cdots + |z_n| + |y_1 - z_1| + \cdots + |y_n - z_n|)\}h^{-n}|\lambda+t|^{-2}dt$$

$$\le \int_{|\lambda+t|\ge Lh^{-m}} M_1 \exp\left\{-\frac{a}{2h}(|y_1| + \cdots + |y_n|)\right\} \times$$

$$\times h^{-n}|\lambda+t|^{-2}dt \sum_{z\in\mathbf{R}_h^n} \exp\left\{-\frac{a}{2h}(|z_1| + \cdots + |z_n|)\right\}.$$

Further,

$$\sum_{z\in\mathbf{R}_h^n} \exp\left\{-\frac{a}{2h}(|z_1| + \cdots + |z_n|)\right\} \le M_2.$$

Therefore,

$$J_2 \le M_3 \exp\left\{-\frac{a}{2h}(|y_1| + \cdots + |y_n|)\right\} h^{-n} \int_{|\lambda+t|\ge Lh^{-m}} |\lambda+t|^{-2}dt.$$

Since $|\arg\lambda| \le \pi - \phi_0$, $|\lambda+t|^2 \ge M(\phi_0)(|\lambda| + t)^2$. Consequently

$$\int_{|\lambda+t|\ge Lh^{-m}} |\lambda+t|^{-2}dt \le \frac{1}{M(\phi_0)} \int_{|\lambda+t|\ge Lh^{-m}} (|\lambda| + t)^{-2}dt = \frac{h^m}{LM(\phi_0)},$$

whence

$$J_2 \le M_4 \exp\left\{-\frac{a}{2h}(|y_1| + \cdots + |y_n|)\right\} h^{m-n} \le$$

$$M_4 \exp\left\{-\frac{a}{4h}(|y_1| + \cdots + |y_n|)\right\} h^{m-n}(1+|k|)^{m-n}\times$$

$$\times \exp\left\{-\frac{a}{4}(|k_1| + \cdots + |k_n|)\right\} (1 + |k|)^{n-m}$$

$$\leq M_5 \exp\left\{-\frac{a}{4h}(|y_1| + \cdots + |y_n|)\right\} (h + |y|)^{m-n}.$$

Under the assumptions of the lemma, $h^{-1} > L^{-1}|\lambda|^{\frac{1}{m}}$; hence,

$$J_2 \leq M_6 \exp\{-a_1|\lambda|^{\frac{1}{m}}|y|\}(h + |y|)^{m-n}.$$

Combining the estimates of the integrals J_1 and J_2 we complete the proof of Lemma 1.14.

In this way we have established the following result

Theorem 1.3. *Suppose that for $|\xi_k h| \leq \pi$ the following conditions are satisfied:* a) $|A(\xi h, h)| \geq M_0|\xi|^m$, *and* b) $|\arg A(\xi h, h)| \leq \phi < \phi_0 < \pi$. *Let* $0 < |\lambda| \leq Lh^{-m}$ *and* $|\arg \lambda| \leq \pi - \phi_0$. *Then the following estimates hold:*

1) *If $m > n$,*

$$|\mathcal{G}^h(y, \lambda)| \leq M \exp\{-a|\lambda|^{\frac{1}{m}}(|y_1| + \cdots + |y_n|)\}|\lambda|^{\frac{n}{m}-1}.$$

2) *If $m = n$,*

$$|\mathcal{G}^h(y, \lambda)| \leq M \exp\{-a|\lambda|^{\frac{1}{m}}(|y_1| + \cdots + |y_n|)\}[1 + \ln\{(|\lambda|^{\frac{1}{m}}(h + |y|))^{-1} + 1\}].$$

3) *If $m < n$,*

$$|\mathcal{G}^h(y, \lambda)| \leq M \exp\{-a|\lambda|^{\frac{1}{m}}|y|\}(h + |y|)^{m-n}.$$

7. Point estimates of difference derivatives of the fundamental solution of the resolvent equation.

As the difference derivative of order r with respect to the j-th variable we shall use in what follows the operators $h^{-r}\Delta_{j\pm}^r = \mathcal{D}_{jh}^r$, since any difference derivative with respect to the j-th variable can be represented as a linear combination of translates of the operator $\mathcal{D}_{jh}^r$.

The estimates of the mixed difference derivatives do not differ from those of the derivatives with respect to a single variable since in both cases one uses only upper bounds on the symbol of the difference derivative, which depend on the order of the derivative but not on the number of the variable. For this reason, from now on the index j in the operator $\mathcal{D}^r_{jh}$ will be omitted.

Using formula (1.23), we obtain

$$\mathcal{D}^r_h \mathcal{G}^h(y, \lambda) = \pm h^{-r} \int_{|\xi_1 h| \leq \pi} \cdots \int_{|\xi_n h| \leq \pi} (\exp\{\pm i\xi_j h\} - 1)^r \times$$

$$\times (A(\xi h, h) + \lambda)^{-1} \exp\{i(\xi, y)\} d\xi_1 \cdots d\xi_n$$

$$= \int_{|\xi_1 h| \leq \pi} \cdots \int_{|\xi_n h| \leq \pi} B^r(\xi h, h)(A(\xi h, h) + \lambda)^{-1} \exp\{i(\xi, y)\} d\xi_1 \cdots d\xi_n, \quad (1.64)$$

where $B^r(\xi h, h)$ denotes the symbol of the difference derivative $\mathcal{D}^r_h$. The latter is an analytic $2\pi h^{-1}$-periodic function, bounded for $|\xi_k h| \leq \pi$, $k = 1, \cdots, n$. From (1.18) we obtain the estimate

$$|B^r(zh, h)| \leq M|z|^r, \tag{1.65}$$

which holds for all $z \in \Omega$. The next three lemmas are generalizations of the estimates obtained in Theorems 1.1 and 1.3.

Lemma 1.15. Let $|\lambda| \geq Lh^{-m}$. Then

$$|\mathcal{D}^r_h \mathcal{G}^h(y, \lambda)| \leq M \exp\{-ah^{-1}(|y_1| + \cdots + |y_n|)\} h^{m-r-n}(|\lambda| h^m + 1)^{-1}. \tag{1.66}$$

Proof. As in the proof of Lemma 1.7, we can write

$$\mathcal{D}^r_h \mathcal{G}^h(y, \lambda) = \int_{|\xi_1 h| \leq \pi} \cdots \int_{|\xi_n h| \leq \pi} B^r(\bar{\xi} h, h)(A(\bar{\xi} h, h) + \lambda)^{-1} \exp\{i(\bar{\xi}, y)\} d\xi_1 \cdots d\xi_n$$

where $\bar{\xi}_k = \xi_k + iah^{-1}\mathrm{sign}\, y_k$. Using this expression and the estimates (1.18) and (1.65), we obtain

$$|\mathcal{D}^r_h \mathcal{G}^h(y, \lambda)| \leq M \exp\{-ah^{-1}(|y_1| + \cdots + |y_n|)\} \times$$

$$\times \int_{|\xi_1 h| \leq \pi} \cdots \int_{|\xi_n h| \leq \pi} |B^r(\bar{\xi} h, h)|(|\lambda| - |A(\bar{\xi} h, h)|)^{-1} d\xi_1 \cdots d\xi_n$$

$$\leq M_1 \exp\{-ah^{-1}(|y_1| + \cdots + |y_n|)\}(|\lambda| - M_2 h^{-m})^{-1} h^{-r}(2\pi h^{-1})^n$$

$$\leq M_3 \exp\{-ah^{-1}(|y_1| + \cdots + |y_n|)\} h^{m-r-n}(|\lambda| h^m + 1)^{-1}.$$

Lemma 1.15 is proved.

Lemma 1.16. *Suppose the conditions* a) *and* b) *of Lemma 1.9 are satisfied. Let* $0 < |\lambda| \leq Lh^{-m}$ *and* $|\arg \lambda| \leq \pi - \phi_0$. *Then for* $m - r > n$ *we have the estimate*

$$|\mathcal{D}_h^r \mathcal{G}^h(y, \lambda)| \leq M \exp\{-a|\lambda|^{\frac{1}{m}}(|y_1| + \cdots + |y_n|)\}|\lambda|^{\frac{n+r}{m}-1}. \tag{1.67}$$

Proof. The substitution $\xi_k = |\lambda|^{\frac{1}{m}} \overline{\xi}_k$ in the integral of formula (1.64) yields

$$\mathcal{D}_h^r \mathcal{G}^h(y, \lambda) = |\lambda|^{\frac{n+r}{m}-1} \int_{|\overline{\xi}_1| \leq \pi |\lambda^{\frac{1}{m}} h|^{-1}} \cdots \int_{|\overline{\xi}_n| \leq \pi |\lambda^{\frac{1}{m}} h|^{-1}} |\lambda|^{-\frac{r}{m}} B^r(|\lambda|^{\frac{1}{m}} \overline{\xi} h, h) \times$$

$$\times (A(|\lambda|^{\frac{1}{m}} \overline{\xi} h, h)|\lambda|^{-1} + \omega)^{-1} \exp\{i|\lambda|^{\frac{1}{m}}(\overline{\xi}, y)\} d\overline{\xi}_1 \cdots d\overline{\xi}_n.$$

Then, repeating the proof of Lemma 1.9 and using the estimates (1.65) and (1.38), we obtain

$$|\mathcal{D}_h^r \mathcal{G}^h(y, \lambda)| \leq M |\lambda|^{\frac{n+r}{m}-1} \exp\{-a|\lambda|^{\frac{1}{m}}(|y_1| + \cdots + |y_n|)\}| \int_{\mathbf{R}^n} |\overline{\xi}|^r (1 + |\overline{\xi}|^{\overset{\bullet}{m}})^{-1} d\overline{\xi}.$$

Since $m - r > n$, the last integral is bounded. Lemma 1.16 is proved.

Lemma 1.17. *Suppose the conditions of Lemma 1.12 are satisfied. Then for* $m - r \leq n$ *and* $0 < \alpha < 1$ *we have the estimate*

$$|\mathcal{D}_h^r \mathcal{G}^h(y, \lambda)| \leq M\alpha^{-1}|\lambda|^{-\frac{\alpha}{m}}(h + |y|)^{m-n-r-\alpha} \exp\{-a|\lambda|^{\frac{1}{m}}(|y_1| + \cdots + |y_n|)\}. \tag{1.68}$$

Proof. By analogy to representation (1.40), we can write

$$\mathcal{D}_h^r \mathcal{G}^h(y, \lambda) = |\lambda|^{\frac{n+r}{m}-1} \exp\{-a|\lambda|^{\frac{1}{m}}(|y_1| + \cdots + |y_n|)\} \times$$

$$\times \int_{|\xi_1| \leq \pi |\lambda^{\frac{1}{m}} h|^{-1}} \cdots \int_{|\xi_n| \leq \pi |\lambda^{\frac{1}{m}} h|^{-1}} \exp\{i|\lambda|^{\frac{1}{m}}(\xi, y)\} \varphi(\xi) d\xi_1 \cdots d\xi_n. \tag{1.69}$$

Here $\varphi(\xi) = |\lambda|^{-\frac{r}{m}} B(|\lambda|^{\frac{1}{m}}(\xi + ia \operatorname{sign} y)h, h) f(\xi)$. By the estimates (1.65) and (1.38),

$$|\varphi(\xi)| \leq M_1(1 + |\xi|^{m-r})^{-1} \tag{1.70}$$

for $|\xi_k| \leq \pi |\lambda^{\frac{1}{m}} h|^{-1}$. In the same way that we established the estimate (1.44), we can show that

$$\left| \left(\frac{\partial}{\partial \xi_k} \right)^s \varphi(\xi) \right| \leq M_2(1 + |\xi|^{m-r+s})^{-1} \tag{1.71}$$

for $|\xi_k| \le \pi |\lambda^{\frac{1}{m}} h|^{-1}$. As in the proof of Lemma 1.12, we may assume for $y \ne 0$ that $y_n = \max_{1 \le k \le n} |y_k|$. Then, upon integrating by parts $n - m + r$ times with respect to the n-th variable in the integral appearing in the right-hand side of (1.69), we obtain

$$\mathcal{D}_h^r \mathcal{G}^h(y, \lambda) = |\lambda|^{\frac{n+r}{m}-1} \exp\{-a|\lambda|^{\frac{1}{m}}(|y_1| + \cdots + |y_n|)\}(i|\lambda|^{\frac{1}{m}} y_n)^{m-n-r} \times$$

$$\times \int_{|\xi_1| \le \pi |\lambda^{\frac{1}{m}} h|^{-1}} \cdots \int_{|\xi_n| \le \pi |\lambda^{\frac{1}{m}} h|^{-1}} \exp\{i|\lambda|^{\frac{1}{m}}(\xi, y)\} \theta(\xi) d\xi_1 \cdots d\xi_n,$$

where $\theta(\xi) = (\varphi(\xi))^{(n-m+r)}$. By (1.71),

$$|\theta(\xi)| \le M(1 + |\xi|^n)^{-1} \tag{1.72}$$

for $|\xi_k| \le \pi |\lambda^{\frac{1}{m}} h|^{-1}$. Repeating the proof of Lemma 1.12 with $\varphi(\xi)$ replaced by $\theta(\xi)$ we obtain the assertion of Lemma 1.17.

In the case $m = n + r$, by choosing α optimally, as we did in Lemma 1.13, we obtain the estimate

$$|\mathcal{D}_h^r \mathcal{G}^h(y, \lambda)| \le M \exp\{-a|\lambda|^{\frac{1}{m}}(|y_1| + \cdots + |y_n|)\} \times$$

$$\times \left(1 + \ln\{(|\lambda|^{\frac{1}{m}}(h + |y|))^{-1} + 1\}\right). \tag{1.73}$$

Let us generalize Lemma 1.14.

Lemma 1.18. *Suppose the conditions of Lemma 1.14 are satisfied. Then for $m - r < n$ we have the inequality*

$$|\mathcal{D}_h^r \mathcal{G}^h(y, \lambda)| \le M \exp\{-a|\lambda|^{\frac{1}{m}} |y|\}(h + |y|)^{m-n-r}. \tag{1.74}$$

Proof. By formula (1.63), we can write

$$\mathcal{D}_h^r \mathcal{G}^h(y, \lambda) = \int_0^\infty (2\pi)^{-n} \sum_{z \in \mathbf{R}_h^n} \mathcal{D}_h^r \mathcal{G}^h(y - z, \lambda + t) \mathcal{G}^h(z, \lambda + t) h^n dt. \tag{1.75}$$

Let $0 < r < m$ and $m \le n$. Then, using formula (1.75) and Lemmas 1.9, 1.14, 1.15, and 1.17, we obtain

$$|\mathcal{D}_h^r \mathcal{G}^h(y, \lambda)| \le \int_{|\lambda+t| \le Lh^{-m}} M^2(\alpha) \sum_{z \in \mathbf{R}_h^n} |\lambda + t|^{-\frac{2\alpha}{m}} \exp\{-a|\lambda + t|^{\frac{1}{m}}(|y - z| + |z|)\} \times$$

$$\times (h + |y - z|)^{m-n-r-\alpha}(h + |z|)^{m-n-\alpha}h^n dt$$

$$+ \int_{|\lambda+t|\geq Lh^{-m}} M^2(\alpha) \sum_{z\in\mathbf{R}_h^n} |\lambda + t|^{-1}(|\lambda + t|h^m + 1)^{-1} \times$$

$$\times h^{-n}h^{m-r-n} \exp\{-ah^{-1}(|y - z| + |z|)\}h^n dt = J_1 + J_2.$$

Let us estimate J_1 following our procedure in Lemma 1.14:

$$J_1 \leq M_1 \int_{|\lambda+t|\leq Lh^{-m}} |\lambda + t|^{-\frac{2\alpha}{m}} \exp\left\{-\frac{a}{2}|\lambda + t|^{\frac{1}{m}}|y|\right\} \left(\sum_{|y-z|\leq|z|} + \sum_{|y-z|>|z|}\right) \times$$

$$\times \exp\left\{-\frac{a}{2}|\lambda + t|^{\frac{1}{m}}(|y - z| + |z|)\right\}(h + |y - z|)^{m-n-r-\alpha}(h + |z|)^{m-n-\alpha}h^n dt$$

$$\leq M_2 \int_{|\lambda+t|\leq Lh^{-m}} |\lambda + t|^{-\frac{2\alpha}{m}} \exp\left\{-\frac{a}{2}|\lambda + t|^{\frac{1}{m}}(h + |y|)\right\} \times$$

$$\times \left(\sum_{|y-z|\leq|z|} (h + |y - z|)^{m-n-r-\alpha}(h + |y/2|)^{m-n-\alpha} \exp\{-a|\lambda + t|^{\frac{1}{m}}|y - z|\}\right.$$

$$\left.+ \sum_{|y-z|>|z|} (h + |y/2|)^{m-n-\alpha}(h + |z|)^{m-n-r-\alpha} \exp\{-a|\lambda + t|^{\frac{1}{m}}|z|\}\right)h^n dt$$

$$\leq M_3 \int_{|\lambda+t|\leq Lh^{-m}} |\lambda + t|^{-\frac{2\alpha}{m}} \exp\left\{-\frac{a}{2}|\lambda + t|^{\frac{1}{m}}(h + |y|)\right\}(h + |y|)^{m-n-\alpha} \times$$

$$\times \sum_{z\in\mathbf{R}_h^n} \exp\{-a|\lambda + t|^{\frac{1}{m}}|z|\}(h + |z|)^{m-n-r-\alpha}h^n dt.$$

Since $m - n - r - \alpha > -n$, it follows that

$$\sigma \equiv \sum_{z\in\mathbf{R}_h^n} \exp\{-a|\lambda + t|^{\frac{1}{m}}|z|\}(h + |z|)^{m-n-r-\alpha}h^n \leq M|\lambda + t|^{\frac{r+\alpha-m}{m}}. \qquad (1.76)$$

Therefore,

$$J_1 \leq M_4 \int_{|\lambda+t|\leq Lh^{-m}} |\lambda + t|^{\frac{r-\alpha-m}{m}} \exp\left\{-\frac{a}{2}|\lambda + t|^{\frac{1}{m}}(h + |y|)\right\}(h + |y|)^{m-n-\alpha}dt$$

$$\leq M_5 \exp\{-a_1|\lambda|^{\frac{1}{m}}(h + |y|)\}(h + |y|)^{m-n-\alpha} \int_0^\infty \exp\{-a_1 t^{\frac{1}{m}}(h + |y|)\}t^{\frac{r-\alpha-m}{m}}dt$$

$$\leq M_6(h + |y|)^{m-n-\alpha} \exp\{-a_1|\lambda|^{\frac{1}{m}}|y|\} \int_0^\infty \exp\{-a_1 s\}s^{r-\alpha-1}ds\,(h + |y|)^{\alpha-r} \leq$$

$$M_7 \exp\{-a_1|\lambda^{\frac{1}{m}}y|\}(h+|y|)^{m-n-r}.$$

Now let us estimate J_2 following the same procedure as in Lemma 1.14:

$$J_2 \leq M_1 \int_{|\lambda+t|\geq Lh^{-m}} \exp\left\{-\frac{a}{2h}|y|\right\}|\lambda+t|^{-1}\times$$

$$\times(|\lambda+t|h^m+1)^{-1}h^{m-n-r}\sum_{z\in\mathbf{R}_h^n}\exp\left\{-\frac{a}{2h}|z|\right\}dt$$

$$\leq M_2\exp\left\{-\frac{a}{2h}|y|\right\}h^{m-n-r}\int_{|\lambda+t|\geq Lh^{-m}}|\lambda+t|^{-1}(|\lambda+t|h^m+1)^{-1}dt$$

$$\leq M_3 h^{-r-n}\exp\left\{-\frac{a}{2h}|y|\right\}\int_{|\lambda|+t\geq L_1 h^{-m}}(|\lambda|+t)^{-2}dt$$

$$\leq M_4\exp\left\{-\frac{a}{2h}|y|\right\}h^{m-n-r} \leq M_5\exp\left\{-\frac{a}{4h}|y|\right\}(h+|y|)^{m-r-n}$$

$$\leq M_6\exp\{-a_1|\lambda^{\frac{1}{m}}y|\}(h+|y|)^{m-n-r}.$$

Combining the estimates for J_1 and J_2 we obtain the assertion of the lemma in the case where $0 < r < m$ and $m \leq n$.

Now let $0 < r < m$ and $m > n$. Using formula (1.75) and Lemmas 1.9, 1.10, 1.15, and 1.17, we obtain

$$|\mathcal{D}_h^r\mathcal{G}^h(y,\lambda)| \leq \int_{|\lambda+t|\leq Lh^{-m}} M^2(\alpha)\sum_{z\in\mathbf{R}_h^n}|\lambda+t|^{-\frac{a}{m}}\times$$

$$\times\exp\{-a|\lambda+t|^{\frac{1}{m}}(|y-z|+|z|)\}(h+|y-z|)^{m-n-r-\alpha}|\lambda+t|^{\frac{n}{m}-1}h^n dt + J_2 = J_3 + J_2.$$

We already estimated the integral J_2. Let us estimate J_3. We have

$$J_3 \leq \int_{|\lambda+t|\leq Lh^{-m}} M_1|\lambda+t|^{\frac{n-\alpha}{m}-1}\exp\left\{-\frac{a}{2}|\lambda+t|^{\frac{1}{m}}|y|\right\}\times$$

$$\times\sum_{z\in\mathbf{R}_h^n}\exp\left\{-\frac{a}{2}|\lambda+t|^{\frac{1}{m}}|z|\right\}h^n(h+|z|)^{m-n-r-\alpha}dt$$

$$\leq M_2\int_{|\lambda+t|\leq Lh^{-m}}|\lambda+t|^{\frac{n-\alpha}{m}-1}\exp\left\{-\frac{a}{2}|\lambda+t|^{\frac{1}{m}}(h+|y|)\right\}|\lambda+t|^{\frac{r+\alpha-m}{m}}dt$$

$$= M_2\int_{|\lambda+t|\leq Lh^{-m}}\exp\left\{-\frac{a}{2}|\lambda+t|^{\frac{1}{m}}(h+|y|)\right\}|\lambda+t|^{\frac{n+r}{m}-2}dt$$

$$\leq M_3 \exp\{-a_1|\lambda|^{\frac{1}{m}}(h+|y|)\} \int_0^\infty \exp\{-a_1 t^{\frac{1}{m}}(h+|y|)\} t^{\frac{n+r}{m}-2} dt$$

$$\leq M_4 \exp\{-a_1|\lambda^{\frac{1}{m}}y|\} \int_0^\infty \exp\{-a_1 s\} s^{n+r-m-1} ds \, (h+|y|)^{m(1-\frac{n+r}{m})}$$

$$\leq M_5 \exp\{-a_1|\lambda^{\frac{1}{m}}y|\}(h+|y|)^{m-n-r}.$$

Combining the estimates of the integrals J_2 and J_3, we obtain the assertion of the lemma in the case where $0 < r < m$ and $m > n$.

Now let $r = m$. Then $m - n - r - \alpha < -n$, and hence the sum σ appearing in the left-hand side of inequality (1.76) does not admit an estimate that does not depend on h. Thus, to prove the assertion of the lemma in this case one cannot argue as above. Instead, we shall use the fact that in (1.75) part of the derivatives under the integral sign can be moved from the first factor to the second. Indeed, let us substitute $y - z \to z$ in (1.75). Then, since $z \in \mathbf{R}_h^n$,

$$\mathcal{D}_h^{r_1} \mathcal{G}^h(y, \lambda) = \int_0^\infty \sum_{z \in \mathbf{R}_h^n} (2\pi)^{-n} \mathcal{D}_h^{r_1} \mathcal{G}^h(z, \lambda + t) \mathcal{G}^h(y - z, \lambda + t) h^n dt.$$

It follows that

$$\mathcal{D}^{r_1 + r_2} \mathcal{G}^h(y, \lambda) = \int_0^\infty \sum_{z \in \mathbf{R}_h^n} (2\pi)^{-n} \mathcal{D}_h^{r_1} \mathcal{G}^h(z, \lambda + t) \mathcal{D}_h^{r_2} \mathcal{G}^h(y - z, \lambda + t) h^n dt. \quad (1.77)$$

Let $r = m$. We need to examine two cases. First, let $m < 2n - 1$. Then we can take r_1 and r_2 such that $r_1 + r_2 = m$, $m - r_1 < n$, and $m - r_2 < n$. Using identity (1.77), the present lemma for $r < n$, and Lemma 1.15, we obtain

$$|\mathcal{D}_h^m \mathcal{G}^h(y, \lambda)| \leq \int_{|\lambda + t| \leq Lh^{-m}} M^2 \sum_{z \in \mathbf{R}_h^n} \exp\{-a|\lambda + t|^{\frac{1}{m}}(|y - z| + |z|)\} \times$$

$$\times (h + |y - z|)^{m-r_1-n}(h + |z|)^{m-r_2-n} h^n dt$$

$$+ \int_{|\lambda + t| \geq Lh^{-m}} M^2 \sum_{z \in \mathbf{R}_h^n} \exp\{-ah^{-1}(|y - z| + |z|)\} \times$$

$$\times h^{m-r_1-n} h^{m-r_2-n}(|\lambda + t|h^m + 1)^{-2} h^n dt = J_1 + J_2.$$

Let us estimate the integral J_1. We have

$$J_1 \leq M_1 \int_{|\lambda + t| \leq Lh^{-m}} \exp\left\{-\frac{a}{2}|\lambda + t|^{\frac{1}{m}}|y|\right\} \left(\sum_{|y-z| \leq |z|} + \sum_{|y-z| > |z|}\right) \times$$

$$\times \exp\left\{-\frac{a}{2}|\lambda + t|^{\frac{1}{m}}(|y-z|+|z|)\right\}(h+|y-z|)^{m-r_1-n}(h+|z|)^{m-r_2-n}h^n dt$$

$$\leq M_2 \int_{|\lambda+t|\leq Lh^{-m}} \exp\left\{-\frac{a}{2}|\lambda+t|^{\frac{1}{m}}|y|\right\} \times$$

$$\times \left(\sum_{|y-z|\leq|z|} \exp\{-a|\lambda+t|^{\frac{1}{m}}|y-z|\}(h+|y-z|)^{-n+1}(h+|z|)^{m-n-1}h^n\right.$$

$$+ \sum_{|y-z|>|z|} \exp\{-a|\lambda+t|^{\frac{1}{m}}|z|\}(h+|y-z|)^{m-n-1}(h+|z|)^{-n+1}h^n\bigg) dt$$

$$\leq M_3 \int_{|\lambda+t|\leq Lh^{-m}} \exp\left\{-\frac{a}{2}|\lambda+t|^{\frac{1}{m}}|y|\right\}(h+|y|)^{m-n-1}\times$$

$$\times \sum_{z\in \mathbf{R}_h^n} \exp\{-a|\lambda+t|^{\frac{1}{m}}|z|\}(h+|z|)^{-n+1}h^n dt.$$

Since $-n+1 > -n$,

$$\sigma \equiv \sum_{z\in \mathbf{R}_h^n} \exp\{-a|\lambda+t|^{\frac{1}{m}}|z|\}(h+|z|)^{-n+1}h^n \leq M|\lambda+t|^{-\frac{1}{m}}. \qquad (1.78)$$

Using (1.78) and proceeding in a similar way to estimate the integral J_1 for $r < m$, we obtain

$$J_1 \leq M \exp\{-a|\lambda|^{\frac{1}{m}}|y|\}(h+|y|)^{-n}.$$

The estimate for J_2 is similar to that for the case $r < m$ and has the same form as for the integral J_1. This yields the assertion of the lemma for the case $r = m$, $m < 2n - 1$.

Now let $r = m$ and $m \geq 2n - 1$, and set $r_1 = m - 1$ and $r_2 = 1$. Then $m - r_1 < n$ and $m - r_2 > n$. Using identity (1.77), the present lemma for $r < m$, and Lemma 1.16, we obtain

$$|\mathcal{D}_h^m \mathcal{G}^h(y,\lambda)| \leq \int_{|\lambda+t|\leq Lh^{-m}} M^2 \sum_{z\in \mathbf{R}_h^n} \exp\{-a|\lambda+t|^{\frac{1}{m}}(|y-z|+|z|)\}\times$$

$$\times (h+|y-z|)^{-n+1}|\lambda+t|^{\frac{n+1}{m}-1}h^n dt + J_2 = J_3 + J_2.$$

The estimation of the integral J_3 is carried out by analogy to the case $r < m$ and yields

$$J_3 \leq M \exp\{-a|\lambda^{\frac{1}{m}}y|\}(h+|y|)^{-n}.$$

Combining this estimate and the estimate of the integral J_2, we obtain the assertion of the lemma in the case $r = m$, $m \geq 2n - 1$. This completes the proof for the case $r = m$.

The case $m < r < 2m - 2$ is dealt with in a similar manner.

If $j(m - 1) < r \leq (j + 1)(m - 1)$, with $j \geq 2$, then to prove the lemma we have to use formula (1.61), in which we put $r = j - 1$. Lemma 1.18 is proved.

To conclude this subsection let us formulate a theorem on the point estimates for derivatives of the fundamental solution of the resolvent equation.

Theorem 1.4. *Suppose the following conditions are satisfied:* a) $|A(\xi h, h)| \geq M_0 |\xi|^m$, *and* b) $|\arg A(\xi h, h)| \leq \phi < \phi_0 < \pi$. *If* $|\lambda| \geq Lh^{-m}$, *then for any* $r \geq 0$ *we have*

$$|\mathcal{D}_h^r \mathcal{G}^h(y, \lambda)| \leq M \exp\{-ah^{-1}(|y_1| + \cdots + |y_n|)\} h^{m-n-r} (|\lambda| h^m + 1)^{-1}.$$

If $0 < |\lambda| \leq Lh^{-m}$ *and* $|\arg \lambda| \leq \pi - \phi_0$, *then the following estimates hold:*

1) *If* $m - r > n$,

$$|\mathcal{D}_h^r \mathcal{G}^h(y, \lambda)| \leq M \exp\{-a|\lambda|^{\frac{1}{m}}(|y_1| + \cdots + |y_n|)\} |\lambda|^{\frac{n+r}{m}-1}.$$

2) *If* $m - r = n$,

$$|\mathcal{D}_h^r \mathcal{G}^h(y, \lambda)| \leq M \exp\{-a|\lambda|^{\frac{1}{m}}|y|\} \left(1 + \ln\{(|\lambda|^{\frac{1}{m}}(h + |y|))^{-1} + 1\}\right).$$

3) *If* $m - r < n$,

$$|\mathcal{D}_h^r \mathcal{G}^h(y, \lambda)| \leq M \exp\{-a|\lambda|^{\frac{1}{m}}|y|\} (h + |y|)^{m-r-n}.$$

Corollary 1.1. *Suppose the conditions of Theorem 1.4 are satisfied and* $0 < r < m$. *Then*

$$\|\mathcal{D}_h^r (\lambda + A_h)^{-1}\|_{C_h \to C_h} \leq M|\lambda|^{\frac{r}{m}-1}, \tag{1.79}$$

where M *does not depend on* h.

Proof. By (1.22), we have

$$\|\mathcal{D}_h^r (\lambda + A_h)^{-1}\|_{C_h \to C_h} \leq (2\pi)^{-n} \sum_{y \in \mathbf{R}_h^n} |\mathcal{D}_h^r \mathcal{G}^h(y, \lambda)| h^n.$$

Hence, to prove the corollary it suffices to establish the inequality

$$M_h(\lambda) \equiv \sum_{y \in \mathbf{R}_h^n} |\mathcal{D}_h^r \mathcal{G}^h(y, \lambda)| h^n \leq M|\lambda|^{\frac{r}{m}-1},$$

with an M that does not depend on h.

In accordance with the four estimates of the derivatives of the fundamental solution given in Theorem 1.4, we will consider four cases of estimation of the quantity $M_h(\lambda)$.

In the case where $|\lambda| \geq Lh^{-m}$ we obtain

$$M_h(\lambda) \leq M \sum_{y \in \mathbf{R}_h^n} \exp\{-ah^{-1}|y|\} h^{m-r-n} (|\lambda| h^m + 1)^{-1} h^n$$

$$\leq Mh^{-r} h^{-n} |\lambda|^{-1} \sum_{y \in \mathbf{R}_h^n} \exp\{-ah^{-1}|y|\} h^n$$

$$\leq M_1 |\lambda|^{-1} h^{-r} \sum_{k \in \mathbf{R}_1^n} \exp\{-a|k|\} \leq M_2 |\lambda|^{-1} h^{-r} \left(\sum_{k=0}^{\infty} \exp\{-ak\} \right)^n$$

$$\leq M_3 |\lambda|^{-1} h^{-r} \leq M_4 |\lambda|^{\frac{r}{m}-1}.$$

In the case where $0 < |\lambda| \leq Lh^{-m}$ and $m - r > n$ we obtain

$$M_h(\lambda) \leq M \sum_{y \in \mathbf{R}_h^n} \exp\{-a|\lambda|^{\frac{1}{m}}|y|\} |\lambda|^{\frac{n}{m}} h^n$$

$$\leq M|\lambda|^{\frac{r}{m}-1} \left(\sum_{k=0}^{\infty} \exp\{-a|\lambda|^{\frac{1}{m}} kh\} |\lambda|^{\frac{1}{m}} h \right)^n$$

$$\leq M_1 |\lambda|^{\frac{r}{m}-1} \left(\int_0^{\infty} \exp\{-a|\lambda|^{\frac{1}{m}} t\} |\lambda|^{\frac{1}{m}} dt \right)^n \leq M_2 |\lambda|^{\frac{r}{m}-1}.$$

In the case where $0 < |\lambda| \leq Lh^{-m}$ and $m - r = n$ we obtain

$$M_h(\lambda) \leq M \sum_{y \in \mathbf{R}_h^n} \exp\{-a|\lambda|^{\frac{1}{m}}|y|\} \left(1 + \ln\{(|\lambda|^{\frac{1}{m}}(h + |y|))^{-1} + 1\} \right) h^n$$

$$\leq M|\lambda|^{\frac{r}{m}-1} \sum_{y \in \mathbf{R}_h^n} \exp\{-a|\lambda|^{\frac{1}{m}}|y|\} \left(1 + \ln\{(|\lambda|^{\frac{1}{m}}(h + |y|))^{-1} + 1\} \right) |\lambda|^{\frac{n}{m}} h^n$$

$$= M|\lambda|^{\frac{r}{m}-1} \sum_{z \in \mathbf{R}_{h_1}^n} \exp\{-a|z|\} \left(1 + \ln\{(h_1 + |z|)^{-1} + 1\} \right) h_1^n$$

$$\leq M_1 |\lambda|^{\frac{r}{m}-1} \sum_{z \in \mathbf{R}^n_{h_1}} \exp\left\{-\frac{a}{2}|z|\right\} h_1^n \leq M_2 |\lambda|^{\frac{r}{m}-1}.$$

Finally, in the case where $0 < |\lambda| \leq Lh^{-m}$ and $m - r < n$ we obtain

$$M_h(\lambda) \leq M \sum_{y \in \mathbf{R}^n_h} \exp\{-a|\lambda|^{\frac{1}{m}}|y|\}(h + |y|)^{m-r-n} h^n$$

$$= M|\lambda|^{\frac{r}{m}-1} \sum_{y \in \mathbf{R}^n_h} \exp\{-a|\lambda|^{\frac{1}{m}}|y|\}(|\lambda|^{\frac{1}{m}}(h + |y|))^{m-r-n} |\lambda|^{\frac{n}{m}} h^n$$

$$= M|\lambda|^{\frac{r}{m}-1} \sum_{z \in \mathbf{R}^n_{h_1}} \exp\{-a|z|\}(h_1 + |z|)^{m-r-n} h_1^n$$

$$\leq M_1 |\lambda|^{\frac{r}{m}-1} \left(\sum_{k=0}^{\infty} \exp\{-akh_1\}(h_1 + kh_1)^{\frac{m-r-n}{n}} h_1 \right)^n$$

$$\leq M_2 |\lambda|^{\frac{r}{m}-1} \left(\int_0^{\infty} \exp\{-at\} t^{\frac{m-r-n}{n}} dt \right)^n \leq M_3 |\lambda|^{\frac{r}{m}-1},$$

since $-1 < (m - r - n)/n < 0$. Thus, in all cases $M_h(\lambda) \leq M|\lambda|^{\frac{r}{m}-1}$. Corollary 1.1 is proved.

2.   FRACTIONAL SPACES IN THE CASE OF AN ELLIPTIC DIFFERENCE OPERATOR

In Chapters 1–3 the differential and difference Cauchy problems were investigated in fractional spaces generated by an abstract positive operator A acting in an abstract Banach space E. In this section we will clarify the structure of the spaces $E'_{\alpha,p}(L_{ph}, A_h)$, $1 \leq p \leq \infty$, generated by the difference operator A_h acting in the Banach spaces L_{ph}.

1.   The fractional spaces $E'_{\alpha,\infty}(C_h, A_h)$.

The study of the structure of these spaces (as well as of other spaces) relies on certain properties of the fundamental solution $\mathcal{G}^h(x, \lambda)$ of the resolvent equation (1.21). As in Lemma 1.5, we shall assume that $A(\xi h, h) + \lambda \neq 0$ for $|\xi_k h| \leq \pi$.

Then for any function $f^h(x) \in C_h$ equation (1.21) is uniquely solvable and formula (1.22) holds for its solution $u^h(x)$.

Lemma 2.1. *The following identity holds:*

$$\mathcal{G}^h(x, \lambda) = \lambda^{\frac{n}{m}-1} \mathcal{G}^{\lambda^{\frac{1}{m}}h}(\lambda^{\frac{1}{m}}x, 1). \tag{2.1}$$

Proof. In (1.22) let us set

$$f^h(x) = \delta^h(x) = \begin{cases} (2\pi)^n h^{-n}, & \text{if } x = 0, \\ 0, & \text{if } x \neq 0. \end{cases} \tag{2.2}$$

Then

$$u^h(x) = (2\pi)^{-n} \sum_{y \in \mathbf{R}_h^n} \mathcal{G}^h(x - y, \lambda) \delta^h(y) h^n = \mathcal{G}^h(x, \lambda).$$

Thus, the grid function $\mathcal{G}^h(x, \lambda)$ is a solution of equation (1.21) with the right-hand side $f^h(x) = \delta^h(x)$. This means that we have the identity

$$\lambda^{-1} A_h \mathcal{G}^h(x, \lambda) + \mathcal{G}^h(x, \lambda) = \lambda^{-1} \delta^h(x).$$

Setting $h_1 = \lambda^{\frac{1}{m}}h$, $\tilde{x} = \lambda^{\frac{1}{m}}x$, we obtain

$$A_{h_1} \mathcal{G}^h(\lambda^{-\frac{1}{m}}\tilde{x}, \lambda) + \mathcal{G}^h(\lambda^{-\frac{1}{m}}\tilde{x}, \lambda) = \lambda^{-1} \delta^h(\lambda^{-\frac{1}{m}}x).$$

Notice that in this identity involves a difference operator with step h_1 with respect to the variable $\tilde{x}$ with the same step h_1. Hence, by formula (1.22), we have

$$\mathcal{G}^h(x, \lambda) = \mathcal{G}^h(\lambda^{-\frac{1}{m}}\tilde{x}, \lambda) = (2\pi)^{-n} \sum_{\tilde{y} \in \mathbf{R}_{h_1}^n} \mathcal{G}^{h_1}(\tilde{x} - \tilde{y}, 1) \lambda^{-1} \delta^h(\lambda^{-\frac{1}{m}}\tilde{y}) h_1^n.$$

Using (2.2), we obtain

$$\mathcal{G}^h(x, \lambda) = \mathcal{G}^{h_1}(\tilde{x}, 1) \lambda^{-1} h^{-n} h_1^n = \mathcal{G}^{\lambda^{\frac{1}{m}}h}(\lambda^{\frac{1}{m}}x, 1) \lambda^{-1+\frac{n}{m}}.$$

Lemma 2.1 is proved.

A direct consequence of formula (1.21) is

Lemma 2.2. *The following identity holds:*

$$(2\pi)^{-n} \sum_{y \in \mathbf{R}_h^n} \mathcal{G}^h(y, \lambda) h^n = \frac{1}{\lambda}. \tag{2.3}$$

Let us denote by $C_h^\beta = C^\beta(\mathbf{R}_h^n)$, $0 < \beta < 1$, the set of all bounded grid functions $u^h(x)$ equipped with the norm

$$\|u^h\|_{C_h^\beta} = \|u^h\|_{C_h} + \sup_{x,y \in \mathbf{R}_h^n,\, x \neq y} \frac{|u^h(x) - u^h(y)|}{|x - y|^\beta}.$$

It is readily seen that for fixed h and different β's these norms are equivalent. However, there are no equivalence constants uniform in h. Recall that $E'_{\alpha,\infty}(C_h, A_h)$ is the set of all bounded grid functions $u^h(x)$ for which the following norm is finite:

$$\|u^h\|_{E'_{\alpha,\infty}(C_h,A_h)} = \sup_{\lambda>0} \lambda^\alpha \|A_h(\lambda + A_h)^{-1} u^h\|_{C_h}.$$

Since for fixed h the operators A_h are bounded, this norm is finite for all grid functions.

Theorem 2.1. *For $0 < \alpha < m^{-1}$ the norms of the spaces $E'_{\alpha,\infty}(C_h, A_h)$ and $C^{2m\alpha}(\mathbf{R}_h^n)$ are equivalent uniformly in h, $0 < h \leq h_0$.*

Proof. For any $t > 0$ we have the obvious equality

$$A_h(t + A_h)^{-1} f^h(x) = t\left[\frac{1}{t} - (t + A_h)^{-1}\right] f^h(x).$$

By formula (1.22) and identity (2.3), we can write

$$A_h(t + A_h)^{-1} f^h(x) = (2\pi)^{-n} t \sum_{y \in \mathbf{R}_h^n} \mathcal{G}^h(x - y, t)[f^h(x) - f^h(y)]h,$$

whence

$$|t^\alpha A_h(t+A_h)^{-1} f^h(x)| \leq t^{\alpha+1} \sum_{y \in \mathbf{R}_h^n} |\mathcal{G}^h(x-y, t)|\, |x-y|^{m\alpha} h^n \|f^h\|_{C_h^{m\alpha}} = J\|f^h\|_{C_h^{m\alpha}}.$$

Now let us estimate

$$J = t^{\alpha+1} \sum_{y \in \mathbf{R}_h^n} |\mathcal{G}^h(x - y, t)|\, |x - y|^{m\alpha} h^n.$$

First let $t \geq Lh^{-m}$. Then from the estimates (1.34) it follows that

$$J \leq M \sum_{y \in \mathbf{R}_h^n,\, y \neq x} t^{1+\alpha} \frac{|x - y|^{m\alpha} \exp\{-ah^{-1}|x - y|\} h^n}{t^{1+m\alpha} h^n h^{m\alpha(m-1)} |x - y|^{m\alpha}}$$

$$\le \frac{M}{(th^m)^{(m-1)\alpha}} \sum_{z\in\mathbf{R}_h^n} \exp\{-ah^{-1}|z|\} \le M_1 \sum_{k\in\mathbf{R}_1^n} \exp\{-a|k|\} \le M_2.$$

Now let $t \le Lh^{-m}$. We shall consider three cases: $m > n$, $m = n$, and $m < n$. In the first case, using Theorem 1.1 we obtain

$$J \le M \sum_{y\in\mathbf{R}_h^n} t^{1+\alpha} \exp\{-at^{\frac{1}{m}}|x-y|\} t^{\frac{n}{m}-1}|x-y|^{m\alpha} h^n.$$

The change of variables $x - y = z$, $z = hr$ yields

$$J \le M t^{\alpha+\frac{n}{m}} \sum_{z\in\mathbf{R}_h^n} \exp\{-at^{\frac{1}{m}}|z|\}|z|^{m\alpha} h^n$$

$$= M t^{\alpha+\frac{n}{m}} \sum_{r\in\mathbf{R}_1^n} \exp\{-at^{\frac{1}{m}}|r|h\}|r|^{m\alpha} h^{m\alpha} h^n.$$

Letting $t^{\frac{1}{m}}h = h_1$, we have

$$J \le M \sum_{r\in\mathbf{R}_1^n} \exp\{-a|r|h_1\}(|r|h_1)^{m\alpha} h_1^n$$

$$= M \sum_{r\in\mathbf{R}_1^n} \exp\{-\frac{a}{2}|r|h_1\} \exp\left\{-\frac{a}{2}|r|h_1\right\} (|r|h_1)^{m\alpha} h_1^n$$

$$\le M_1 \sum_{r\in\mathbf{R}_1^n} \exp\left\{-\frac{a}{2}|r|h_1\right\} h_1^n \le M_1 \left(\sum_{r\in\mathbf{R}_1^1} \exp\left\{-\frac{a}{2\sqrt{n}}|r|h_1\right\} h_1\right)^n$$

$$\le M_2 \left(\int_0^\infty \exp\left\{-\frac{a}{2\sqrt{n}}s\right\} ds\right)^n \le M_3.$$

In the case $m = n$ we let $t^{\frac{1}{m}}h = h_1{}'$ and use Theorem 1.3 to obtain

$$J \le M t^{1+\alpha} \sum_{z\in\mathbf{R}_h^n} \exp\{-at^{\frac{1}{m}}|z|\}|z|^{m\alpha} \ln\left|\left[t^{\frac{1}{m}}(h+|z|)\right]^{-1}+1\right| h^n$$

$$= M \sum_{\tilde{z}=t^{\frac{1}{m}}z\in\mathbf{R}_{h_1}^n} \exp\{-a|\tilde{z}|\} \ln\left|\left[h_1+|\tilde{z}|\right]^{-1}+1\right| |\tilde{z}|^{m\alpha} h_1^n t^{1-\frac{n}{m}}$$

$$= M \sum_{\tilde{z}\in\mathbf{R}_{h_1}^n} \exp\left\{-\frac{a}{2}|\tilde{z}|\right\} \exp\left\{-\frac{a}{2}|\tilde{z}|\right\} |\tilde{z}|^{m\alpha} \ln\left|\left[h_1+|\tilde{z}|\right]^{-1}+1\right| h_1^n$$

$$\leq M_1 \sum_{\bar{z} \in \mathbf{R}^n_{h_1}} \exp\left\{-\frac{a}{2}|\bar{z}|\right\} h_1^n \leq M_1 \left(\sum_{r \in \mathbf{R}^1_1} \exp\left\{-\frac{a}{2\sqrt{n}}|r|\right\} h_1\right)^n$$

$$\leq M_2 \left(\int_0^\infty \exp\left\{-\frac{a}{2\sqrt{n}}z\right\} dz\right)^n \leq M_3.$$

In the case $m < n$ we let $t^{\frac{1}{m}} h = h_1$ and use Theorem 1.3 to obtain

$$J \leq M t^{1+\alpha} \sum_{z \in \mathbf{R}^n_h} \exp\{-at^{\frac{1}{m}}|z|\}(h+|z|)^{m-n}|z|^{m\alpha} h^n$$

$$= M t^{1+\alpha} \sum_{\bar{z}=t^{\frac{1}{m}}z \in \mathbf{R}^n_{h_1}} \exp\{-a|\bar{z}|\}(h_1+|\bar{z}|)^{m-n+m\alpha} h_1^n t^{-\frac{m-n+m\alpha}{m}} t^{-\frac{n}{m}}$$

$$= M \sum_{\bar{z} \in \mathbf{R}^n_{h_1}} \exp\{-a|\bar{z}|\}(h_1+|\bar{z}|)^{m-n+m\alpha} h_1^n \leq M_1 \int_{\bar{z} \in \mathbf{R}^n} \exp\{-a|\bar{z}|\}|\bar{z}|^{m-n+m\alpha} d\bar{z}.$$

Since $2-n \leq m-n+m\alpha < 1$, the last integral converges. Consequently, $J \leq M_2$.

Thus, for any $t \geq 0$ and $x \in \mathbf{R}^n_h$ we established the inequality

$$|t^\alpha A_h(t+A_h)^{-1} f^h(x)| \leq M\|f^h\|_{C^{m\alpha}_h}.$$

This means that

$$\|f^h\|_{E'_{\alpha,\infty}(C_h,A_h)} \leq M\|f^h\|_{C^{m\alpha}_h}.$$

Let us prove the opposite inequality. For any positive operator A we can write

$$v = \int_0^\infty A(t+A)^{-2} v \, dt.$$

From this relation and formula (1.22) it follows that

$$f^h(x) = \int_0^\infty (t+A_h)^{-1} A_h(t+A_h)^{-1} f^h(x) \, dt$$

$$= (2\pi)^{-n} \int_0^\infty \sum_{y \in \mathbf{R}^n_h} \mathcal{G}^h(x-y,t) A_h(t+A_h)^{-1} f^h(y) h^n \, dt.$$

Consequently,

$$f^h(x_1) - f^h(x_2)$$

$$= (2\pi)^{-n} \int_0^\infty t^{-\alpha} \sum_{y \in \mathbf{R}^n_h} [\mathcal{G}^h(x_1-y,t) - \mathcal{G}^h(x_2-y,t)] t^\alpha A_h(t+A_h)^{-1} f^h(y) h^n \, dt,$$

whence

$$|f^h(x_1) - f^h(x_2)|$$

$$\leq \int_0^\infty t^{-\alpha} \sum_{y \in \mathbf{R}_h^n} |\mathcal{G}^h(x_1 - y, t) - \mathcal{G}^h(x_2 - y, t)| h^n \, dt \, \|f^h\|_{E'_{\alpha,\infty}(C_h, A_h)}.$$

Let

$$T = |x_1 - x_2|^{-m\alpha} \int_0^\infty t^{-\alpha} \sum_{y \in \mathbf{R}_h^n} |\mathcal{G}^h(x_1 - y, t) - \mathcal{G}^h(x_2 - y, t)| h^n \, dt.$$

Then

$$\frac{|f^h(x_1) - f^h(x_2)|}{|x_1 - x_2|^{m\alpha}} \leq T \|f^h\|_{E'_{\alpha,\infty}(C_h, A_h)}.$$

To estimate T let us perform the change of variables $x_1 - y = z$; we obtain

$$T = |x_1 - x_2|^{-m\alpha} \int_0^\infty t^{-\alpha} \sum_{z \in \mathbf{R}_h^n} |\mathcal{G}^h(z, t) - \mathcal{G}^h(x_2 - x_1 + z, t)| h^n \, dt.$$

From identity (2.1) it follows that

$$T = |x_1 - x_2|^{-m\alpha} \int_0^\infty t^{-\alpha - 1 + \frac{n}{m}} \times$$

$$\times \sum_{z \in \mathbf{R}_h^n} h^n |\mathcal{G}^{t^{\frac{1}{m}} h}(t^{\frac{1}{m}} z, 1) - \mathcal{G}^{t^{\frac{1}{m}} h}(t^{\frac{1}{m}}(x_2 - x_1 + z), 1)| dt.$$

If we now put $t^{\frac{1}{m}} h = h_1$, then

$$T = |x_1 - x_2|^{-m\alpha} \int_0^\infty t^{-\alpha - 1} \sum_{\tilde{z} \in \mathbf{R}_{h_1}^n} |\mathcal{G}^{h_1}(\tilde{z}, 1) - \mathcal{G}^{h_1}(t^{\frac{1}{m}}(x_2 - x_1) + \tilde{z}, 1)| h_1^n dt.$$

Introducing the new variable $t^{\frac{1}{m}} |x_2 - x_1| = \lambda$, we obtain

$$T = \frac{m|x_2 - x_1|^{m(1+\alpha)}}{|x_1 - x_2|^{m\alpha} |x_2 - x_1|^m} \int_0^\infty \lambda^{-m\alpha - 1} \times$$

$$\times \sum_{\tilde{z} \in \mathbf{R}_{h_1}^n} \left| \mathcal{G}^{h_1}(\tilde{z}, 1) - \mathcal{G}^{h_1}\left(\frac{x_2 - x_1}{|x_2 - x_1|} \lambda + \tilde{z}, 1 \right) \right| h_1^n d\lambda$$

$$= m \int_0^\infty \lambda^{-m\alpha - 1} \sum_{\tilde{z} \in \mathbf{R}_{h_1}^n} \left| \mathcal{G}^{h_1}(\tilde{z}, 1) - \mathcal{G}^{h_1}\left(\frac{x_2 - x_1}{|x_2 - x_1|} \lambda + \tilde{z}, 1 \right) \right| h_1^n d\lambda$$

$$= m \left\{ \int_0^{L^{\frac{1}{m}}|x_2-x_1|/h} + \int_{L^{\frac{1}{m}}|x_2-x_1|/h}^{\infty} \right\} \lambda^{-m\alpha-1} \times$$

$$\times \sum_{\tilde{z} \in \mathbf{R}_{h_1}^n} \left| \mathcal{G}^{h_1}(\tilde{z},1) - \mathcal{G}^{h_1}\left(\frac{x_2-x_1}{|x_2-x_1|}\lambda + \tilde{z}, 1 \right) \right| h_1^n d\lambda = m(T_1 + T_2).$$

Let us estimate T_1 and T_2 separately. Clearly, $\lambda \geq L^{\frac{1}{m}}|x_2 - x_1|/h$ implies $t \geq Lh^{-m}$. Hence, using Theorem 1.4 we obtain

$$T_2 \leq M \int_{L^{\frac{1}{m}}|x_2-x_1|/h}^{\infty} \lambda^{-m\alpha-1} \times$$

$$\times \sum_{\tilde{z} \in \mathbf{R}_{h_1}^n} \left[\exp\{-ah_1^{-1}|\tilde{z}|\} + \exp\{-ah_1^{-1}|\tilde{z} + (x_2-x_1)\lambda/|x_2-x_1||\}\right] d\lambda$$

$$= M \int_{L^{\frac{1}{m}}|x_2-x_1|/h}^{\infty} \lambda^{-m\alpha-1}[I_1(\lambda) + I_2(\lambda)]d\lambda.$$

Let us estimate $I_k(\lambda)$, $k = 1, 2$. First,

$$I_1(\lambda) = \sum_{\tilde{z} \in \mathbf{R}_{h_1}^n} \exp\{-ah_1^{-1}|\tilde{z}|\} \leq \left(\sum_{z \in \mathbf{R}_1^1} \exp\left\{ -\frac{a}{\sqrt{n}}|k| \right\} \right)^n \leq M_1.$$

Second,

$$I_2(\lambda) = \sum_{\tilde{z} \in \mathbf{R}_{h_1}^n} \exp\left\{ -a\left| \frac{\tilde{z}}{h_1} + \frac{x_2-x_1}{|x_2-x_1|} \cdot \frac{\lambda}{h_1} \right| \right\}$$

$$= \sum_{\tilde{z} \in \mathbf{R}_{h_1}^n} \exp\left\{ -a\left| \frac{\tilde{z}}{h_1} + \left[\frac{x_2-x_1}{|x_2-x_1|} \cdot \frac{\lambda}{h_1} \right] - \left(\left[\frac{x_2-x_1}{|x_2-x_1|} \cdot \frac{\lambda}{h_1} \right] - \frac{x_2-x_1}{|x_2-x_1|} \cdot \frac{\lambda}{h_1} \right) \right| \right\}$$

$$\leq \exp\{na\} \sum_{\tilde{z} \in \mathbf{R}_{h_1}^n} \exp\left\{ -a\left| \frac{\tilde{z}}{h_1} + \left[\frac{x_2-x_1}{|x_2-x_1|} \cdot \frac{\lambda}{h_1} \right] \right| \right\} \leq M_2 \sum_{k \in \mathbf{R}_1^n} \exp\{-a|k+m|\}.$$

Here $[b] = \{[b_k]\}_{k=1}^n \in \mathbf{R}^n$, where $[b_k]$ denotes the integer part of the number b_k for $k = 1, \cdots, n$, and $m = [\frac{(x_2-x_1)}{|x_2-x_1|} \cdot \frac{\lambda}{h_1}] \in \mathbf{Z}_1^n$ is a fixed vector with integer coordinates. Letting $k + m = \tilde{k}$, we have

$$I_2(\lambda) \leq M_2 \sum_{\tilde{k} \in \mathbf{R}_1^n} \exp\{-a|\tilde{k}|\} \leq M_3.$$

This yields

$$T_2 \le M_4 \int_{L^{\frac{1}{m}}|x_2-x_1|/h}^{\infty} \lambda^{-m\alpha-1}d\lambda = \frac{1}{m\alpha} \frac{M_4}{(L^{\frac{1}{m}}|x_1-x_2|/h)^{m\alpha}} \le \frac{M_5}{\alpha}.$$

Now let us estimate T_1. First let us note that from identity

$$\mathcal{G}^{h_1}(\tilde{z},1) - \mathcal{G}^{h_1}(\tilde{x}_2 - \tilde{x}_1 + \tilde{z}, 1)$$

$$= \sum_{k=1}^{n} \sum_{i=\tilde{x}_{1,k}h_1^{-1}}^{\tilde{x}_{2,k}h_1^{-1}-1} \left[\mathcal{G}^{h_1}\left(\sum_{j=1}^{k-1}\tilde{x}_{1,j}e_j + ie_kh_1 + \sum_{j=k+1}^{n}\tilde{x}_{2,j}e_j - \tilde{x}_1 + \tilde{z}, 1 \right) \right.$$

$$\left. -\mathcal{G}^{h_1}\left(\sum_{j=1}^{k-1}\tilde{x}_{1,j}e_j + (i+1)e_kh_1 + \sum_{j=k+1}^{n}\tilde{x}_{2,j}e_j - \tilde{x}_1 + \tilde{z}, 1 \right) \right]$$

it follows that (see Section 1)

$$I(\tilde{z},\lambda) = \left| \mathcal{G}^{h_1}(\tilde{z},1) - \mathcal{G}^{h_1}\left(\frac{x_2-x_1}{|x_2-x_1|}\lambda + \tilde{z}, 1 \right) \right| \le \sum_{k=1}^{n} \sum_{i=x_{1,k}h_1^{-1}}^{x_{2,k}h_1^{-1}-1} \frac{\lambda h_1}{|x_2-x_1|} \times$$

$$\times \left| \mathcal{D}_{h_1}\mathcal{G}^{h_1}\left(\left(\sum_{j=1}^{k-1}x_{1,j}e_j + ie_kh_1 + \sum_{j=k+1}^{n}x_{2,j}e_j - x_1 \right) \frac{\lambda}{|x_2-x_1|} + \tilde{z}, 1 \right) \right|.$$

Let

$$P(\lambda) = \sum_{\tilde{z}\in\mathbf{R}_{h_1}^n} I(\tilde{z},\lambda)h_1^n.$$

Clearly, $\lambda \le L^{\frac{1}{m}}|x_2-x_1|/h$ implies $t \le Lh^{-m}$. We need to consider three cases: $m < n$, $m = n$, and $m > n$. In the case $m < n$ Theorem 1.3 yields

$$P(\lambda) \le M \sum_{\tilde{z}\in\mathbf{R}_{h_1}^n} \exp\{-a|\tilde{z}|\}(h_1 + |\tilde{z}|)^{m-n}h_1^n$$

$$+M \sum_{\tilde{z}\in\mathbf{R}_{h_1}^n} \exp\{-a|\tilde{z}+(x_2-x_1)\lambda/|x_2-x_1||\}(h_1+|\tilde{z}+(x_2-x_1)\lambda/|x_2-x_1||)^{m-n}h_1^n$$

$$\le MP_1(\lambda) + M_1 \sum_{\tilde{z}\in\mathbf{R}_{h_1}^n} \exp\{na\}\exp\{-a|\tilde{z} + h_1[(x_2-x_1)\lambda/|x_2-x_1|]|\} \times$$

$$\times (h_1 + |\tilde{z} + h_1[(x_2-x_1)\lambda/|x_2-x_1|]|)^{m-n}h_1^n$$

$$\leq M P_1(\lambda) + M_2 \sum_{\tilde{\tilde{z}} \in \mathbf{R}^n_{h_1}} \exp\{-a|\tilde{\tilde{z}}|\}(h_1 + |\tilde{\tilde{z}}|)^{m-n} h_1^n = (M + M_2) P_1(\lambda),$$

where $\tilde{\tilde{z}} = \tilde{z} + h_1[(x_2 - x_1)\lambda/|x_2 - x_1|]$. Here

$$P_1(\lambda) = \sum_{z \in \mathbf{R}^n_{h_1}} \exp\{-a|z|\}(h_1 + |z|)^{m-n} h_1^n.$$

Thus, $P(\lambda)$ obeys the estimate

$$P(\lambda) \leq M_3 \sum_{z \in \mathbf{R}^n_{h_1}} \exp\{-a|z|\}(h_1 + |z|)^{m-n} h_1^n. \tag{2.4}$$

On the other hand, using the inequality (1.67) with $r = 1$ we deduce that

$$P(\lambda) \leq M \sum_{\tilde{z} \in \mathbf{R}^n_{h_1}} \sum_{k=1}^{n} \sum_{i=x_{1,k} h_1^{-1}}^{x_{2,k} h_1^{-1} - 1} h_1^{n+1} \frac{\lambda}{|x_2 - x_1|} \times$$

$$\times \exp\left\{-a\left|\tilde{z} + (ih_1 - x_{1,k})\frac{\lambda e_k}{|x_2 - x_1|} + \sum_{j=k+1}^{n} (x_{2,j} - x_{1,j})\frac{\lambda e_j}{|x_2 - x_1|}\right|\right\} \times$$

$$\times \left(h_1 + \left|\tilde{z} + (ih_1 - x_{1,k})\frac{\lambda e_k}{|x_2 - x_1|} + \sum_{j=k+1}^{n} (x_{2,j} - x_{1,j})\frac{\lambda e_j}{|x_2 - x_1|}\right|\right)^{m-n-1}$$

$$= M \sum_{k=1}^{n} \sum_{i=x_{1,k} h_1^{-1}}^{x_{2,k} h_1^{-1} - 1} h_1 \frac{\lambda}{|x_2 - x_1|} \times$$

$$\times \sum_{\tilde{z} \in \mathbf{R}^n_{h_1}} \exp\left\{-a\left|\tilde{z} + (ih_1 - x_{1,k})\frac{\lambda e_k}{|x_2 - x_1|} + \sum_{j=k+1}^{n} (x_{2,j} - x_{1,j})\frac{\lambda e_j}{|x_2 - x_1|}\right|\right\} \times$$

$$\times \left(h_1 + \left|\tilde{z} + (ih_1 - x_{1,k})\frac{\lambda e_k}{|x_2 - x_1|} + \sum_{j=k+1}^{n} (x_{2,j} - x_{1,j})\frac{\lambda e_j}{|x_2 - x_1|}\right|\right)^{m-n-1} h_1^n$$

$$= M \sum_{k=1}^{n} \sum_{i=x_{1,k} h_1^{-1}}^{x_{2,k} h_1^{-1} - 1} h_1 \frac{\lambda}{|x_2 - x_1|} \tilde{P}(\lambda).$$

First of all note that

$$\tilde{P}(\lambda) \le \tilde{M} \sum_{\tilde{z} \in \mathbf{R}^n_{h_1}} h_1^n \exp \left\{ - a \left| \tilde{z} + e_k h_1 \left[\frac{(ih_1 - x_{1,k})\lambda}{|x_2 - x_1|h_1} \right] \right. \right.$$

$$\left. + \sum_{j=k+1}^{n} e_j h_1 \left[\frac{(x_{2,j} - x_{1,j})\lambda}{|x_2 - x_1|} \right] \right| \right\} \times$$

$$\times \left(h_1 + \left| \tilde{z} + e_k h_1 \left[\frac{(ih_1 - x_{1,k})\lambda}{|x_2 - x_1|h_1} \right] + \sum_{j=k+1}^{n} e_j h_1 \left[\frac{(x_{2,j} - x_{1,j})\lambda}{|x_2 - x_1|h_1} \right] \right| \right)^{m-n-1}.$$

Set

$$\tilde{\tilde{z}} = \tilde{z} + e_k h_1 \left[\frac{(ih_1 - x_{1,k})\lambda}{|x_2 - x_1|h_1} \right] + \sum_{j=k+1}^{n} e_j h_1 \left[\frac{(x_{2,j} - x_{1,j})\lambda}{|x_2 - x_1|h_1} \right].$$

Then

$$\tilde{P}(\lambda) \le \tilde{M} \sum_{\tilde{\tilde{z}} \in \mathbf{R}^n_{h_1}} \exp\{-a|\tilde{\tilde{z}}|\}(h_1 + |\tilde{\tilde{z}}|)^{m-n-1} h_1^n.$$

Consequently,

$$P(\lambda) \le M_1 \sum_{k=1}^{n} \sum_{i=x_{1,k}h_1^{-1}}^{x_{2,k}h_1^{-1}-1} h_1 \frac{\lambda}{|x_2 - x_1|} \sum_{z \in \mathbf{R}^n_{h_1}} \exp\{-a|z|\}(h_1 + |z|)^{m-n-1} h_1^n$$

$$\le M_2 \sum_{z \in \mathbf{R}^n_{h_1}} \exp\{-a|z|\}(h_1 + |z|)^{m-n-1} \lambda h_1^n. \qquad (2.5)$$

Suppose first that $L^{\frac{1}{m}}|x_2 - x_1|/h \le 1$. Then (2.5) yields

$$T_1 \le \int_0^1 P(\lambda)\lambda^{-m\alpha-1} d\lambda \le M_2 \int_0^1 \lambda^{-m\alpha} \sum_{z \in \mathbf{R}^n_{h_1}} \exp\{-a|z|\}(h_1 + |z|)^{m-n-1} h_1^n$$

$$\le \frac{M_3}{1 - m\alpha} \int_{z \in \mathbf{R}^n} \exp\{-a|z|\}|z|^{m-n-1} dz.$$

Since $n - m + 1 \le n - 1$,

$$\int_{z \in \mathbf{R}^n} \exp\{-a|z|\}|z|^{m-n-1} dz \le M.$$

Therefore, $T_1 \le M_4/(1 - m\alpha)$. Now suppose that $L^{\frac{1}{m}}|x_2 - x_1|/h \ge 1$. Then from the last inequality and (2.4) it follows that

$$T_1 \le \int_0^1 P(\lambda)\lambda^{-m\alpha-1} d\lambda + \int_1^{L^{\frac{1}{m}}|x_2-x_1|/h} P(\lambda)\lambda^{-m\alpha-1} d\lambda$$

$$\leq \frac{M_4}{1-m\alpha} + M_5 \int_1^\infty \lambda^{-m\alpha-1} d\lambda \sum_{z \in \mathbf{R}_{h_1}^n} \exp\{-a|z|\}(h_1 + |z|)^{m-n} h_1^n$$

$$\leq \frac{M_4}{1-m\alpha} + \frac{M_5}{m\alpha} \int_{z \in \mathbf{R}^n} \exp\{-a|z|\}|z|^{m-n} dz,$$

whence $T_1 \leq M_6/\{m\alpha(1-m\alpha)\}$.

In the case $m = n$ Theorem 1.3 yields

$$P(\lambda) \leq M \sum_{\tilde{z} \in \mathbf{R}_{h_1}^n} \exp\{-a|\tilde{z}|\}(1 + \ln[(h_1 + |\tilde{z}|)^{-1} + 1])h_1^n$$

$$+ M \sum_{\tilde{z} \in \mathbf{R}_{h_1}^n} \exp\left\{ -a \left| \tilde{z} + \frac{(x_2 - x_1)\lambda}{|x_2 - x_1|} \right| \right\} \times$$

$$\times \left(1 + \ln \left[\left(h_1 + \left| \tilde{z} + \frac{(x_2 - x_1)\lambda}{|x_2 - x_1|} \right| \right)^{-1} + 1 \right] \right) h_1^n$$

$$\leq MP_1(\lambda) + M_1 \sum_{\tilde{z} \in \mathbf{R}_{h_1}^n} \exp\left\{ -a \left| \tilde{z} + h_1 \left[\frac{(x_2 - x_1)\lambda}{|x_2 - x_1|h_1} \right] \right| \right\} \times$$

$$\times \left(1 + \ln \left[\left(h_1 + \left| \tilde{z} + h_1 \left[\frac{(x_2 - x_1)\lambda}{|x_2 - x_1|h_1} \right] \right| \right)^{-1} + 1 \right] \right) h_1^n$$

$$= MP_1(\lambda) + M_1 \sum_{\tilde{\tilde{z}} \in \mathbf{R}_{h_1}^n} \exp\{-a|\tilde{\tilde{z}}|\}(1 + \ln((h_1 + |\tilde{\tilde{z}}|)^{-1} + 1))h_1^n = (M + M_1)P_1(\lambda),$$

where

$$\tilde{\tilde{z}} = \tilde{z} + h_1 \left[\frac{(x_2 - x_1)\lambda}{|x_2 - x_1|h_1} \right]$$

and

$$P_1(\lambda) = \sum_{z \in \mathbf{R}_{h_1}^n} \exp\{-a|z|\}(1 + \ln((h_1 + |z|)^{-1} + 1))h_1^n.$$

Therefore,

$$P(\lambda) \leq M_2 \sum_{z \in \mathbf{R}_{h_1}^n} \exp\{-a|z|\}(1 + \ln((h_1 + |z|)^{-1} + 1))h_1^n. \tag{2.6}$$

Suppose that $L^{\frac{1}{m}}|x_2 - x_1/h \geq 1$. Then from (2.5) and (2.6) it follows that

$$T_1 \leq \int_0^1 P(\lambda)\lambda^{-m\alpha-1} d\lambda + \int_1^{L^{\frac{1}{m}}|x_2-x_1/h|} P(\lambda)\lambda^{-m\alpha-1}$$

$$\leq M_2 \int_0^1 \lambda^{-m\alpha} d\lambda \sum_{z \in \mathbf{R}_{h_1}^n} \exp\{-a|z|\}(h_1 + |z|)^{-1} h_1^n$$

$$+ M_2 \int_1^\infty \lambda^{-m\alpha-1} d\lambda \sum_{z \in \mathbf{R}_{h_1}^n} \exp\{-a|z|\}(1 + \ln((h_1 + |z|)^{-1} + 1)) h_1^n$$

$$\leq \frac{M_3}{1 - m\alpha} + \frac{M_2}{m\alpha} \sum_{z \in \mathbf{R}_{h_1}^n} \exp\left\{-\frac{a}{2}|z|\right\} (h_1 + |z|)^{-1} \times$$

$$\times \exp\left\{-\frac{a}{2}|z|\right\} (1 + \ln((h_1 + |z|)^{-1} + 1)) h_1^n$$

$$\leq \frac{M_3}{1 - m\alpha} + \frac{M_4}{m\alpha} \sum_{z \in \mathbf{R}_{h_1}^n} \exp\left\{-\frac{a}{2}|z|\right\} (h_1 + |z|)^{-1} h_1^n \leq \frac{M_5}{m\alpha(1 - m\alpha)}.$$

The estimate for T_1 in the case $L^{\frac{1}{m}}|x_2 - x_1|/h \leq 1$ is plain.

Finally, when $m > n$ we consider two separate cases: $m - 1 = n$ and $m - 1 > n$. First of all, using inequality (1.74) with $r = 1$ and inequality (1.67) with $r = 0, 1$, respectively, one can obtain the estimates

$$P(\lambda) \leq M \sum_{z \in \mathbf{R}_{h_1}^n} \exp\{-a|z|\}(1 + \ln((h_1 + |z|)^{-1} + 1))\lambda h_1^n, \quad \text{if } m - 1 = n, \quad (2.7)$$

$$P(\lambda) \leq M \sum_{z \in \mathbf{R}_{h_1}^n} \exp\{-a|z|\} h_1^n, \quad \text{if } m - 1 \geq n, \quad (2.8)$$

and

$$P(\lambda) \leq M \sum_{z \in \mathbf{R}_{h_1}^n} \exp\{-a|z|\}\lambda h^n, \quad \text{if } m - 1 > n. \quad (2.9)$$

The proof of these estimates follows the scheme given above.

Let $m - 1 = n$. Then for $L^{\frac{1}{m}}|x_2 - x_1|/h \geq 1$, (2.7) and (2.8) yield

$$T_1 \leq M \int_0^1 \lambda^{-m\alpha} d\lambda \sum_{z \in \mathbf{R}_{h_1}^n} \exp\{-a|z|\}(1 + \ln((h_1 + |z|)^{-1} + 1))\lambda h_1^n$$

$$+ M \int_1^\infty \lambda^{-m\alpha-1} d\lambda \sum_{z \in \mathbf{R}_{h_1}^n} \exp\{-a|z|\} h_1^n$$

$$\leq \frac{M_1}{1 - m\alpha} \sum_{z \in \mathbf{R}_{h_1}^n} \exp\left\{-\frac{a}{2}|z|\right\} (h_1 + |z|)^{-1} + \frac{M_1}{m\alpha}$$

$$\leq \frac{M_2}{1 - m\alpha} + \frac{M_1}{m\alpha} \leq \frac{M_3}{m\alpha(1 - m\alpha)}.$$

Clearly, a similar estimate for T_1 in the case $L^{\frac{1}{m}} |x_2 - x_1|/h \leq 1$ follows from (2.7).

Finally, let $m - 1 > n$. For $L^{\frac{1}{m}} |x_2 - x_1|/h \geq 1$ the estimates (2.8) and (2.9) yield

$$T_1 \leq M \int_0^1 \lambda^{-m\alpha} d\lambda \sum_{z \in \mathbf{R}_{h_1}^n} \exp\{-a|z|\} h_1^n$$

$$+ M \int_1^\infty \lambda^{-m\alpha-1} d\lambda \sum_{z \in \mathbf{R}_{h_1}^n} \exp\{-a|z|\} h_1^n \leq \frac{M_1}{m\alpha(1 - m\alpha)}.$$

Clearly, a similar estimate for T_1 in the case $L^{\frac{1}{m}} |x_2 - x_1|/h \leq 1$ follows from (2.8).

Thus, for any $x_1, x_2 \in \mathbf{R}_{h_1}^n$ we have established the inequality

$$\frac{|f^h(x_1) - f^h(x_2)|}{|x_1 - x_2|^{m\alpha}} \leq \frac{\tilde{M}}{m\alpha(1 - m\alpha)} \|f^h\|_{E'_{\alpha,\infty}(C_h, A_h)}.$$

This means that the following inequality holds:

$$\|f^h\|_{C_h^{m\alpha}} \leq \frac{\tilde{M}}{m\alpha(1 - m\alpha)} \|f^h\|_{E'_{\alpha,\infty}(C_h, A_h)}.$$

Theorem 2.1 is proved.

Since the operator A_h is strongly positive in C_h, it is also strongly positive in $E'_{\alpha,\infty}(C_h, A_h)$. Hence, by Theorem 2.1, A_h is strongly positive in $C_h^{m\alpha}$ for any $0 < \alpha < 1/m$.

2.   Positivity of the elliptic difference operator in L_{1h}. The fractional spaces $E'_{\alpha,1}(L_{1h}, A_h)$.

Let us introduce the space of grid functions $L_{1h} = L_1(\mathbf{R}_h^n)$, defined by the norm

$$\|u^h\|_{L_{1h}} = \sum_{x \in \mathbf{R}_h^n} |u^h(x)| h^n.$$

To study the positivity of the operator A_h in the space L_{1h} we need to study the resolvent equation (1.2) in this space. From the inequality $|u^h(x_0)| \leq \sum_{x \in \mathbf{R}_h^n} |u^h(x)|$ for any $x_0 \in \mathbf{R}_h^n$ it follows that L_{1h} is continuously embedded (for fixed $h > 0$) in C_h. Hence, for any grid function $f^h(x) \in L_{1h}$ equation (1.21) has a unique solution $u^h(x)$ in C_h, and this solution is given by formula (1.22). In the proof of Theorem 2.2 below it will be shown that this solution belongs to L_{1h}. Since $L_{1h} \subset C_h$, it follows that equation (1.22) is uniquely solvable in L_{1h}, i.e., the inverse operator $(\lambda + A_h)^{-1}$, given by formula (1.22), is defined.

Theorem 2.2. *Suppose that the assumptions of Theorem 1.1 hold. Then the operator $A_h + \varepsilon$ is strongly positive on L_{1h} for $\varepsilon > 0$.*

Proof. We have the inequality

$$\|(\lambda + A_h)^{-1}\|_{L_{1h} \to L_{1h}} \le \sum_{z \in \mathbf{R}_h^n} |\mathcal{G}^h(z, \lambda)| h^n. \tag{2.10}$$

Indeed, by (1.22),

$$|(\lambda + A_h)^{-1} f^h(x)| \le \sum_{z \in \mathbf{R}_h^n} |\mathcal{G}^h(z, \lambda)| \, |f^h(x + z)| h^n,$$

whence

$$\|(\lambda + A_h)^{-1} f^h\|_{L_{1h}} = \sum_{x \in \mathbf{R}_h^n} |(\lambda + A_h)^{-1} f^h(x)| h^n$$

$$\le \sum_{x \in \mathbf{R}_h^n} \sum_{z \in \mathbf{R}_h^n} |\mathcal{G}^h(z, \lambda)| \, |f^h(x + z)| h^{2n} = \sum_{z \in \mathbf{R}_h^n} |\mathcal{G}^h(z, \lambda)| \sum_{x \in \mathbf{R}_h^n} |f^h(x + z)| h^n h^n.$$

Making the substitution $x + z = y$, we obtain

$$\|(\lambda + A_h)^{-1} f^h\|_{L_{1h}} \le \sum_{z \in \mathbf{R}_h^n} |\mathcal{G}^h(z, \lambda)| h^n \sum_{y \in \mathbf{R}_h^n} |f^h(y)| h^n$$

$$= \sum_{z \in \mathbf{R}_h^n} |\mathcal{G}^h(z, \lambda)| h^n \, \|f^h\|_{L_{1h}},$$

which yields (2.10).

In Subsection 5 of Section 1 we have shown that $\sum_{z \in \mathbf{R}_h^n} |\mathcal{G}^h(z, \lambda)| h^n \le M|\lambda|^{-1}$. The assertion of the theorem follows from this inequality and (2.10).

In this subsection we shall investigate the structure of the fractional spaces $E'_{\alpha,1}(L_{1h}, A_h)$. To formulate our result we need to introduce the space of grid functions $W_1^\beta(\mathbf{R}_h^n)$, $0 < \beta < 1$, defined by the norm

$$\|u^h\|_{W_1^\beta(\mathbf{R}_h^n)} = \sum_{x \in \mathbf{R}_h^n} \sum_{z \in \mathbf{R}_h^n, \, z \ne 0} \frac{|u^h(x) - u^h(x + z)|}{|z|^{n+\beta}} h^{2n} + \|u^h\|_{L_{1h}}.$$

Theorem 2.3. *The spaces $E'_{\alpha,1}(L_{1h}, A_h)$ and $W_1^{m\alpha}(\mathbf{R}_h^n)$ are identical for $0 < \alpha < 1/m$, and their norms are equivalent uniformly in h for $0 < h \le h_0$.*

Proof. Using the identity (given in Subsection 1 of Section 2)

$$t^\alpha A_h(t+A_h)^{-1} f^h(x) = (2\pi)^{-n} \sum_{z \in \mathbf{R}_h^n,\, z \ne 0} t^{1+\alpha} \mathcal{G}^h(z,t)[f^h(x) - f^h(x-z)] h^n, \quad (2.11)$$

we obtain

$$\sum_{x \in \mathbf{R}_h^n} t^\alpha |A_h(t+A_h)^{-1} f^h(x)| \le t^{1+\alpha} \sum_{x \in \mathbf{R}_h^n} \sum_{z \in \mathbf{R}_h^n,\, z \ne 0} |\mathcal{G}^h(z,t)[f^h(x) - f^h(x-z)]| h^{2n}$$

$$= t^{1+\alpha} \sum_{z \in \mathbf{R}_h^n,\, z \ne 0} |z|^{n+m\alpha} |\mathcal{G}^h(z,t)| h^n \sum_{x \in \mathbf{R}_h^n} \frac{|f^h(x) - f^h(x-z)|}{|z|^{n+m\alpha}} h^n.$$

Consequently,

$$\int_0^\infty t^\alpha \sum_{x \in \mathbf{R}_h^n} |A_h(t+A_h)^{-1} f^h(x)| h^n \frac{dt}{t}$$

$$\le \sum_{z \in \mathbf{R}_h^n,\, z \ne 0} |z|^{n+m\alpha} \int_0^\infty t^\alpha |\mathcal{G}^h(z,t)| dt\, h^n \sum_{x \in \mathbf{R}_h^n} \frac{|f^h(x) - f^h(x-z)|}{|z|^{n+m\alpha}} h^n$$

$$\le \sup_{z \in \mathbf{R}_h^n,\, z \ne 0} |z|^{n+m\alpha} \int_0^\infty t^\alpha |\mathcal{G}^h(z,t)| dt \sum_{z \in \mathbf{R}_h^n,\, z \ne 0} \sum_{x \in \mathbf{R}_h^n} \frac{|f^h(x) - f^h(x-z)|}{|z|^{n+m\alpha}} h^n$$

$$\le J \|f^h\|_{W_1^{m\alpha}(\mathbf{R}_h^n)}.$$

Now let us estimate the expression

$$J = \sup_{z \in \mathbf{R}_h^n,\, z \ne 0} |z|^{n+m\alpha} \int_0^\infty t^\alpha |\mathcal{G}^h(z,t)| dt.$$

By (1.34), we have

$$|z|^{n+m\alpha} \int_{Lh^{-m}}^\infty t^\alpha |\mathcal{G}^h(z,t)| dt$$

$$\le M \int_{Lh^{-m}}^\infty |z|^{n+m\alpha} \exp\{-ah^{-1}|z|\} \frac{t^\alpha dt}{|z|^{m\alpha} h^n t^{1+m\alpha} h^{m\alpha(m-1)}}.$$

Since $|z|^n \exp\{-ah^{-1}|z|\} \le M_1 h^n$, it follows that

$$|z|^{n+m\alpha} \int_{Lh^{-m}}^\infty t^\alpha |\mathcal{G}^h(z,t)| dt \le M_2 \int_{Lh^{-m}}^\infty \frac{dt}{h^{m\alpha(m-1)} t^{1+(m-1)\alpha}}$$

$$\leq \frac{M_3}{(m-1)\alpha L^{\alpha(m-1)}} \leq \frac{M_4}{\alpha}. \tag{2.12}$$

Next, let us establish the estimate

$$|z|^{n+m\alpha} \int_0^{Lh^{-m}} t^\alpha |\mathcal{G}^h(z,t)| dt \leq M. \tag{2.13}$$

We need to examine separately the three cases $m > n$, $m = n$, and $m < n$. In the case $m > n$ Theorem 1.1 yields

$$|z|^{n+m\alpha} \int_0^{Lh^{-m}} t^\alpha |\mathcal{G}^h(z,t)| dt \leq |z|^{n+m\alpha} M \int_0^{Lh^{-m}} t^\alpha \exp\{-at^{\frac{1}{m}}|z|\} t^{\frac{n}{m}-1} dt.$$

Performing the change of variables $t^{\frac{1}{m}}|z| = s$, we obtain

$$|z|^{n+m\alpha} \int_0^{Lh^{-m}} t^\alpha |\mathcal{G}^h(z,t)| dt$$

$$\leq M|z|^{n+m\alpha} \int_0^{L^{\frac{1}{m}}h^{-1}|z|} \exp\{-as\} \left(\frac{s}{|z|}\right)^{m\alpha+n-m} m \frac{s^{m-1} ds}{|z|^m}$$

$$\leq M_1 \int_0^\infty s^{n-1+m\alpha} \exp\{-as\} ds \leq M_2. \tag{2.14}$$

In the case $m = n$ Theorem 1.3 yields

$$|z|^{m+m\alpha} \int_0^{Lh^{-m}} t^\alpha |\mathcal{G}^h(z,t)| dt$$

$$\leq M|z|^{m+m\alpha} \int_0^{Lh^{-m}} t^\alpha \exp\{-at^{\frac{1}{m}}|z|\} \left(1 + \ln((t^{\frac{1}{m}}(h+|z|))^{-1})\right) dt$$

$$\leq M_1|z|^{m+m\alpha} \int_0^{Lh^{-m}} t^\alpha \exp\left\{-\frac{a}{2}t^{\frac{1}{m}}|z|\right\} dt.$$

Performing the change of variables $t^{\frac{1}{m}}|z| = s$ we obtain

$$|z|^{m+m\alpha} \int_0^{Lh^{-m}} t^\alpha |\mathcal{G}^h(z,t)| dt \leq M_1|z|^{m+m\alpha} \int_0^{L^{\frac{1}{m}}h^{-1}|z|} \exp\left\{-\frac{a}{2}s\right\} \frac{s^{m\alpha} m s^{m-1} ds}{|z|^{m\alpha}|z|^m}$$

$$\leq M_2 \int_0^\infty s^{m-1+m\alpha} \exp\left\{-\frac{a}{2}s\right\} ds \leq M_3, \tag{2.15}$$

because $m - n$. Finally, in the case $m < n$ Theorem 1.3 yields

$$|z|^{n+m\alpha} \int_0^{Lh^{-m}} t^\alpha |\mathcal{G}^h(z,t)| dt \le M|z|^{n+m\alpha} \int_0^{Lh^{-m}} t^\alpha \exp\{-at^{\frac{1}{m}}|z|\}(h+|z|)^{m-n} dt$$

$$\le M_1 |z|^{m+m\alpha} \int_0^{Lh^{-m}} t^\alpha \exp\{-at^{\frac{1}{m}}|z|\} dt.$$

Performing the change of variables $t^{\frac{1}{m}}|z| = s$ we obtain

$$|z|^{n+m\alpha} \int_0^{Lh^{-m}} t^\alpha |\mathcal{G}^h(z,t)| dt \le M_1 |z|^{m+m\alpha} \int_0^{L^{\frac{1}{m}}h^{-1}|z|} \frac{s^{m\alpha}}{|z|^{m\alpha}} \exp\{-as\} \frac{ms^{m-1} ds}{|z|^m}$$

$$\le M_2 \int_0^\infty s^{m-1+m\alpha} \exp\{-as\} ds \le M_3. \tag{2.16}$$

From the estimates (2.14), (2.15), and (2.16) we obtain inequality (2.13). By (2.12) and (2.13), $J \le M/\alpha$, whence

$$\|f^h\|_{E'_{\alpha,1}(L_{1,h}, A_h)} \le \frac{M}{\alpha} \|f^h\|_{W_1^{m\alpha}(\mathbf{R}_h^n)}. \tag{2.17}$$

Now let us prove the opposite inequality:

$$\|f^h\|_{W_1^{m\alpha}(\mathbf{R}_h^n)} \le \frac{M}{m\alpha(1-m\alpha)} \|f^h\|_{E'_{\alpha,1}(L_{1,h}, A_h)}. \tag{2.18}$$

Arguing exactly as in the first subsection, we write

$$\frac{f^h(x) - f^h(x-z)}{|z|^{n+m\alpha}}$$

$$= (2\pi)^{-n} \int_0^\infty t^{-\alpha} \frac{\mathcal{G}^h(x-y,t) - \mathcal{G}^h(x-z-y,t)}{|z|^{n+m\alpha}} t^\alpha A_h(t+A_h)^{-1} f^h(y) h^n dt.$$

Making the substitution $x - y = \tilde{y}$ and using the triangle inequality, we obtain

$$\sum_{z \in \mathbf{R}_h^n,\ z \ne 0} \sum_{x \in \mathbf{R}_h^n} \frac{|f^h(x) - f^h(x-z)|}{|z|^{n+m\alpha}} h^{2n}$$

$$\le \int_0^\infty t^{-\alpha} \sum_{z \in \mathbf{R}_h^n,\ z \ne 0} \sum_{x \in \mathbf{R}_h^n} \sum_{\tilde{y} \in \mathbf{R}_h^n} \frac{|\mathcal{G}^h(\tilde{y},t) - \mathcal{G}^h(\tilde{y}-z,t)|}{|z|^{n+m\alpha}} \times$$

$$\times t^\alpha |A_h(t+A_h)^{-1} f^h(x-\tilde{y})| h^{3n} dt$$

$$= \int_0^\infty t^{-\alpha+1} \sum_{y\in\mathbf{R}_h^n} \sum_{z\in\mathbf{R}_h^n,\ z\neq0} \frac{|\mathcal{G}^h(y,t)-\mathcal{G}^h(y-z,t)|}{|z|^{n+m\alpha}} h^{2n} \times$$

$$\times \sum_{x\in\mathbf{R}_h^n} t^\alpha |A_h(t+A_h)^{-1}f^h(x)|h^n \frac{dt}{t}$$

$$\leq \sup_{t\geq0} t^{-\alpha+1} \sum_{y\in\mathbf{R}_h^n} \sum_{z\in\mathbf{R}_h^n,\ z\neq0} \frac{|\mathcal{G}^h(y,t)-\mathcal{G}^h(y-z,t)|}{|z|^{n+m\alpha}} h^{2n} \times$$

$$\times \int_0^\infty t^\alpha \sum_{x\in\mathbf{R}_h^n} |A_h(t+A_h)^{-1}f^h(x)|h^n \frac{dt}{t} \leq T\|f^h\|_{E'_{\alpha,1}(L_{1,h},A_h)}.$$

Let us estimate the expression

$$T = \sup_{t\geq0} t^{-\alpha+1} \sum_{y\in\mathbf{R}_h^n} \sum_{z\in\mathbf{R}_h^n,\ z\neq0} \frac{|\mathcal{G}^h(y,t)-\mathcal{G}^h(y-z,t)|}{|z|^{n+m\alpha}} h^{2n}.$$

Set $h_1 = t^{\frac{1}{m}}h$. Using the identity (1.1), we obtain

$$T = \sup_{t\geq0} t^{-\alpha+1} \sum_{y\in\mathbf{R}_h^n} \sum_{z\in\mathbf{R}_h^n,\ z\neq0} h^{2n}t^{-1+\frac{n}{m}} \frac{|\mathcal{G}^{t^{\frac{1}{m}}h}(t^{\frac{1}{m}}y,1)-\mathcal{G}^{t^{\frac{1}{m}}h}(t^{\frac{1}{m}}(y-z),1)|}{|z|^{n+m\alpha}}$$

$$= \sup_{h_1} \sum_{\tilde{y}\in\mathbf{R}_{h_1}^n} \sum_{\tilde{z}\in\mathbf{R}_{h_1}^n,\ \tilde{z}\neq0} \frac{|\mathcal{G}^{h_1}(\tilde{y},1)-\mathcal{G}^{h_1}(\tilde{y}-\tilde{z},1)|}{|\tilde{z}|^{n+m\alpha}} h_1^{2n}$$

$$\leq \sup_{h_1} \sum_{\tilde{y}\in\mathbf{R}_{h_1}^n} \sum_{\tilde{z}\in\mathbf{R}_{h_1}^n,\ |\tilde{z}|\geq1} \frac{|\mathcal{G}^{h_1}(\tilde{y},1)-\mathcal{G}^{h_1}(\tilde{y}-\tilde{z},1)|}{|\tilde{z}|^{n+m\alpha}} h_1^{2n}$$

$$+ \sup_{h_1} \sum_{\tilde{y}\in\mathbf{R}_{h_1}^n} \sum_{\tilde{z}\in\mathbf{R}_{h_1}^n,\ 0<|\tilde{z}|<1} \frac{|\mathcal{G}^{h_1}(\tilde{y},1)-\mathcal{G}^{h_1}(\tilde{y}-\tilde{z},1)|}{|\tilde{z}|^{n+m\alpha}} h_1^{2n}$$

$$= \sup_{h_1} T_{1h_1} + \sup_{h_1} T_{2h_1}.$$

First let us estimate T_{1h_1}. The triangle inequality yields

$$T_{1h_1} \leq \sum_{\tilde{y}\in\mathbf{R}_{h_1}^n} \sum_{\tilde{z}\in\mathbf{R}_{h_1}^n,\ |\tilde{z}|\geq1} \frac{[|\mathcal{G}^{h_1}(\tilde{y},1)|+|\mathcal{G}^{h_1}(\tilde{y}-\tilde{z},1)|]h_1^{2n}}{|\tilde{z}|^{n+m\alpha}}$$

$$= \sum_{\tilde{y}\in\mathbf{R}_{h_1}^n} \sum_{\tilde{z}\in\mathbf{R}_{h_1}^n,\ |\tilde{z}|\geq1} \frac{|\mathcal{G}^{h_1}(\tilde{y},1)|}{|\tilde{z}|^{n+m\alpha}} h_1^{2n} + \sum_{\tilde{y}\in\mathbf{R}_{h_1}^n} \sum_{\tilde{z}\in\mathbf{R}_{h_1}^n,\ |\tilde{z}|\geq1} \frac{|\mathcal{G}^{h_1}(\tilde{y}-\tilde{z},1)|}{|\tilde{z}|^{n+m\alpha}} h_1^{2n}.$$

In the second sum, let us change the order of summation and change the variables to $\tilde{y} - \tilde{z} = \tilde{\tilde{y}}$. Then

$$\sum_{\tilde{y}\in\mathbf{R}^n_{h_1}} \sum_{\tilde{z}\in\mathbf{R}^n_{h_1},\ |\tilde{z}|\geq 1} \frac{|\mathcal{G}^{h_1}(\tilde{y}-\tilde{z},1)|}{|\tilde{z}|^{n+m\alpha}} h_1^{2n} = \sum_{\tilde{z}\in\mathbf{R}^n_{h_1},\ |\tilde{z}|\geq 1} \sum_{\tilde{y}\in\mathbf{R}^n_{h_1}} \frac{|\mathcal{G}^{h_1}(\tilde{y}-\tilde{z},1)|}{|\tilde{z}|^{n+m\alpha}} h_1^{2n}$$

$$= \sum_{\tilde{z}\in\mathbf{R}^n_{h_1},\ |\tilde{z}|\geq 1} \sum_{\tilde{\tilde{y}}\in\mathbf{R}^n_{h_1}} \frac{|\mathcal{G}^{h_1}(\tilde{\tilde{y}},1)|}{|\tilde{z}|^{n+m\alpha}} h_1^{2n} = \sum_{\tilde{\tilde{y}}\in\mathbf{R}^n_{h_1}} \sum_{\tilde{z}\in\mathbf{R}^n_{h_1},\ |\tilde{z}|\geq 1} \frac{|\mathcal{G}^{h_1}(\tilde{\tilde{y}},1)|}{|\tilde{z}|^{n+m\alpha}} h_1^{2n}.$$

Therefore,

$$T_{1h_1} \leq 2 \sum_{\tilde{y}\in\mathbf{R}^n_{h_1}} \sum_{\tilde{z}\in\mathbf{R}^n_{h_1},\ |\tilde{z}|\geq 1} \frac{|\mathcal{G}^{h_1}(\tilde{y},1)|}{|\tilde{z}|^{n+m\alpha}} h_1^{2n}.$$

Since

$$\sum_{\tilde{z}\in\mathbf{R}^n_{h_1},\ |\tilde{z}|\geq 1} \frac{h_1^n}{|\tilde{z}|^{n+m\alpha}} \leq \frac{M}{m\alpha},$$

the last inequality yields the estimate

$$T_{1h_1} \leq \frac{2M}{m\alpha} \sum_{\tilde{y}\in\mathbf{R}^n_{h_1}} |\mathcal{G}^{h_1}(\tilde{y},1)| h_1^n.$$

By Theorem 1.2,

$$\sum_{\tilde{y}\in\mathbf{R}^n_{h_1}} |\mathcal{G}^{h_1}(\tilde{y},1)| h_1^n \leq \tilde{M}.$$

Hence,

$$T_{1h_1} \leq \frac{M_1}{m\alpha}. \tag{2.19}$$

Now let us estimate T_{2h_1}. From the identity

$$\mathcal{G}^{h_1}(\tilde{y},1) - \mathcal{G}^{h_1}(\tilde{y}+\tilde{z},1)$$

$$= -\sum_{k=1}^{n} \sum_{i=0}^{|\tilde{z}_k|h_1^{-1}-1} \left[\mathcal{G}^{h_1}\left(\tilde{y} + \sum_{j=1}^{k-1} \tilde{z}_j e_j + \operatorname{sign} \tilde{z}_k (i+1) e_k h_1, 1 \right) \right.$$

$$\left. - \mathcal{G}^{h_1}\left(\tilde{y} + \sum_{j=1}^{k-1} \tilde{z}_j e_j + \operatorname{sign} \tilde{z}_k i e_k h_1, 1 \right) \right]$$

it follows (see Section 2) that

$$J(\tilde{z},\tilde{y}) = |\mathcal{G}^{h_1}(\tilde{y},1) - \mathcal{G}^{h_1}(\tilde{y}+\tilde{z},1)|$$

$$\le \sum_{k=1}^{n} \sum_{i=0}^{|\tilde{z}_k|h_1^{-1}-1} \left| \mathcal{D}_h^1 \mathcal{G}^{h_1} \left(\tilde{y} + i \operatorname{sign} \tilde{z}_k e_k h_1 + \sum_{j=1}^{k-1} \tilde{z}_j e_j, 1 \right) \right|.$$

Clearly, $t \ge Lh^{-m}$ implies $h_1 \ge L^{\frac{1}{m}}$. Hence, using the estimate (1.66) with $r = 1$, we obtain

$$T_{2h_1} = \sum_{\tilde{y} \in \mathbf{R}_{h_1}^n} \sum_{\tilde{z} \in \mathbf{R}_{h_1}^n, 0<|\tilde{z}|<1} \sum_{k=1}^{n} \sum_{i=0}^{|\tilde{z}_k|h_1^{-1}-1} \frac{J(\tilde{z}, \tilde{y}) h_1^{2n}}{|\tilde{z}|^{n+m\alpha}}$$

$$\le M \sum_{\tilde{z} \in \mathbf{R}_{h_1}^n, 0<|\tilde{z}|<10} \sum_{k=1}^{n} \sum_{i=0}^{|\tilde{z}_k|h_1^{-1}-1} \sum_{\tilde{y} \in \mathbf{R}_{h_1}^n} h_1 \times$$

$$\times \exp\left\{ -a \left| \tilde{y} i \operatorname{sign} \tilde{z}_k e_k h_1 + \sum_{j=1}^{k} \tilde{z}_j e_j \right| h_1^{-1} \right\} h_1^{-1-n} h_1^{2n}$$

$$= M \sum_{\tilde{z} \in \mathbf{R}_{h_1}^n, 0<|\tilde{z}|<1} \frac{1}{|\tilde{z}|^{n+m\alpha}} \sum_{k=1}^{n} \sum_{i=0}^{|\tilde{z}_k|h_1^{-1}-1} h_1 \sum_{\tilde{y} \in \mathbf{R}_{h_1}^n} \exp\{-a|\tilde{y}|h_1^{-1}\} h_1^{-1-n} h_1^{2n}$$

$$= M \sum_{\tilde{z} \in \mathbf{R}_{h_1}^n, 0<|\tilde{z}|<1} \frac{1}{|\tilde{z}|^{n+m\alpha}} \sum_{k=1}^{n} |\tilde{z}_k| \sum_{y \in \mathbf{R}_1^n} \exp\{-a|y|\} h_1^{-1+n}$$

$$\le M_1 \sum_{\tilde{z} \in \mathbf{R}_{h_1}^n, 0<|\tilde{z}|<1} \frac{h_1^{n-1}}{|\tilde{z}|^{n-1+m\alpha}} \sum_{y \in \mathbf{R}_1^n} \exp\{-a|y|\} \le \frac{M_2}{h_1} \le \frac{M_2}{L^{\frac{1}{m}}} \le M_3.$$

If $t \le Lh^{-m}$, then $h_1 \le L^{\frac{1}{m}}$, and we have to consider three cases: $m < n$, $m = n$, and $m > n$. In the first case, inequality (1.67) with $r = 1$ yields

$$T_{2h_1} \le \sum_{\tilde{y} \in \mathbf{R}_{h_1}^n} \sum_{\tilde{z} \in \mathbf{R}_{h_1}^n, 0<|\tilde{z}|<1} \sum_{k=1}^{n} \sum_{i=0}^{|\tilde{z}_k|h_1^{-1}-1} \frac{J(\tilde{z}, \tilde{y}) h_1^{2n}}{|\tilde{z}|^{n+m\alpha}}$$

$$\le M \sum_{\tilde{z} \in \mathbf{R}_{h_1}^n, 0<|\tilde{z}|<1} \frac{1}{|\tilde{z}|^{n+m\alpha}} \sum_{k=1}^{n} \sum_{i=0}^{|\tilde{z}_k|h_1^{-1}-1} \sum_{\tilde{y} \in \mathbf{R}_{h_1}^n} h_1^{1+2n} \times$$

$$\times \exp\left\{ -a \left| \tilde{y} + i \operatorname{sign} \tilde{z}_k e_k h_1 + \sum_{j=1}^{k-1} \tilde{z}_j e_j \right| \right\} \times$$

$$\times \left(h_1 + \left| \tilde{y} + i \operatorname{sign} \tilde{z}_k e_k h_1 + \sum_{j=1}^{k} \tilde{z}_j e_j \right| \right)^{m-n-1}$$

$$= M \sum_{\tilde{z} \in \mathbf{R}_{h_1}^n, 0 < |\tilde{z}| < 1} \frac{1}{|\tilde{z}|^{n+m\alpha}} \sum_{k=1}^{n} |\tilde{z}_k| \sum_{\tilde{\tilde{y}} \in \mathbf{R}_{h_1}^n} \exp\{-a|\tilde{\tilde{y}}|\}(h_1 + |\tilde{\tilde{y}}|)^{m-n-1} h_1^{2n}$$

$$\leq M_1 \sum_{\tilde{z} \in \mathbf{R}_{h_1}^n, 0 < |\tilde{z}| < 1} \frac{h_1^n}{|\tilde{z}|^{n-1+m\alpha}} \sum_{\tilde{\tilde{y}} \in \mathbf{R}_{h_1}^n} \exp\{-a|\tilde{\tilde{y}}|\}(h_1 + |\tilde{\tilde{y}}|)^{m-n-1} h_1^n.$$

Since $n - m + 1 \leq n - 1$,

$$\sum_{\tilde{\tilde{y}} \in \mathbf{R}_{h_1}^n} \exp\{-a|\tilde{\tilde{y}}|\}(h_1 + |\tilde{\tilde{y}}|)^{m-n-1} h_1^n$$

$$= h_1^{m-1} + \sum_{\tilde{\tilde{y}} \in \mathbf{R}_{h_1}^n, \tilde{\tilde{y}} \neq 0} \exp\{-a|\tilde{\tilde{y}}|\}(h_1 + |\tilde{\tilde{y}}|)^{m-n-1} h_1^n$$

$$\leq L^{\frac{m-1}{m}} + M \sum_{\tilde{\tilde{y}} \in \mathbf{R}_{h_1}^n} \exp\{-a(h_1 + |\tilde{\tilde{y}}|)\}(h_1 + |\tilde{\tilde{y}}|)^{m-n-1} h_1^n$$

$$\leq L^{\frac{m-1}{m}} + M \int_{z \in \mathbf{R}^n} \exp\{-a|z|\}|z|^{m-n-1} dz.$$

Consequently,

$$T_{2h_1} \leq M_1 \sum_{\tilde{z} \in \mathbf{R}_{h_1}^n, h_1 \leq |\tilde{z}| \leq 1} \frac{h_1^n}{|\tilde{z}|^{n-1+m\alpha}} \leq \frac{M_2}{1 - m\alpha}.$$

In the cases $m = n$ and $m \geq n$ the needed estimates for T_{2h_1} are established following the same scheme as above, using the estimates (1.73) and (1.67), respectively.

Combining the estimates for T_{2h_1}, we obtain

$$T_{2h_1} \leq \frac{M_1}{m\alpha(1 - m\alpha)},$$

which in conjunction with the estimate (2.19) yields the bound

$$T \leq \frac{M}{m\alpha(1 - m\alpha)}.$$

This completes the proof of Theorem 2.3.

By Theorem 2.2, $A_h + \varepsilon$, $\varepsilon > 0$, is a strongly positive operator in L_{1h}. This implies that $A_h + \varepsilon$ is a strongly positive operator in $E'_{\alpha,1}(L_{1h}, A_h)$. Hence, by Theorem 2.3, $A_h + \varepsilon$ is a strongly positive operator in $W_1^{m\alpha}(\mathbf{R}_h^n)$ for any α, $0 < \alpha < 1/m$.

3. Positivity of elliptic difference operators in L_{ph}. The fractional spaces $E'_{\alpha,p}(L_{ph}, A_h)$.

Let us introduce the spaces of grid functions $L_{ph} = L_p(\mathbf{R}_h^n)$, $1 \le p \le \infty$, defined by the norm

$$\|u^h\|_{L_{ph}} = \left(\sum_{x \in \mathbf{R}_h^n} |u^h(x)|^p h^n \right)^{\frac{1}{p}}.$$

Theorem 2.4. *Suppose that the conditions of Theorem 1.1 are satisfied. Then the operator $A_h + \varepsilon$ is strongly positive in L_{ph} for $\varepsilon > 0$.*

Proof. The unique solvability of the resolvent equation (2.1) is established as in the case $p = 1$. Hence, to prove our theorem it suffices to prove the inequality

$$\|(\lambda + A_h)^{-1}\|_{L_{ph} \to L_{ph}} \le \sum_{z \in \mathbf{R}_h^n} |\mathcal{G}^h(z, \lambda)| h^n. \qquad (2.20)$$

Using formula (1.22) and the Minkowski inequality (in the variable x), we obtain

$$\|(\lambda + A_h)^{-1} f^h\|_{L_{ph}} = \left(\sum_{x \in \mathbf{R}_h^n} |(\lambda + A_h)^{-1} f^h(x)|^p h^n \right)^{\frac{1}{p}}$$

$$\le \sum_{z \in \mathbf{R}_h^n} |\mathcal{G}^h(z, \lambda)| \left(\sum_{x \in \mathbf{R}_h^n} |f^h(x + z)|^p h^n \right)^{\frac{1}{p}} h^n.$$

The change of variables $x + z = y$ yields

$$\|(\lambda + A_h)^{-1} f^h\|_{L_{ph}} \le \sum_{z \in \mathbf{R}_h^n} |\mathcal{G}^h(z, \lambda)| \left(\sum_{y \in \mathbf{R}_h^n} |f^h(y)|^p h^n \right)^{\frac{1}{p}} h^n$$

$$= \sum_{z \in \mathbf{R}_h^n} |\mathcal{G}^h(z, \lambda)| h^n \|f^h\|_{L_{ph}}.$$

This proves (2.20) and completes the proof of Theorem 2.4.

In this subsection we shall investigate the structure of the fractional spaces $E'_{\alpha,p}(L_{ph}, A_h)$. To formulate our result we need to introduce the space of grid functions $W_p^\beta(\mathbf{R}_h^n)$, $0 < \beta < 1$, defined by the norm

$$\|u^h\|_{W_p^\beta(\mathbf{R}_h^n)} = \left(\sum_{x \in \mathbf{R}_h^n} \sum_{z \in \mathbf{R}_h^n,\, z \neq 0} \frac{|u^h(x) - u^h(x+z)|^p}{|z|^{n+\beta p}} h^{2n} + \sum_{x \in \mathbf{R}_h^n} |u^h(x)|^p h^n \right)^{\frac{1}{p}}.$$

Theorem 2.5. *The spaces $E'_{\alpha,p}(L_{ph}, A_h)$ and $W_p^{m\alpha}(\mathbf{R}_h^n)$ coincide for $0 < \alpha < 1/m$, and their norms are equivalent uniformly in h, $0 < h \leq h_0$.*

Proof. Using the identity (2.11) and the Minkowski inequality (in the variable x), we obtain

$$\left(\int_{Lh^{-m}}^\infty \sum_{x \in \mathbf{R}_h^n} |t^\alpha A_h (t + A_h)^{-1} f^h(x)|^p h^n \frac{dt}{t} \right)^{\frac{1}{p}}$$

$$\leq \sum_{z \in \mathbf{R}_h^n} h^n \left(\int_{Lh^{-m}}^\infty |t^{1+\alpha} \mathcal{G}^h(z,t)|^p \frac{dt}{t} \right)^{\frac{1}{p}} \left(\sum_{x \in \mathbf{R}_h^n} |f^h(x) - f^h(x+z)|^p h^n \right)^{\frac{1}{p}}$$

$$= \sum_{z \in \mathbf{R}_h^n} h^n \left(\int_{Lh^{-m}}^\infty |z|^{n+m\alpha p} |t^{1+\alpha} \mathcal{G}^h(z,t)|^p \frac{dt}{t} \right)^{\frac{1}{p}} \times$$

$$\times \left(\sum_{x \in \mathbf{R}_h^n} \frac{|f^h(x) - f^h(x+z)|^p}{|z|^{n+m\alpha p}} h^n \right)^{\frac{1}{p}}.$$

Next, using the Hölder inequality (in z), we obtain $(1/p + 1/q = 1)$

$$\left(\int_{Lh^{-m}}^\infty \sum_{x \in \mathbf{R}_h^n} |t^\alpha A_h (t + A_h)^{-1} f^h(x)|^p h^n \frac{dt}{t} \right)^{\frac{1}{p}}$$

$$\leq \left(\sum_{z \in \mathbf{R}_h^n,\, z \neq 0} \left(\int_{Lh^{-m}}^\infty |z|^{n+m\alpha p} |t^{1+\alpha} \mathcal{G}^h(z,t)|^p \frac{dt}{t} \right)^{\frac{q}{p}} h^n \right)^{\frac{1}{q}} \times$$

$$\times \left(\sum_{z \in \mathbf{R}_h^n,\, z \neq 0} \sum_{x \in \mathbf{R}_h^n} \frac{|f^h(x) - f^h(x+z)|^p}{|z|^{n+m\alpha p}} h^{2n} \right)^{\frac{1}{p}} \leq J \|f^h\|_{W_p^{m\alpha}(\mathbf{R}_h^n)},$$

where

$$J = \left(\sum_{z \in \mathbf{R}_h^n,\, z \neq 0} \left(\int_{Lh^{-m}}^\infty |z|^{n+m\alpha p} |t^{1+\alpha} \mathcal{G}^h(z,t)|^p \frac{dt}{t} \right)^{\frac{q}{p}} h^n \right)^{\frac{1}{q}}.$$

Let us estimate J. From (1.34) it follows that

$$J \leq M \left(\sum_{z \in \mathbf{R}_h^n,\, z \neq 0} \left(\int_{Lh^{-m}}^\infty |z|^{n+m\alpha p} \left| t^{1+\alpha} \exp\{-ah^{-1}|z|\} \times \right. \right.$$

$$\left. \left. \times \frac{t^{-1-m\alpha}}{|z|^{m\alpha} h^n h^{m\alpha(m-1)}} \right|^p \frac{dt}{t} \right)^{\frac{q}{p}} h^n \right)^{1/q}$$

$$\leq M \left(\sum_{z \in \mathbf{R}_h^n,\, z \neq 0} h^{-n(q-1)} |z|^{n\frac{q}{p}} \exp\{-aqh^{-1}|z|\} \right)^{\frac{1}{q}} \left(\int_{Lh^{-m}}^\infty \frac{h^{-pm\alpha(m-1)} dt}{t^{1+p\alpha(m-1)}} \right)^{\frac{1}{p}}$$

$$= \left(\sum_{k \in \mathbf{R}_1^n} |k|^{n(q-1)} \exp\{-aq|k|\} \right)^{\frac{1}{q}} \frac{ML^{-(m-1)\alpha}}{(p\alpha(m-1))^{1/p}} \leq \frac{M_1}{\alpha^{\frac{1}{p}}}.$$

Therefore,

$$\left(\int_{Lh^{-m}}^\infty \sum_{x \in \mathbf{R}_h^n} |t^\alpha A_h(t + A_h)^{-1} f^h(x)|^p h^n \frac{dt}{t} \right)^{\frac{1}{p}} \leq \frac{M_1}{\alpha^{1/p}} \|f^h\|_{W_p^{m\alpha}(\mathbf{R}_h^n)}. \quad (2.21)$$

Let $0 \leq t \leq Lh^{-m}$ and $n > m$. Then, by formula (1.22) and the estimate (1.62),

$$|t^\alpha A_h(t + A_h)^{-1} f^h(x)|$$

$$\leq M \sum_{z \in \mathbf{R}_h^n} t^{1+\alpha} \exp\{-at^{\frac{1}{m}}|z|\}(h + |z|)^{m-n+1} |f^h(x) - f^h(x-z)| h^n.$$

Since the last factor vanishes at $z = 0$,

$$|t^\alpha A_h(t + A_h)^{-1} f^h(x)|$$

$$\leq M_1 \sum_{z \in \mathbf{R}_h^n} t^{1+\alpha} \exp\left\{-\frac{a}{2} t^{\frac{1}{m}} (h + |z|)\right\} (h + |z|)^{m-n} |f^h(x) - f^h(x - z)| h^n \equiv \mathcal{I}.$$

Let us represent $\mathcal{I}$ as an integral over $\mathbf{R}^n$. To this end we extend the grid functions appearing in $\mathcal{I}$ to piecewise-linear functions, which yields

$$\mathcal{I} = M_1 \int_{\mathbf{R}^n} t^{1+\alpha} \exp\left\{-\frac{a}{2} t^{\frac{1}{m}} \left(\left|\left[\frac{z}{h}\right] h\right| + h\right)\right\} \times$$

$$\times \left(\left|\left[\frac{z}{h}\right] h\right| + h\right)^{m-n} \left|f^h(x) - f^h\left(x - \left[\frac{z}{h}\right] h\right)\right| dz.$$

Here, for a vector $y = (y_1, \cdots, y_n)$ we denote by $[y]$ the vector with coordinates $[y_k]$, where $[\cdot]$ stands for the integer part. Clearly, $|[\frac{y_k}{h}] h| + h \geq |y_k|$ for any real y_k, and hence $|[\frac{z}{h}] h| + h \geq \frac{1}{\sqrt{n}} |z|$. Therefore, we have the inequality

$$\mathcal{I} \leq M_1 \int_{z \in \mathbf{R}^n} t^{1+\alpha} \exp\left\{-\frac{a}{2\sqrt{n}} t^{\frac{1}{m}} |z|\right\} |z|^{m-n} \left|f^h(x) - f^h\left(x - \left[\frac{z}{h}\right] h\right)\right| dz.$$

Performing the change of variables $y = t^{\frac{1}{m}} z$ we obtain the inequality

$$|t^\alpha A_h(t + A_h)^{-1} f^h(x)|$$

$$\leq M_1 \int_{y \in \mathbf{R}^n} \exp\left\{-\frac{a}{2\sqrt{n}} |y|\right\} |y|^{n-m} t^\alpha \left|f^h(x) - f^h\left(x - \left[\frac{y}{t^{\frac{1}{m}} h}\right] h\right)\right| dt.$$

Applying the Minkowski inequality we obtain

$$\mathcal{I}_1 \equiv \left(\int_0^{Lh^{-m}} \sum_{x \in \mathbf{R}_h^n} |t^\alpha A_h(t + A_h)^{-1} f^h(x)|^p h^n \frac{dt}{t}\right)^{\frac{1}{p}}$$

$$\leq M_1 \int_{y \in \mathbf{R}^n} \exp\left\{-\frac{a}{2\sqrt{n}} |y|\right\} \times$$

$$\times \left(\int_0^\infty \sum_{x \in \mathbf{R}_h^n} \left|t^\alpha \left(f^h(x) - f^h\left(x - \left[\frac{y}{t^{\frac{1}{m}} h}\right] h\right)\right)\right|^p h^n \frac{dt}{t}\right)^{\frac{1}{p}} dy.$$

Changing the variables in the inner integral to $t = |y|^m \rho^{-m}$ so that $dt/t = -m d\rho/\rho$, we obtain

$$\mathcal{I}_1 \leq M_1 m^{\frac{1}{p}} \int_{y \in \mathbf{R}^n} \exp\left\{-\frac{a}{2\sqrt{n}} |y|\right\} |y|^{m\alpha+m-n} \times$$

$$\times \left(\int_0^\infty \sum_{x \in \mathbf{R}_h^n} \frac{\left| f^h(x) - f^h \left(x - \left[\frac{y\rho}{|y|h} \right] h \right) \right|^p h^n}{\rho^{m\alpha+1}} d\rho \right)^{\frac{1}{p}} dy.$$

Let us pass to spherical coordinates in the outer integral, putting $r = |y|$ and $\sigma = y/|y|$ for $y \in \mathbf{R}^n$, $y \neq 0$; here σ is the projection of y on the unit sphere Σ. Then

$$\mathcal{I}_1 \leq M_1 m^{\frac{1}{p}} \int_0^\infty \exp \left\{ -\frac{a}{2\sqrt{n}} r \right\} r^{(1+\alpha)m-n} r^{n-1} dr \times$$

$$\times \int_\Sigma \left(\int_0^\infty \sum_{x \in \mathbf{R}_h^n} \frac{\left| f^h(x) - f^h \left(x - \left[\frac{\rho\sigma}{h} \right] h \right) \right|^p h^n}{\rho^{m\alpha p+1}} d\rho \right)^{\frac{1}{p}} d\sigma.$$

Calculating the integral with respect to r and applying the Hölder inequality to the integral over Σ, we obtain

$$\mathcal{I}_1 \leq M_1 m^{\frac{1}{p}} \Gamma((1+\alpha)m) \left(\frac{2\sqrt{n}}{a} \right)^{(1+\alpha)m} |\Sigma|^{\frac{1}{q}} \times$$

$$\times \left(\int_\Sigma d\sigma \int_0^\infty \sum_{x \in \mathbf{R}_h^n} \frac{\left| f^h(x) - f^h \left(x - \left[\frac{\rho\sigma}{h} \right] h \right) \right|^p h^n}{\rho^{m\alpha p+1}} d\rho \right)^{\frac{1}{p}}.$$

Here $1/q = 1 - 1/p$ and $|\Sigma|$ is the area of the unit sphere Σ in $\mathbf{R}^n$. Since

$$\int_\Sigma \int_0^\infty (\cdots) \rho^{n-1} d\rho \, d\sigma = \int_{y \in \mathbf{R}^n} (\cdots) dy,$$

the last inequality can be recast in the form

$$\mathcal{I}_1 \leq M_2 \left(\sum_{x \in \mathbf{R}_h^n} \int_{y \in \mathbf{R}^n} \frac{\left| f^h(x) - f^h \left(x - \left[\frac{y}{h} \right] h \right) \right|^p}{|y|^{n+m\alpha p}} h^n dy \right)^{\frac{1}{p}}.$$

By the definition of the function $\left[\frac{y}{h} \right] h$, it follows that

$$\mathcal{I}_1 \leq M_2 \left(\sum_{x \in \mathbf{R}_h^n} \int_{y \in \mathbf{R}^n, \, |y|>h} \frac{\left| f^h(x) - f^h \left(x - \left[\frac{y}{h} \right] h \right) \right|^p}{|y|^{n+m\alpha p}} h^n dy \right)^{\frac{1}{p}}.$$

In the domain of integration of the inner integral the inequality $|y| \geq \delta \left| \left[\frac{y}{h} \right] h \right|$ holds with some δ that does not depend on h. Consequently,

$$\mathcal{I}_1 \leq M_3 \left(\sum_{x \in \mathbf{R}_h^n} \int_{|y|>h} \frac{\left| f^h(x) - f^h \left(x - \left[\frac{y}{h} \right] h \right) \right|^p}{\left| \left[\frac{y}{h} \right] h \right|} h^n dy \right)^{\frac{1}{p}}.$$

Since now the integrand involves a piecewise-constant function y, we have the estimate

$$\mathcal{I}_1 \leq M_3 \left(\sum_{x \in \mathbf{R}_h^n} \sum_{y \in \mathbf{R}_h^n, \, |y| \neq 0} \frac{|f^h(x) - f^h(x-y)|^p}{|y|^{n+m\alpha p}} h^{2n} \right)^{\frac{1}{p}}.$$

Thus, in the case $n > m$ we have established the inequality

$$\left(\int_0^{Lh^{-m}} \sum_{x \in \mathbf{R}_h^n} |t^\alpha A_h (t + A_h)^{-1} f^h(x)|^p h^n \frac{dt}{t} \right)^{\frac{1}{p}} \leq M_3 \|f^h\|_{W_p^{m\alpha}(\mathbf{R}_h^n)}. \qquad (2.22)$$

The cases $n = m$ and $n < m$ are dealt with in the same manner, using the estimates (1.54) and (1.35), respectively, for the modulus of the fundamental solution $\mathcal{G}^h(z, \lambda)$. Now let us combine the estimates (2.21) and (2.22). To this end let $\chi(t)$ denote the indicator function of the segment $[0, Lh^{-m}]$. Then

$$t^\alpha A_h (t + A_h)^{-1} f^h(x) = \chi(t) t^\alpha A_h (t + A_h)^{-1} f^h(x) + (1 - \chi(t)) t^\alpha A_h (t + A_h)^{-1} f^h(x).$$

Using the triangle inequality and the estimates (2.21) and (2.22), we obtain the inequality

$$\|f^h\|_{E'_{\alpha,p}(L_{ph}, A_h)} \leq \frac{M}{\alpha^{\frac{1}{p}}} \|f^h\|_{W_p^{m\alpha}(\mathbf{R}_h^n)}, \qquad (2.23)$$

where M does not depend on h and α.

Now let us establish the opposite inequality:

$$\|f^h\|_{W_p^{m\alpha}(\mathbf{R}_h^n)} \leq \frac{M}{m\alpha(1 - m\alpha)} \|f^h\|_{E'_{\alpha,p}(L_{ph}, A_h)} \qquad (2.24)$$

First let us prove the inequality

$$\|f^h\|_{W_p^{m\alpha}(\mathbf{R}_h^n)}$$

$$\leq M \left[\|f^h\|_{L_{ph}} + \sum_{k=1}^n \left\{ \sum_{x \in \mathbf{R}_h^n} \int_h^\infty \frac{|f^h(x) - f^h\left(x + \left[\frac{\rho}{h}\right] e_k h\right)|^p}{\rho^{1+m\alpha p}} h^n d\rho \right\}^{\frac{1}{p}} \right]. \qquad (2.25)$$

Set $z = z_1 e_1 + \cdots + z_n e_n$. Then the triangle inequality yields

$$\left(\sum_{x \in \mathbf{R}_h^n} \sum_{z \in \mathbf{R}_h^n, \, z \neq 0} \frac{|f^h(x) - f^h(x+z)|^p}{|z|^{n+m\alpha p}} h^{2n} \right)^{\frac{1}{p}}$$

$$= \left(\sum_{x\in\mathbf{R}_h^n} \sum_{z\in\mathbf{R}_h^n,\, z\neq 0} \frac{|\sum_{k=1}^n (f^h(x+\sum_{i=1}^k z_i e_i) - f^h(x+\sum_{i=1}^{k-1} z_i e_i))|^p}{|z|^{n+map}} h^{2n} \right)^{\frac{1}{p}}$$

$$\leq \sum_{k=1}^n \left(\sum_{x\in\mathbf{R}_h^n} \sum_{z\in\mathbf{R}_h^n,\, z\neq 0} \frac{|f^h(x+\sum_{i=1}^k z_i e_i) - f^h(x+\sum_{i=1}^{k-1} z_i e_i)|^p}{|z|^{n+map}} h^{2n} \right)^{\frac{1}{p}}$$

$$= \sum_{k=1}^n \left(\sum_{\tilde{x}\in\mathbf{R}_h^n} \sum_{z\in\mathbf{R}_h^n,\, z\neq 0} \frac{|f^h(\tilde{x}+z_k e_k) - f^h(\tilde{x})|^p}{|z|^{n+map}} h^{2n} \right)^{\frac{1}{p}},$$

where $\tilde{x} = x + \sum_{i=1}^{k-1} z_i e_i$. To prove (2.25) it suffices to establish the inequality

$$\sum_{z\in\mathbf{R}_h^n,\, z\neq 0} \frac{|f^h(x+z_k e_k) - f^h(x)|^p}{|z|^{n+map}} h^n \leq M \int_h^\infty \frac{|f^h\left(x+\left[\frac{\rho}{h}\right]e_k h\right) - f^h(x)|^p}{\rho^{1+map}} d\rho$$

$$(2.26)$$

for all $k = 1, \cdots, n$ and $x \in \mathbf{R}_h^n$. It suffices to take $k = 1$ (the proof is the same for all values of k). Write $z = (z_1, z')$, $z' = (z_2, \cdots, z_n)$; then $|z|^2 = |z_1|^2 + |z'|^2$. We have

$$\sum_{z\in\mathbf{R}_h^n,\, z_1\neq 0} \frac{|f^h(x+z_1 e_1) - f^h(x)|^p}{|z|^{n+map}} h^n$$

$$= \sum_{z_1\in\mathbf{R}_h^1,\, z_1\neq 0} \sum_{z'\in\mathbf{R}_h^{n-1}} \frac{|f^h(x+z_1 e_1) - f^h(x)|^p}{(|z_1|^2 + |z'|^2)^{\frac{n+map}{2}}} h^n$$

$$= \sum_{z_1\in\mathbf{R}_h^1,\, z_1\neq 0} h|f^h(x+z_1 e_1) - f^h(x)|^p \sum_{z'\in\mathbf{R}_h^{n-1}} \frac{h^{n-1}}{(|z_1|^2 + |z'|^2)^{\frac{n+map}{2}}}. \qquad (2.27)$$

Further,

$$\sum_{z'\in\mathbf{R}_h^{n-1}} \frac{h^{n-1}}{(|z_1|^2 + |z'|^2)^{\frac{n+map}{2}}} = \frac{h^{n-1}}{|z_1|^{n+map}} + \sum_{z'\in\mathbf{R}_h^{n-1},\, z'\neq 0} \frac{h^{n-1}}{(|z_1|^2 + |z'|^2)^{\frac{n+map}{2}}}$$

$$\leq \frac{1}{|z_1|^{1+map}} + M \sum_{z'\in\mathbf{R}_h^{n-1},\, z'\neq 0} \frac{h^{n-1}}{(|z_1|^2 + |z'|^2 + h^2)^{\frac{n+map}{2}}}$$

$$= \frac{1}{|z_1|^{1+map}} + M \int_{z'\in\mathbf{R}^{n-1}} \frac{dz'}{(|z_1|^2 + |[\frac{z'}{h}]h|^2 + h^2)^{\frac{n+map}{2}}}$$

$$\leq \frac{1}{|z_1|^{1+map}} + M_1 \int_{z'\in\mathbf{R}^{n-1}} \frac{dz'}{(|z_1|^2 + +|z'|^2)^{\frac{n+map}{2}}}.$$

Let us change the variables in the last integral to $z' = |z_1|y'$. Then

$$\int_{z' \in \mathbf{R}^{n-1}} \frac{dz'}{(|z_1|^2 + +|z'|^2)^{\frac{n+map}{2}}} = \int_{y' \in \mathbf{R}^{n-1}} \frac{dy'}{(1+|y'|^2)^{\frac{n+map}{2}}|z_1|^{1+map}}.$$

Since

$$\int_{y' \in \mathbf{R}^{n-1}} \frac{dy'}{(1+|y'|^2)^{\frac{n+map}{2}}} \leq \tilde{M},$$

it follows that

$$\sum_{z' \in \mathbf{R}_h^{n-1}} \frac{h^{n-1}}{(|z_1|^2 + |z'|^2)^{\frac{n+map}{2}}} \leq \frac{M_2}{|z_1|^{1+map}}.$$

Consequently,

$$\sum_{z \in \mathbf{R}_h^n,\, z \neq 0} \frac{|f^h(x + z_1 e_1) - f^h(x)|^p h^n}{|z|^{n+map}}$$

$$\leq M_2 \sum_{z_1 \in \mathbf{R}_h^1,\, z_1 \neq 0} \frac{|f^h(x + z_1 e_1) - f^h(x)|^p h}{|z_1|^{1+map}}. \qquad (2.28)$$

Since

$$\sum_{z_1 \in \mathbf{R}_h^1,\, z_1 \neq 0} \frac{|f^h(x + z_1 e_1) - f^h(x)|^p h}{|z_1|^{1+map}} = \int_h^\infty \frac{|f^h(x + [\frac{\rho}{h}]he_1) - f^h(x)|^p}{|[\frac{\rho}{h}]h|^{1+map}} d\rho$$

$$\leq \tilde{\tilde{M}} \int_h^\infty \frac{|f^h(x + [\frac{\rho}{h}]he_1) - f^h(x)|^p}{\rho^{1+map}} d\rho,$$

inequalities (2.27) and (2.28) yields (2.26). Next, from inequality (2.25) it follows that in order to prove (2.24) it suffices to establish the inequality

$$\left\{ \sum_{x \in \mathbf{R}_h^n} \int_h^\infty \frac{|f^h(x + [\frac{\rho}{h}]he_1) - f^h(x)|^p}{\rho^{1+map}} h^n d\rho \right\}^{\frac{1}{p}}$$

$$\leq \frac{M}{ma(1 - ma)} \|f^h\|_{E'_{\alpha,p}(L_{ph}, A_h)}. \qquad (2.29)$$

Arguing as in the first subsection of Section 1, we have

$$f^h(x + [\frac{\rho}{h}]he_1) - f^h(x)$$

$$= (2\pi)^{-n} \int_0^\infty \sum_{y \in \mathbf{R}_h^n} [\mathcal{G}^h(y + [\frac{\rho}{h}]he_1, t) - \mathcal{G}^h(y, t)] A_h(t + A_h)^{-1} f^h(x - y) h^n dt$$

$$= (2\pi)^{-n} \left(\int_0^{\rho^{-m}} + \int_{\rho^{-m}}^{Lh^{-m}} + \int_{Lh^{-m}}^{\infty} \right) \sum_{y \in \mathbf{R}_h^n} [\mathcal{G}^h(y + [\tfrac{\rho}{h}]he_1, t) - \mathcal{G}^h(y, t)] \times$$

$$A_h(t + A_h)^{-1} f^h(x - y) h^n dt = \mathcal{I}_1(x, \rho) + \mathcal{I}_2(x, \rho) + \mathcal{I}_3(x, \rho).$$

With no loss of generality we may assume that $L > 1$. Let us estimate $\mathcal{I}_j$, $j = 1, 2, 3$, in the norm given by the left-hand side of inequality (2.29). We start with $\mathcal{I}_3$. Using the estimate (1.27), we obtain

$$|\mathcal{I}_3(x, \rho)| \le M \int_{Lh^{-m}}^{\infty} \sum_{y \in \mathbf{R}_h^n} \left(\exp \left\{ -a \left| \tfrac{y}{h} \right| \right\} + \exp \left\{ -a \left| \tfrac{y}{h} + \left[\tfrac{\rho}{h} \right] e_1 \right| \right\} \right) \times$$

$$\times h^{-n} |A_h(t + A_h)^{-1} f^h(x - y)| h^n \frac{dt}{t}.$$

The change of variables $t = r^m / \rho^m$ yields

$$|\mathcal{I}_3(x, \rho)| \le Mm \int_{L^{1/m}(\rho/h)^m}^{\infty} \sum_{k \in \mathbf{R}_1^n} \left(\exp \left\{ -a|k| \right\} + \exp \left\{ \left| k + \left[\tfrac{\rho}{h} \right] e_1 \right| \right\} \right) \times$$

$$\times |A_h((r^m / \rho^m) + A_h)^{-1} f^h(x - y)| \frac{dr}{r}$$

$$\le Mm \left[\int_{L^{1/m}}^{\infty} \sum_{k \in \mathbf{R}_1^n} \exp\{-a|k|\} |A_h((r^m / \rho^m) + A_h)^{-1} f^h(x - kh)| \frac{dr}{r} \right.$$

$$+ \int_{L^{1/m}}^{\infty} \sum_{k \in \mathbf{R}_1^n} \exp\{-a|k|\} |A_h((r^m / \rho^m) + A_h)^{-1} f^h(x - (k - [\rho/h]e_1)h)| \frac{dr}{r} \right].$$

Using the Minkowski inequality, we obtain

$$\left(\sum_{x \in \mathbf{R}_h^n} \int_h^{\infty} \frac{|\mathcal{I}_3(x, \rho)|^p}{\rho^{1+m\alpha p}} d\rho \, h^n \right)^{\frac{1}{p}} \le Mm \int_{L^{1/m}}^{\infty} \sum_{k \in \mathbf{R}_1^n} \exp\{-a|k|\} \times$$

$$\times \left[\left(\sum_{x \in \mathbf{R}_h^n} \int_h^{\infty} \frac{1}{\rho^{m\alpha p}} |A_h((r^m / \rho^m) + A_h)^{-1} f^h(x - kh)|^p h^n \frac{d\rho}{\rho} \right)^{\frac{1}{p}} \right.$$

$$+ \left(\sum_{x \in \mathbf{R}_h^n} \int_h^{\infty} \frac{1}{\rho^{m\alpha p}} |A_h((r^m / \rho^m) + A_h)^{-1} f^h(x - (k - [h/\rho]e_1)h)|^p h^n \frac{d\rho}{\rho} \right)^{\frac{1}{p}} \right] \frac{dr}{r}$$

$$= 2mM \int_{L^{1/m}}^{\infty} \sum_{k \in \mathbf{R}_1^n} \exp\{-a|k|\} \times$$

$$\times \left(\sum_{x \in \mathbf{R}_h^n} \int_h^{\infty} \frac{1}{\rho^{m\alpha p}} |A_h((r^m/\rho^m) + A_h)^{-1} f^h(x)|^p h^n \frac{d\rho}{\rho} \right)^{\frac{1}{p}} \frac{dr}{r}.$$

Since $\sum_{k \in \mathbf{R}_1^n} \exp\{-a|k|\} \le M_1$, it follows that

$$\left(\sum_{x \in \mathbf{R}_h^n} \int_h^{\infty} \frac{|\mathcal{I}_3(x,\rho)|^p}{\rho^{1+m\alpha p}} d\rho\, h^n \right)^{\frac{1}{p}}$$

$$\le M_2 \int_{L^{1/m}}^{\infty} \left(\sum_{x \in \mathbf{R}_h^n} \int_h^{\infty} \frac{1}{\rho^{m\alpha p}} |A_h((r^m/\rho^m) + A_h)^{-1} f^h(x)|^p h^n \frac{d\rho}{\rho} \right)^{\frac{1}{p}} \frac{dr}{r}$$

$$\le M_3 \int_{L^{1/m}}^{\infty} \frac{dr}{r^{1+m\alpha}} \|f^h\|_{E'_{\alpha,p}(L_{ph}, A_h)} = \frac{M_3}{m\alpha L^{\alpha}} \|f^h\|_{E'_{\alpha,p}(L_{ph}, A_h)}.$$

Thus, we have established the inequality

$$\left(\sum_{x \in \mathbf{R}_h^n} \int_h^{\infty} \frac{|\mathcal{I}_3(x,\rho)|^p}{\rho^{1+m\alpha p}} d\rho\, h^n \right)^{1/p} \le \frac{M_3}{m\alpha} \|f^h\|_{E'_{\alpha,p}(L_h, A_h)}. \tag{2.30}$$

Now let us consider the second term, $\mathcal{I}_2(x, \rho)$. Let $n > m$. Then, using the estimate (1.62) we obtain

$$|\mathcal{I}_2(x,\rho)| \le M \int_{\rho^{-m}}^{Lh^{-m}} \sum_{y \in \mathbf{R}_h^n} \left[\exp\{-at^{\frac{1}{m}}|y|\}(h+|y|)^{m-n+1} \right.$$

$$\left. + \exp\{-at^{\frac{1}{m}}|y+[\rho/h]he_1|\}(h+[\rho/h]he_1+|y|)^{m-n+1} \right] |A_h(t+A_h)^{-1} f^h(x-y)| h^n dt$$

$$= M \int_{\rho^{-m}}^{Lh^{-m}} \sum_{y \in \mathbf{R}_h^n} h^n \exp\{-at^{\frac{1}{m}}|y|\}(h+|y|)^{m-n+1} \times$$

$$\times \left[|A_h(t+A_h)^{-1} f^h(x-y)| + |A_h(t+A_h)^{-1} f^h(x-y-[\rho/h]he_1)| \right] dt$$

$$= M \int_{\rho^{-m}}^{Lh^{-m}} \int_{y \in \mathbf{R}^n} \exp\{-at^{\frac{1}{m}}|[y/h]h|\}(h+|[y/h]h|)^{m-n+1} \times$$

$$\times \left[|A_h(t + A_h)^{-1} f^h(x - [y/h]h)| \right.$$

$$+ |A_h(t + A_h)^{-1} f^h(x - [y/h]h - [\rho/h]he_1)| \Big] dy\, dt.$$

Since $|[y/h]h| + h \geq \frac{1}{\sqrt{n}}|y|$ and $t^{\frac{1}{m}}h \leq L^{\frac{1}{m}}$, it follows that

$$|\mathcal{I}_2(x, \rho)| \leq M_1 \int_{\rho^{-m}}^{Lh^{-m}} \int_{y \in \mathbf{R}^n} \exp\{-at^{\frac{1}{m}}|y|\}|y|^{m-n+1} \times$$

$$\times \left[|A_h(t + A_h)^{-1} f^h(x - [y/h]h)| \right.$$

$$+ |A_h(t + A_h)^{-1} f^h(x - [y/h]h - [\rho/h]he_1)| \Big] dy\, dt.$$

The change of variables $z = t^{\frac{1}{m}} y$ yields

$$|\mathcal{I}_2(x, \rho)| \leq M_1 \int_{\rho^{-m}}^{Lh^{-m}} \int_{z \in \mathbf{R}^n} \exp\{-a|z|\}|z|^{m-n} \times$$

$$\times \left[|A_h(t + A_h)^{-1} f^h(x - [t^{-\frac{1}{m}}z/h]h)| \right.$$

$$+ |A_h(t + A_h)^{-1} f^h(x - [t^{-1\frac{1}{m}}z/h]h - [\rho/h]he_1)| \Big] dz\, \frac{dt}{t}.$$

Next, performing the change of variables $t = r^m/\rho^m$ in the outer integral, we obtain

$$|\mathcal{I}_2(x, \rho)| \leq M_1 m \int_1^{L(\rho/h)^m} \int_{z \in \mathbf{R}^n} \exp\{-a|z|\}|z|^{m-n} \times$$

$$\times \left[|A_h((r^m/\rho^m) + A_h)^{-1} f^h(x - [\rho z/(rh)]h)| \right.$$

$$+ |A_h((r^m/\rho^m) + A_h)^{-1} f^h(x - [\rho z/(rh)]h - [\rho/h]he_1)| \Big] dz\, \frac{dr}{r}$$

$$\leq M_1 m \int_1^{\infty} \int_{z \in \mathbf{R}^n} \exp\{-a|z|\}|z|^{m-n} \times$$

$$\times \left[|A_h((r^m/\rho^m) + A_h)^{-1} f^h(x - [\rho z/(rh)]h)| \right.$$

$$+ |A_h((r^m/\rho^m) + A_h)^{-1} f^h(x - [\rho z/(rh)]h - [\rho/h]he_1)| \Big] dz\, \frac{dr}{r}.$$

With the help of the Minkowski inequality, this yields

$$\left(\sum_{x \in \mathbf{R}_h^n} \int_h^{\infty} \frac{|\mathcal{I}_2(x, \rho)|^p}{\rho^{1+m\alpha p}} d\rho\, h^n \right)^{\frac{1}{p}} \leq M_1 m \int_1^{\infty} \int_{z \in \mathbf{R}^n} \exp\{-a|z|\}|z|^{m-n} \times$$

$$\times \left\{ \sum_{x \in \mathbf{R}_h^n} \int_h^\infty \frac{1}{\rho^{m\alpha p}} \left[|A_h((r^m/\rho^m) + A_h)^{-1} f^h(x - [\rho z/(rh)]h)| \right.\right.$$

$$\left.\left. + |A_h((r^m/\rho^m) + A_h)^{-1} f^h(x - [\rho z/(rh)]h - [\rho/h]he_1)| \right]^p h^n \frac{d\rho}{\rho} \right\}^{\frac{1}{p}} dz \frac{dr}{r}$$

$$\leq 2^{1/p} M_1 m \int_1^\infty \int_{z \in \mathbf{R}^n} \exp\{-a|z|\}|z|^{m-n} dz \times$$

$$\times \left\{ \sum_{x \in \mathbf{R}_h^n} \int_h^\infty \frac{1}{\rho^{m\alpha p}} |A_h((r^m/\rho^m) + A_h)^{-1} f^h(x)|^p h^n \frac{d\rho}{\rho} \right\}^{\frac{1}{p}} \frac{dr}{r}.$$

Since $\int_{z \in \mathbf{R}^n} \exp\{-a|z|\}|z|^{m-n} dz \leq \tilde{M}$, we conclude that

$$\left(\sum_{x \in \mathbf{R}_h^n} \int_h^\infty \frac{|\mathcal{I}_2(x,\rho)|^p}{\rho^{1+m\alpha p}} d\rho\, h^n \right)^{\frac{1}{p}}$$

$$\leq M_2 \int_1^\infty \left\{ \sum_{x \in \mathbf{R}_h^n} \int_0^\infty \frac{r^{m\alpha p}}{\rho^{m\alpha p}} |A_h((r^m/\rho^m) + A_h)^{-1} f^h(x)|^p h^n \frac{d\rho}{\rho} \right\}^{\frac{1}{p}} \frac{dr}{r^{1+m\alpha}}$$

$$\leq M \int_1^\infty \frac{dr}{r^{1+m\alpha}} \|f^h\|_{E'_{p,\alpha}(L_{ph}, A_h)} = \frac{M_2}{m\alpha} \|f^h\|_{E'_{p,\alpha}(L_{ph}, A_h)}.$$

Thus, in the case $n > m$ we have established the inequality

$$\left(\sum_{x \in \mathbf{R}_h^n} \int_h^\infty \frac{|\mathcal{I}_2(x,\rho)|^p d\rho}{\rho^{1+m\alpha p}} h^n \right)^{1/p} \leq \frac{M}{m\alpha} \|f^h\|_{E'_{\alpha,p}(L_{ph}, A_h)}. \tag{2.31}$$

The cases $n = m$ and $n < m$ are dealt with in a similar manner, using the estimates (1.54) and (1.35), respectively, for the modulus of the function $\mathcal{G}^h(y, \lambda)$.

Finally, let us estimate the term $\mathcal{I}_1(x, \rho)$. Let $n + 1 > m$. Before embarking on the proof of the needed inequality for $\mathcal{I}_1(x, \rho)$ we must introduce some notations and establish an inequality for a difference of fundamental solutions. Let $\varphi^h(x)$ be a grid function on $\mathbf{R}_h^1$ and $\tilde{\varphi}(s)$ be the piecewise-linear extension of $\varphi^h(x)$: by definition, $\tilde{\varphi}(s)|_{s=0,\pm h,\cdots} = \varphi^h(s)$ and for any point s, $(l-1)h < s < lh$, $l = 0, \pm 1, \cdots$, there exists the derivative

$$\tilde{\varphi}'(s) = \frac{\varphi^h(lh) - \varphi^h((l-1)h)}{h}.$$

Then
$$\mathcal{G}^h(y + [\rho/h]he_1, t) - \mathcal{G}^h(y, t) = \int_0^1 \frac{d}{ds}\tilde{\mathcal{G}}^h(y + s[\rho/h]he_1, t)ds.$$

From the definition of the piecewise-linear extension it follows that
$$\frac{d}{ds}\tilde{\mathcal{G}}^h(y + s[\rho/h]he_1, t) = \mathcal{D}_h^1 \mathcal{G}^h(y + [\rho/h]he_1, t)$$

for $(l-1)h \le s \le lh$. Hence, by (1.74), we have the estimate
$$\left|\frac{d}{ds}\tilde{\mathcal{G}}^h(y + s[\rho/h]he_1, t)\right|$$

$$\le M \exp\{-at^{\frac{1}{m}}|y + (l-1)he_1|\}(h + |y + (l-1)he_1|)^{m-n-1}$$

for all s, $(l-1)h \le s \le lh$, $l = 1, 2, \cdots$. Since $t^{\frac{1}{m}}h \le L$ and $|y + (l-1)he_1| \ge |y + [s[\rho/h]]he_1| - h$, it follows that
$$\left|\frac{d}{ds}\tilde{\mathcal{G}}^h(y + s[\rho/h]he_1, t)\right|$$

$$\le M_1 \exp\{-at^{\frac{1}{m}}|y + [s[\rho/h]]he_1|\}(h + |y + [s[\rho/h]]he_1|)^{m-n-1}.$$

Consequently,
$$|\mathcal{G}^h(y + [\rho/h]he_1, t) - \mathcal{G}^h(y, t)|$$

$$\le M_1 \int_0^1 \rho \exp\{-at^{\frac{1}{m}}|y + [s[\rho/h]]he_1|\}(h + |y + [s[\rho/h]]he_1|)^{m-n-1}ds. \quad (2.32)$$

Now let us pass to the estimation of $\mathcal{I}_1(x, \rho)$. By (2.32),
$$|\mathcal{I}_1(x, \rho)| \le M_1 \int_0^{\rho^{-m}} \sum_{y \in \mathbf{R}_h^n} \int_0^1 \rho \exp\{-at^{\frac{1}{m}}|y + [s[\rho/h]]he_1|\} \times$$

$$\times (h + |y + [s[\rho/h]]he_1|)^{m-n-1}ds \, |A_h(t + A_h)^{-1}f^h(x - y)|h^n dt$$

$$= M_1 \int_0^1 \int_0^{\rho^{-m}} \rho \sum_{y \in \mathbf{R}_h^n} \exp\{-at^{\frac{1}{m}}|y|\}(h + |y|)^{m-n-1} \times$$

$$\times |A_h(t + A_h)^{-1}f^h(x - y - [s[\rho/h]]he_1)|h^n dt \, ds.$$

Replacing the sum by an integral, we obtain
$$|\mathcal{I}_1(x, \rho)| \le M_2 \int_0^1 \int_0^{\rho^{-m}} \int_{y \in \mathbf{R}_h^n} \rho \exp\{-at^{\frac{1}{m}}|y|\}|y|^{m-n-1} \times$$

$$\times |A_h(t + A_h)^{-1} f^h(x - [y/h]h - [s[\rho/h]]|he_1)| \, dy \, dt \, ds.$$

Making the change of variables $y = t^{-\frac{1}{m}} z$ in the inner integral we obtain

$$|\mathcal{I}_1(x, \rho)| \leq M_2 \int_0^1 \int_0^{\rho^{-m}} \int_{z \in \mathbf{R}^n} \rho t^{\frac{1}{m}} \exp\{-a|z|\} |z|^{m-n-1} \times$$

$$\times |A_h(t + A_h)^{-1} f^h(x - [z/(t^{\frac{1}{m}} h)]h - [s[\rho/h]]|he_1)| dz \, \frac{dt}{t} \, ds.$$

Further, the change of variables $t = r^m / \rho^m$ yields

$$|\mathcal{I}_1(x, \rho)| \leq M_2 m \int_0^1 \int_0^1 \int_{z \in \mathbf{R}^n} \exp\{-a|z|\} |z|^{m-n-1} \times$$

$$\times |A_h((r^m/\rho^m) + A_h)^{-1} f^h(x - [z/(t^{\frac{1}{m}} h)]h - [s[\rho/h]]|he_1)| dz \, dr \, ds,$$

whence, via the Minkowski inequality,

$$\left(\sum_{x \in \mathbf{R}_h^n} \int_h^\infty \frac{|\mathcal{I}_1(x, \rho)|^p d\rho}{\rho^{1+m\alpha p}} h^n \right)^{\frac{1}{p}} \leq M_2 m \int_0^1 \int_0^1 \int_{z \in \mathbf{R}^n} \exp\{-a|z|\} |z|^{m-n-1} \times$$

$$\times \left\{ \sum_{x \in \mathbf{R}_h^n} h^n \int_h^\infty \frac{1}{\rho^{m\alpha p}} |A_h((r^m/\rho^m) + A_h)^{-1} \times \right.$$

$$\times f^h(x - [z/(t^{\frac{1}{m}} h)]h - [s[\rho/h]]|he_1)|^p \frac{d\rho}{\rho} \Bigg\}^{\frac{1}{p}} \, dr \, ds \, dz$$

$$= M_2 m \int_0^1 \int_{z \in \mathbf{R}^n} \exp\{-a|z|\} |z|^{m-n-1} dz \times$$

$$\times \left\{ \sum_{x \in \mathbf{R}_h^n} \int_h^\infty \frac{1}{\rho^{m\alpha p}} |A_h((r^m/\rho^m) + A_h)^{-1} f^h(x)|^p h^n \frac{d\rho}{\rho} \right\}^{\frac{1}{p}} \, dr.$$

Since

$$\int_{z \in \mathbf{R}_h^n} \exp\{-a|z|\} |z|^{m-n-1} dz \leq \tilde{M},$$

it follows that

$$\left(\sum_{x \in \mathbf{R}_h^n} \int_h^\infty \frac{|\mathcal{I}_1(x, \rho)|^p d\rho}{\rho^{1+m\alpha p}} h^n \right)^{\frac{1}{p}}$$

$$\leq M_3 \int_0^1 \frac{1}{r^{m\alpha}} \left\{ \sum_{x \in \mathbf{R}_h^n} \int_0^\infty \frac{r^{m\alpha p}}{\rho^{m\alpha p}} |A_h((r^m/\rho^m) + A_h)^{-1} f^h(x)|^p h^n \frac{d\rho}{\rho} \right\}^{1/p} dr$$

$$= M_3 \int_0^\infty \frac{1}{r^{m\alpha}} \|f^h\|_{E'_{\alpha,p}(L_{ph},A_h)} = \frac{M_3}{1-m\alpha} \|f^h\|_{E'_{\alpha,p}(L_{ph},A_h)}.$$

Thus, in the case $n + 1 > m$ we have established the inequality

$$\left(\sum_{x \in \mathbf{R}_h^n} \int_h^\infty \frac{|\mathcal{I}_1(x,\rho)|^p d\rho}{\rho^{1+m\alpha p}} h^n \right)^{\frac{1}{p}} \leq \frac{M}{1-m\alpha} \|f^h\|_{E'_{\alpha,p}(L_{ph},A_h)}. \qquad (2.33)$$

The cases $n = m-1$, m and $n < m-1$ are dealt with in the same manner, using the estimates (1.34), (1.73), and (1.67), respectively, for the modulus of the difference derivative of the fundamental solution $\mathcal{G}^h(y, \lambda)$. From (2.30), (2.31), (2.33), and the triangle inequality we obtain inequality (2.29). Theorem 2.5 is proved.

By Theorem 2.4, $A_h + \varepsilon$, $\varepsilon > 0$, is a strongly positive operator in L_{ph}. This implies that $A_h + \varepsilon$ is strongly positive in $E'_{\alpha,p}(L_{ph}, A_h)$, and hence, by Theorem 2.3, $A_h + \varepsilon$ is strongly positive in $W_p^{m\alpha}(\mathbf{R}_h^n)$ for all α, $0 < \alpha < 1/m$.

4.   The coercivity inequality for an elliptic difference operator in $C^{m\alpha}(\mathbf{R}_h^n)$ and $W_p^{m\alpha}(\mathbf{R}_h^n)$.

Let us show that the estimates of the derivatives of order m of the Green function of the resolvent of the operator $-A_h$ imply the coercivity inequality for an elliptic difference operator in the norms of $C^{m\alpha}(\mathbf{R}_h^n)$ and $W_p^{m\alpha}(\mathbf{R}_h^n)$ for $0 < \alpha < 1/m$.

Theorem 2.6. *Suppose that for $|\xi_k h| \leq \pi$ the conditions of Lemma 1.9 are fulfilled:* a) $|A(\xi h, h)| \geq M_0 |\xi|^m$, *where $M_0 > 0$, and* b) $|\arg A(\xi h, h)| \leq \phi < \phi_0 \leq \pi$. *Let $|\arg \lambda| \leq \pi - \phi_0$. Then the solutions of problem (1.21) obey the coercivity inequality*

$$\sum_{|r|=m} \|\mathcal{D}_h^r u^h\|_{C^{m\alpha}(\mathbf{R}_h^n)} \leq \frac{M}{(m\alpha(1-m\alpha))^2} \|f^h\|_{C^{m\alpha}(\mathbf{R}_h^n)}, \quad 0 < \alpha < 1/m, \quad (2.34)$$

where M does not depend on α, h, and f^h.

Proof. It follows from the definition of the solution of problem (1.21) that

$$u^h(x) = (\lambda + A_h)^{-1} f^h(x) = (2\pi)^{-n} \sum_{y \in \mathbf{R}_h^n} \mathcal{G}^h(x - y, \lambda) f^h(y) h^n.$$

Here we cannot directly use the formula

$$\mathcal{D}_h^r u^h(x) = (2\pi)^{-n} \sum_{y \in \mathbf{R}_h^n} \mathcal{D}_h^r \mathcal{G}^h(x - y, \lambda) f^h(y) h^n,$$

since for $|r| = m$ the difference derivative of the fundamental solution admits only a (uniform in h) singular estimate. For this reason we shall use the relation

$$A_h^{-1} = 4A_h(A_h + A_h)^{-2}.$$

Applying the Cauchy-Riesz formula, we obtain the representation

$$\lambda^\alpha A_h(\lambda + A_h)^{-1} A_h^{-1} f^h(x) = \frac{1}{2\pi i} \int_{S_1 \cup S_2} \frac{\lambda^\alpha 4z}{\lambda + z} (z + A_h)^{-2} A_h(z - A_h)^{-1} f^h(x) dz. \tag{2.35}$$

Let $\mathcal{D}_h^m$ denote the difference derivative of order m. Since m is even, we can write $\mathcal{D}_h^m = \tilde{\mathcal{D}}_h^{m/2} \tilde{\tilde{\mathcal{D}}}_h^{m/2}$. Now, A_h is a difference operator with constant symbol; hence, A_h as well as its resolvent commute with the operators $\tilde{\mathcal{D}}_h^{m/2}$ and $\tilde{\tilde{\mathcal{D}}}_h^{m/2}$. Consequently,

$$\lambda^\alpha A_h(\lambda + A_h)^{-1} \mathcal{D}_h^m A_h^{-1} f^h(x)$$

$$= \frac{1}{2\pi i} \int_{S_1 \cup S_2} \frac{\lambda^\alpha 4z}{\lambda + z} \tilde{\mathcal{D}}_h^{m/2} (z + A_h)^{-1} \tilde{\tilde{\mathcal{D}}}_h^{m/2} (z + A_h)^{-1} A_h(z - A_h)^{-1} f^h(x) dz. \tag{2.36}$$

Note that in the right-hand side of (2.36) contains the difference derivatives of order $m/2$ of the resolvent, which admit estimates in terms of the estimates of the difference derivatives of the fundamental solution. Applying (1.79), we obtain the estimates

$$\|\tilde{\mathcal{D}}_h^{m/2} (z + A_h)^{-1}\|_{C_h \to C_h} \le \frac{M}{(1 + |z|)^{1/2}},$$

$$\|\tilde{\tilde{\mathcal{D}}}_h^{m/2} (z + A_h)^{-1}\|_{C_h \to C_h} \le \frac{M}{(1 + |z|)^{1/2}}. \tag{2.37}$$

The estimates (2.37) and formula (2.36) yield the inequality

$$\|\lambda^\alpha A_h(\lambda + A_h)^{-1} \mathcal{D}_h^m A_h^{-1} f^h\|_{C_h} \le M \int_0^\infty \frac{\lambda^\alpha}{\lambda + \rho} \|A_h(\rho + A_h)^{-1} f^h\|_{C_h} d\rho.$$

Further, proceeding as in Chapter 1, we obtain the inequality

$$\sup_{\lambda>0} \|\lambda^\alpha A_h(\lambda+A_h)^{-1}\mathcal{D}_h^m A_h^{-1}f^h\|_{C_h} \leq M\frac{\pi}{\sin\pi\alpha}\sup_{\rho>0}\|\rho^\alpha A_h(\rho+A_h)^{-1}f^h\|_{C_h}.$$

Thus, we have shown that

$$\|\mathcal{D}_h^m A_h^{-1}f^h\|_{E'_{\alpha,\infty}(C_h,A_h)} \leq M\frac{\pi}{\sin\pi\alpha}\|f^h\|_{E'_{\alpha,\infty}(C_h,A_h)}.$$

Since the operator $A_h(\lambda+A_h)^{-1}$ is bounded in C_h uniformly in λ, it is bounded in the space $E'_{\alpha,\infty}(C_h,A_h)$. Consequently, the operator $\mathcal{D}_h^m(\lambda+A_h)^{-1}$ is bounded in $E'_{\alpha,\infty}(C_h,A_h)$ uniformly in λ. Finally, recalling (Theorem 2.1) that the spaces $E'_{\alpha,\infty}(C_h,A_h)$ and $C_h^{m\alpha}$ coincide (for $m\alpha < 1$), we obtain the conclusion of Theorem 2.6.

In a similar manner one proves

Theorem 2.7. *Under the conditions of Theorem 2.6 the solution of problem* (1.21) *satisfies the inequality*

$$\sum_{|r|=m}\|\mathcal{D}_h^r u^h\|_{W_p^{m\alpha}(\mathbf{R}_h^n)} \leq \frac{M}{(m\alpha(1-m\alpha))^2}\|f^h\|_{W_p^{m\alpha}(\mathbf{R}_h^n)}. \qquad (2.38)$$

5.　　Elliptic difference operators in $\mathcal{L}_{2h}$.

In the preceding subsection we have established coercivity inequalities for solutions of the resolvent equation in the fractional spaces $E'_{\alpha,p}(L_{ph},A_h)$ for any $1 \leq p \leq \infty$, $0 < \alpha < 1$. In the case $p = 2$ such an inequality holds in the original space L_{2h} $(\alpha = 0)$.

Theorem 2.8. *Suppose that the symbol $A(\xi h,h)$ $(0 \leq |\xi_k h| \leq \pi)$ of the operator A_h satisfies the conditions of Lemma 1.9:* a) $|A(\xi h,h)| \geq M_0|\xi|^m$, *where $M_0 > 0$;* b) $|\arg A(\xi h,h)| \leq \phi < \phi_0 < \pi$. *Let $|\arg\lambda| \leq \pi - \phi_0$. Then the solution u^h of the resolvent equation* (1.21) *satisfies the inequality*

$$\sum_{|r|=m}\|\mathcal{D}_h^r u^h\|_{L_{2h}} \leq M\|f^h\|_{L_{2h}}, \qquad (2.39)$$

where M does not depend on f^h.

Proof. Let us recall (see Section 1) the connections between grid functions and their Fourier transforms and between difference operators and their symbols. For a given grid function $f^h(x)$ the formula

$$\tilde{f}(\xi) = (2\pi)^{-n} \sum_{x \in \mathbf{R}_h^n} \exp\{-i(x, \xi)\} f^h(x) h^n$$

defines its Fourier transform, which is a $2\pi h^{-1}$-periodic function. The Parseval equality holds:

$$\int_{|\xi_k h| \leq \pi} |\tilde{f}(\xi)|^2 d\xi = (2\pi)^{-n} \sum_{x \in \mathbf{R}_h^n} |f^h(x)|^2 h^n. \tag{2.40}$$

The symbol of the operator A_h is defined by the rule

$$(\widetilde{A_h f^h})(\xi) = A(\xi h, h)\tilde{f}(\xi).$$

Hence, taking the Fourier transformation in equation (1.21) we obtain the relation

$$\lambda \tilde{u}(\xi) + A(\xi h, h)\tilde{u}(\xi) = \tilde{f}(\xi). \tag{2.41}$$

Under the assumptions of the theorem the estimate

$$|\lambda + A(\xi h, h)| \geq \delta(|\lambda| + |\xi|^m) \tag{2.42}$$

holds for some $\delta > 0$. Consequently, (2.21) can be recast as

$$\tilde{u}(\xi) = [\lambda + A(\xi h, h)]^{-1} \tilde{f}(\xi). \tag{2.43}$$

A difference derivative of order m of the grid function u^h has the form

$$h^{-m} \Delta_{1-}^{s_1} \Delta_{1+}^{s_2} \cdots \Delta_{n-}^{s_{2n-1}} \Delta_{n+}^{s_{2n}} u^h(x),$$

where the s_k are nonnegative integers such that $\sum_{k=1}^{2n} s_k = m$. The Fourier transform of such a derivative is then given by the formula

$$h^{-n} [1 - \exp(-i\xi_1)]^{s_1} [\exp(i\xi_1) - 1]^{s_2} \cdots [1 - \exp(-i\xi_n)]^{s_{2n-1}} [\exp(i\xi_n) - 1]^{s_{2n}} \tilde{u}(\xi).$$

From this and (2.42), (2.43) we obtain the inequality

$$\left|(\widetilde{\mathcal{D}_h^m u^h})(\xi)\right| \leq \frac{M}{\delta} \frac{|\xi_1|^{s_1} |\xi_2|^{s_2} \cdots |\xi_n|^{s_{2n-1}} |\xi_n|^{s_{2n}}}{|\lambda| + |\xi|^m} |\tilde{f}(\xi)| \leq M_1 |\tilde{f}(\xi)|.$$

Therefore,

$$\int_{\xi_k h|\leq\pi} \left|\mathcal{D}_h^m\widetilde{u^h})(\xi)\right|^2 d\xi \leq M_1 \int_{|\xi_k h|\leq\pi} |\tilde{f}(\xi)|^2 d\xi.$$

Applying the Parseval equality (2.40) we obtain inequality (2.39). Theorem 2.8 is proved.

3.   STABILITY AND COERCIVITY ESTIMATES

In this section the abstract theorem given in Chapter 3 are applied in the investigation of difference schemes of higher order of accuracy with respect to the set of all variables for approximate solution of the Cauchy problem for parabolic equations

$$\frac{\partial v(t,x)}{\partial t} + \sum_{|r|=m} a_r \frac{\partial^r v(t,x)}{\partial x_1^{r_1}\cdots\partial x_n^{r_n}} = f(t,x), \tag{3.1}$$

$$0 < t \leq 1, \quad v(0,x) = v_0(x), \quad x \in \mathbf{R}^n.$$

The discretization of problem (3.1) is carried out in two steps.

I.   Approximation with respect to the space variables.

Let us give the difference operator A_h by the formula

$$A_h u^h(x) = \sum_{m\leq|r|\leq S} b_r(h)\mathcal{D}_h^r u^h(x). \tag{3.2}$$

The coefficients are chosen in such a way that the operator A_h approximates in a specified way the operator

$$\sum_{|r|=m} a_r \frac{\partial^r}{\partial x_1^{r_1}\cdots\partial x_n^{r_n}}.$$

We shall assume that for $|\xi_k h| \leq \pi$ the symbol $A(\xi h, h)$ of the operator A_h satisfies the inequalities

$$|A(\xi h, h)| \geq M_1|\xi|^m, \quad |\arg A(\xi h, h)| \leq \phi < \phi_0 \leq \pi. \tag{3.3}$$

With the help of A_h we arrive at the Cauchy problem

$$(v^h(t,x))' + A_h v^h(t,x) = f^h(t,x), \quad 0 \le t \le 1, \tag{3.4}$$

$$v^h(0,x) = v_0^h(x) = v_0(x), \ x \in \mathbf{R}_h^n,$$

for an infinite system of ordinary differential equations.

II. Approximation with respect to the time variable.

We replace problem (3.4) by the Padé difference scheme

$$\tau^{-1}(u_k^h(x) - u_{k-1}^h(x)) + \tau^{-1}[I - R_{j,l}(\tau A_h)]u_{k-1}^h(x) = \varphi_k^{j,l}(x), \tag{3.5}$$

$$1 \le k \le N, \quad u_0^h(x) = v_0^h(x), \ x \in \mathbf{R}_h^n.$$

Let us give a number of corollaries of the abstract theorems of Chapter 3.

Theorem 3.1. *The solutions of the difference schemes (3.5) satisfy the following stability estimates:*

  1) *If $j = l - 2, l - 1$,*

$$\|u^{\tau,h}\|_{C(\tau,C_h)} \le M \left[\|u_0^h\|_{C_h} + \|\varphi^{\tau,h}\|_{C(\tau,C_h)} \right]; \tag{3.6}$$

  2) *If $j = l$,*

$$\|u^{\tau,h}\|_{C(\tau,C_h)} \le M \ln \frac{1}{\tau + h} \left[\|u_0^h\|_{C_h} + \|\varphi^{\tau,h}\|_{C(\tau,C_h)} \right] \tag{3.7}$$

and

$$\|u^{\tau,h}\|_{C(\tau,C_h)} \le M \left[\|(I + \tau A_h)u_0^h\|_{C_h} + \|(I + \tau A_h)\varphi^{\tau,h}\|_{C(\tau,C_h)} \right]. \tag{3.8}$$

Here M does not depend on τ, h, u_0^h, and $\varphi^{\tau,h}$.

The proof of this theorem is based on the abstract theorems 1.5, 1.27, and 1.10 of Chapter 3, the positivity of the operator A_h in C_h, and the estimate

$$\min \left\{ \ln \frac{1}{\tau}, 1 + |\ln \|A_h\|_{C_h \to C_h}| \right\} \le M \ln \frac{1}{\tau + h}. \tag{3.9}$$

Theorem 3.2. *The solutions of the difference schemes* (3.5) *satisfy the following coercive stability estimates:*

1) *If* $j = l - 2$, $l - 1$ *or if* $j = l$ *is an even number,*

$$\|\mathcal{D}_\tau^1 \Pi(u_0) u^{\tau,h}\|_{C(\tau,C_h)} \leq M \left[\|A_h u_0^h\|_{C_h} + \ln \frac{1}{\tau + h} \|\varphi^{\tau,h}\|_{C(\tau,C_h)} \right]; \qquad (3.10)$$

2) *If* $j = l$ *is an odd number,*

$$\|\mathcal{D}_\tau^1 \Pi(u_0) u^{\tau,h}\|_{C(\tau,C_h)}$$

$$\leq M \left[\|A_h u_0^h\|_{C_h} + \ln \frac{1}{\tau + h} \|(I + \tau A_h)\varphi^{\tau,h}\|_{C(\tau,C_h)} \right]. \qquad (3.11)$$

Here M *does not depend on* τ, h, u_0^h, *and* $\varphi^{\tau,h}$.

The proof of this theorem is based on the abstract theorems 1.6, 1.9, and 1.11 of Chapter 3, the positivity of the operator A_h in C_h, and the estimates (3.9) and

$$\|\mathcal{D}_h^m A_h^{-1}\|_{C_h \to C_h} \leq M \ln \frac{1}{h}. \qquad (3.12)$$

The last estimate is a consequence of (2.34).

Theorem 3.3. *Let* $j = l - 2$, $l - 1$. *Then the solutions of the difference schemes* (3.5) *satisfy the stability estimates*

$$\|u^{\tau,h}\|_{C_0^\alpha(\tau,C_h)} \leq M \left[\|u_0^h\|_{C_h} + \|\varphi^{\tau,h}\|_{C_0^\alpha(\tau,C_h)} \right].$$

The proof of this theorem is based on the abstract theorem 2.1 of Chapter 3 and the positivity of the operator A_h in C_h.

Theorem 3.4. *Let* $j = l - 2$, $l - 1$ *or let* $j = l$ *be an even number. Then the solutions of the difference schemes* (3.5) *satisfy the coercivity estimates*

$$\|\mathcal{D}_\tau^1 \tau \Pi(u_0) u^{\tau,h}\|_{C_0^\alpha(\tau,C_h^{m\alpha})}$$

$$\leq \frac{M}{\alpha^3(1 - m\alpha)^2} \left[\sum_{|r|=m} \|\mathcal{D}_h^r u_0^h\|_{C_h^{m\alpha}} + \|\varphi^{\tau,h}\|_{C_0^\alpha(\tau,C_h^{m\alpha})} \right], \qquad 0 < \alpha < 1/m.$$

Here M *does not depend on* τ, h, α, u_0^h, *and* $\varphi^{\tau,h}$.

The proof of this theorem is based on the abstract theorems 2.2–2.6 of Chapter 3, the positivity of the operator A_h in $C_h^{m\alpha}$, and the coercivity inequality of Theorem 3.6.

Theorem 3.5. *Let $j = l-2$, $l-1$, l. Then the solutions of the difference schemes* (3.5) *satisfy the stability estimates*

$$\|u^{\tau,h}\|_{C(\tau,C_h^{m\alpha})} \leq M(\alpha) \left[\|u_0^h\|_{C_h^{m\alpha}} + \|\varphi^{\tau,h}\|_{C(\tau,C_h^{m\alpha})} \right],$$

$0 < \alpha < 1/m$, *where M does not depend on τ, h, u_0^h, and $\varphi^{\tau,h}$.*

The proof of this theorem is based on the abstract theorem 4.1 of Chapter 3 and the positivity of the operator A_h in $C_h^{m\alpha}$.

Theorem 3.6. *Let $j = l-2$, $l-1$ or let $j = l$ be an even number. Then the solutions of the difference schemes* (3.5) *satisfy the coercivity estimates*

$$\|\mathcal{D}_t^1 a u \Pi(u_0) u^{\tau,h}\|_{C(\tau,C_h^{m\alpha})} \leq M(\alpha) \left[\sum_{|r|=m} \|\mathcal{D}_h^r u_0^h\|_{C_h^{m\alpha}} + \|\varphi^{\tau,h}\|_{C(\tau,C_h^{m\alpha})} \right],$$

$0 < \alpha < 1/m$, *where $M(\alpha)$ does not depend on τ, h, u_0^h, and $\varphi^{\tau,h}$.*

The proof of this theorem is based on the abstract theorems 4.2 and 4.4 of Chapter 3 and the positivity of the operator A_h in $C_h^{m\alpha}$.

Theorem 3.7. *Suppose that inequalities* (3.3) *hold with $\varphi_0 \leq \pi/(2l)$ and $j = l-2$, $l-1$. Then the solutions of the difference schemes* (3.5) *satisfy the coercivity estimates*

$$\|\mathcal{D}_\tau^1 \Pi(u_0) u^{\tau,h}\|_{L_p(\tau,L_{2h})} \leq M(p) \left[\|u_0^h\|_{W_{2h}^{1-\frac{1}{p}}(\mathbf{R}_h^n)} + \|\varphi^{\tau,h}\|_{L_p(\tau,L_{2h})} \right],$$

$1 < p < m/(m-1)$, *and*

$$\|\mathcal{D}_h^1 \tau \Pi(u_0) u^{\tau,h}\|_{L_p(\tau,W_q^{m\alpha}(\mathbf{R}_h^n))}$$

$$\leq M(\alpha,p,q) \left[\|u_0^h\|_{E_{1-\frac{1}{p}}(A_h,W_q^{m\alpha}(\mathbf{R}_h^n))} + \|\varphi^{\tau,h}\|_{L_p(\tau,W_q^{m\alpha}(\mathbf{R}_h^n))} \right],$$

$1 < p,q < \infty$, $0 < \alpha < 1/m$, *where $M(p)$ and $M(\alpha,p,q)$ do not depend on τ, h, u_0^h, and $\varphi^{\tau,h}$.*

The proof of this theorem is based on the abstract theorems 3.6 and 4.5 of Chapter 3, the positivity of the operator A_h in the spaces L_{2h} and $W_q^{m\alpha}(\mathbf{R}_h^n)$, and Theorems 1.7 and 2.8.

COMMENTS ON THE LITERATURE

Chapter 1

The role played by coercive inequalities in the study of boundary value problems for elliptic and parabolic partial differential equations is well known (see, e.g., [33, 34, 65]).

Coercivity inequalities for the solutions of an abstract differential equation of parabolic type were established for the first time in [54a]. The Cauchy problem for such an equation was considered in various spaces $F(E)$ of functions defined on the interval $[0, 1]$ with values in a Banach space E, and it was established that a necessary condition for coercivity is the strong positivity of the operator coefficient A. This condition is also sufficient in the spaces $C_0^\alpha(E)$. In the case of the spaces $L_p(E)$ with $p \in (1, \infty)$ the following extrapolation result was established: for the Cauchy problem to be coercive in $L_p(E)$ for all such p it suffices that it be coercive for some $p_0 \in (1, \infty)$. The application of this results to parabolic partial differential equations allowed one to obtain a series of new coercivity inequalities involving different norms with respect to t and the space variables.

As noted above, in contrast to $L_p(E)$, in the spaces $C_0^\alpha(E)$ an unconditional coercivity inequality holds, and in connection with this there arises the question of whether the same is valid for the ordinary Hölder space $C^\alpha(E)$. In [6g] this was shown to be true for the two-parameter series of spaces $C_0^{\beta,\gamma}(E)$ with $0 < \beta < 1$, $0 \le \gamma \le \beta$, under the additional assumption that $Av_0 - f(0) \in E_{\beta,\infty}$. Earlier an unconditional coercivity inequality was established in the spaces $W_p^\alpha(E)$ with $1 \le p < \infty$, $0 < \alpha < 1$ [4, 24], as well as in the spaces $L_p(E_{\alpha,p})$ with $0 < \alpha < 1$, $1 \le p \le \infty$ [54e, 17a, 17b].

Finally, let us mention the paper [18], in which an unconditional coercivity inequality in $L_p(E)$ was established under the assumption that the imaginary powers $A^{i\beta}$ are bounded (in some restrictions of the space E).

The material in **Section 1** was written on the basis of the papers [54a, 54g]. Here we give elements of the theory of operator semigroups. A detailed exposition of this material can be found, for example, in the references [31, 32, 58]. We discuss the connection between the well-posedness in $C(E)$ of the Cauchy problem (1) and the membership of the semigroup $\exp\{-tA\}$ in the class of analytic semigroups.

In **Section 2** we give a complete proof of Theorem 2.1 of [54a].

In **Section 3** we provide a complete proof of the theorems on well-posedness of the Cauchy problem (1) in $L_p(E)$ given in [54a].

The material in **Section 4** is written on the basis of the papers [54e, 17a, 17b].

Section 5: Theorems 5.1 and 5.2 for the case $\alpha = \beta$ and $\gamma = 0$ were stated without proofs in [7j]. In the general case these theorems are presented here for the first time. Theorems 5.3 and 5.4 are taken from [6g]. Finally, we state without a proof Theorem 5.5 of [64]. Earlier results of the type of Theorem 5.5 were obtained in [25].

Chapter 2

The construction of the discretization with respect to time for parabolic equations was initiated in Rothe's paper [46]. When the study of the discretization with respect to t for the abstract Cauchy problem is applied to parabolic partial differential equations it also covers the case of the full discretization with respect to time and space variables.

To prove stability, in a number of works (see [35, 36, 48a, 48b, 49, 63a, 63b] and the references given therein) difference schemes were treated as operator equations in a Hilbert space, and the investigation was based on the symmetry properties of the operator coefficient. This led to L_2-stability estimates.

Of great interest is the study of stability in the C-norm. In [2, 5a, 5b, 51a-e, 59a, 59b, 66] the corresponding estimates were obtained for the simplest difference schemes approximating the Cauchy problem for parabolic equations and systems with constant coefficients and one space variable. The proof is based on estimates of parabolic difference Green functions (fundamental solutions). In the case of parabolic equations with variable coefficients and more than one space variable, stability in the C-norm is known to hold only for difference schemes that satisfy a maximum principle (see [36, 48b]).

An important type of stability is the coercive stability (well-posedness) of difference schemes. There are no coercive estimates in the C-norm, because they do not hold for the corresponding differential equations. The coercive stability of difference schemes of first order of accuracy was apparently studied for the first time in [56a, 56b] for the first boundary value problem for second-order parabolic equations in L_2. A more general sharp-angle inequality for parabolic difference operators was established. This inequality follows from the corresponding difference inequality for elliptic difference operators, which is discussed in the comments to Chapter 4. Then followed the works [2, 19, 23, 29, 56d, 56e], where coercive inequalities in the L_2- and L_p-norms were given. In [54b] difference schemes are treated as operator equations in Banach spaces. In a Banach space there is no notion of symmetry and for this reason the study of stability relies on the positivity property of the operator coefficients, expressed in terms of properties of their resolvents. In this way in [54b] coercive estimates were established in difference analogues of Hölder spaces with a weight with respect to the time variable. The application of these results to the initial-boundary value problem for second-order parabolic equations without mixed derivatives in rectangular domains allowed one to obtain a series of new coercivity inequalities in difference analogues of various norms with respect to t and x for the solutions of difference schemes of first order of accuracy in time and second order of accuracy in the space variables. Enlisting the theory of positive operators turned out to be effective in the investigation of such difference schemes in difference analogues of Bochner spaces (see [42a, 42d]). The coercivity inequality in the difference analogue of the Bochner space $L_p(\tau, E)$ holds for any $p \in (1, \infty)$ whenever it holds for at least one $p_0 \in (1, \infty)$, i.e., here we have a result analogous to that in the differential case [54a].

An unconditional coercivity inequality in the spaces $L_q(\tau, E'_{\alpha,q})$ $(0 < \alpha < 1,\ 1 \le q \le \infty)$ was obtained in [7f, 7g]. One is led to asking whether a coercivity inequality holds for the Rothe difference scheme in the difference analogue of the usual Hölder space $C^\alpha(\tau, E)$. The answer turns out to be affirmative under the additional restrictions $Au_0 \in E'_{\alpha,\infty}$ and $\varphi_1 \in E'_{\alpha,\infty}$. This result is a particular case of the assertion that the difference problem is coercive in the two-parameter family of Banach spaces $C_0^{\beta,\gamma}(E)$. In this book we do not discuss results on the Rothe scheme in the case of a t-dependent operator, for which the reader is referred to the papers [7e, 41, 42b, 54f, 54g].

Section 0 is based on the papers [54b, 54g]. We give the definition of stability of difference schemes and a criterion for stability of such schemes in the difference analogues of the spaces $C(E)$, $C_0^\alpha(E)$, and $L_p(E)$.

Section 1 is based on the papers [54b, 54g]. Theorem 1.1 was formulated in [54c]. The real-field analyticity criterion was established in [54g]. Theorem 1.2 is also taken from [54g].

In Section 2 we give a complete proof of the theorem on the well-posedness of the difference scheme in $C_0^\alpha(\tau, E)$ found in [54f].

Section 3 is based on the papers [42a, 42d].

In Section 4 we reproduce our results from [7f, 7g].

The study of the difference scheme in difference analogues of the spaces of smooth functions, as carried out in Section 5, appears here for the first time.

Chapter 3

The utilization of Padé aproximants for computational purposes has a long history. The approximants $R_{1,1}(z)$ and $R_{2,2}(z)$ were first applied by Euler to calculate the scalar exponential (see [3]). The fact that these approximants are attributed to Padé is based on his 1892 dissertation, in which he studied and tabulated them. In recent years we have seen a sharp rise in the interest in classical methods of rational approximation of analytical functions, primarily in Padé approximants and their application to physics and engineering problems. For a detailed exposition of these investigation we refer the reader to [10].

Apparently the first study that uses rational approximation of $\exp\{-\tau A\}$ in the case of a positive (unbounded) selfadjoint operator A was carried out in the paper [63a], which deals with approximation on the positive half-line. That study was continued by a number of authors [15, 48a, 63b]. In applications to problems of mathematical physics this resulted in stability and convergence estimates for difference schemes in L_2.

The rational function $R_{1,1}^k(\tau A)$ was used to approximate the semigroup $\exp\{-tA\}$, where A is an unbounded strongly positive operator acting in a Banach space E, for the first time in [54c, 54h]. Further, in [1a, 1b, 42e], the approximants $R_{1,2}^k(\tau A)$, $R_{2,2}^k(\tau A)$, and several other were studied. When applied to various problems for parabolic equations these studies led to difference schemes second-, third-, or fourth-order accurate in time that are stable in the C and L_p norms.

The application of the approximants $R_{j,j}^k(\tau A)$ in the construction of difference schemes of a high order of accuracy in the time variable in the case of an arbitrary Banach space was studied independently by several authors [1d, 12, 13,

27]. In [12, 13, 27] difference schemes were studied for a class of operators that generate not only analytic, but also strongly continuous semigroups. In the general case it was not possible to establish stability, i.e., boundedness of the powers $R_{j,l}^k(\tau A)$ for $j = l - 2, l$. However, it was shown that the norms of these operators grow sufficiently slowly when $k \to \infty$, which allowed, based on approximation, to establish convergence.

For the more restricted class of strongly positive operators. stability with a factor depending logarithmically on the grid step was established in [1d]. This allowed one to prove the convergence of difference schemes in the case $j = l$ with almost the same order as the order of their approximation.

The stability of difference schemes with $j = l$ in the case of the fractional spaces $E_{\alpha,p}$ was established only later in [7h, 7i].

Subsequently the stability of such schemes in the space E was established for a more restricted class of operators in [9a, 9c]. Unfortunately, complete proofs of the announced results have not been published.

Considerable progress has been made recently. The stability in E is now established for a wide class of difference schemes, generated by the rational functions $R_{j,l}(z)$ with $j \le l$ [70].

Finally, let us mention a large cycle of works on difference schemes for parabolic partial differential equations (see, e.g., [14, 40] and the references given therein), in which stability in the C-norm was established under the assumption that the magnitudes of the grid steps with respect to the time and space variables are connected. In abstract terms this means, in particular, that the condition $\tau \|A\| \le M$ is satisfied.

The study of Padé difference schemes in the case of a strongly positive operator allowed to establish not only the stability, but also the coercive stability in difference analogues of spaces of smooth functions [6c, 7a, 7c, 7f–i].

This book does not touch upon the results of [6d, 6h, 7b, 7e–h, 9d, 30, 50a, 50b, 67] on the theory of highly accurate two-step difference schemes for the Cauchy problem (1) of Chapter 1 with a t-dependent operator.

The investigation of the stability and well-posedness of Padé difference schemes relies on a number of properties of the rational functions $R_{j,l}(z)$ that generate them. In **Section 0** we give known (Lemmas 0.1–0.3 and 0.7; see [55a, 55b]) as well as new (Lemmas 0.4–0.6 and 0.8) facts concerning $R_{j,l}(z)$. We construct Padé difference schemes and give the definition of stability as well as stability criteria for these schemes in difference analogues of the spaces $C(E)$, $C_0^\alpha(E)$, and $L_p(E)$.

In **Section 1** we generalize the results of Section 1 of Chapter 2 to a wide class of Padé difference schemes. Theorem 1.1 appears here for the first time. Theorem 1.1 and Theorem 1.2 of [12, 27] allow one to obtain a new real-field criterion for an operator to generate a strongly continuous semigroup. Theorem 1.3 follows from the stability and approximation properties of difference schemes. It is known that the solution of the homogeneous Cauchy problem (1) has additional smothness for $t > 0$ (see, e.g., [6a, 6b, 6e, 28, 45, 57, 69a, 69b]), and therefore the condition of Theorem 3.1 can be weakened for $t > 0$. By using the method of [54g] one shows that a necessary condition for well-posedness in $C(E)$ of Padé difference schemes is the analyticity of the semigroup. In the general case the necessity of this condition requires a new proof. The estimates (1.22), (1.23), (1.25), and (1.26) are taken from [12] for $j = l - 1$, and appear here for the first time for $j = l - 2$. Theorems 1.5, 1.6 and 1.8, 1.9 are taken from [1d, 7h]. Theorem 1.7 is new. Theorem 1.10 is taken from [7h]. Theorem 1.11 is new.

Section 2 is based on [6i]. Some results of this section for a second-order operator were obtained independently in [9d].

The results in **Section 3** appear here for the first time.

The assertions of Theorems 4.1–4.4 of **Section 4** were announced in [7h].

The results in **Section 5** appear here for the first time.

Chapter 4

The application of the abstract results of Chapter 3 to the Cauchy problem for m-th order parabolic equations relies on the following facts: the strong positivity of an elliptic difference operator A_h in the Banach space E_h, the well-posedness of the resolvent equation of A_h in E_h or in $E'_{\alpha,p}(E_h, A_h)$, and the structure of the fractional spaces $E'_{\alpha,p}(E_h, A_h)$.

The strong positivity of elliptic difference operators in Hilbert norms was studied in many works (see, e.g., [35, 48a]). Such a study is usually based on the selfadjointness and positive definiteness of the principal part of the operator A_h.

Considerably less studied is the strong positivity or simply the positivity of difference operators in Banach norms. For second-order elliptic operators and the simplest difference scheme positivity in the C-norm follows from the maximum principle (see [49, 50a, 50b]).

The strong positivity in difference analogues of weighted Hölder spaces for an elliptic difference operator A_h of second order of accuracy that approximates an elliptic operator without mixed derivatives was established for the first time in [54b].

The strong positivity of the simplest multidimensional second-order elliptic difference operator in the L_p- and C-norms was established in [1a, 1c]. The most thorough study of the strong positivity in the C-norm of a wide class of operators that approximate elliptic operators of arbitrary order in $\mathbf{R}^n$ was carried out in [52a–d]. The situation is different for difference operators that approximate boundary value problems for an elliptic equation. The positivity in the C-norm of difference operators of this type was studied in a number of particular cases [1a, 16a–c].

The study of well-posedness of difference schemes for elliptic equations was initiated in [39]. Further investigations of the simplest difference schemes for the approximate solution of boundary value problems were carried out in [20a–e, 42c, 43a–e, 44, 47a, 47b, 54b, 54d, 54f, 54g, 54i, 56b–d, 56f, 61, 62, 68].

Coercive inequalities in the L_2- and L_p-norms for difference operators generated by general boundary value problems in a half-space were established in [22a, 22b, 37, 38a, 38b].

The Padé difference scheme for the approximate solution of the Dirichlet problem for a second-order operator differential equation of elliptic type was studied in the papers [6f, 7n], where its well-posedness in spaces of smooth functions was established.

Finally, let us comment on the structure of fractional spaces. It is established that for any $0 < \alpha < 1/m$ and $1 \leq p \leq \infty$ the norms in the spaces $E'_{\alpha,p}(L_p(\mathbf{R}^n_h), A_h)$ and $W^{m\alpha}_p(\mathbf{R}^n_h)$ are equivalent uniformly in h. This fact corresponds to the following equality, known in interpolation theory (see [60]):

$$E'_{\alpha,p}(L_p(\mathbf{R}^n), A) = W^{m\alpha}_p(\mathbf{R}^n), \quad 0 < \alpha < 1/m, \quad 1 < p < \infty,$$

which in turn follows from the equality $D(A) = W^m_p(\mathbf{R}^n)$ for an m-th order elliptic operator A in $L_p(\mathbf{R}^n)$, $1 < p < \infty$, via the real interpolation method. The alternative method of investigation adopted in Chapter 4, based on estimates of the fundamental solution of the resolvent equation for the operator A_h, allows us to consider also the cases $p = 1$ and $p = \infty$.

Section 1 is written on the basis of the papers [52a–d].

The results of **Section 2** in the case $p = \infty$ were obtained in [7d, 7f], and for the general case appear here for the first time.

The results of **Section 3** appear here for the first time.

REFERENCES*

[1a]   Alibekov, Kh. A. and Sobolevskiĭ, P.E., *Stability of difference schemes for parabolic equations,* Dokl. Akad. Nauk SSSR **232** (1977), no. 4, 737-740. (Russian). [MR **55** #13023]

[1b]   Alibekov, Kh. A. and Sobolevskiĭ, P.E., *Stability and convergence of high-order difference schemes of approximation for parabolic equations,* Ukrain. Mat. Zh. **31** (1979), no. 6, 627-634. (Russian). [MR **81i:65040**]

[1c]   Alibekov, Kh. A. and Sobolevskiĭ, P.E., *Stability and convergence of difference schemes of a high order for parabolic partial differential equations,* Ukrain. Mat. Zh. **32** (1980), no. 3, 291-300. (Russian). [MR **81j:65100**]

[1d]   Alibekov, Kh. A. and Sobolevskiĭ, P.E., *On a method for constructing and studying schemes of Padé class,* in: Differential Equations and their Applications, Vol. 32, pp. 9-29, Viln'nyus (1982). (Russian). [RZh. Mat. 1983:5 Б1048]

[2]   Andreev, V. B., *On uniform convergence of certain difference schemes,* Zh. Vychisl. Mat. i Mat. Fiz. **6** (1966), no. 2, 238-250. (Russian). [MR **34** #967]

[3]   Andreev, Yu. N., *Control of Finite-Dimensional Linear Objects,* "Nauka", Moscow, 1976. (Russian). [MR **57** #9150]

[4]   Anosov, V. P. and Sobolevskii, P.E., *The coercive solvability of parabolic equations,* Mat. Zametki **11** (1972), no. 2, 409-419. (Russian). [MR **46** #558]

[5a]   Aronson, D. G., *On the correctness of partial differential operators and the von Neumann condition for stability of finite difference operators,* Proc. Amer. Math. Soc. **14** (1963), 948-955. [MR **27** #6407]

[5b]   Aronson, D. G., *On the stability of certain finite difference approximations to parabolic systems of partial differential equations,* Numer. Math. **5** (1963), 118-137; correction, ibid. **5** (1963), 290. [MR **27** #6408]

---

*   For the reader's convenience, Math. Reviews or Ref. Zh. Mat. numbers are provided.

[6a]   Ashyralyev, A. O., *Some difference schemes for parabolic equations with nonsmooth initial data,* Izv. Akad. Nauk Turkmen. SSR Ser. Fiz.-Tekhn. Khim. Geol. Nauk No. 6 (1983), 70-72. (Russian). [MR **86c:65093**]

[6b]   Ashyralyev, A. O., *An estimate of the error of purely implicit difference schemes for parabolic equations with nonsmooth data,* in: Applied Mathematics and Computer Software, pp. 26-28, Moscow, 1985. (Russian). [RZh. Mat. 1986:5 Б1509]

[6c]   Ashyralyev, A. O., *A purely implicit difference scheme of second order of approximation for parabolic equations,* Izv. Akad. Nauk Turkmen. SSR Ser. Fiz.-Tekhn. Khim. Geol. Nauk No. 4 (1987), 3-13. (Russian). [MR **89g:65112**]

[6d]   Ashyralyev, A. O., *Coercive stability of a difference scheme of the second order of approximation for differential equations with a time-dependent operator in a Banach space,* in: Measure and Integral, pp. 27-36, Kuĭbyshev. Gos. Univ., Kuĭbyshev, 1988. (Russian).

[6e]   Ashyralyev, A. O., *Estimation of the convergence of modified Crank-Nicolson difference schemes for parabolic equations with nonsmooth imput data,* Izv. Akad. Nauk Turkmen. SSR Ser. Fiz.-Tekhn. Khim. Geol. Nauk No. 1 (1989), 3-8. (Russian). [MR **90f:65159**]

[6f]   Ashyralyev, A. O., *On difference schemes of higher order of accuracy for elliptic equations,* in: Applications of Functional Analysis to Some Nonclassical Equations of Mathematical Physics, pp. 3-14, Novosibirsk, 1989. (Russian). [RZh. Mat. 1990:10 Г122]

[6g]   Ashyralyev, A. O., *Coercive solvability of parabolic equations in spaces of smooth functions,* Izv. Akad. Nauk Turkmen. SSR Ser. Fiz.-Tekhn. Khim. Geol. Nauk No. 3 (1989), 3-13. (Russian). [MR **90i:35150**]

[6h]   Ashyralyev, A. O., *On difference schemes of higher order of accuracy for parabolic differential equations with variable coefficients,* in: Optimal Control and Differential Equations: Proceedings of All-Union Conference, Akad. Nauk. Turkmen. SSR, Ashkhabad, 1990. (Russian).

[6i]   Ashyralyev, A. O., *Correct solvability of Padé difference schemes for parabolic equations in Hölder spaces,* Ukrain. Mat. Zh. **44** (1992), 1466-1476. (Russian). [MR **94c:65103**]

[7a] Ashyralyev, O. A. and Sobolevskiĭ, P. E., *Correct solvability of the Crank-Nicolson scheme for parabolic equations,* Izv. Akad. Nauk Turkmen. SSR Ser. Fiz.-Tekhn. Khim. Geol. Nauk No. 6 (1981), 10-16. (Russian). [MR **83g:34064**]

[7b] Ashyralyev, O. A. and Sobolevskiĭ, P. E., *The Crank-Nicolson difference scheme for differential equations in a Banach space with a time-dependent operator,* Izv. Akad. Nauk Turkmen. SSR Ser. Fiz.-Tekhn. Khim. Geol. Nauk No. 3 (1982), 3-9. (Russian). [MR **84c:65080**]

[7c] Ashyralyev, O. A. and Sobolevskiĭ, P. E., *Coercive stability of a Crank-Nicolson difference scheme in spaces $\tilde{C}_0^\alpha$,* in: Approximate Methods for Investigating Differential Equations and their Applications, pp. 16-24, Kuĭbyshev. Gos. Univ., Kuĭbyshev, 1982. (Russian). [MR **86e:65117**]

[7d] Ashyralyev, O. A. and Sobolevskiĭ, P. E., *Investigation of the stability of difference schemes in Hölder spaces,* Voronezh. Gosud. Univ., Voronezh, 1983, 63 p. Deposited VINITI 4.12.1983, No. 2745. (Russian). [RZh. Mat. 1983:8 Б1116 Dep.]

[7e] Ashyralyev, O. A. and Sobolevskiĭ, P. E., *On the coercive stability of difference schemes for abstract parabolic equations with variable operator coefficients in interpolation spaces,* in: Investigations on the Theory of Differential Equations, pp. 65-77, Ashkhabad, 1983. (Russian). [RZh. Mat. 1983:12 Б1415]

[7f] Ashyralyev, O. A. and Sobolevskiĭ, P. E., *The theory of interpolation of linear operators and the stability of difference schemes,* Dokl. Akad. Nauk SSSR **275** (1985), no. 6, 1289-1291. (Russian). [MR **85m: 65046**]

[7g] Ashyralyev, O. A. and Sobolevskiĭ, P. E., *Coercive stability of difference schemes of first and second order for parabolic equations with coefficients independent of time,* Izv. Akad. Nauk Turkmen. SSR Ser. Fiz.-Tekhn. Khim. Geol. Nauk No. 6 (1985), 3-11. (Russian). [RZh. Mat. 1986:8 Б1305]

[7h] Ashyralyev, O. A. and Sobolevskiĭ, P. E., *Difference schemes for parabolic equations,* in: Differential Equations and their Applications, Abstracts of All-Union Conference, pp. 39-40, Ashkhabad, 1985. (Russian).

[7i]   Ashyralyev, O. A. and Sobolevskiĭ, P. E., *Stability in Hölder norms of dif-ference schemes for parabolic equations,* in: Numerical Methods for Solving Transport Equations, pp. 20-23, Tartu, 1986. (Russian).

[7j]   Ashyralyev, O. A. and Sobolevskiĭ, P. E., *On a coercive estimate for an ab-stract parabolic equation in Hölder space* in: Abstracts of the 9-th All-Union Conference on the Theory of Operators in Functional Spaces, Chelyabinsk, 1986. (Russian).

[7k]   Ashyralyev, O. A. and Sobolevskiĭ, P. E., *Coercive stability of a multidimen-sional differential elliptic equation of 2m-th order with variable coefficients,* in: Investigations in the Theory of Differential Equations, 31-43, Minvuz Turkmen. SSR, Ashkhabad, 1987. (Russian). [see MR **90d:00022**]

[7l]   Ashyralyev, O. A. and Sobolevskiĭ, P. E., *Difference schemes of a high order of accuracy for parabolic equations with variable coefficients,* Dokl. Akad. Nauk Ukrain. SSR Ser. A No. 6 (1988), 3-7. (Russian). [MR **90a:65190**]

[7m]   Ashyralyev, O. A. and Sobolevskiĭ, P.E., *Stability of difference schemes for parabolic equations in interpolation spaces,* in: Applied Methods of Func-tional Analysis, pp. 9-17, Voronezh, 1985. (Russian). [RZh. Mat. 1986:1 Б1416]

[7n]   Ashyralyev, O. A. and Sobolevskiĭ, P.E., *On a class of three-step difference schemes of higher order of accuracy for elliptic equations in Hilbert space,* in: Numerical Methods for Solving Transport Equations: Seminar Abstracts, Akad. Nauk. Estonsk. SSR, 1990. (Russian).

[8]   Baillon, J.-B., *Charactère borné de certains générateurs de semi-groupes linéaires dans les espaces de Banach,* C. R. Acad. Sci. Paris Sér. A-B **290**, No. 16 (1980), A757-A760. [MR **82a:47038**]

[9a]   Bakaev, N. Yu., *The theory of the stability of difference schemes in arbitrary norms,* Dokl. Akad. Nauk SSSR **297** (1987), no. 2, 275-279. (Russian). [MR **88m:65086**]

[9b]   Bakaev, N. Yu., *Estimates for the stability of difference schemes for a dif-ferential equation with a constant operator. I.,* in: Partial Differential Equa-tions, pp. 3-14, Akad. Nauk SSSR Sibirsk. Otdel., Inst. Mat., Novosibirsk, 1989. (Russian). [MR **93bg:65079**]

[9c]   Bakaev, N. Yu., *Estimates for the stability of difference schemes for a differential equation with a constant operator. II.*, in: Embedding Theorems and their Applications to Problems in Mathematical Physics, pp. 18-37, Akad. Nauk SSSR Sibirsk. Otdel., Inst. Mat., Novosibirsk, 1989. (Russian). [MR **93b:65080**]

[9d]   Bakaev, N. Yu., *On the theory of coercive stability of two-layer difference schemes,* Differentsial'nye Uravneniya **26** (1990), no. 5, 898-900. (Russian). [MR **91g:65188**]

[10]   Baker, G. A., Jr. and Graves-Morris, P. R., *Padé Approximants. Part I. Basic Theory,* and *Part II. Extensions and Applications,* Addison-Wesley, Reading, Mass., 1981. [MR **83a:41009a,b**]

[11]   Benedek, A., Calderón, A. P., and Panzone, R., *Convolution operators on Banach space valued functions,* Proc. Nat. Acad. Sci. U.S.A. **48** (1962), 356-365. [MR **24** #A3479]

[12]   Brenner, Ph. and Thomée, V., *On rational approximations of semigroups,* SIAM J. Numer. Anal. **16** (1979), no. 4, 683-694. [MR **80j:47052**]

[13]   Brenner, Ph., Crouzeix, M., and Thomée, V., *Single-step methods for inhomogeneous linear differential equations in Banach space,* RAIRO Anal. Numér. **16** (1982), no. 1, 5-26. [MR **83d:65268**]

[14]   Brenner, Ph., Thomée, V., and Wahlbin, L. B., *Besov Spaces and Applications to Difference Methods for Initial Value Problems,* Lecture Notes in Mathematics, Vol. 434, Springer-Verlag, Berlin-New York, 1975. [MR **57** #1106]

[15]   Cody, W. J., Meinardus, G, and Varga, R. S., *Chebyshev rational approximations to e^{-x} in $[0, +\infty)$ and applications to heat-conduction problems,* J. Approximation Theory **2** (1969), 50-65. [MR **39** #6536]

[16a]   Danelich, S. I., *Positive difference operators in R_{h1},* Voronezh. Gosud. Univ. 1987, 13 p. Deposited VINITI 3.18.1987, No. 1936-B87. (Russian). [RZh. Mat. 1987:8 Б1336 Dep.]

[16b]   Danelich, S. I., *Positive difference operators with constant coefficients in a half-space,* Voronezh. Gosud. Univ. 1987, 56 p. Deposited VINITI 11.5.1987, No. 7747-B87. (Russian). [RZh. Mat. 1988:2 Б1240 Dep.]

[16c]   Danelich, S. I., *Positive difference operators with variable coefficients on the half-line,* Voronezh. Gosud. Univ. 1987, 16 p. Deposited VINITI 11.9.1987, No. 7713-B87. (Russian). [RZh. Mat. 1988:2 Б1241 Dep.]

[17a]   Da Prato, G. and Grisvard, P., *Sommes d'opérateurs linéaires et équations différentielles opérationnelles,* J. Math. Pures Appl. (9) **54** (1975), no. 3, 305-387. [MR **56** #1129]

[17b]   Da Prato, G. and Grisvard, P., *Équations d'évolution abstraites non linéaires de type parabolique,* C. R. Acad. Sci. Paris Sér. A-B **283** (1976), no. 9, A709-A711. [MR **54** #13647]

[18]   Dore, G. and Venni, A., *On the closedness of the sum of two closed operators,* Math. Z. **196** (1981), no. 2, 189-201. [MR **88m:47072**]

[19]   Dryya, M., *Convergence in C inside the domain of difference schemes with a splitting operator for parabolic systems,* Zh. Vychisl. Mat. i Mat. Fiz. **11** (1971), no. 3, 658-666. (Russian). [MR **44** #7779]

[20a]   D'yakonov, E. G., *On the convergence of a certain iteration process,* Uspekhi Mat. Nauk **21** (1966), no. 1, 179-182 (Russian). [RZh. Mat. 1967:2 Б661]

[20b]   D'yakonov, E. G., *Iteration methods for the solution of difference analogues of boundary value problems for elliptic-type equations,* in: Proceedings of International Summer School on Numerical Methods, Vol. 4, Kiev, 1970 (Russian). [RZh. Mat. 1971:3 Б621]

[20c]   D'yakonov, E. G., *Approximate methods for the solution of operator equations,* Dokl. Akad. Nauk SSSR **198** (1971), no. 3, 516-519 (Russian). [MR **44** #7768]

[20d]   D'yakonov, E. G., *Difference Methods for the Solution of Boundary Value Problems (Stationary Problems),* Moskov. Gosud. Univ., Moscow, 1971. (Russian).

[20e]   D'yakonov, E. G., *On some methods for the solution of systems of equations by difference and projection-difference schemes,* in: Computational Methods in Linear Algebra, pp. 28-58, Vychisl. Tsentr Sib. Otdel. Akad. Nauk SSSR, Novosibirsk, 1972. (Russian). [RZh. Mat. 1972:9 Б742]

[21]   Fedoryuk, M. V. *On stability in C of the Cauchy problem for difference equations and partial differential equations,* Zh. Vychisl. Mat. i Mat. Fiz. **7** (1967), no. 3, 510-540. (Russian). [MR **35** #5147]

[22a]   Frank, L. S. *Convolution difference operators,* Dokl. Akad. Nauk **181** (1968), no. 2, 42-45. (Russian). [MR **43** #752]

[22b]   Frank, L. S. *Coercive boundary value problems for difference operators,* Dokl. Akad. Nauk **192** (1970), no. 1, 286-289. (Russian). [MR **58** #23742]

[23]   Grif, A. G., *On the stability in $W_2^{2,1}$ of difference schemes for parabolic equations,* in: Investigations on the Theory of Difference Schemes for Elliptic and Parabolic Equations, pp. 88-112, Mosk. Gos. Univ., Moscow, 1973. (Russian). [RZh. Mat. 1973:11 Б838]

[24]   Grisvard, P., *Équations différentielles abstraites,* Ann. Sci. École. Norm. Sup. (4) **2** (1969), no. 3, 311-395. [MR **42** #5101]

[25]   Gudkin, V. P., Dyment, D. A., and Matveev, V. A., *Coercive solvability of abstract parabolic equations in weighted Hölder spaces,* in: Proceedings of the Conference of Young Scientists of Departments of the Khabarovsk Inst. of Railway Transp., pp. 47–53, Khabarovsk, 1973. (Russian). [RZh. Mat. 1974:10 Б885]

[26]   Guilinger, W. H., Jr., *The Peaceman-Rachford method for small mesh increments,* J. Math. Anal. Appl. **11** (1965), no. 4, 261-277. [MR **32** #607]

[27]   Hersch, R. and Kato, T., *High-accuracy stable difference schemes for well-posed initial value problems,* SIAM J. Numer. Anal. **16** (1979), no. 4, 670-682. [MR **80h:65036**]

[28]   Huang, Ming You and Thomée, V., *On the backward Euler method for parabolic equations with rough initial data,* SIAM J. Numer. Anal. **19** (1982), no. 3, 599-603. [MR **83f:65143**]

[29]   Ionkin, N. I. and Mokin, Yu. I., *The parabolicity of difference schemes,* Zh. Vychisl. Mat. i Mat. Fiz. **14** (1974), no. 2, 402-417. (Russian). [MR **49** #8383]

[30]   Karakashian, O. A., *On Runge-Kutta methods for parabolic problems with time-dependent coefficients,* Math. Comp. **47** (1986), no. 175, 77-101. [MR **87i:65161**]

[31]   Krasnosel'skiĭ, M. A., Zabreĭko, P. P., Pustyl'nik, E. I., and Sobolevskiĭ, P.
        E., *Integral Operators in Spaces of Summable Functions,* "Nauka", Moscow,
        1966 (Russian). [MR **34** #6568]; English transl.: *Integral Operators in Spaces
        of Summable Functions,* Noordhoff, Leiden, 1976. [MR **52** #6505]

[32]   Kreĭn, S. G., *Linear Differential Equations in a Banach Space,* "Nauka",
        Moscow, 1966 (Russian). [MR **40** #508]; English transl.: *Linear Differential
        Equations in Banach space,* Translations of Mathematical Monographs, Vol.
        29, American Mathematical Society, Providence, RI, 1971. [MR **49** #7548]

[33]   Ladyzhenskaya, O. A., Solonnikov, V. A., and Ural'tseva, N. N., *Linear and
        Quasilinear Equations of Parabolic Type,* "Nauka", Moscow, 1967 (Rus-
        sian). [MR **39** #3159a]; English transl.: *Linear and Quasilinear Equations
        of Parabolic Type,* Translations of Mathematical Monographs, Vol. 23, Amer-
        ican Mathematical Society, Providence, RI, 1968. [MR **39** #3159b]

[34]   Ladyzhenskaya, O. A. and Ural'tseva, N. N., *Linear and Quasilinear Equa-
        tions of Elliptic Type,* Second edition, revised "Nauka", Moscow, 1973 (Rus-
        sian). [MR **58** #23009]; English transl. of first edition: *Linear and Quasi-
        linear Elliptic Equations,* Academic Press, New York-London, 1968. [MR
        39 #5941]

[35]   Marchuk, G. I., *Methods of Numerical Mathematics,* Second edition, "Na-
        uka", Moscow, 1980 (Russian). [MR **81m:65003**]; English transl.: *Methods
        of Numerical Mathematics,* Springer-Verlag, New York-Berlin, 1982. [MR
        83e:65004]

[36]   Marchuk, G. I. and Shaĭdurov, V. V., *Increasing the Accuracy of Solutions of
        Difference Schemes,* "Nauka", Moscow, 1979 (Russian). [MR **81m:65004**]

[37]   Mokin, Yu. I., *Estimations of the L_p-norms of mesh functions in the limiting
        cases,* Differentsial'nye Uravneniya **11** (1975), no. 9, 1652-1663. (Russian).
        [MR **52** #8910]

[38a]   Mokin, Yu. I. and Lazarov, R. D., *On the stability of elliptic difference
        problems,* Zh. Vychisl. Mat. i Mat. Fiz. **13** (1973), no. 2, 488-494. (Russian).
        [MR **48** #3264]

[38b] Mokin, Yu. I. and Lazarov, R. D., *The stability of elliptic difference schemes in the metrics $L_{p,h}$,* in: Investigations on the Theory of Difference Schemes for Elliptic and Parabolic Equations, pp. 40-87, Mosk. Gos. Univ., Moscow, 1973. (Russian). [RZh. Mat. 1973:11 Б844]

[39]   Nitsche, J. A. and Nitsche, J. C. C., *Error estimates for the numerical solution of elliptic differential equations,* Arch. Rational Mech. Anal. **5** (1960), 488-494. [MR **22** #8664]

[40]   Piskarev, S. I., *Error estimates in the approximation of semigroups of operators by Padé fractions,* Izv. Vyssh. Uchebn. Zaved. Mat. (1979), no. 4, 33-38. (Russian). [MR **81j:47032**]

[41]   Polichcka, A. E., *Correct solvability of a difference Cauchy problem with a variable operator in a Bochner space,* Differentsial'nye Uravneniya **13** (1977), no. 9, 1723-1726. (Russian). [MR **57** #10153]

[42a] Polichka, A. E. and Sobolevskiĭ, P. E., *New L_p-estimates for parabolic difference problems,* Zh. Vychisl. Mat. i Mat. Fiz. **16** (1976), no. 5, 1155-1163. (Russian). [MR **57** #18129]

[42b] Polichka, A. E. and Sobolevskiĭ, P. E., *Rothe's method of approximate solution of the Cauchy problem for differential equations in a Banach space with a variable unbounded operator,* Differentsial'nye Uravneniya **12** (1976), no. 9, 1693-1704. (Russian). [MR **57** #13058]

[42c] Polichka, A. E. and Sobolevskiĭ, P. E., *Correct solvability of a difference boundary value problem in a Bochner space,* Ukrain. Mat. Zh. **28** (1976), no. 4, 511-523. (Russian). [MR **55** #795]

[42d] Polichka, A. E. and Sobolevskiĭ, P. E., *Correct solvability of parabolic difference equations in Bochner spaces,* Trudy Mosk. Mat. Obshch. **36** (1978), 29-57. (Russian). [MR **80g:34063**]

[42e] Polichka, A. E. and Sobolevskiĭ, P. E., *Some properties of the Crank-Nicolson scheme,* in: Computations with Sparse Matrices, pp. 115-122, Novosibirsk, 1981. (Russian). [RZh. Mat. 1982:8 Б1048]

[43a] Polichka, A. E. and Tiunchik, M. F., *L_p-estimates of solutions of a difference boundary value problem,* Sibirsk. Mat. Zh. **22** (1981), no. 6, 162-167. (Russian). [MR **83d:34099**]

[43b] Polichka, A. E. and Tiunchik, M. F., *Some estimates of solutions of dif-
      ference schemes of a Neumann problem and a mixed problem,* Zh. Vychisl.
      Mat. i Mat. Fiz. **22** (1982), no. 3, 735-738. (Russian). [MR **84a:65098**]

[43c] Polichka, A. E. and Tiunchik, M. F., *Estimates in L_p for some differential
      boundary value problems,* in: Qualitative Methods of the Theory of Dy-
      namical Systems, pp. 108-114, Dal'nevost. Gos. Univ., Vladivostok, 1982.
      (Russian).

[43d] Polichka, A. E. and Tiunchik, M. F., *Estimates for the solutions of some dif-
      ference schemes for mixed boundary value problems in cylindrical domains,*
      Differentsial'nye Uravneniya **22** (1986), no. 6, 1054-1060. (Russian). [MR
      87j:65139]

[43e] Polichka, A. E. and Tiunchik, M. F., *Estimates for the solutions of a dif-
      ference scheme of first order of approximation for the Neumann problem,* in:
      Problems in Applied Analysis, pp. 81-83, Akad. Nauk SSSR, Dal'nevostochn.
      Nauchn. Tsentr Akad Nauk SSSR, Vladivostok, 1986 (Russian) [MR **90f:
      65188**]

[44] Primakova, S. I. and Sobolevskiĭ, P. E., *The coercive solvability of fourth or-
      der difference schemes,* Differentsial'nye Uravneniya **10** (1974), no. 9, 1699-
      1713. (Russian) [MR **52** #1076]

[45] Rannacher, R. *Discretization of the heat equation with singular initial data,*
      Z. Angew. Math. Mech **62** (1982), no. 5, T346-T348. [see MR **83j:0009**]

[46] Rothe, E. *Zweidimensionale parabolische Randwertaufgaben als eindimen-
      sionaler Randwertaufgaben,* Math. Ann. **102** (1930), 650-670.

[47a] Rukavishnikov, V. A., *On the convergence of coercive difference schemes
      approximating the second boundary value problem,* Dal'nevostochn. Nauchn.
      Tsentr Akad. Nauk SSSR Preprint, Vladivostok, 1982, 29 p. (Russian).
      [RZh. Mat. 1983:2 Б1197]

[47b] Rukavishnikov, V. A., *A coercive estimate of the rate of convergence of an
      approximate solution of the second boundary value problem,* Dokl. Akad.
      Nauk SSSR **271** (1983), no. 4, 798-801. (Russian). [MR **85a:65152**]

[48a] Samarskiĭ, A. A., *Regularization of difference schemes,* Zh. Vychisl. Mat. i
      Mat. Fiz. **7** (1967), no. 1, 62-93. (Russian). [MR **35** #3930]

[48b] Samarskiĭ, A. A., *Introduction to the Theory of Difference Schemes,* "Na-uka", Moscow, 1971 (Russian). [MR **49** #11822]

[49] Samarskiĭ, A. A. and Gulin, A. V., *Stability of Difference Schemes,* "Nauka", Moscow, 1973 (Russian). [RZh. Mat. 1974:4 Б962]

[50a] Sammon, P. H., *Approximations for parabolic equations with time dependent coefficients,* PhD. Thesis, Cornell University, Ithaca, 1978.

[50b] Sammon, P. H., Fully discrete approximation for parabolic problems, SIAM J. Numer. Anal. **20** (1983), 437-469. [MR **85a:65147**]

[51a] Serdyukova, S. I., *Stability in C of linear difference schemes with constant real coefficients,* Zh. Vychisl. Mat. i Mat. Fiz. **6** (1966), no. 3, 477-486. (Russian). [MR **34** # 5329]

[51b] Serdyukova, S. I., *Uniform stability with respect to the initial data of a six-point symmetric scheme for the heat equation,* in: Numerical Methods for Solving Differential and Integral Equations, and Quadrature Formulas, Zh. Vychisl. Mat. i Mat. Fiz. **4** (1964), no. 4, suppl., 212-216. (Russian). [MR **31** # 5353]

[51c] Serdyukova, S. I., *Uniform stability of a six-point scheme of higher order accuracy for the heat equation,* Zh. Vychisl. Mat. i Mat. Fiz. **7** (1967), no. 1, 214-218. (Russian). [MR **35** # 1229]

[51d] Serdyukova, S. I., *On the stability in the uniform metric of systems of difference equations,* Zh. Vychisl. Mat. i Mat. Fiz. **7** (1967), no. 3, 497-509. (Russian). [MR **35** # 7023]

[51e] Serdyukova, S. I., *A necessary and sufficient condition for the stability of a certain class of difference boundary value problems,* Dokl. Akad. Nauk SSSR **208** (1973), no. 1, 52-55. (Russian). [MR **47** # 3868]

[52a] Smirnitskiĭ, Yu. A. and Sobolevskiĭ, P. E., *Positivity of multidimensional difference operators in the C-norm,* Uspekhi Mat. Nauk **36** (1981), no. 4, 202-203. (Russian). [RZh. Mat. 1981:12 Б807]

[52b] Smirnitskiĭ, Yu. A. and Sobolevskiĭ, P. E., *Positivity of difference operators,* in: Spline Methods, Novosibirsk, 1981. (Russian). [RZh. Mat. 1982:12 Б1022]

[52c] Smirnitskiĭ, Yu. A. and Sobolevskiĭ, P. E., *Pointwise estimates of the Green function of a difference elliptic operator,* Chisl. Metody Mekh. Sploshn. Sredy **15** (1982), no. 4, 129-142. (Russian). [MR **85m:65111**]

[52d] Smirnitskiĭ, Yu. A. and Sobolevskiĭ, P. E., *Pointwise estimates of the Green function of the resolvent of a difference elliptic operator with variable co- efficients in* $\mathbf{R}^n$, Voronezh. Gosud. Univ. 1982, 32 p., Deposited VINITI 5.2.1982, No. 1519. (Russian). [RZh. Mat. 1982:8 Б275 Dep.]

[53]   Sobolev, S. L., *Some Applications of Functional Analysis in Mathemati- cal Physics,* Third edition, "Nauka", Moscow, 1988. (Russian). [MR **90m: 46059**]

[54a] Sobolevskiĭ, P. E., *Coerciveness inequalities for abstract parabolic equa- tions,* Dokl. Akad. Nauk SSSR **157** (1964), no. 1, 52-55. (Russian). [MR **29** #3762]

[54b] Sobolevskiĭ, P. E., *The coercive solvability of difference equations,* Dokl. Akad. Nauk SSSR **201** (1971), no. 5, 1063-1066. (Russian). [MR **44** #7375]

[54c] Sobolevskiĭ, P. E., *On the stability and convergence of the Crank-Nicolson scheme,* in: Variational-Difference Methods in Mathematical Physics, pp. 146-151, Vychisl. Tsentr Sibirsk. Otdel. Akad. Nauk SSSR, Novosibirsk, 1973(1974). (Russian). [RZh. Mat. 1974:2 Б513]

[54d] Sobolevskiĭ, P. E., *Difference Methods for the Approximate Solution of Dif- ferential Equations,* Izdat. Voronezh. Gosud. Univ., Voronezh, 1975. (Rus- sian).

[54e] Sobolevskiĭ, P. E., *Some properties of the solutions of differential equations in fractional spaces,* Trudy Nauchn.-Issled. Inst. Mat. Voronezh. Gos. Univ. No. 14 (1974), 68-74. (Russian). [RZh. Mat. 1975:7 Б825]

[54f] Sobolevskiĭ, P. E., *On the well-posedness in C of the first boundary value problem for difference elliptic and parabolic equations in rectangular do- mains,* Vychisl. Tsentr Sibirsk. Otdel. Akad. Nauk SSSR Preprint, Novosi- birsk, 1976, 10 p. (Russian).

[54g] Sobolevskiĭ, P. E., *The theory of semigroups and the stability of difference schemes,* in: Operator Theory in Function Spaces (Proc. School, Novosi- birsk, 1975), pp. 304-337, "Nauka", Sibirsk. Otdel., Novosibirsk, 1977. (Rus- sian). [MR **58** #31877]

[54h] Sobolevskiĭ, P. E., *On the Crank-Nicolson difference scheme for parabolic equations,* in: Nonlinear Oscillations and Control Theory, pp. 98-106, Izhevsk, 1978. (Russian). [RZh. Mat. 1978:8 Б1075]

[54i] Sobolevskiĭ, P. E., *The correct solvability in C of elliptic and parabolic difference boundary value problems,* Voronezh. Gos. Univ. Trudy Nauchn.-Issled. Inst. Mat. No. 17 (1975), 94–95. (Russian). [MR **58**# 19212]

[55a] Sobolevskiĭ, P. E. and Hoàng Văn Lai, *Algorithms of optimal type for the approximate solution of parabolic equations,* Vychisl. Systemy No. 72 (1977), 79-91. (Russian). [MR **80f:65084**]

[55b] Sobolevskiĭ, P. E. and Hoàng Văn Lai, *Difference schemes of optimal type for the approximate solution of parabolic equations (the Banach case),* Ukrain. Mat. Zh. **33** (1981), no. 1, 39-46. (Russian). [MR **82e:65066**]

[56a] Sobolevskiĭ, P. E. and Tiunchik, M. F., *On a comparison method in the theory of approximate methods,* in: Problems on the Accuracy and Efficiency of Computational Algorithms, pp. 138-145 Proc. of Sympos., Vol. 5, Kiev, 1969. (Russian). [RZh. Mat. 1970:4 Б909]

[56b] Sobolevskiĭ, P. E. and Tiunchik, M. F., *On a difference method for approximate solution of quasilinear elliptic and parabolic equations,* Voronezh. Gos. Univ. Trudy Mat. Fak. No. 2 (1970), 82-106. (Russian). [RZh. Mat. 1971:10 Б763]

[56c] Sobolevskiĭ, P. E. and Tiunchik, M. F., *The difference method of approximate solution for elliptic equations,* Voronezh. Gos. Univ. Trudy Mat. Fak. No. 4 (1970), 117-127. (Russian). [MR **55** #6883]

[56d] Sobolevskiĭ, P. E. and Tiunchik, M. F., *On a difference method for approximate solution of boundary value problems for quasilinear elliptic and parabolic equations,* in: Proceedings of the 8-th All-Union Inter-College Far-East Conference on Mathematical Science, pp. 126-129, Khabarovsk. Pedagog. Inst., Khabarovsk, 1970. (Russian).

[56e] Sobolevskiĭ, P. E. and Tiunchik, M. F., *The angle between parabolic operators,* Voronezh. Gos. Univ. Trudy Mat. Fak. No. 5 (1971), 103-113. (Russian). [MR **53** #11229]

[56f] Sobolevskiĭ, P. E. and Tiunchik, M. F., *On the well-posedness of the second boundary value problem for difference equations in weighted Hölder norms,* in: Qualitative Methods of the Theory of Dynamical Systems, pp. 27-37, Dal'nevost. Gos. Univ., Vladivostok, 1982. (Russian).

[57]   Suzuki, T. *Full-discrete finite element approximation of evolution equation $u_t + A(t)u = 0$ of parabolic type,* J. Fac. Sci. Univ. Tokyo Sec. IA **29** (1982), no. 1, 195-240. [MR **83g:65112**]

[58]   Tanabe, H., *Equations of Evolution,* Translated from the Japanese, Pitman, Boston, Mass.-London,1979. [MR **82g:47032**]

[59a] Thomée, V. *Stability of difference schemes in the maximum-norm,* J. Differential Equations **1** (1965), no. 3, 273-292. [MR **31** #515]

[59b] Thomée, V. *On maximum-norm stable difference operators,* in: Numerical Solution of Partial Differential Equations (Proc. Sympos. Univ. Maryland, 1965), pp. 125-151 Academic Press, New York, 1966. [MR **35** #1225]

[60]   Triebel, H., *Interpolation Theory, Function Spaces, Differential Operators,* North-Holland, Amsterdam-New York, 1978. [MR **80i:46032b**]

[61]   Vaĭnniko, G. M., *Analysis of Discretization Methods, Special Course,* Tartu. Gosudarstv. Univ, Tartu, 1976. (Russian). [MR **58** #13699]

[62]   Vaĭnniko, G. M. and É. É. Tamme, *Convergence of the difference method in the problem of periodic solutions of equations of elliptic type,* Zh. Vychisl. Mat. i Mat. Fiz. **16** (1976), no. 3, 652-664. (Russian). [MR **54** #4137]

[63a] Varga, R. S., *On higher order stable implicit implicit methods for solving parabolic partial differential equations,* J. Math. and Phys. **40** (1961), 220-231. [MR **25** #3613]

[63b] Varga, R. S., *Functional Analysis and Approximation Theory in Numerical Analysis,* CBMS Regional Conference Series in Applied Mathematics, No. 3, SIAM, Philadelphia, 1971. [MR **46** #9602]

[64]   Vasil'ev, V. V., *Coercive stability of abstract parabolic equations with a constant operator,* Differentsial'nye Uravneniya **14** (1978), no. 8, 1507-1510. (Russian). [MR **80d:34088**]

[65]   Vishik, M. I., Myshkis, A. D., and Oleĭnik, O. A. *Partial differential equa-
       tions,* in: Mathematics in USSR in the Last 40 Years, 1917–1957, Vol. 1, pp.
       563-599, Fizmatgiz, Moscow, 1959. (Russian). [RZh. Mat. 1960:2 1962]

[66]   Widlund, O. B., *Stability of parabolic difference schemes in the maximum
       norm,* Numer. Math. **18** (1966), no. 2, 186-202. [MR **33** #5149]

[67]   Wu, Wei, *The Calahan method for parabolic equations with time-dependent
       coefficient,* J. Comput. Math. **5** (1987), no. 1, 10-20. [MR **89a:65141**]

[68]   Zarubin, A. G. and Tiunchik, M. F., *Approximate methods for the solution
       of a certain class of nonlinear operator equations,* Zh. Vychisl. Mat. i Mat.
       Fiz. **16** (1976), no. 3, 567-576. (Russian). [MR **55** #6840]

[69a] Zlotnik, A. A., *Estimates of the rate of convergence in $V_2(Q_T)$ of projection-
       difference schemes for parabolic equations,* Vestnik Moskov. Univ. Ser. XV
       Vychisl. Mat. Kibernet. no. 1 (1980), 27-35. (Russian). [MR **81i:65101**]

[69b] Zlotnik, A. A., *An estimate of the method of variable directions for the heat
       equation with nonsmooth data,* Vychisl. Tsentr Sibirsk. Otdel. Akad. Nauk
       SSSR Preprint No. 54, 1978, 14p. (Russian). [RZh. Mat. 1978:3 Б946]

[70]   Crouzeix, M., Larsson, S., Piskarev, S., and Thomée, V. *The stability of
       rational approximations of analytic semigroups,* Preprint, Dept. of Mathe-
       matics, Chalmers Institute of Technology, Göteborg, 1991–28.

Titles previously published in the series

OPERATOR THEORY: ADVANCES AND APPLICATIONS
BIRKHÄUSER VERLAG

1. **H. Bart, I. Gohberg, M.A. Kaashoek:** Minimal Factorization of Matrix and Operator Functions, 1979, (3-7643-1139-8)
2. **C. Apostol, R.G. Douglas, B.Sz.-Nagy, D. Voiculescu, Gr. Arsene** (Eds.): Topics in Modern Operator Theory, 1981, (3-7643-1244-0)
3. **K. Clancey, I. Gohberg:** Factorization of Matrix Functions and Singular Integral Operators, 1981, (3-7643-1297-1)
4. **I. Gohberg** (Ed.) Toeplitz Centennial, 1982, (3-7643-1333-1)
5. **H.G. Kaper, C.G. Lekkerkerker, J. Hejtmanek:** Spectral Methods in Linear Transport Theory, 1982, (3-7643-1372-2)
6. **C. Apostol, R.G. Douglas, B. Sz-Nagy, D. Voiculescu, Gr. Arsene** (Eds.): Invariant Subspaces and Other Topics, 1982, (3-7643-1360-9)
7. **M.G. Krein:** Topics in Differential and Integral Equations and Operator Theory, 1983, (3-7643-1517-2)
8. **I. Gohberg, P. Lancaster, L. Rodman:** Matrices and Indefinite Scalar Products, 1983, (3-7643-1527-X)
9. **H. Baumgärtel, M. Wollenberg:** Mathematical Scattering Theory, 1983, (3-7643-1519-9)
10. **D. Xia:** Spectral Theory of Hyponormal Operators, 1983, (3-7643-1541-5)
11. **C. Apostol, C.M. Pearcy, B. Sz.-Nagy, D. Voiculescu, Gr. Arsene** (Eds.): Dilation Theory, Toeplitz Operators and Other Topics, 1983, (3-7643-1516-4)
12. **H. Dym, I. Gohberg** (Eds.). Topics in Operator Theory Systems and Networks, 1984, (3-7643-1550-4)
13. **G. Heinig, K. Rost:** Algebraic Methods for Toeplitz-like Matrices and Operators, 1984, (3-7643-1643-8)
14. **H. Helson, B. Sz.-Nagy, F.-H. Vasilescu, D.Voiculescu, Gr. Arsene** (Eds.): Spectral Theory of Linear Operators and Related Topics, 1984, (3-7643-1642-X)
15. **H. Baumgärtel:** Analytic Perturbation Theory for Matrices and Operators, 1984, (3-7643-1664-0)
16. **H. König:** Eigenvalue Distribution of Compact Operators, 1986, (3-7643-1755-8)
17. **R.G. Douglas, C.M. Pearcy, B. Sz.-Nagy, F.-H. Vasilescu, D. Voiculescu, Gr. Arsene** (Eds.) Advances in Invariant Subspaces and Other Results of Operator Theory, 1986, (3-7643-1763-9)
18. **I. Gohberg** (Ed) I. Schur Methods in Operator Theory and Signal Processing, 1986, (3-7643-1776-0)

19. **H. Bart, I. Gohberg, M.A. Kaashoek** (Eds.): Operator Theory and Systems, 1986, (3-7643-1783-3)
20. **D. Amir:** Isometric characterization of Inner Product Spaces, 1986, (3-7643-1774-4)
21. **I. Gohberg, M.A. Kaashoek** (Eds.). Constructive Methods of Wiener-Hopf Factorization, 1986, (3-7643-1826-0)
22. **V.A. Marchenko:** Sturm-Liouville Operators and Applications, 1986, (3-7643-1794-9)
23. **W. Greenberg, C. van der Mee, V. Protopopescu:** Boundary Value Problems in Abstract Kinetic Theory, 1987, (3-7643-1765-5)
24. **H. Helson, B. Sz.-Nagy, F.-H. Vasilescu, D. Voiculescu, Gr. Arsene** (Eds.): Operators in Indefinite Metric Spaces, Scattering Theory and Other Topics, 1987, (3-7643-1843-0)
25. **G.S. Litvinchuk, I.M. Spitkovskii:** Factorization of Measurable Matrix Functions, 1987, (3-7643-1883-X)
26. **N.Y. Krupnik:** Banach Algebras with Symbol and Singular Integral Operators, 1987, (3-7643-1836-8)
27. **A. Bultheel:** Laurent Series and their Pade Approximation, 1987, (3-7643-1940-2)
28. **H. Helson, C.M. Pearcy, F.-H. Vasilescu, D. Voiculescu, Gr. Arsene** (Eds.): Special Classes of Linear Operators and Other Topics, 1988, (3-7643-1970-4)
29. **I. Gohberg** (Ed.): Topics in Operator Theory and Interpolation, 1988, (3-7634-1960-7)
30. **Yu.I. Lyubich:** Introduction to the Theory of Banach Representations of Groups, 1988, (3-7643-2207-1)
31. **E.M. Polishchuk:** Continual Means and Boundary Value Problems in Function Spaces, 1988, (3-7643-2217-9)
32. **I. Gohberg** (Ed.): Topics in Operator Theory. Constantin Apostol Memorial Issue, 1988, (3-7643-2232-2)
33. **I. Gohberg** (Ed.): Topics in Interplation Theory of Rational Matrix-Valued Functions, 1988, (3-7643-2233-0)
34. **I. Gohberg** (Ed.): Orthogonal Matrix-Valued Polynomials and Applications, 1988, (3-7643-2242-X)
35. **I. Gohberg, J.W. Helton, L. Rodman** (Eds.): Contributions to Operator Theory and its Applications, 1988, (3-7643-2221-7)
36. **G.R. Belitskii, Yu.I. Lyubich:** Matrix Norms and their Applications, 1988, (3-7643-2220-9)
37. **K. Schmüdgen:** Unbounded Operator Algebras and Representation Theory, 1990, (3-7643-2321-3)
38. **L. Rodman:** An Introduction to Operator Polynomials, 1989, (3-7643-2324-8)
39. **M. Martin, M. Putinar:** Lectures on Hyponormal Operators, 1989, (3-7643-2329-9)
40. **H. Dym, S. Goldberg, P. Lancaster, M.A. Kaashoek** (Eds.): The Gohberg Anniversary Collection, Volume I, 1989, (3-7643-2307-8)
41. **H. Dym, S. Goldberg, P. Lancaster, M.A. Kaashoek** (Eds.): The Gohberg Anniversary Collection, Volume II, 1989, (3-7643-2308-6)

42. **N.K. Nikolskii** (Ed.): Toeplitz Operators and Spectral Function Theory, 1989, (3-7643-2344-2)

43. **H. Helson, B. Sz.-Nagy, F.-H. Vasilescu, Gr. Arsene** (Eds.) Linear Operators in Function Spaces, 1990, (3-7643-2343-4)

44. **C. Foias, A. Frazho:** The Commutant Lifting Approach to Interpolation Problems, 1990, (3-7643-2461-9)

45. **J.A. Ball, I. Gohberg, L. Rodman:** Interpolation of Rational Matrix Functions, 1990, (3-7643-2476-7)

46. **P. Exner, H. Neidhardt** (Eds) Order, Disorder and Chaos in Quantum Systems, 1990, (3-7643-2492-9)

47. **I. Gohberg** (Ed.): Extension and Interpolation of Linear Operators and Matrix Functions, 1990, (3-7643-2530-5)

48. **L. de Branges, I. Gohberg, J. Rovnyak** (Eds): Topics in Operator Theory. Ernst D. Hellinger Memorial Volume, 1990, (3-7643-2532-1)

49. **I. Gohberg, S. Goldberg, M.A. Kaashoek:** Classes of Linear Operators, Volume I, 1990, (3-7643-2531-3)

50. **H. Bart, I. Gohberg, M.A. Kaashoek** (Eds.): Topics in Matrix and Operator Theory, 1991, (3-7643-2570-4)

51. **W. Greenberg, J. Polewczak** (Eds.): Modern Mathematical Methods in Transport Theory, 1991, (3-7643-2571-2)

52. **S. Prössdorf, B. Silbermann:** Numerical Analysis for Integral and Related Operator Equations, 1991, (3-7643-2620-4)

53 **I. Gohberg, N. Krupnik:** One-Dimensional Linear Singular Integral Equations, Volume I, Introduction, 1992, (3-7643-2584-4)

54 **I. Gohberg, N. Krupnik:** One-Dimensional Linear Singular Integral Equations, Volume II, General Theory and Applications, 1992, (3-7643-2796-0)

55. **R.R. Akhmerov, M.I. Kamenskii, A.S. Potapov, A.E. Rodkina, B.N. Sadovskii:** Measures of Noncompactness and Condensing Operators, 1992, (3-7643-2716-2)

56. **I. Gohberg** (Ed.) Time-Variant Systems and Interpolation, 1992, (3-7643-2738-3)

57. **M. Demuth, B. Gramsch, B.W. Schulze** (Eds.). Operator Calculus and Spectral Theory, 1992, (3-7643-2792-8)

58. **I. Gohberg** (Ed.): Continuous and Discrete Fourier Transforms, Extension Problems and Wiener-Hopf Equations, 1992, (3-7643-2809-6)

59. **T. Ando, I. Gohberg** (Eds.): Operator Theory and Complex Analysis, 1992, (3-7643-2824-X)

60. **P.A. Kuchment:** Floquet Theory for Partial Differential Equations, 1993, (3-7643-2901-7)

61. **A. Gheondea, D. Timotin, F.-H. Vasilescu** (Eds.): Operator Extensions, Interpolation of Functions and Related Topics, 1993, (3-7643-2902-5)

62. **T. Furuta, I. Gohberg, T. Nakazi** (Eds.): Contributions to Operator Theory and its Applications. The Tsuyoshi Ando Anniversary Volume, 1993, (3-7643-2928-9)

63. **I. Gohberg, S. Goldberg, M.A. Kaashoek:** Classes of Linear Operators, Volume 2, 1993, (3-7643-2944-0)

64. **I. Gohberg** (Ed.): New Aspects in Interpolation and Completion Theories, 1993, (3-7643-2948-3)

65. **M.M. Djrbashian:** Harmonic Analysis and Boundary Value Problems in the Complex Domain, 1993, (3-7643-2855-X)

66. **V. Khatskevich, D. Shoiykhet:** Differentiable Operators and Nonlinear Equations, 1993, (3-7643-2929-7)

67. **N.V. Govorov** †: Riemann's Boundary Problem with Infinite Index, 1994, (3-7643-2999-8)

68. **A. Halanay, V. Ionescu:** Time-Varying Discrete Linear Systems Input-Output Operators. Riccati Equations. Disturbance Attenuation, 1994, (3-7643-5012-1)

69. **A. Ashyralyev, P.E. Sobolevskii:** Well-Posedness of Parabolic Difference Equations, 1994, (3-7643-5024-5)

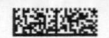